Electricity and Modern Physics

Cover picture: A flame-probe being used to measure the potential near an electrically charged sphere (the flame is too small to be visible). See p. 205 and cf. Fig. 11.5, p. 186.

Photo: Bromhead (Bristol) Ltd.

Electricity and Modern Physics 2ND EDITION

G. A. G. BENNET
*Formerly Head of the Physics Department,
Clifton College*

Edward Arnold

© G. A. G. Bennet 1974

First published 1968 by Edward Arnold (Publishers) Limited
41 Bedford Square, London WC1B 3DQ

Edward Arnold (Australia) Pty Ltd,
80 Waverley Road, Caulfield East,
Victoria 3145, Australia.

Edward Arnold, 3 East Read Street,
Baltimone, Maryland 21202, USA.

Reprinted 1968, 1969, 1970, 1972

Second edition 1974
Reprinted 1976
Reprinted with corrections 1978
Reprinted 1980
Reprinted 1981
Reprinted 1983, 1986

ISBN: 0 7131 2459 8

Printed in Hong Kong by
Wing King Tong Co., Ltd.

'And He exists before everything, and all things are held together in Him'

Colossians 1 : 17, *New English Bible*

The theoretical structure of electromagnetism

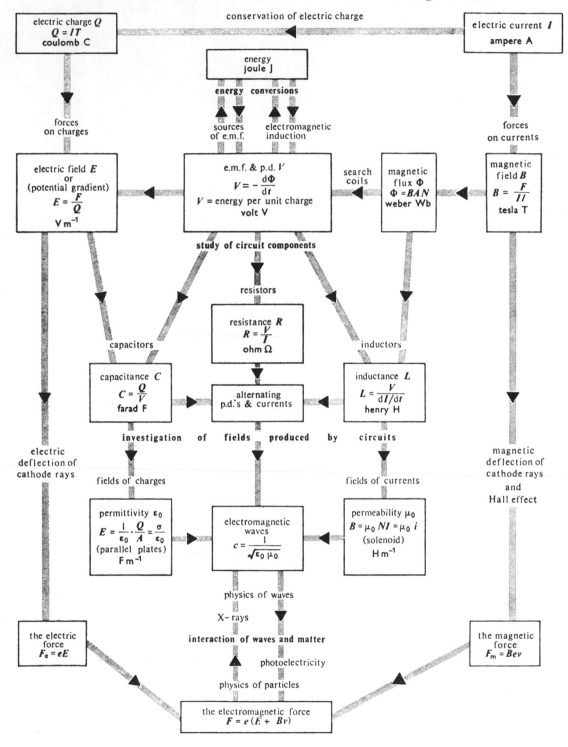

Preface

SI and other Standards

The need for a second edition of this book has been created by the general adoption in the last few years of SI for scientific and other purposes. In addition the work of the Symbols Committee of the Royal Society has led to a standardization of the symbols, nomenclature and typography to be employed in scientific writing. This was based on international agreement, and has been adapted for school work by the Education (Research) Committee of the Association for Science Education (A.S.E.). The British Standards Institution has also been cooperating in this process, and has itself been working towards agreed standards for the *diagrams* to be used in electrical and electronic work (British Standard 3939: *Graphical Symbols*). This edition is fully up to date in all these matters, incorporating wherever appropriate the recommendations to be found in the publications of the bodies mentioned above.

Order of treatment

When the first edition of this book was published, a number of differing m.k.s. approaches to the teaching of Electricity were represented in our schools. Since then there has been a process of convergence and simplification towards an agreed presentation of the subject. On the opposite page is shown in diagrammatic form the theoretical structure of electromagnetism as presented in this book. The diagram displays the main concepts involved together with their defining formulae and units; it also shows the areas of experimental investigation by which the different concepts are linked together, the order of treatment being in the direction of the arrows.

This structure still allows for a good deal of flexibility in the order of teaching. For instance, electric fields can be handled before magnetic fields or vice versa. Or the study of electric fields can stop short at capacitors, and of magnetic fields at inductors, leaving the harder theoretical work associated with ε_0 and μ_0 to a later stage of the course after the treatment of alternating currents and electromagnetic waves. But the use of this book *does* require both electric and magnetic fields to be approached via circuits and current electricity, and some traditional orders of treatment are thereby precluded. However, I believe this is in accord with a general consensus among physics teachers at the present time.

The theory of fields is developed entirely in terms of the vectors B and E. Most of this work is handled using the *uniform fields* in solenoids or between capacitor plates. Fields with cylindrical or spherical symmetry are introduced at later stages in the chapters concerned.

The revolution in physics teaching

The last six years have witnessed not only a change in the content of syllabuses but also a radical reappraisal of the *methods* of a school science course. To call it a revolution is no understatement. I offer in this edition a consistent and up to date account of the subject with (I believe) a minimum of unnecessary material from a former era.

The essence of the new approach is the way simple experiments are employed to string together the theoretical ideas of the subject. Formerly such experiments were used largely for illustration, and the backbone of the subject was provided by a chain of logical definitions and theoretical deductions. This sort of theoretical synthesis we are now content to leave to the university stage. Definitions and mathematical deductions there must still be; but experiments now have an essential part to play, at every stage, in establishing the theoretical structure. Even where a mathematical proof alone might be sufficient to link together two concepts we prefer rather to base the work on an experimental bridge. This is both more satisfying for the student in the age range 16 to 20 for whom this book is intended, but also gives him a better insight into the true nature of the scientific process. The well-trained hunch, the rough experiment, and the hesitant speculation are essential ingredients of scientific discovery; and it is good that students should experience these as they themselves work towards a clear understanding of the subject. The logical synthesis comes only at a much later stage—a stage at which the subject becomes a closed one and of much less real scientific interest. Too often we have presented Physics, and particularly Electromagnetism, as a branch of Mathematics, and have done our students no service in doing so.

Physics teaching owes much at this point to the work of the Nuffield Foundation, who have sponsored research into these new ideas and their practical realization through the Nuffield Advanced Physics Team. New technology and improvements of older methods, both from the work of this Team and from many other sources, have provided us with a fresh range of experimental techniques to put into the hands of a student. With their aid whole areas of electromagnetism can be made the object of intelligent exploration by the student himself.

New apparatus

It is worth listing some of the most important of the new instruments that we now have available, and which a student will need to use if he is to get the best value from this book. In the study of magnetic fields the simple wire-loop *current balance* in its various forms gives substance to the force-on-a-current approach to the definition of *B*—generally recognized now as the simplest route into magnetic field theory. The *Hall probe* provides us with a simple *B*-meter for measuring and comparing fields. The range of magnetic field investigations can be completed with *a.c. search coils* and an oscilloscope, enabling us to study the nature of fields that could otherwise only be derived by mathematical analysis.

With electric fields the modern range of compact and robust *mirror galvanometers* allow us to detect and measure the movements of charge between conductors under circumstances that would have been unthinkable a few years ago. An even more sensitive instrument for measuring small charges and currents is the modern transistorized measuring amplifier or *electrometer*. By making 'visible' the movements of electric charge these two instruments enable us to recast the subject and to dispose of many of the difficulties of traditional Electrostatics.

For these reasons the chapters dealing with electromagnetic field theory (Chapters 5, 6, 7, 10, 11) have been completely rewritten in this edition. Chapter 4 has also been largely rewritten, since I feel that the time has come to demote the straight-wire potentiometer from its former pre-eminent place in the school laboratory. Potentiometer circuits can be constructed with good radio-type components; these bear a closer resemblance to commercial potentiometer-type measuring instruments and are just as useful from the teaching point of view. Likewise the metre-wire form of bridge has been eliminated from this edition, though no doubt it may continue to appear in school practical courses as a particular example of a Wheatstone network.

The transistorized electrometer also simplifies elementary experiments in radioactivity, rendering the pulse electrometer obsolete. With a suitable ionization chamber the modern type of electrometer can be used to detect all three kinds of radiation. The study of the radiations can then be extended with a G-M tube; and, if this is of the special very thin-walled type, again all three radiations can be observed with the one instrument.

Another piece of apparatus that has more than proved its worth in school work is the *3-cm wave transmitter and receiver*, providing valuable insights into wave processes of all kinds. Its particular value for the purposes of this book has been to provide analogue experiments in X-ray crystallography. Real experiments with X-rays will probably continue to be banned in school laboratories for safety reasons; and except for those schools fortunate enough to have access to a university crystallographic laboratory the 3-cm wave analogue experiment is the only thing available in this important field.

Electronics

Since the publication of the first edition of this book the silicon transistor has come into its own. Its high gain and negligible leakage current make it eminently suitable for elementary electronics experiments. The sections of the book dealing with rectifiers, amplifiers, oscillators, etc., have therefore been rewritten using the n–p–n silicon transistor (and this conveniently enables $+$ve to be drawn at the top in all electronics diagrams). Thermionic tubes have been relegated to a subsidiary section. The chief function of the thermionic effect in a physics course is now to provide evidence on the nature of the electron.

A substantial section has been devoted to the nature of transistorized switching circuits, since such logic circuits are likely to play an increasing part in school physics and mathematics courses at all levels. Also a short introductory section has been included on operational amplifiers; these are becoming very simple and cheap devices, and can provide excellent value in the study of oscillatory processes of all kinds.

Flexibility

In all this rewriting and re-arranging I have sought to preserve those features of the first edition that made the book adaptable to the varying backgrounds and differing needs of our students. In each chapter the advanced work is arranged to *follow* the more elementary sections in which fundamental concepts are presented. Any student should thus be able to read up to a certain point in a chapter and find everything within his grasp. All the hardest mathematical and theoretical work is safely confined to the later sections of the chapters, and can be left to a second reading. Occasionally I have been forced to include harder topics in an elementary section. This is indicated by the use of smaller type with a ★ at the beginning.

I am aware that a textbook must often be used primarily as a work of reference. I have therefore preserved the same comprehensive type of index as before (with over 1000 entries), so that any topic can quickly be located. Within the text there are numerous cross-references to other parts of the book; and the page headlines are designed to help in finding any required section quickly.

Questions

Questions (with Answers) on the subject matter of each chapter are gathered together under their respective chapter headings at the end of the book. These have been sorted into three grades of difficulty—**A**, **B** and **C**. Questions in the **A** grade are straightforward introductory applications of new ideas; **C** grade Questions are probably rather too hard for the average student. At the end of each group of questions cross-references are given to questions of similar type under other chapters.

I have taken the opportunity to prune away a number of older questions from the previous edition and to incorporate some more up to date ones instead. In response to many requests I have increased the proportion of **A** grade problems so as to provide more practice material in the elementary handling of basic concepts.

Many of the questions, old and new, are from past examination papers of various examining bodies, as acknowledged below. These bodies have allowed me to make such alterations in their questions as will bring them into conformity with the current standards of printing and nomenclature.

I have not attempted to provide practice questions of the Multiple Choice, Multiple Completion or other types increasingly being used in modern examining. To give an adequate selection of these would have taken up too much space. Besides, there are comprehensive books of practice questions of these types already available.*

Slide-rules

An efficient physicist should be able to use a slide-rule with speed and accuracy. I have therefore attached some hints on this as an appendix to the text. I recommend any student to set aside an hour or so to work through the method with the examples given; it will certainly save many hours of hard labour later in his course. The appendix also includes a list of mathematical short-cuts that should form part of the mental equipment of any physicist.

---

* e.g. *Multiple Choice Questions for 'A' Level Physics*, Marshall, Akrill and Cosh, Edward Arnold, 1972.

Acknowledgements

I acknowledge with gratitude the help I have received from various busy people in finding suitable photographs of particle tracks and for permission to reproduce these: The Director of the Science Museum; Professor Lord Blackett, F.R.S., Professor C. F. Powell, F.R.S., and Professor A. B. Pippard, F.R.S. I am also grateful to the Director of the the National Physical Laboratory for providing the photograph of their current balance (on which Crown Copyright is reserved), and for giving permission for this to be used.

I should like to express my gratitude to my former colleagues at Clifton College who have given endless advice and assistance in creating this second edition: Dr I. B. Hopley, Mr T. B. Akrill, and Dr J. Q. Cosh; also to Mr K. W. Watson whose unsurpassed technical skill was always available to help in trying out new ideas in the laboratory. I have gained many fresh insights and much inspiration from a spell of work with Dr G. E. Foxcroft of Rugby School and Mr G. W. Dorling of Worcester College of Education, and I am more than grateful to them both.

I am grateful to the various examining bodies listed below for permission to make use of questions from their past examination papers, and for allowing me to make alterations to them to bring them into conformity with current standards of printing and nomenclature. The copyrights of such questions remain the property of the bodies concerned, while the responsibility for any alterations is entirely mine. Each question is individually acknowledged as indicated below.

Oxford and Cambridge Schools
   Examination Board   (*O & C*)
Oxford Local Examinations Board   (*O*)
Cambridge Local Examinations Syndicate:

   Advanced level papers   (*C*)
   Papers set for H.M. Forces   (*Cam. Forces*)
   Cambridge Oversea H.S.C.   (*Cam. Oversea*)

London School Examination Council   (*L*)
Joint Matriculation Board   (*N*)
Southern Universities Joint Board   (*S*)
Welsh Joint Education Committee   (*W*)
Associated Examining Board   (*A.E.B.*)
Oxford Entrance Scholarships   (*Ox. Schol.*)
Cambridge Entrance Scholarships

                      (*Cam. Schol.*)

G. A. G. BENNET

Motcombe,
March 1974

Contents

★ = advanced topic that may be omitted at a first reading

      ★ = advanced topic that may be omitted at a first reading

1 Atoms and Electric Currents

1.1 The atomic theory

To the present generation it is no doubt a commonplace that matter consists of *atoms*—very small, almost indestructible particles, whose combinations and interactions give rise to the properties of matter in bulk. However, it is only in this century that the atomic theory has entered the realm of well-accepted scientific truth; and it is important for the student to appreciate the kind of evidence on which our belief in atoms is based. None of the ordinary techniques used for making measurements on small pieces of matter in a school laboratory can possibly deal with single atoms. The diameter of an atom is far less than the resolving limit of any kind of microscope. No balance is sensitive enough to weigh a single atom. It is true that there are experiments in which we claim to be observing events involving single atoms. We may for instance 'observe' the disintegration of an atom in a cloud chamber (p. 319). But what we actually see are chains of water droplets intersecting at a point in the cloud chamber; each droplet must contain many millions of atoms. We say that each chain of droplets shows the track of a particle emerging from an atomic disintegration; but this is a human interpretation which we ourselves impose on the observations. Its validity depends on our correctly reading into the experiment a great deal more than is immediately obvious from this experiment alone.

Most of the evidence for the atomic structure of matter is even more indirect than this. We start by suggesting an imaginary model for the composition of matter. We then work out what behaviour to expect from matter constructed according to this model. In some respects the behaviour of actual matter may agree with this; in other respects it does not. The model is then varied and modified in ever greater and greater detail, until the agreement between the properties of the theoretical model and of real matter is

exact. This is a process which still continues at the present day. No single piece of experimental evidence for the atomic theory is by itself conclusive; but the accumulation of evidence from many branches of Physics and Chemistry, always consistent with the same picture of the structure of matter, leaves very little room in the end for doubting the correctness of the theory.

An atomic theory of a kind first arose as a philosophical speculation amongst the ancient Greeks; but it was only in the early years of the nineteenth century that the first clear evidence for the theory was assembled, when Dalton pointed out that the laws of chemical combination were readily explained on an atomic basis. Subsequent developments showed that the atomic picture of chemical processes could be made wholly consistent. The chemical evidence for the atomic theory, compelling though it is, gives little indication of the properties of atoms; all it can do is to provide a table of the relative masses of the different kinds of atoms. Chemistry cannot provide any figure for the actual mass or size of an atom.

Little physical evidence in support of the atomic theory could be offered until about the middle of the nineteenth century. By then the work of Joule had established the true nature of heat as a form of energy. This paved the way for rapid developments in the atomic theory; this particular branch of it became known as the *kinetic theory of matter*. It was supposed that the atoms in a piece of matter are in a state of continuous random motion; in a solid this motion must be limited to oscillations about fixed mean positions if the material is to retain its shape permanently; but in a fluid the motion can be more general. The *heat energy* carried by the matter is supposed to be the *sum total* of the mechanical energy of its atoms. Its *temperature* is taken to be a measure of the *average energy* of the atoms. A simple confirmation of the theory is provided by observation

1

of the movements of small particles suspended in a gas or liquid—the *Brownian movement*, as it is called. For instance, if a drop of water containing a little black water-colour pigment is placed on a slide and observed with a medium-power microscope, the black particles are found to be in a state of ceaseless agitation. It is easy to suppose that we are observing the particles sharing in the random motions of the atoms adjoining them.

Later developments of the kinetic theory showed that detailed observation of the Brownian movement in a gas enables the number of molecules per unit volume to be calculated; it comes to the enormous figure of 2.7×10^{25} molecules per cubic metre at *standard temperature and pressure* (s.t.p.). By dividing this into the mass per cubic metre (i.e. the density) of the gas, we obtain the mass of a single molecule. This gives the mass of a hydrogen molecule as 3.4×10^{-27} kg; if each molecule of hydrogen contains two atoms, the mass of a single hydrogen atom is then 1.7×10^{-27} kg.

The kinetic theory was developed first for gases, since the mathematical difficulties are in this case less than for solids or liquids. It is supposed that in a gas the average distance of a molecule from its nearest neighbours is very much greater than the diameter of the molecule. The interactions between molecules are then assumed to have negligible effects on their motions, except actually at the moments of collisions. The pressure of the gas is supposed to arise through the continual impacts of its molecules on the walls of the containing vessel. Using the known laws of mechanics, it is possible to predict how the pressure exerted by such a molecular 'model' will depend on the volume and temperature. The predictions are found to be in excellent agreement with the basic gas laws (*Boyle's law and Charles's law*). Furthermore, the theory provides a formula connecting the velocities of the molecules with the pressure and density of the gas. For oxygen at room temperature the average velocity is 0.5 km s^{-1}, for hydrogen about 2 km s^{-1}.

Other types of measurement enable estimates to be made of molecular diameters. The kinetic theory shows that certain properties of gases (e.g. their thermal conductivity, viscosity, etc.) should depend on the average distance travelled between collisions—a quantity known as the *mean free path*. This in turn depends on the diameter of the molecules; the larger they are, the more frequent must be the collisions between them. The measurements show that for air at s.t.p. the mean free path is 10^{-7} m, which gives the diameter of the molecules as about 3×10^{-10} m.

It is also possible to estimate the dimensions of atoms by a number of other methods. For instance, using the known values of the masses of atoms we can calculate how many atoms there must be in a piece of solid material of known mass and volume. It is reasonable to assume that in the solid state the atoms are closely packed together; and so we can use the figures to calculate the volume occupied by a single atom. Always the results obtained for atomic dimensions are of the same order of magnitude. There are, however, significant variations between the estimates made by different methods; and it is clear that the elementary concept of atoms as hard elastic spheres is inadequate. When an atom is fixed in a stable configuration, such as a chemical compound or the regularly spaced lattice of a crystal, the forces on it are in equilibrium; and the distance between the centres of two atoms is a precisely fixed quantity, which cannot be varied much by elastic stresses or changes of temperature. But when the molecules are free to move at random in a gas, they do not appear to behave as though they had sharply defined surfaces. If we must form mental pictures of atoms, it is better not to think of billiard balls but of puffs of cotton wool with hard centres! But such efforts of the imagination are of dubious scientific value. The point is that the molecular model only agrees with the observed properties of gases if we suppose that the closest distance of approach of two molecules depends on the violence of the collision.

There are many other lines of evidence that have been considered in the development of the theory, and these have led to endless refinements of our ideas about the properties of atoms. But enough has been said here to show the basis of the modern atomic theory; without ever observing individual atoms we have developed a single consistent picture of an atomic structure of matter, which suffices to give detailed explanations of a vast body of experimental facts. At the present time there certainly seems to be no possible ground for doubting the validity of the theory.

Even so it is hard enough to appreciate the minute scale and vast rapidity of atomic processes. It may be useful therefore to consider some of the implications of the figures we have given. The volume of a molecule of diameter 3×10^{-10} m must be about 1.5×10^{-29} m^3. In 1 m^3 of a gas at s.t.p. the total volume occupied by the molecules themselves is thus about 4×10^{-4} m^3—about 1 part in 2500 of the actual volume of the gas. With a mean free path of 10^{-7} m each molecule travels about 300 times its own diameter between collisions. But since it moves at an average speed of 5×10^2 m s^{-1}, it must make some 5×10^9 collisions with other molecules every second! In spite of this it spends less than 1% of its time being affected in any way by molecular collisions. The mean free path is inversely proportional to the pressure of the gas. Thus at the lowest pressures readily attainable in the laboratory (about 10^{-6} N m^{-2}) the mean free path is nearly 10 km. Some idea of the smallness of molecules can be gained by considering that even at this pressure there are over 10^{14} molecules per cubic metre along the path. At the lowest densities believed to exist in parts of outer space there are still several molecules per cubic metre. But the mean free paths are then to be measured in light-years and collisions only occur at intervals of several centuries!

1.2   The structure of the atom

Our knowledge of the internal structure of atoms has been arrived at in much the same way as our belief in the atoms themselves. The ambition of physicists is to develop a theory of the atom that shall be able to explain all the observed properties of matter. For instance any satisfactory theory must be able to account for the varied behaviour of atoms in chemical processes. It must also explain the emission and absorption of light by matter.

In this book we are chiefly concerned with electric and magnetic effects, which, as we shall see, arise from the properties of the fundamental particles that go to make up the atom itself. In due course we shall consider some of the detailed evidence. But for the time being we must be content with a bare sketch of the atomic model that has been found to fit the facts.

Although the effective diameter of an atom is about 3×10^{-10} m, nearly all its mass appears to be concentrated in a very small central region, called the *nucleus*. This is rather less than 10^{-14} m in diameter, and itself contains fundamental particles of two kinds—*protons* and *neutrons*—packed closely together; these two particles have nearly the same mass. The rest of the space occupied by the atom contains particles of a third type, called *electrons*, which are in motion round the nucleus. The number of electrons in the outer parts of the atom is normally exactly the same as the number of protons in the nucleus, but each electron is only about 1/2000 as massive as the proton or the neutron.

The atom is held together by two different types of forces:

1 *Nuclear forces.*   All the particles in the nucleus (known collectively as *nucleons*) are attracted to one another by very short-range forces; that is, the forces are negligible except when the particles are about 10^{-15} m apart; but at this separation the forces become very large, sufficient to hold the nucleons together in a closely packed structure.

2 *Electrical forces.*   A different type of force is found to operate between electrons and protons. These forces are effective at much greater separations than the nuclear forces. The force acts as a repulsion between like particles (i.e. between electron and electron or between proton and proton), and as an attraction between unlike particles (i.e. between electron and proton). These forces are described by saying that the particles carry *electric charge*, the proton a positive charge and the electron an equal negative charge; the neutron is an uncharged particle. Thus, like charges repel, and unlike charges attract. Normally atoms are electrically neutral, since the number of electrons in them is equal to the number of protons and the total charge is thereby zero. In this condition the atom exerts little or no electrical force on charged particles in its neighbourhood.

In certain circumstances an atom may lose electrons from the electron-cloud surrounding its nucleus, or it may gain surplus electrons from its

The twelve lightest elements and their most important isotopes

Element	Symbol	Atomic number Z	Relative atomic mass A_r	Proton number Z	Neutron number N	Mass number A	or	Nucleon number = Z+N
Hydrogen	H	1	1.0078	1	0			1
Deuterium	D		2.0141	1	1			2
Helium	He	2	3.0161	2	1			3
			4.0026	2	2			4
Lithium	Li	3	6.015	3	3			6
			7.016	3	4			7
Beryllium	Be	4	9.012	4	5			9
			10.012	4	6			10
Boron	B	5	10.013	5	5			10
			11.009	5	6			11
Carbon	C	6	**12.000**	6	6			12
			13.003	6	7			13
			*14.003	6	8			14
Nitrogen	N	7	14.003	7	7			14
			15.000	7	8			15
Oxygen	O	8	15.995	8	8			16
			16.999	8	9			17
			17.999	8	10			18
Fluorine	F	9	18.998	9	10			19
Neon	Ne	10	19.992	10	10			20
			20.994	10	11			21
			21.991	10	12			22
Sodium	Na	11	22.990	11	12			23
Magnesium	Mg	12	23.985	12	12			24
			24.986	12	13			25
			25.983	12	14			26

* This isotope is radioactive. Short-lived radioactive isotopes have not been included in the table.

surroundings. The whole atom then acquires a net positive or negative charge and behaves in many respects like a single charged particle. An atom in this charged condition is referred to as an *ion*. Hydrogen and metals have a tendency to lose electrons, forming positive ions; while many non-metals have a tendency to gain extra electrons, forming negative ions. Some simple chemical compounds are produced by the forces of attraction between such pairs of oppositely charged ions. In fact many of the processes of inorganic chemistry are concerned at the atomic level with transferring electrons from one atom or ion to another. In chemical terms, adding an electron to an atom or ion is described as *reduction*, and removing an electron as *oxidation*—a broadening of the original meanings of these

terms. In a given atom only a small number of the outermost electrons are involved in any chemical effects. These are known as the *valence electrons*.

The chemical behaviour of an atom is decided by the number of electrons it contains when unionized. This in turn is fixed by the number of protons in the nucleus, which is called the *atomic number Z*. Classifying the atoms according to their atomic numbers reveals the sequence familiar in the periodic table of the chemical elements. The first few members of this sequence are shown in the above table.

The mass of an atom is almost equal to that of the nucleons it contains; the electrons contribute only about 1/4000 of the total mass. The modern practice is to measure atomic masses relative to that of the commonest sort of *carbon* atom (whose nucleus contains 12 nucleons); this is arbitrarily given a relative atomic mass of 12 precisely. Thus the *relative atomic mass* A_r of a species of nucleus is defined as

$$A_r = \frac{m_a}{m_C/12}$$

where m_a = the mass of one atom of the species, and m_C = the mass of one atom of carbon-12. The quantity $m_C/12$ is referred to as the *unified atomic mass constant* (m_u). Thus we can write

$$A_r = \frac{m_a}{m_u}$$

It is sometimes convenient to define m_u as a special unit of mass. This is then called the *unified atomic mass unit* (u). Measurement of the actual atomic mass of carbon-12 gives the connection between the u and the kg. Thus

$$m_u = \frac{m_C}{12} = 1\ u = 1.66 \times 10^{-27}\ kg$$

This is *nearly* equal to the atomic mass of hydrogen; it would be precisely so if the relative atomic mass of hydrogen were exactly 1. But in fact we have for hydrogen

$$A_r = 1.0078$$

Therefore the mass m_H of one atom of hydrogen is given by

$$\begin{aligned} m_H &= 1.0078\ u \\ &= 1.0078 \times 1.66 \times 10^{-27}\ kg \\ &= 1.67 \times 10^{-27}\ kg \end{aligned}$$

Expressed like this to only three significant figures, this quantity is also the mass m_p of the proton (the hydrogen nucleus).

Likewise for other species of nucleus the relative atomic mass is always close to a whole number; this whole number is the number of nucleons in the atom, and is referred to as its *mass number A*. [N.B. *Mass number* has no units and is necessarily a whole number. *Atomic mass* must be expressed in appropriate units—kg or u—and, when expressed in u, is *nearly* equal to the mass number.]

Thus the *mass number A* is the sum of the *proton number Z* (which is the same as the atomic number) and the *neutron number N*.

$$\therefore\ A = Z + N$$

The meanings of these quantities are clearly shown in the Table on p. 4.

The simplest possible atom is that of *hydrogen* consisting of a single proton round which orbits a single electron. The nucleus of *deuterium* contains one proton and one neutron; and since one electron only is required to balance this electrically, it has the same chemical properties as ordinary hydrogen. It is sometimes called *heavy hydrogen*. It occurs as about 1 part in 4000 of natural hydrogen.

The atoms of a given chemical element must all contain the same number of protons. With the light elements in the table the number of neutrons is also approximately equal to this; but in most cases there are several possible alternatives. Groups of atoms having the same atomic number but differing atomic masses are known as *isotopes*. Samples of a given element derived from natural sources consist of mixtures of the possible isotopes of the element; the proportions are found to vary little from one sample to another. The measurement of relative atomic mass by chemical methods therefore gives a figure that is an average for the naturally occurring mixture of isotopes. This figure is not necessarily close to a whole number. For instance natural *magnesium* is a mixture of isotopes of mass numbers 24, 25 and 26, the lightest providing nearly 80% of the total. The mixture is in such proportions that the relative atomic mass is 24.32. It is this quantity that must be used in chemical calculations involving magnesium.

1.3 Electric currents

There are many substances in which there are charged atomic particles that can wander freely through the material. In such a substance electric charge can be transferred from one point to another by a general drift of the charged particles within it. Such a movement of electric charge is called an *electric current*. Materials through which an electric current will pass are known as *conductors*. Those substances in which there are no charged particles that are free to move are known as *non-conductors* or *insulators*.

The most important class of conductors is formed by the *metals*. In any solid the atoms themselves are necessarily fixed at permanent sites in the crystal. But the outermost (valence)

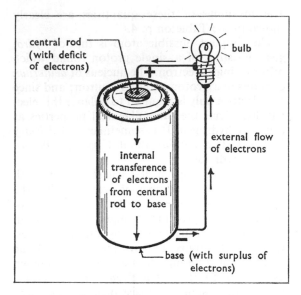

Fig. 1.1   The flow of electrons in a simple circuit

electrons of each atom are partly shared with neighbouring atoms. It is this feature that gives a solid its cohesive strength. In most non-metals each valence electron is attached to a particular group of atoms; and no general movement of electrons through the material is possible. But with metals the bonds between atoms are such that some of the valence electrons are free to move from one atom to the next right through the crystal lattice.

Another important type of conductor is the class of liquids known as *electrolytes*. Many chemical compounds when they go into solution break up into positively and negatively charged ions, which then move independently in the liquid, and are therefore available to conduct an electric current through it.

The electrical forces between charged particles are so great that it is impossible to unbalance the distribution of charge in a piece of matter to any appreciable extent. Any tendency for charge of one sign to pile up at some point generates forces of repulsion that soon bring the process to a halt. A continuous flow of electric current can thus only happen in a closed path or *circuit*, as it is called. The circuit must also include some device such as an electric battery or dynamo, whose function is to maintain the circulation of electric charge. This it does by an internal transference of electrons

from one of its terminals to the other; in a battery this is brought about by chemical processes. One terminal thus gains a surplus of electrons and so carries a negative charge. At the other terminal there is a corresponding deficit of electrons and a net positive charge. When a complete conducting path is provided externally between the terminals, the electrical forces tend to even out the distribution of charge by a suitable flow of current; the battery or dynamo then acts to maintain the initial distribution of charge by a fresh transference of electrons between the terminals *inside* the device. A continuous circulation is thus created (Fig. 1.1).

When an electric current flows in a circuit, there are three different types of effect that may arise:

(*i*) *A heating effect.*   In all forms of electric conduction the movements of the charged particles are continually impeded by their interactions with stationary particles of matter in their paths. The energy lost in collisions increases the random motion of the particles of the conductor through which the current passes., and therefore raises its temperature. The heating effect of a current in the 'element' of an electric fire or the filament of a light bulb is too well known to require any demonstration. A further discussion of this process will be found later in this chapter and in Chapter 2.

(*ii*) *A chemical effect.*   The movement of electrons through a metal does not cause any chemical effect in it. But in an electrolyte the position is different. In this case the conduction takes place through the simultaneous movement of

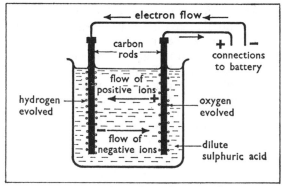

Fig. 1.2   The chemical effect of an electric current—carbon electrodes in dilute sulphuric acid

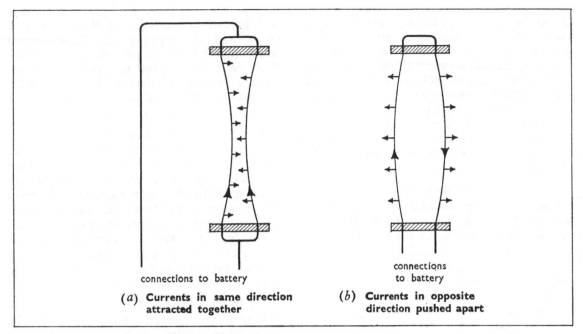

(a) **Currents in same direction attracted together**

(b) **Currents in opposite direction pushed apart**

Fig. 1.3 The magnetic force between neighbouring electric currents

positive and negative ions in opposite directions. Where the current enters and leaves the solution through suitable conducting plates (called *electrodes*) some of the ions are discharged; and chemical changes are therefore observed as long as the current flows. The particular reactions that occur depend on the materials used. The effect may be demonstrated by joining two carbon rods to the terminals of a battery by means of copper wires; the circuit may then be completed by dipping the rods into a beaker of dilute sulphuric acid—which is an electrolyte (Fig. 1.2). It will then be noticed that bubbles of hydrogen are evolved at the surface of the negative rod and bubbles of oxygen on the positive rod. The action is clearly caused by the electric current, since it ceases when one of the copper wires is disconnected from the battery. If instead the rods are dipped into a solution of copper sulphate, oxygen is still evolved at the positive rod, but metallic copper is now found to be deposited on the negative one. This process is considered in detail in Chapter 3.

(*iii*) *A magnetic effect*. This may be demonstrated by causing an electric current to flow in the vicinity of a small pivoted magnet (i.e. a compass). In most positions close to the circuit the compass will be deflected slightly from its usual north-pointing direction. This effect is due to yet another type of force that acts between the fundamental particles of matter; it occurs between *charged particles in motion*, and acts in addition to the electrical forces already considered. Since charges in motion constitute an electric current, we may say that magnetic effects are due to forces acting between electric currents. This may be shown by passing a large current through two parallel flexible conductors (Fig. 1.3). If the currents are in the same direction, the magnetic force is seen to act so as to draw the conductors together. When the currents are in opposite directions, the wires are repelled from one another. Such effects are considered in more detail in Chapters 5 and 6.

It is not perhaps obvious how the properties of *magnets* can arise from the interactions of electric currents. But a little thought will show that on an atomic scale there are never-ending circulations of electric charge taking place. An electron in rapid motion round a nucleus constitutes an electric current. Indeed it is now understood that electron, proton and neutron must all be regarded as being permanently in a state of 'spin', which causes a magnetic effect even when they are not otherwise in motion. We should therefore expect all matter to show magnetic effects; and such is found to be the case, though with most materials the effects are very weak. It so happens that in iron and a few other substances the interactions between atoms are such as to magnify the effects considerably. This aspect of magnetism is dealt with more fully in Chapter 7.

The strength of an electric current could well be measured by the intensity of any one of the three effects to which it gives rise. However, by far the most convenient is the magnetic effect.

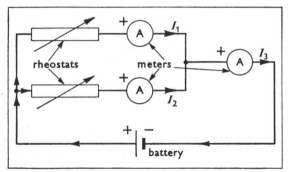

Fig. 1.4   The distribution of currents in a branching circuit

An instrument for measuring an electric current usually consists of a small coil pivoted between the poles of a magnet; the current to be measured passes through this coil. The instrument registers the moment of the couple that acts on the coil due to the magnetic forces. The details of the construction of *moving-coil meters*, as they are called, is considered in Chapter 8.

There is nothing obvious to indicate that a flow of *anything* is taking place in an electric circuit. The flow of charge is a theoretical explanation of the process that we believe in because it proves to be entirely consistent. So we must review briefly the evidence to show that an electric current consists of a flow of electric charge.

(*i*) The measurements of the currents in branching circuits is consistent with this explanation. Thus, in the circuit shown in Fig. 1.4, the current from the positive terminal of the battery

divides into two parts, I_1 through one meter with its current controller (or *rheostat*, as it is called), and I_2 through the other. These currents recombine to form the current I_3 through the third meter. The currents may be varied in any manner we wish with the two rheostats; but for all settings of the rheostats we find that

$$I_3 = I_1 + I_2$$

A similar result holds in other kinds of flow situations. For instance in a branching system of water pipes the quantity of water passing per second in a pipe that divides is equal to the sum of the quantities flowing per second in the pipes that it branches into. This is because the water is not lost from the system; or, as we say, the water is *conserved*. Likewise the above evidence suggests that an electric current, as measured by a moving-coil meter, is a *rate of flow of something that is conserved within the circuit*.

(*ii*) In the chemical effect of an electric current there is an actual transport through the electrolyte of matter, which then appears at the electrodes. In the examples above the *rate* of evolution of gaseous hydrogen or oxygen or the *rate* of deposition of copper is found to be proportional to the current through the electrolyte, as measured by a moving-coil meter joined in series. This is consistent with the idea that each particle (*ion*) of matter in the electrolyte carries a definite quantity of electric charge. A certain rate of deposition of matter on the electrodes will then be associated with a definite rate of passage of electric charge through the electrolyte, and one should be proportional to the other.

(*iii*) In Chapters 10 and 11 we shall consider the effects due to isolated electric charges (of both signs) produced from the terminals of a battery. We shall analyse the forces between these charges and show how these forces cause the charges to be carried through conductors, giving rise in the process to all the effects we associate with electric currents.

(*iv*) We may cause beams of electrons or ions to pass through a vacuum, and we may then measure the effect of electric and magnetic forces on the beam. The experimental results are consistent with the view that each electron or ion carries a definite quantity of electric charge; indeed it is such experiments that provide most of the detailed evidence that enables us to describe electrons and protons as particles possessing definite charge, mass, etc. This evidence is considered in detail in Chapter 13. An electron beam exhibits all the properties of an electric current. The beam can exist only when it forms part of an electric circuit; a moving-coil meter joined in series with the electrodes between which the beam travels registers a current only when the beam is passing.

There is thus good evidence for the working hypotheses that we adopt.

> An electric current consists of electric charges in motion.
> A moving-coil meter indicates the rate of passage of electric charge through the instrument.

The same electric current may be carried in various ways in different parts of the same circuit; at one point there may be a movement of positively charged particles in one direction; at another there may be an equal movement of negatively charged particles in the opposite direction—the net transference of charge could be the same in both cases; or there may be a simultaneous movement of positive and negative charges in opposite directions (as in electrolytic conduction). Any or all of these processes may occur in any one circuit.

By general agreement the *positive* direction of electric current is taken to be that in which a positive charge at that point of the circuit would flow. An arrow drawn on a circuit diagram indicates this positive direction; it implies that, if the current at that point is actually a flow of negatively charged particles, then the movement of the particles is opposite to that of the arrow. Fig. 1.4 shows how this is done on a conventional circuit diagram with a single-celled battery. Note that in the conventional symbol for a battery the *longer stroke* is taken to be its *positive* terminal.

The unit of electric current is called the *ampere* (A) after the French scientist who established the laws of electromagnetism. It is defined in terms of the magnetic forces that act between current-carrying conductors. The definition of the ampere could in principle be cast in any number of different forms. The laws of electromagnetism enable us to work out expressions for the magnetic forces acting between electric currents flowing in any imaginable configuration (though the calculations can well be of extreme complexity in some cases). These expressions each contain a numerical constant, whose value depends on our choice of unit for measuring electric currents. To fix the ampere it is only necessary to specify the value of this constant for some stated arrangement of conductors. The corresponding constants for all other arrangements can then be derived from this one. The form of definition given below is that adopted in 1948 by the International Conference on Weights and Measures. It was established in 1960 as one of the base units of the International System of Units, known as SI units (standing for *Système Internationale*). Historically, the ampere was fixed originally in a very different way. The numerical constant 2×10^{-7} that appears in the modern definition was chosen so as to keep the magnitude of the ampere the same as formerly.

> The **ampere** is that constant current which, if maintained in two straight parallel conductors of infinite length, of negligible circular cross-section, and placed 1 metre apart in vacuum, would produce between these conductors a force equal to 2×10^{-7} newton per metre of length.

[The *newton* (N) is the SI unit of force. It is the force that produces an acceleration of 1 m s^{-2} in a mass of 1 kg. Thus

$$1 \text{ N} = 1 \text{ m kg s}^{-2}]$$

The prefixes used in the metric system are applied to electrical units also so as to provide convenient multiples and submultiples of the basic units. For example, the *milliampere* (mA) and *microampere* (μA) are derived units in regular use for recording small currents. Instruments for measuring currents in amperes are called *ammeters*, or perhaps *milliammeters* or *microammeters*, according to how they are calibrated.

The instrument used in the standardizing laboratories for fixing the value of the ampere is the *current balance*. One such instrument is shown in Fig. 1.5. The beam of the balance, supported on its knife-edges at the centre, carries at each end a cylindrical coil. These coils are joined in series, and the connection to them is by means of wires arranged to exert

Fig. 1.5   The National Physical Laboratory current balance. In this picture the outer, fixed pair of coils has been lowered on the left-hand side so that one of the inner suspended coils may be seen. The weights are lifted onto or off the small scale pans above the coils by means of rods controlled from outside the case.

Photograph: National Physical Laboratory, Crown Copyright

negligible torque on the beam. A set of fixed, single-layer, cylindrical coils surrounds the suspended coils, concentric with them. All the coils, suspended and fixed, are joined in series with one another in such a way that the forces operating between them will combine to deflect the beam in one direction. The beam is maintained in equilibrium when the current is switched on by adding weights to the small scale pans. In this way the forces acting between the coils are measured in terms of a known weight, and so the current in the coils is found. The accuracy attainable at present is about 5 parts in 10^6.

If an electric current is a rate of flow of electric charge, fixing the unit of current automatically establishes also the unit of charge. This is called the *coulomb* (C), after the French scientist who first investigated the forces acting between electric charges.

> The **coulomb** is the quantity of electric charge that passes a given point in a circuit when a current of 1 ampere flows for 1 second.

Thus $1\,\text{C} = 1\,\text{A s}$

Or we can say $1\,\text{A} = 1\,\text{C s}^{-1}$

If a current I flows in a circuit for time t, the quantity Q of electric charge that passes is given by

$$Q = It$$

The charge on one electron turns out to be -1.60×10^{-19} C. Or we may say that one coulomb is numerically the charge on 6.2×10^{18} electrons. [Sometimes the phrase 'quantity of electricity' is used where we have written 'quantity of electric charge'. This is an old-fashioned usage, and it will not be adopted in this book. We shall keep the word 'electricity' to refer to the subject under study. However, the student may well come across the older phrase in books and examination questions.]

The velocities of the current carriers

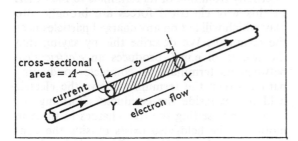

Fig. 1.6   Calculating the velocity of the charged particles in a conductor

Consider a length of conductor of cross-sectional area A containing n electrons per cubic metre (Fig. 1.6). The number of electrons in the conductor remains constant; when a current flows in it, electrons enter it at one end and the same number leave at the other. The motion of the individual electrons is doubtless somewhat irregular, but suppose their average velocity is v. Then the number of electrons passing a point such as Y is the number contained in the part of the conductor between X and Y of length v.

$$\text{volume between X and Y} = Av$$
$$\therefore \text{ number of electrons contained} = nAv$$

The electric current I flowing in the conductor is equal to the quantity of electric charge passing a point such as Y in one second. If the charge carried by one electron is $-e$, then

$$I = nAve$$
$$\therefore v = \frac{I}{nAe}$$

The resistance to the motion of the electrons is roughly proportional to their average velocity v. A good conducting material is one for which n is large and v therefore small. Copper is one of the best conductors, having about 10^{29} free electrons per cubic metre.

In a simple circuit of conductors joined in series the current I is the same at all points. However, the velocity of the electrons and therefore the resistance to the flow may vary from one part of the circuit to another. In a copper wire of relatively large cross-section v is small and very little resistance is presented to the flow of current. But if the circuit includes a length of fine wire such as the filament of a light bulb, v is there much greater, and most of the available energy is dissipated in driving the current through this wire—which thus becomes white hot.

The actual values of the velocity v are amazingly small. Consider for instance a current of 1 amp flowing through a copper wire of cross-sectional area 2.0×10^{-7} m^2 (about 0.5 mm diameter).

$$v = \frac{I}{nAe} = \frac{1}{10^{29} \times 2 \times 10^{-7} \times 1.6 \times 10^{-19}}$$
$$= 3.1 \times 10^{-4} \text{ m s}^{-1}$$

Although the electrons move so slowly, the electric forces that set them in motion are propagated round a circuit with great rapidity—in fact very nearly with the speed of light (3×10^8 m s^{-1}). If a current of 1 ampere flowed steadily along such a wire in a transatlantic cable (3000 km long), it would take about 300 years for a given electron to traverse the ocean. And yet a small displacement of the electrons at the London end sends a wave of electrical forces along the cable, and a similar displacement of the electrons in New York occurs within 0.01 s! Even in the circuit of a small electric torch it is doubtful whether a given electron could pass right round the circuit before the battery is run down.

The same analysis applies to conduction in electrolytes, except that the different types of ions may well move with different velocities. In the nature of things the cross-sectional areas of electrolyte are usually greater than those of metallic wires, and the velocities are correspondingly less (p. 38).

In an electron beam the order of magnitude of the velocities is vastly greater. For instance in a television picture tube such a beam is used to provide the moving spot of light that builds up the picture. In this case the electrons move along the tube unimpeded by any collisions, and travel at about a quarter the speed of light. The entire energy of the electrons is given up on impact with the front wall of the tube. Because the speed is so high the number of electrons in the beam is relatively small. Suppose there are N electrons per metre length of the beam moving at a speed v. The number of electrons passing a given point is then Nv per second; and the current I in the beam is given by

$$I = Nve$$

With electrons moving at, say, 8×10^7 m s^{-1} and a total beam current of 100 μA, we have

$$N = \frac{I}{ve} = \frac{10^{-4}}{8 \times 10^7 \times 1.6 \times 10^{-19}}$$
$$= 8 \times 10^6 \text{ electrons per metre}$$

This is a very small fraction of the number of free electrons contained in a metre length of even the finest metal wire that might carry this current.

1.4   Electrical energy

The circulation of an electric current leads to the continual dissipation of energy throughout the circuit in the form of heat; it may also result in the production of other forms of energy—e.g. mechanical energy in an electric motor. All this energy must be continuously supplied by the battery, dynamo, etc., that maintains the flow of current. Such devices for generating electric current are therefore primarily means of converting energy to electrical form from some other form. Thus an electric battery is a means of producing electrical energy from chemical potential energy; a dynamo produces it from kinetic energy, a microphone from the vibrational energy of a sound wave, a photocell from light energy, and so on.

In the electric circuit this production of energy is made apparent by the existence of electrical forces acting on the charged particles in the circuit, giving rise to the current. This is described by saying that the battery, dynamo, etc. produces an *electromotive force*, usually abbreviated to *e.m.f.* The electromotive force is measured in terms of the energy imparted to the complete circuit per unit charge driven round the circuit by the source of e.m.f. (This includes the energy used in driving the current through the source of e.m.f. itself.)

> The **electromotive force (e.m.f.)** of a battery (or dynamo, etc.) is equal to the energy converted to electrical form per unit charge passing through the battery.

Potential difference

A source of e.m.f. imparts energy to a circuit through its ability to set up an uneven distribution of electric charge, by transferring electrons internally from one of its terminals to the other. In this way electrical forces are brought into play which will act on any charged particles near the terminals. We describe this by saying that the source of e.m.f. produces an *electric field* between its terminals. A complete conducting path between the terminals allows the electric field to act inside the conductors forming the circuit, thus setting the free charges in them in motion. (The field also exists outside the conductors; but here it is not effective in causing a current—provided the insulation between the terminals does not break down!) Inside the conductors the field does work on the moving charged particles, causing the electrical energy provided by the source of e.m.f. to be converted into heat and perhaps other forms of energy.

A useful way of describing this field is to attach to each point of the circuit a figure, called its *potential*, in such a way that the flow of current (in the conventional sense) is from points at higher potential to points at lower potential. The difference of potential between two given points is an indication of the strength of the electric field acting to drive current from one point to the other. It is defined as the work done by the forces of the electric field per unit charge passing between the points.

> The **potential difference (p.d.)** between two points acts so as to drive electric current from the point at higher potential to that at lower potential; it is equal to the energy converted from electrical to other forms per unit electric charge that passes.

The SI unit of energy is called the *joule* (J), after the British scientist who established the relationship between heat and mechanical forms of energy. It is the quantity of energy supplied (or the quantity of work done) when a force of 1 newton moves its point of application 1 metre in the direction of the force. Thus

$$1 \, J = 1 \, N \, m$$

The SI unit of potential difference is called the *volt* (V), after the Italian scientist Volta, who first investigated the action of electric batteries. The *millivolt* (mV) and *kilovolt* (kV) are derived units that are also in regular use.

> The **volt** is the potential difference between two points such that the energy converted from electrical to other forms is **1** joule per coulomb of electric charge that passes from one point to the other.

Thus $$1 \, V = 1 \, J \, C^{-1}$$

If a charge Q flows in a part of a circuit across which there is a p.d. V, the electrical energy supplied is given by

$$\text{energy} = QV$$

Since both p.d. and e.m.f. are defined in terms of the energy converted per unit electric charge, it is clear that the same unit is to be used to measure both quantities. Thus the volt is also the unit of e.m.f. If a charge Q flows in a circuit containing a source of e.m.f. E, the electrical energy supplied to the whole circuit is given by

$$\text{energy} = QE$$

It is important to understand the distinction between *electromotive force* and *potential difference*, and to use the two concepts correctly. A source of e.m.f. is an energy converting device, the e.m.f. being the electrical manifestation of the conversion process. The e.m.f. acts *in* the circuit so as to drive current round it. The action of the e.m.f. is to create an electric field between the terminals of the source of e.m.f.; we describe this by saying that the e.m.f. produces a potential difference between the terminals. The potential decreases as we pass round the circuit from the positive to the negative terminal. The potential difference *across* any part of the circuit acts so as to drive current through that part of the circuit and supply electrical energy to it. But this derives from the source of e.m.f., and there could be no p.d. *across* any part of the circuit without an e.m.f. acting *in* the circuit. Because some of the energy generated by the source of e.m.f. is wasted as heat driving current through the source itself, the p.d. between its terminals is usually less than the e.m.f.

Although only *differences* of potential are of any significance in electric circuits, it is sometimes convenient to have a conventional *zero of potential* with reference to which we calculate the potentials of other points. Many pieces of electrical apparatus are operated with some part connected to the earth. It is then usual to regard the earth as being at zero potential. But the choice is quite arbitrary, and we are always free to fix the zero (if we need one at all) to suit our convenience (p. 188).

Batteries

Electric *batteries* consist of several electric *cells* joined together so as to provide greater p.d.'s or greater currents than would be possible with a single cell. There are two ways of doing this:

(i) Cells in series. In this case the same current, and therefore the same charge, flows in each cell. Suppose a charge Q passes through the battery. The total energy imparted to the circuit is the sum of the quantities of energy provided by each cell. In Fig. 1.7a the total energy produced

$$= E_1 \, Q + E_2 \, Q + E_3 \, Q$$

$$\therefore \text{ total e.m.f.} = \frac{\text{total energy}}{Q} = E_1 + E_2 + E_3$$

Most electric cells have e.m.f.'s between 1 and 2 volts. By joining a number of them in series in this way batteries of large e.m.f. may be produced.

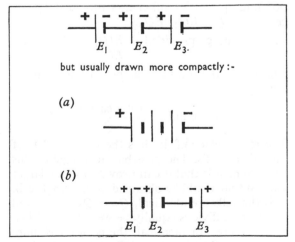

Fig. 1.7   Cells in series: (*a*) all acting in the same direction; (*b*) one cell reversed and acting in opposition to the others

If one of the cells is reversed so that its e.m.f. acts in opposition to the others, it must be counted as negative in assessing the resultant e.m.f. of the combination. Thus in Fig. 1.7*b*

$$\text{resultant e.m.f.} = E_1 + E_2 - E_3$$

The effect of driving a current through a cell in opposition to its e.m.f. is to reverse the energy conversion that usually takes place in it; i.e. chemical energy is formed again from the electrical energy made available by the other cells.

(*ii*) *Cells in parallel.* Suppose we have 3 cells of equal e.m.f. *E* joined in parallel as in Fig. 1.8.

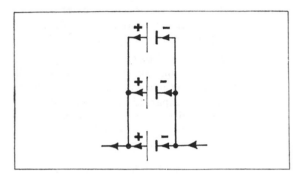

Fig. 1.8   Cells in parallel

In this case the current through the battery divides, part of it passing to each cell. If the quantities of charge passing in a given time through the cells are Q_1, Q_2, Q_3, the corresponding quantities of energy provided by the cells are EQ_1, EQ_2, EQ_3.

$$\therefore \text{ total energy provided} = E(Q_1 + Q_2 + Q_3)$$
$$= EQ$$

where Q is the total charge passing through the battery.

$$\therefore \text{ total e.m.f.} = \frac{\text{total energy}}{Q} = E$$

The resultant e.m.f. is thus the same as that of one of the cells. The possible advantage of this arrangement is that it can provide a much larger current than could be obtained from a single cell.

If the cells are identical, $Q_1 = Q_2 = Q_3$, and the current divides equally between them. When the cells are not identical, the position is more complex. It then becomes difficult to predict how the current will be apportioned between them; and if their e.m.f.'s differ also, it is even possible for some of the cells to drive current backwards through the others. Such arrangements have no practical value.

Power

The rate at which energy is converted from one form to another in some mechanical or electrical device is called its *power*.

The **power** of a device is the rate at which energy is converted in it.

The unit of power is called the *watt* (W) after the British engineer who developed the steam engine. Other units of power in regular use are the *kilowatt* (kW) and the *megawatt* (MW).

Thus $1 \text{ W} = 1 \text{ J s}^{-1}$

A device of power P operating for time t converts a quantity of energy given by

$$\text{energy} = Pt$$

When an electric current is flowing through a circuit component, the p.d. V across it is the energy converted within it per unit charge that passes. The current I is the quantity of electric charge passing per unit time. Therefore the product VI is the energy converted per unit time, i.e. the *power P* of the device.

$$\therefore P = VI$$

If the component is a simple conductor, the energy is dissipated as heat within it and lost to the surroundings. But if the device converts energy in some other way, there may be other forms of energy produced as well; e.g. in an electric motor the power P calculated from the product VI is the total rate of production of energy, both mechanical energy and heat in the motor.

By the same reasoning the power P_{tot} generated by a source of e.m.f. E is given by

$$P_{tot} = EI$$

This is the total power converted to electrical form by the source of e.m.f.; part of it is lost as heat within the source of e.m.f. (e.g. inside a battery), the rest is delivered to the circuit external to the source of e.m.f., and reappears there as heat and perhaps other forms of energy.

★   *Other units of energy and power*

The progress of physics and engineering has been littered with a trail of other units of energy of various kinds.

Some of these remain in use for particular purposes for which they have some special convenience.

(i) *The kilowatt-hour* (kW h).   The calculation of quantities of electrical energy for commercial purposes is conveniently carried out by means of this unit. In Great Britain it is defined by Act of Parliament as the *Board of Trade Unit* of electrical energy, often colloquially called 'the unit', for short. As the name of the unit implies, to calculate quantities of energy in kW h it is only necessary to multiply the power consumption in kilowatts by the time in hours for which it is used.

Since $\quad\quad\quad\quad 1 \text{ kW} = 10^3 \text{ J s}^{-1}$
and $\quad\quad\quad\quad 1 \text{ hour} = 3.6 \times 10^3 \text{ s}$
$\therefore\ 1 \text{ kW h} = 3.6 \times 10^6 \text{ J}$

(ii) *The British Thermal Unit* (B.T.U.).   This is used to some extent by British heating engineers.

$$1 \text{ B.T.U.} = 1.054 \times 10^3 \text{ J}$$

(iii) *The therm*.   This is used for calculating quantities of energy in the gas industry.

$$1 \text{ therm} = 10^5 \text{ B.T.U.} = 1.054 \times 10^8 \text{ J}$$

(iv) *The calorie* (cal).   This was a unit originally devised to measure quantities of heat energy. In fact no fewer than three different versions of the calorie are in existence; (a) the thermochemical calorie (4.184 J); (b) the International Steam Table calorie (4.1868 J); (c) the 15 °C calorie (4.1855 J). To 2 significant figures we have in all three cases:

$$1 \text{ cal} \approx 4.2 \text{ J}$$

(v) *The erg*.   This is an older scientific unit of energy.

$$1 \text{ erg} = 10^{-7} \text{ J}$$

(vi) *The electron-volt* (eV).   This is a convenient unit for the calculation of particle energies in atomic and nuclear physics. It is the quantity of energy gained by an electron in accelerating through a p.d. of 1 V (p. 188). Since the charge on one electron is -1.60×10^{-19} C, we have

$$1 \text{ eV} = 1.60 \times 10^{-19} \text{ J}$$

This is the only additional unit of energy that we shall employ in this book, and then only for its own rather specialized purposes in atomic and nuclear physics.

Each of the above units of energy can of course be combined with a unit of time to form a unit of power, and this is sometimes done. But there is also one other unit of power that we should mention here for the sake of completeness, and as a reference for the student.

The horsepower (h.p.).   This was adopted by British engineers as an advertising stunt at the time when they were seeking to persuade the eighteenth-century mine owners to replace the horse for haulage purposes by the steam engine; the power of the steam engines was therefore quoted in horsepower. For all its absurdity it has proved a remarkably persistent part of the British engineering tradition!

$$1 \text{ h.p.} = 746 \text{ W}$$
$$\therefore\ 1 \text{ h.p.} \approx \tfrac{3}{4} \text{ kW}$$

2 Resistance

2.1 Ohm's law

An electric current is caused to flow through a conductor by applying a potential difference across it. Ohm was the first person to distinguish clearly between the two concepts, p.d. and current, and to investigate the relation between them.

The current in a metallic conductor depends on a number of factors besides the p.d. across it; the most important of these is the temperature. But such things as the elastic strain to which it is subjected, the illumination of its surface, or indeed almost any of its physical conditions, may also affect the current. However, Ohm found that if all these were kept constant, the current in a given conductor is directly proportional to the p.d. across it.

For electrolytic conduction the connection between p.d. and current is complicated by e.m.f.'s that arise at the electrodes, accompanying the conversion of electrical energy to chemical energy (p. 40). The same situation arises in other cases where the conductor is itself the site of an e.m.f.—for instance, the armature coil of a motor, or a coil in which a rapidly changing current induces an e.m.f. by self-induction (p. 81). For these reasons *Ohm's law* only applies strictly to steady currents in metallic conductors which are not themselves the site of any e.m.f.

Ohm's law. A steady current flowing through a metallic conductor, which is not itself the site of an e.m.f., is proportional to the potential difference between its ends provided the temperature and other physical conditions are constant.

A consequence of Ohm's law is that the quantity

$$\frac{\text{potential difference}}{\text{current}}$$

is a constant for a given metallic conductor under steady physical conditions. It is known as its *resistance*. Even when we are dealing with a circuit component for which Ohm's law does not apply, it is still often convenient to define this quantity as the resistance of the component; in such a case the resistance depends on the current.

The **resistance** of a conductor which is not itself the site of an e.m.f. is

$$\frac{\text{the potential difference across it}}{\text{the current through it}}$$

when a steady current is flowing; for a conductor that obeys Ohm's law this quantity is constant for all currents if the temperature and other physical conditions are steady.

A knowledge of the resistances of the conductors making up a circuit enables us to calculate the distribution of currents and p.d.'s in it. The unit of resistance is called the *ohm* (Ω—the Greek letter 'omega').

The **ohm** is the resistance of a conductor through which a steady current of 1 amp passes when a potential difference of 1 volt exists across it, the conductor not itself being the site of an e.m.f.

Thus $$1 \ \Omega = 1 \ V \ A^{-1}$$

If a current I flows through a resistor when the p.d. across it is V, its resistance R is given by

$$R = \frac{V}{I}$$

It equally follows from Ohm's law that the reciprocal quantity

$$\frac{\text{current}}{\text{potential difference}}$$

is a constant for a given metallic conductor; this is known as its *conductance G*. The unit is called the Ω^{-1} or sometimes the *siemens* (S) after another German scientist.

Thus $$1 \ S = 1 \ \Omega^{-1} = 1 \ A \ V^{-1}$$

and we have

$$G = \frac{1}{R} = \frac{I}{V}$$

The relationship between current and applied p.d. for a coil of wire (or any other suitable component) may be investigated with the simple

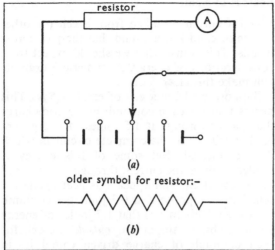

Fig. 2.1   (a) Investigating the relationship between current and p.d. for a resistor; (b) the older symbol for a resistor

arrangement shown in Fig. 2.1a. [Note the symbol for the coil of wire or *resistor*, as we call it. An older symbol for a resistor was the zig-zag line shown below in Fig. 2.1b. This symbol is still widely used. But in this book we shall employ only the internationally agreed forms of circuit symbols, incorporated in British Standard 3969.]

The applied p.d. may be varied in equal steps by selecting different tapping points on the battery, and in each case the current through the resistor is noted. If Ohm's law applies, a graph of p.d. against current will be a straight line through the origin. If the e.m.f. of each cell is also known, then the p.d.'s can be expressed in volts, and the gradient of the graph line then gives the resistance in ohms. The battery must be of a type that itself introduces negligible resistance into the circuit. A set of lead-acid accumulators is ideal (p. 43), e.g. a car battery; but dry cells can be used provided their internal resistance is small compared with that of the resistor. The measured resistance inevitably includes the small resistance of the ammeter, and this must be allowed for if necessary (p. 49).

Variable resistors

The arrangement in Fig. 2.1 only enables the current and p.d. to be adjusted in steps. If we

want a finer control than this we must employ a variable resistor. The simple ones used in radio circuits consist of a strip of carbon composition in the form of a broken circle to which a sliding connection is made by means of a rotating arm; fixed connections are made to the two ends of the strip. A similar wire-wound design has the wire wound on thin card, which is bent into an arc of a circle; the rotating arm makes contact with the turns along the edge of the card.

For dealing with larger currents a straight ceramic tube is wound with the resistance wire, and the sliding contact is carried on a metal bar parallel with the tube.

A variable resistor can be used in two ways in circuits:

(a) *As a rheostat*—for controlling the current in a low-resistance device; only one of the end connections is used in this case (Fig. 2.2a), and often the circuit symbol for this is simplified as shown in the alternative diagram alongside. To provide a large range of adjustment of current the rheostat must have a maximum resistance much greater than that of the rest of the circuit in series with it.

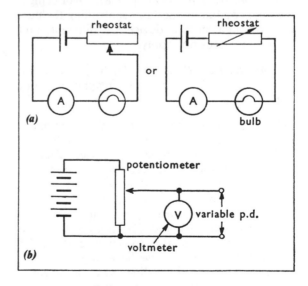

Fig. 2.2   The uses of a variable resistor (a) as a rheostat to control a current—two alternative symbols are shown; (b) as a potentiometer to provide a variable p.d.

(*b*) *As a potentiometer*—for controlling the p.d. across some device. Fig. 2.2*b* shows how the variable resistor is connected up for this purpose. The battery is joined to the two end terminals of the resistor, so that the battery p.d. is applied across the whole length of resistance wire. Any required fraction of the total p.d. can then be tapped off between the sliding contact and one end of the wire. This arrangement enables the p.d. to be varied continuously right down to zero.

The latter arrangement can be used to provide a continuously variable p.d. to apply across the resistor and ammeter in the experiment of Fig. 2.1. With this improved arrangement it is instructive to repeat this experiment not only for coils of wire at steady temperatures for which Ohm's law holds, but also for a number of components for which it does not hold. For example:

(*i*) *A fine wire*, such as an electric bulb filament. In this case the temperature changes considerably with the current, and Ohm's law does not apply. This is even more obviously the case with the special temperature dependent resistors, known as *thermistors,* made of semiconducting substances (p. 25).

(*ii*) *A semiconductor diode* (p. 258).   Not being a metallic conductor the laws controlling the current through such a device are different, and Ohm's law does not apply.

(*iii*) *An electric cell* (p. 39).   Such a device is the site of an e.m.f. and again Ohm's law does not apply.

With each of the last two the experiment should be performed with the current flowing in both directions through the device. There is no single constant that describes the properties of such components. Rather we must display them by plotting a graph of p.d. *V* against current *I*; and, if the properties are unsymmetrical—as with (*ii*) and (*iii*) above—then the graph must extend also to negative values of *V* and *I*. In some cases we have to plot a range of curves of this sort for different values of some other quantity on which the current depends, e.g. the temperature. Such sets of curves are called the *characteristics* of the component concerned.

Sources of e.m.f.

When a current is taken from a cell (or other source of e.m.f.), it is found that the p.d. across it falls. This is just what we should expect from considerations of energy. A numerical example will make this clear:

Suppose we have a cell of e.m.f. 1.5 V. This means that for each coulomb of electric charge that passes through it 1.5 joules of energy are made available to the complete circuit in which it is connected. But some of this energy is needed to drive the current through the material of the cell itself, and appears as heat inside the cell. Let us suppose that 0.3 joule per coulomb is so used. This means that 1.2 joules of energy are available in the circuit *outside* the cell for each coulomb of charge driven round it—in other words, that the p.d. across it is 1.2 volts.

The way in which the p.d. across a cell varies with the current through it may be investigated with the arrangement shown in Fig. 2.3.

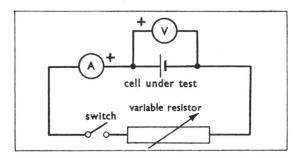

Fig. 2.3   Finding how the p.d. across a cell varies with the current through it

With most types of cell the results are not at all uniform; the chemical changes occurring in a cell when a current flows lead to changes in its e.m.f.—an effect known as *polarization* (p. 40). The p.d. across the cell therefore depends not only on the current taken, but also on the length of time for which it has been flowing and on the previous treatment of the cell. However, a graph of p.d. against current is of the general form shown in Fig. 2.4. For zero current the p.d. is equal to the e.m.f. (provided the voltmeter draws negligible current); the subsequent *drop in p.d.* is approximately proportional to the current, shown by the initial straight line part of the graph.

A similar result is found for most other sources

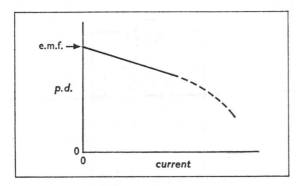

Fig. 2.4   The variation of the p.d. across a cell with the current through it

of e.m.f. The p.d. falls when a current is taken, and, as with a cell, it is often difficult to disentangle how much of this fall of p.d. is due to the internal resistance of the source and how much to a change in the e.m.f. For instance when current is taken from a d.c. dynamo, as well as the drop in p.d. due to the resistance of the dynamo coils there is also likely to be a change in the e.m.f. caused by a change in speed of the machine. However, when allowance is made for such effects, it is found that the difference between the e.m.f. E of the source and the p.d. V across it is proportional to the current I flowing through it. Thus

$$(E - V) \propto I$$

When $I = 0$, $V = E$, and only with zero current is the p.d. measured with a high resistance voltmeter equal to the e.m.f.

A source of e.m.f. behaves as though it included an *internal resistance* R_{int} joined in series with it (Fig. 2.5); it is as though the observed drop in p.d. $(E - V)$ occurred across

this. The internal resistance R_{int} is of course an integral part of the source of e.m.f. that can never be separated from it. But for the purpose of circuit analysis we may represent it separately as in Fig. 2.5, provided we remember that the p.d. V across the cell is actually that across cell and internal resistance combined. We therefore have

$$R_{int} = \frac{(E - V)}{I}$$
$$\therefore E = V + IR_{int}$$

But V is also the p.d. across the *external resistance* R_{ext} of the circuit, and we have

$$R_{ext} = \frac{V}{I}$$
$$\therefore V = IR_{ext}$$
$$\therefore E = I(R_{ext} + R_{int})$$

Now $(R_{ext} + R_{int})$ is the *total resistance* of the circuit. If we represent this by R_{tot}, we have

$$R_{tot} = \frac{E}{I}$$

The three relationships printed here in bold letters are of essentially the same form; they show the application of the concept of resistance to the analysis of the electrical behaviour of the different parts of a circuit: (*a*) of the source of e.m.f. (*b*) of the external circuit joined to the source, and (*c*) of the complete circuit in which the source of e.m.f. is included.

EXAMPLE   A dry cell of e.m.f. 1.5 V and internal resistance 2.0 Ω is connected to a torch bulb whose resistance at its operating temperature is 5.5 Ω. What is the p.d. across the cell?

The total resistance R_{tot} of the circuit is given by

$$R_{tot} = 5.5 + 2.0 \ \Omega$$
$$= 7.5 \ \Omega$$

$$\therefore \ \text{current} = \frac{E}{R_{tot}} = \frac{1.5}{7.5} \ \text{A}$$

$$= 0.2 \ \text{A}$$

Now considering the cell only,

$$\text{drop in p.d.} = IR_{int} = 0.2 \times 2.0 \ \text{V}$$
$$= 0.4 \ \text{V}$$

$$\therefore \ \text{p.d. across cell} = 1.5 - 0.4 \ \text{V}$$
$$= 1.1 \ \text{V}$$

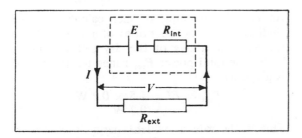

Fig. 2.5   The internal resistance of a cell represented by a separate resistance R_{int} in series

Alternatively we could reason:

$$\text{p.d. across bulb} = IR_{ext} = 0.2 \times 5.5 \text{ V}$$
$$= 1.1 \text{ V}$$

and this is necessarily the same as the p.d. across the cell.

A simple method of measuring the internal resistance of a source of e.m.f. is that shown in Fig. 2.6. With the switch open the high resistance voltmeter indicates the e.m.f. E of the source. Then when the switch is closed the p.d. across the source falls, and the reading V is taken. We then have

for the complete circuit

$$E = I(R_{ext} + R_{int})$$

and for the external circuit

$$V = IR_{ext}$$

Dividing these equations,

$$\frac{E}{V} = \frac{R_{ext} + R_{int}}{R_{ext}} = 1 + \frac{R_{int}}{R_{ext}}$$

$$R_{int} = R_{ext}\left(\frac{E}{V} - 1\right)$$

If there is any significant polarization in a cell, this will be shown up by a steady fall in the voltmeter reading after the switch has been closed. The reading required to give the internal resistance is that immediately after closing the switch.

Power dissipation in resistors

The electrical power P delivered to a resistor (or any other circuit component) is given by

$$P = VI \quad \text{(p. 14)}$$

where $V =$ the p.d. across it,
and $I =$ the current through it.

But $R = \dfrac{V}{I}$

substituting for V,

$$P = I^2R$$

This expresses the result first discovered experimentally by Joule:

Joule's law. The rate of production of heat by an electric current in a given resistance is proportional to the square of the current flowing in it. ($P \propto I^2$)

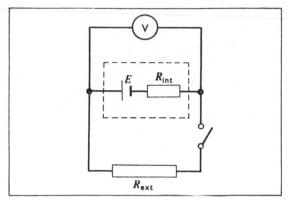

Fig. 2.6   Measuring the internal resistance of a cell

Substituting also for I,

$$P = \frac{V^2}{R}$$

Thus we have three alternative expressions for the power production P:

$$P = VI = I^2R = \frac{V^2}{R}$$

The last two of these formulae can apply only to the production of *heat* in resistors. But the first formula, not involving the resistance R, may be used to give the rate of production of energy in any form in a circuit component in which a current I flows at a potential difference V. For example, if a current of 0.5 A is flowing through the armature of a small motor at a p.d. of 20 V, the total power supplied is 10 W; this includes heat and mechanical energy. If the armature resistance is 12 Ω then

the rate of production of *heat* $= I^2R = 3$ W
$\therefore$ the rate of production of *mechanical energy*

$$= 10 - 3 \text{ W}$$
$$= 7 \text{ W}$$

As a further example of such calculations, we shall consider the quantities of power used in the example above of the dry cell and torch bulb. The total power P_{tot} generated in the cell is given by

$$P_{tot} = EI = 1.5 \times 0.2 \text{ W}$$
$$= 0.3 \text{ W}$$

This is the rate at which energy is converted from chemical to electrical form in the cell. Some of this energy is dissipated forthwith as heat in the internal resistance R_{int} of the cell itself. The power P_{int} lost in this way is given by

$$P_{int} = I^2 R_{int} = (0.2)^2 \times 2.0 \text{ W}$$
$$= 0.08 \text{ W}$$

The power P_{ext} delivered to the external resistance R_{ext}, in this case the resistance of the torch bulb, is given by

$$P_{ext} = I^2 R_{ext} = (0.2)^2 \times 5.5 \text{ W}$$
$$= 0.22 \text{ W}$$

[Equally we could say,

$$P_{ext} = VI = 1.1 \times 0.2 \text{ W}$$
or, $$P_{ext} = V^2/R_{ext} = (1.1)^2/5.5 \text{ W}]$$

And of course we observe in passing that

$$P_{tot} = P_{int} + P_{ext}$$

★ *The maximum power theorem*

A source of e.m.f. E delivering a current I generates a total power P_{tot} given by

$$P_{tot} = EI$$

Some of this power is wasted producing heat in the source of e.m.f. and the remainder is delivered for use in the rest of the circuit (often called the *load*). It is of importance to know under what conditions the power P_{ext} delivered to the load will be a maximum.

Suppose the source of e.m.f. has internal resistance R_{int} and the load is a resistance R_{ext}. Then

$$I = \frac{E}{R_{ext} + R_{int}}$$

$$\therefore P_{ext} = I^2 R_{ext} = \frac{E^2 R_{ext}}{(R_{ext} + R_{int})^2}$$

To find the value of R_{ext} that makes this expression a maximum we need to differentiate it with respect to R_{ext} (taking R_{int} as constant), and equate to zero.

$$\therefore \frac{dP_{ext}}{dR_{ext}} = \frac{E^2(R_{int} - R_{ext})}{(R_{ext} + R_{int})^3}$$

This is zero, when

$$R_{ext} = R_{int}$$

It is easily verified that this stationary value is indeed a maximum.

A source of e.m.f. delivers maximum power when the resistance of the load is equal to the internal resistance of the source.

The condition for maximum *efficiency* is rather different. Thus

$$\text{efficiency} = \frac{P_{ext}}{P_{tot}} = \frac{R_{ext}}{R_{ext} + R_{int}}$$

To make this a maximum we need R_{ext} to be large compared with R_{int}. In the maximum *power* condition the efficiency is 50%, i.e. half the power generated is wasted

as heat in the source of e.m.f. When large power outputs are concerned, this situation is clearly unacceptable, and we shall rather aim at maximum efficiency—chiefly by making the internal resistance of the source small. However, when we need to extract the maximum possible value from a small source of power, it may be desirable to aim at the maximum power condition. An example of this situation for alternating currents (where similar considerations apply) is an amplifier delivering power to a loudspeaker.

2.2 Combinations of resistors

Many circuits contain networks of resistors and cells; and we must be able to calculate the combined effective resistance of various arrangements.

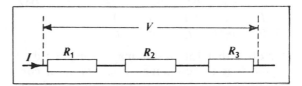

Fig. 2.7   Resistors in series

(*i*) *Resistors in series* (Fig. 2.7).   In this arrangement the same current I flows through all the resistors, R_1, R_2 and R_3.

The combined series resistance R_{ser} is given by—

$$R_{ser} = \frac{V}{I}$$

where V is the total p.d. across the combination. V is the sum of the separate p.d.'s across the individual resistors.

$$\therefore V = IR_1 + IR_2 + IR_3$$
$$= I(R_1 + R_2 + R_3)$$
$$\therefore R_{ser} = R_1 + R_2 + R_3$$

(*ii*) *Resistors in parallel* (Fig. 2.8).   In this arrangement the same p.d. V exists across all the resistors, R_1, R_2, R_3.

The combined effective resistance R_{par} is again given by—

$$R_{par} = \frac{V}{I}$$

where I is the total current through the combination. I is the sum of the separate currents through the individual resistors.

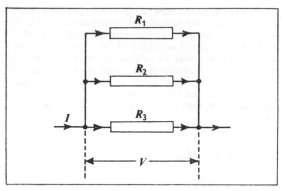

Fig. 2.8   Resistors in parallel

$$\therefore I = \frac{V}{R_1} + \frac{V}{R_2} + \frac{V}{R_3} = V\left(\frac{1}{R_1} + \frac{1}{R_2} + \frac{1}{R_3}\right)$$

$$\therefore \frac{I}{V} = \frac{1}{R_{\text{par}}} = \frac{1}{R_1} + \frac{1}{R_2} + \frac{1}{R_3}$$

Notice that this result may be expressed by saying that the combined *conductance* of conductors in parallel is equal to the sum of their separate *conductances*, since conductance is the reciprocal of resistance. Thus

$$G_{\text{par}} = G_1 + G_2 + G_3$$

In two particular cases the last formula may be simplified, making for greater speed of calculation:

(*a*) The combined resistance of *n equal resistances R* in parallel is given by—

$$R_{\text{par}} = \frac{R}{n}$$

(*b*) For *two resistors* only in parallel, re-arrangement gives—

$$R_{\text{par}} = \frac{R_1 R_2}{R_1 + R_2} = \left(\frac{\text{product}}{\text{sum}}\right)$$

The same rules may be used to calculate the combined effective internal resistance of an arrangement of cells. The rules for obtaining the combined e.m.f. of a complete battery have already been discussed (p. 13). Consider for instance a battery of three identical cells each of e.m.f. 1.5 V and internal resistance 0.6 Ω.

Joined (*a*) in series, their combined e.m.f. is 4.5 V and the internal resistance is 1.8 Ω.

Joined (*b*) in parallel, the e.m.f. is the same as that of one cell, 1.5 *V*, and the internal resistance

is equal to that of three 0.6 Ω resistances in parallel, namely 0.2 Ω.

With these rules it is possible to deduce the effective resistance of most arrangements of conductors, and to calculate the currents in many types of circuits. Cross-connected arrangements (such as unbalanced forms of the Wheatstone networks and potentiometer circuits discussed in Chapter 4) cannot generally be handled in this way, but must be approached from first principles individually.

2.3 Resistivity

The resistance of a conductor at a given temperature depends on its length and cross-sectional area and on the material of which it is made. It is easily shown by experiment that the resistance of a uniform conductor made of any given material is directly proportional to its length and inversely proportional to its cross-sectional area. On theoretical grounds this can be seen to follow from the results of the previous section. A conductor of length *l* can be regarded as a set of conductors in series each of unit length; suppose each such conductor has a resistance *r*. Then the total resistance is *lr*, which is proportional to the length. Likewise a conductor of cross-sectional area *A* can be regarded as a collection of *A* conductors in parallel each of unit cross-sectional area; suppose each conductor has a resistance *r'*. Then the total resistance is *r'/A*, which is inversely proportional to the cross-sectional area. Thus the resistance *R* of a conductor of length *l* and cross-sectional area *A* may be written in the form

$$R = \frac{\rho l}{A}$$

where ρ is a property of the material of the conductor, called its *resistivity*.

Resistivity. The resistance *R* of a uniform conductor is directly proportional to its length *l* and inversely proportional to its cross-sectional area *A* ;

$$\therefore R \propto \frac{l}{A}$$

$$\text{or} \quad R = \frac{\rho l}{A}$$

where ρ is a constant (for fixed temperature and other physical conditions) for the material of the conductor known as its resistivity.

Re-arranging this formula,

$$\rho = \frac{RA}{l}$$

This shows us that the dimensions of resistivity are those of [resistance × length]. The unit of resistivity is accordingly the Ω m. The reciprocal of resistivity is known as the *conductivity* σ of the material. Its unit is the Ω^{-1} m^{-1}.

$$\therefore\ \sigma = \frac{1}{\rho}$$

In order to measure the resistivity of a material we have first to select a wire of suitable quality. Modern wire is manufactured to very high standards of uniformity of diameter, but we cannot rely on this in conducting the measurements. The diameter should be measured at several positions along the length of it—in case it tapers; and at each position in two directions at right angles—in case the cross-section is not exactly circular. If the figures differ very little, their mean is taken as the effective diameter. The accuracy of the value of resistivity obtained is limited chiefly by the measurement of the diameter. For example, suppose the wire used has a diameter of about 1 mm; using a micrometer screw-gauge, this should be found with a maximum error of 0.005 mm—i.e. $\frac{1}{2}\%$ error. But this figure will be squared in the calculation of the area of cross-section. Therefore the resulting final error will be twice this, namely 1%. There is plainly no great difficulty in finding the other quantities needed to at least this accuracy; provided the resistance R is more than 1 ohm, a bridge method will suffice (p. 50), and, if the length l is more than 0.5 m, an error of 2 mm in measuring this will be relatively unimportant. If the material under test is a pure metal, its temperature must be controlled and recorded; an uncertainty of 1 K could be significant. However, in the case of the special resistance alloys—constantan, eureka and manganin (p. 25)—this precaution is unnecessary, since their resistivities scarcely change at all with temperature.

Silver is the best conductor, closely followed by copper. The resistivity of a pure metal is considerably increased by even small traces of impurity; for this reason the copper used for electrical connecting wire is always electrolytically purified (p. 34). Aluminium has a resistivity about twice that of copper. But since its density is about a third as much, an aluminium conductor can be much lighter and cheaper than a copper one. The overhead power lines of the national grid system are made of aluminium; the lighter cables enable less expensive pylons to be used to support them. The cables carry a steel core to give them the tensile strength to enable very long loops to be suspended between pylons; the aluminium strands are twisted around this core.

Resistivities of various common metals
(293.15 K or 20°C)

Metal	Resistivity 10^{-8} Ω m	Temp. coeff. K^{-1}	
Pure metals			
Silver	1.63		0.0040
Copper	1.69		0.0043
Aluminium	3.21		0.0038
Gold	2.42		0.0040
Platinum	11.0		0.0038
Iron (soft)	14.0		0.0062
Lead	20.8		0.0043
Mercury	94.1		0.0009
Alloys			
Nichrome	130		0.00017
Eureka, Constantan	49	about	0.00002
Manganin	44	about	0.00002

The table shows the resistivities of a number of pure metals and alloys. The resistivities of non-metallic substances are vastly greater. Almost all of them are more than 10^4 Ω m; and those materials used for insulation of electrical circuits have resistivities between 10^{13} and 10^{16} Ω m. Intermediate between these extremes comes the important class of materials known as *semiconductors*—germanium, silicon, etc. In these substances only a very small proportion of the valence electrons are 'free' to move through the crystal lattice; the resistivity of germanium is 0.65 Ω m, and of silicon 2.3 × 10^3 Ω m. However, these figures are obtained only with the purest specimens; and the addition of minute traces of impurities may increase the number of free charges in the crystal lattice considerably, and so *reduce* the resistivity (p. 257). By contrast, impurities in a metal do not much affect the number of conduction electrons, but have a marked effect on their freedom of movement and therefore *increase* the resistivity.

The student should notice the use of suffixes in the words introduced in this chapter—

resist*or*   resist*ance*   resist*ivity*
conduct*or*   conduct*ance*   conduct*ivity*

This usage is common in scientific terminology, and its significance should be apparent from these examples.

'*-or*' signifies the *body* that possesses the property considered—in the above examples it is perhaps a metal wire.

'*-ance*' signifies the *property* that the *body* possesses.

'*-ivity*' signifies the corresponding *property* of the *material* of which the body is composed.

The physical basis of Ohm's law

For a conductor that obeys Ohm's law we have

$$\frac{V}{I} = R = \frac{\rho l}{A}$$

where the resistivity ρ is a constant for the material at given temperature. Re-arranging, we can write this as

$$\frac{V}{l} = \rho \frac{I}{A}$$

The quantity I/A is the current per unit area of cross-section; it is known as the *current density j* in the material.

$$\therefore j = \frac{I}{A}$$

Thus if Ohm's law holds for the material we have

$$j \propto \frac{V}{l}$$

This quantity V/l is the potential gradient measured along the length of the conductor; we can regard it as a measure of the *electric field strength E* acting on the charged particles in the conductor. Thus

$$E = \frac{V}{l} \quad (= \text{the potential gradient})$$

If a material obeys Ohm's law we have

$$j \propto E$$

or

$$j = \frac{1}{\rho}E = \sigma E$$

where σ is the conductivity of the material.

We have already deduced an expression giving the velocity v of the charged particles in a conductor of cross-sectional area A carrying a current I.

$$v = \frac{I}{nAe} \quad \text{(p. 11)}$$

where n is the number of 'free' charged particles per unit volume and e is their charge. In terms of the current density j this can be written

$$v = \frac{j}{ne}$$

$$\therefore v = \frac{\sigma E}{ne}$$

At the atomic level we can see that Ohm's law will hold for the material if the average velocity v of the charge carriers is proportional to the electric field strength E in which they are placed.

$$v \propto E$$

We can write this as: $v = u_e E$

where u_e is a constant known as the *electric mobility* of the charge carriers. Comparing with the previous expression, u_e is given by

$$u_e = \frac{\sigma}{ne}$$

$$\therefore \sigma = u_e ne$$

We shall see in due course a method of measuring the concentration n of charge carriers in a material (p. 70). The measurement of conductivity then provides us with a direct measurement of the mobility of the charge carriers u_e.

The same analysis applies to conduction in an electrolyte, except that in this case there may be several types of ion present in the solution each with its own mobility. The result above shows that the conductivity of the electrolyte should be proportional to the number of ions n per unit volume. This is confirmed by experiment— at least for very dilute solutions in which the solute is completely dissociated into its constituent ions (p. 38). If the concentration of solute is established by chemical means, a measurement of conductivity gives the average mobility of the ions. Conversely, if the mobility is known, a measurement of conductivity gives a quick and accurate indication of the concentration.

When the temperature of a metal is raised, the thermal vibrations of the atoms increase; this increases the interaction of the electrons with the crystal lattice, and therefore reduces u_e. On this account the resistance of a metal increases with temperature. On the other hand in electrolytes raising the temperature decreases the viscosity of the liquid, allowing the ions to move more freely. A temperature rise therefore increases the conductivity of an electrolyte (and so decreases its resistivity).

Semiconductors show a very considerable decrease of resistivity when the temperature rises. In these materials greater thermal vibration of the atoms 'frees' more electrons in the crystal lattice; and n is thus greatly increased. In some semiconductors the number of conduction electrons can also be increased by raising the p.d. These materials do not then obey Ohm's law; indeed the current in them may vary as the third or fourth power of the p.d. Substances with this property are used for making *surge suppressors*. These are devices for protecting equipment from sudden surges of current such as may occur at the moments of switching on and off. The surge suppressor is joined in parallel with the equipment to be protected; very little current flows through it at the normal p.d. But a sudden increase in p.d. causes an enormous rise in the suppressor current, and diverts the surge from the equipment.

The resistivity of any pure metal is found to increase nearly uniformly with temperature. This variation is described with the aid of a temperature coefficient similar to the coefficients of expansion used in dealing with changes of length and volume.

The **temperature coefficient of resistivity** of a substance is the increase in its resistivity per unit rise of temperature divided by its resistivity at 0°C.

If this temperature coefficient is α, and the resistivity at $t\,°C$ is ρ_t we have

$$\alpha = \frac{\rho_t - \rho_0}{\rho_0 t}$$

Re-arranging $\rho_t = \rho_0(1 + \alpha t)$

There is surprisingly little variation between the temperature coefficients of resistivity of many pure metals, as may be seen from the table on p. 23. Also the values of these coefficients occupy a range near the figure 3.7×10^{-3} K^{-1}, which is the same as the coefficient of expansion of a gas ($1/273$ K^{-1}). It follows that the resistivity of a pure metal is approximately proportional to its absolute temperature. This provides a very useful way of doing rough calculations of changes of resistance with temperature.

When measurements of resistivity are made over large temperature ranges it is found that the variation is not quite uniform. But for all pure metals a slight modification of the above expression can be made to agree very closely with the experimental values:

$$\rho_t = \rho_0(1 + \alpha t + \beta t^2)$$

where β is a coefficient that is very small in comparison with α.

For alloys the temperature coefficients of resistivity are generally less than for pure metals. For example, for nichrome (80% nickel, 20% chromium) it is about 0.17×10^{-3} K^{-1}. This alloy is used for heating elements of electric fires, cookers, etc.; it has the property of forming a protective oxide layer when heated, which partly protects it against further oxidation at high temperatures and provides insulation between the coils of wire in the element. It is also commonly used in wire-wound resistors for electronic equipment, since its resistivity is unusually high. Eureka or constantan (60% copper, 40% nickel) and manganin (84% copper, 12% manganese, 4% nickel) are two alloys whose temperature coefficients of resistivity are very small; since they also have quite high resistivities, they are useful for constructing standard resistors. Semiconductors have large *negative* temperature coefficients of resistivity.

At temperatures near to the absolute zero some metals are found to lose the property of resistance altogether and become perfect conductors. They are then said to be *superconducting*. The onset of superconductivity occurs at a sharply defined transition temperature, which is lowered if the specimen is placed in a magnetic field. Electric currents have been maintained in rings of superconducting material for periods of several months, conclusively demonstrating the complete absence of resistance in metals in this condition. Above the transition temperature the same currents would have ceased in a few seconds in even the largest rings. Fig. 2.9 shows how the resistivity of such a metal varies from the absolute zero upwards.

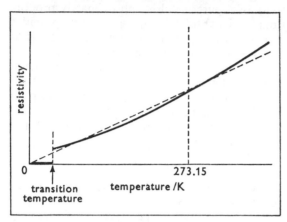

Fig. 2.9   The variation of resistivity with temperature for a typical pure metal that becomes superconducting near the absolute zero

2.4   Types of resistor

Many different methods have been developed for constructing fixed and variable resistors suitable for the numerous purposes for which they are required in radio engineering and science. The most important types are described briefly below.

(*i*) *Carbon composition resistors*. By mixing carbon, which is a conductor, with non-conducting materials it is possible to produce a mixture which can have almost any value of resistivity. The usual mixture is of carbon black, resin binder and a refractory powder (of the kind used in making fire bricks); this is milled and sifted, and then pressed into short cylindrical blocks. Heating in a kiln solidifies the blocks. The ends are sprayed with metal, to which tinned copper connecting wires are soldered (Fig. 2.10).

The resistors are then measured and sorted into resistance 'groups'. They are sold as having 5%, 10% or 20% tolerance. Their stability is very poor; the values are likely to drift by about 5% in the course of a year, and any stress (such

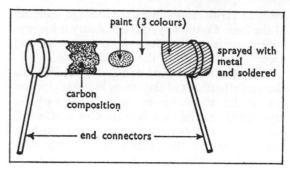

Fig. 2.10   A carbon composition resistor

as excessive current or heating in other ways) may cause a permanent change in resistance of 30% or more. A further defect is that the resistance may vary with the p.d. applied, which is another way of saying that Ohm's law does not hold for this composition. However, these defects are quite unimportant for many radio purposes, and their cheapness and small size lead to their being widely used.

★   The values of low accuracy resistors are usually marked on them with the Standard Colour Code:

The table shows the digits denoted by each colour. (The sequence of colours is approximately that of the spectrum.)

Colour	Digit	Colour	Digit
Black	0	Green	5
Brown	1	Blue	6
Red	2	Violet	7
Orange	3	Grey	8
Yellow	4	White	9

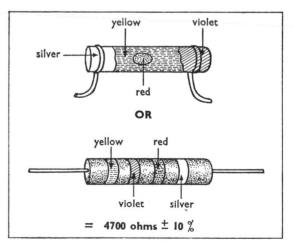

Fig. 2.11   The use of the Standard Colour Code

The colour of the body of the resistor or of the first ring or dot from the end indicates the first digit; the colour of the end or of the second ring or dot gives the second digit; and a single spot in the middle or the third ring or dot shows the number of ensuing noughts (Fig. 2.11). A fourth colour is used to show the tolerance; gold 5%, silver 10%, and no colour 20%.

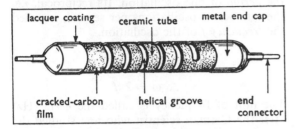

Fig. 2.12   A cracked-carbon resistor

(*ii*) *Cracked-carbon resistors.*   By heating small ceramic rods to a temperature of about 1300 K in methane vapour, decomposition or 'cracking' of the vapour occurs, and a uniform layer of carbon is deposited. The resistance between the ends at this stage is between 10 and 1000 ohms, depending on the thickness of the layer. This is then increased by anything up to 10 000 times by cutting a helical groove in the film, thereby altering the shape of the actual conducting path. In this way, the resistance may be adjusted to the required value to within 2% or better. Connection to the ends is made by metal caps fitted over them (Fig. 2.12). These resistors maintain their values to within about 2% under most ordinary stresses.

(*iii*) *Wire-wound resistors.*   For precision purposes this is the only form of resistor ever used.

Two common designs are shown in Fig. 2.13. Enamelled nichrome wire is used in the ordinary commercial type, since it has a very high resistivity and may be used in fine gauges without much danger of damage by corrosion; the overall accuracy and stability of such resistors is about 1%. Manganin is used for high precision work, since it has a very low temperature coefficient of resistance, and also produces very low thermal e.m.f.'s (p. 45) when connected in a circuit with copper wires. The accuracy and stability can be better than 0.01% in ordinary use, and rather better than this by careful control of the temperature. The value of a *standard resistor* is sometimes adjusted to the exact value required by incorporating a large resistance in parallel with it; large variations may be made in the length of wire in the latter resistor to effect only small changes in the total resistance.

With any type of resistor, and particularly with standards, great care should be taken not to exceed the maximum power dissipation for which it is designed. Excessive heating invariably changes the resistance value and may damage the insulation or even fuse the wire. Even if the power dissipation is kept within the rated amount, adequate ventilation must be allowed to avoid undue rise of temperature.

★   Wire-wound resistors for use in circuits with varying currents have to be wound *non-inductively* (p. 95); this means that the magnetic field produced by the coil must be a minimum. With high-frequency alternating currents it is also necessary to reduce the effective capacitance between the ends of the coil (p. 165); this means that the electric field between the ends must be a minimum, or in other words that the ends of the coil (between which there

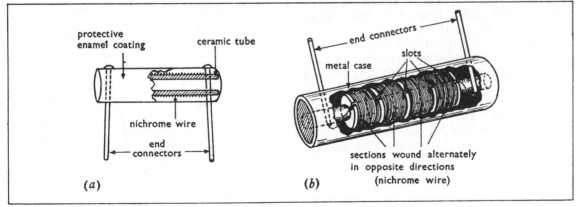

Fig. 2.13   Wire-wound resistors

Fig. 2.14   A simple non-inductive winding for a resistance coil

may be a high p.d.) must be kept well separated from each other.

The simplest form of non-inductive winding is made by doubling the wire to be used, and then winding the doubled piece on a suitable core (Fig. 2.14). The two halves of the coil then produce equal and opposite magnetic effects. But this arrangement is not suitable for high frequencies, since the capacitance between the ends is large. Fig. 2.13b shows a non-inductive winding in sections; the sections are wound alternately in opposite directions.

2.5   Alternating currents

We have dealt so far only with cases in which the currents are steady and in one direction: this is called *direct current* (d.c.). For large-scale power distribution there are however many advantages in using *alternating current* (a.c.). In this case the direction of the e.m.f. acting in the circuit reverses many times a second and so therefore does the current.

The variation of current with time (the waveform) may follow almost any repetitive pattern; but the most important for practical purposes is the sinusoidal waveform shown in Fig. 2.15. This is the waveform to which the public electricity supply approximates closely. The sinusoidal waveform is so called because the variation of current I with time t is according to the *sine function*:

$$I = I_0 \sin \omega t$$

where I_0 and ω are constants. Since the maximum and minimum value of $\sin \omega t$ are $+1$ and

-1, I_0 in this equation is the maximum value (or *peak value*) of the current in either direction. Also, a sine function goes through a complete cycle of variation when the angle advances by 2π radians; in the above equation ωt advances by 2π when t advances by $2\pi/\omega$. This is called the *period* of one oscillation. Its reciprocal, i.e. the number of oscillations per second, is called the *frequency f* of the oscillation.

Thus
$$f = \frac{\omega}{2\pi}$$
$$\therefore \; \omega = 2\pi f$$

The unit of frequency is called the *hertz* (Hz) after the German scientist who first studied the properties of oscillatory electric discharges. Thus

$$1 \text{ Hz} = 1 \text{ s}^{-1}$$

The frequency of the public supply in Great Britain (and in most of Europe) is 50 Hz; in the U.S.A. a frequency of 60 Hz is used.

For many applications it is quite immaterial whether alternating or direct current is used. Electric heating and lighting for instance are unaffected by the direction of flow of the current. The power dissipation in a resistor fluctuates (100 times per second with the 50 Hz mains); but this is too rapid to allow much oscillation of the temperature of an electric fire element or of the filament of a bulb, though the variation of temperature is quite detectable. A slight hum can usually be heard from an electric fire at 100 oscillations per second (a note close to A flat on the musical scale); this is due to the slight alternate contrac-

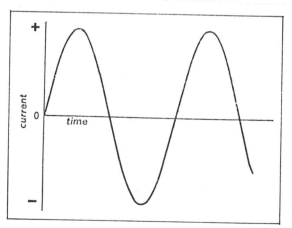

Fig. 2.15   A sinusoidal alternating current

tions and expansions of the element. Likewise the small variations in light intensity from an electric bulb can be detected by moving some object rapidly from side to side in the light; instead of seeing a continuous 'blur', a series of separate impressions is made on the eye, these being 0.01 s apart in time. The effect is even more marked with fluorescent lighting; the discharge through the mercury vapour actually stops when the applied p.d. falls below a certain value, and the only light then left is the weak continuing glow from the fluorescent coating of the tube. This causes much greater flicker than a filament lamp. But these effects are rarely of sufficient importance to make any practical difference. For some purposes direct current is essential; e.g. for electro-plating, and other electrolytic processes, and for electronic equipment—radio, television, etc. But in such cases simple and efficient devices have been developed for *rectifying* the alternating current to give direct current at any desired p.d. Alternating current motors are less efficient than d.c. ones; but this is not important with small and medium-sized motors. Only with the largest motors, such as those used in electric trains, has the d.c. motor any marked advantage. But the generation and distribution of alternating current is so much cheaper and more efficient than direct current that its use for large-scale power purposes is now universal.

There are several ways in which we can express the *magnitude* of an alternating current:

(*i*) *The peak value* of the current is its maximum value (in either direction); in the equation

$$I = I_0 \sin \omega t$$

I_0 is the peak value.

(*ii*) *The mean value* or *half-cycle average* of the current is its average value taken over *half a cycle* while the current flows in one direction only (the average current over the whole cycle is of course zero).

(*iii*) *The root-mean-square* (*r.m.s.*) *value* of an alternating current I_{rms} is the square-root of the mean value of the square of the current taken over a whole cycle.

$$\therefore I_{rms} = \sqrt{(\text{mean value of } I^2)}$$

At first sight this may seem an unnecessarily complicated way of specifying an alternating current, but it is in fact much the most useful way. The reason for this may be seen as follows. The power P dissipated in a resistance is proportional to the square of the current

$$P = I^2 R$$

With alternating current the value of I and so of P varies from moment to moment.

$\therefore$ The mean value of P

$$= (\text{mean value of } I^2) \times R$$
$$= I_{rms}^2 R$$

This means that if a steady direct current of magnitude I_{rms} were passing through the resistance R, the same power $I_{rms}^2 R$ would be dissipated in it. For calculations of power the r.m.s. value is the most convenient way of specifying an alternating current, since it enables us to use the same methods of calculation as in d.c. situations.

Alternating potential differences may be analysed in the same way: the *peak value*, the *mean value* and the *r.m.s. value* of an alternating p.d. are all employed on occasions, but the r.m.s. value is generally the most useful for the same reason as before. The power P dissipated in a resistance R when the p.d. across it is V is given by

$$P = \frac{V^2}{R}$$

$\therefore$ the mean value of P

$$= \frac{(\text{the mean value of } V^2)}{R}$$
$$= \frac{V_{rms}^2}{R}$$

if $V_{rms} = \sqrt{(\text{mean value of } V^2)}$

Again, if a steady p.d. of magnitude V_{rms} were placed across this resistance, the power dissipated in it would be given by the same expression.

The practical result of all this is then very simple: provided we use r.m.s. values of current and p.d., the analysis of the currents flowing in networks of resistances and the calculations of power dissipated can be performed in exactly the same way as with direct currents. Likewise light bulbs and heating elements designed for a given d.c. supply will work correctly on alternating supplies of the same r.m.s. value. Only when a circuit contains devices (e.g. transformers) whose functioning depends on the use of alternating currents, must different methods of calculation be employed. This is considered in detail in Chapter 12.

Fig. 2.16 shows a method of calibrating an a.c. ammeter to read r.m.s. values by comparison with a d.c. ammeter. The two-way switch enables the light bulb to be lit either by an alternating current, indicated by the a.c. meter, or by a direct current measured by the d.c. meter (which could be a moving-coil instrument). The bulb is contained in a light-tight box together with a photocell (e.g. the selenium photovoltaic kind, p. 238). The galvanometer joined to the photocell gives a reading that depends on the brightness of the bulb.

Using first the alternating supply, the potentiometer across this is adjusted to bring the meter reading to the point of its scale at which the calibration is required; and the galvanometer reading is taken. Then the supply is switched to d.c., and the potentiometer across the battery is adjusted until the galvanometer once more gives the same reading. The brightness of the bulb and the mean power dissipated in it are now the same as they were before and the d.c. meter is recording the value of the r.m.s. current that previously passed through the a.c. meter. This procedure can then be repeated for a number of points of the scale.

The relation between the peak and r.m.s. values of a current or p.d. depends on the waveform involved. For the case of a sinusoidal waveform it may be found as follows. The current I is given by

$$I = I_0 \sin \omega t$$

By definition, the r.m.s. current I_{rms} is given by

$$I_{rms} = \sqrt{(\text{mean value of } I^2)}$$
$$= I_0 \sqrt{(\text{mean value of } \sin^2 \omega t)}$$

Using a well-known trigonometrical identity we have

$$\sin^2 \omega t = \tfrac{1}{2} - \tfrac{1}{2} \cos 2\omega t$$

The mean value of the second term on the right-hand side is zero, since the positive and negative parts of a sine or cosine curve are equal.

$$\therefore \text{ the mean value of } \sin^2 \omega t = \tfrac{1}{2}$$

$$\therefore I_{rms} = I_0 \sqrt{\tfrac{1}{2}} = \frac{I_0}{\sqrt{2}} = 0.707 \, I_0$$

Fig. 2.17 shows the relation graphically. The full line is the graph of $\sin \omega t$; it oscillates both sides of zero between $+1$ and -1. The graph of $\sin^2 \omega t$ is shown dotted; it is always positive and oscillates between 0 and 1 at *twice* the frequency of $\sin \omega t$. Its mean value is $\tfrac{1}{2}$.

Similarly, the r.m.s. value V_{rms} and the peak value V_0 of the p.d. are connected by

$$V_{rms} = \frac{V_0}{\sqrt{2}} = 0.707 \, V_0$$

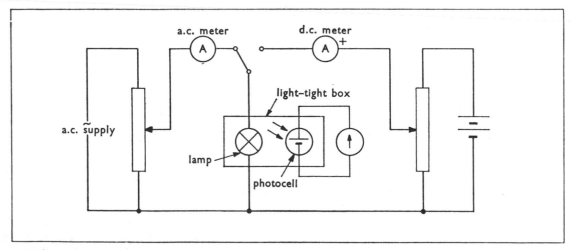

Fig. 2.16   Calibrating an a.c. meter to read r.m.s. values of current by comparison with a d.c. meter

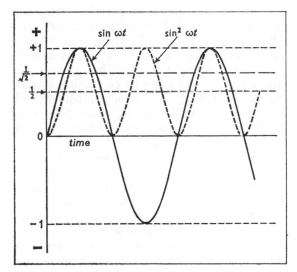

Fig. 2.17   Deducing the relation between peak and r.m.s. values

★ The connection between the peak and mean values of current or p.d. again depends on the waveform concerned. For the particular case of a sinusoidal current with a period of oscillation T, we have

$$\text{half-cycle average} = \frac{\int_0^{\frac{1}{2}T} I\,dt}{\frac{1}{2}T}$$

$$= \frac{I_0 \int_0^{\pi/\omega} \sin \omega t\,dt}{\pi/\omega}$$

$$= \frac{\omega I_0}{\pi}\left[-\frac{1}{\omega}\cos \omega t\right]_0^{\pi/\omega}$$

$$= \frac{2}{\pi}I_0$$

$$\therefore \text{ mean current} = \frac{2}{\pi}I_0 = 0.637\,I_0$$

Similarly,   mean p.d. $= \dfrac{2}{\pi}V_0 = 0.637\,V_0$

Combining the two results we have

$$\frac{\text{r.m.s. value}}{\text{mean value}} = \frac{0.707}{0.637} = 1.11$$

2.6 Fuses and filaments

When an electric current flows in a wire, heat is evolved in it and its temperature rises until the rate of loss of heat from its surface is equal to the rate at which the heat is supplied. The temperature then remains steady. The tungsten filament of an electric lamp is so fine that this final temperature is well in excess of 2300 K. On the other hand the wires used for joining up an electric circuit must be thick enough so that the temperature they reach is normally very little above that of the surroundings. The function of a fuse is to protect the circuit against the effects of excessive currents such as can arise from short circuits. It consists of a short length of relatively fine wire contained in a porcelain holder or a glass tube; its diameter is selected so that it quickly reaches its melting point and breaks the circuit if the current exceeds the maximum safe value.

The filament of a small electric lamp is of such low heat capacity that it reaches its final steady temperature in a fraction of a second. It is important in a lighting circuit that the fuse should heat up much more slowly than the lamp filaments; otherwise the fuse might blow every time the lights are switched on! When a lamp is cold, the filament has less than a tenth of its normal resistance at its working temperature. Therefore for an instant after switching on there is likely to be a surge of current that may well be in excess of the normal maximum for the circuit. Fortunately the wire of the fuse takes longer to heat up than the much finer filaments and usually survives the initial surge.

The behaviour of fuse wires and filaments may be discussed mathematically as follows. Consider a wire of radius r and length l through which a current I is flowing. If its resistivity is ρ, its resistance R is given by

$$R = \frac{\rho l}{\pi r^2}$$

and the rate of production of heat P in it given by

$$P = \frac{I^2 \rho l}{\pi r^2}$$

When the wire has reached a steady temperature, P is also the rate of loss of heat from its surface by conduction, radiation, etc. Assuming this is proportional to the surface area we can write

$$P = 2\pi r l \cdot p$$

where p is the rate of loss of heat per unit area of surface; it is a function of the temperature and is the same for all wires of the same material losing heat in similar circumstances.

$$\therefore \quad \frac{I^2 \rho l}{\pi r^2} = 2\pi r l . p$$

$$\therefore \quad I^2 = 2\pi^2 r^3 \times \frac{p}{\rho}$$

Without assuming anything yet about the variation of p and ρ with temperature, this expression leads to two important conclusions:

(*i*) The temperature reached by a given wire depends only on the current through it, and for a given current is not dependent on the length l. This last statement needs some qualification. The ends of a wire can lose heat by conduction through its supports, and are usually cooler than the centre. The temperature of the central part is only independent of the length provided the wire is long enough for end effects not to matter.

(*ii*) For wires of a given kind heated to the same temperature

$$I^2 \propto r^3$$

or

$$r \propto I^{2/3}$$

Thus if we know the current at which a given fuse wire melts, this result may be used to determine what diameter wire of the same metal should be used as a fuse at any other current. The same expression can also be used to compare the radii of the filaments of lamps to be run at the same temperature but with differing currents.

To obtain maximum luminous efficiency the temperature of a lamp filament should be as high as possible. Tungsten is therefore used, since this has a melting point of about 3700 K. However, even tungsten cannot be used in a vacuum above 2400 K if serious evaporation of the filament is to be avoided. Most lamps are nowadays filled with an inert gas. This reduces evaporation and enables temperatures of 2600 K or more to be reached. Also the tungsten that does evaporate is now carried away by convection and deposited on the top of the bulb instead of producing a uniform blackening of the glass, as in the vacuum type. However, the heat loss from the filament is so increased by the presence of the gas that a straight wire would need to be excessively fine and fragile to reach the required temperature. To get over this difficulty a coiled filament is used. This has the effect of reducing the exposed surface area, and so enables the same temperature to be reached with a more robust wire. The same design also makes for a more compact source of light—a matter of some importance in projector and headlamp bulbs. Such bulbs are for preference low-p.d., high-current devices so that the filament can be as short and thick as possible.

An even higher filament temperature, and therefore greater luminous efficiency, is obtained in the *quartz-iodine lamp*, in which an ingenious chemical process causes the evaporated tungsten to be recycled. These lamps are filled with iodine vapour, and the bulb itself is made of fused quartz. The tungsten evaporated from the filament diffuses to the walls of the bulb, and in contact with the quartz forms the compound tungstate(VI) iodide, which is a vapour. This compound diffuses back to the filament, where it is decomposed by the high temperature, depositing the tungsten once more on the filament. The life of the filament is thus considerably prolonged in spite of the high temperature employed.

3 Electrolysis and Thermoelectricity

3.1 Conduction in liquids

Except for the case of molten metals (such as mercury at room temperature), the passage of an electric current through a liquid causes chemical changes; the process is called *electrolysis*. Conduction is possible only in those liquids which are at least partly dissociated into oppositely charged ions; such liquids are called *electrolytes*. Solutions of many inorganic chemical compounds (common salt, sulphuric acid, etc.) are examples of this type of liquid. There are however many substances (sugar, for instance) which dissolve without splitting up into ions. Solutions of these do not conduct electric currents, and are called *non-electrolytes*.

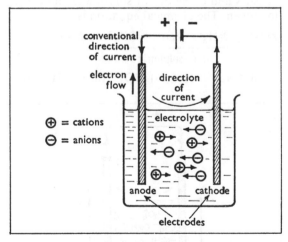

Fig. 3.1 The terms used in describing electrolysis

The plates through which the current enters and leaves an electrolyte are known as *electrodes*. That towards which the positive ions travel is called the *cathode*; and the other, towards which the negative ions travel, is called the *anode* (Fig. 3.1). The flow of current through an electrolyte takes place by the movement of all the ions in it

—the positive ions in one direction and the negative in the other. The positive ions are referred to as *cations*; these are mostly formed from metals or hydrogen. The negative ions are referred to as *anions*. Thus in a circuit diagram (Fig. 3.1) the arrows which show the direction of the current (in the conventional sense) actually indicate the direction of movement of the cations; the anions move in the opposite direction. An electrode must normally be made of metal (or perhaps graphite or a semiconductor); and in such a material conduction of current can take place only by the movement of the free electrons it contains (p. 6); in the electrolyte, however, there are no free electrons, and conduction takes place only by the movement of the ions. At the surface of the electrodes, where the current passes between electrode and electrolyte, the mechanism of conduction must change. This can happen in one of two ways; either some of the ions are discharged, electrons being transferred between ions and electrode; or else fresh ions are formed from the material of the electrode, and these pass into solution. The discharging of an ion usually causes the substance concerned to come out of solution and be liberated either as a deposit on the electrode or in bubbles of gas, as the case may be. The chemical changes that occur in electrolysis are thus seen to arise at the surface of the electrodes. Two examples will show the way in which these processes operate.

(i) The electrolysis of copper sulphate solution with inert electrodes (carbon or platinum). The copper sulphate in solution consists of the ions $Cu(H_2O)_4{}^{2+}$ and $SO_4{}^{2-}$. Also the water itself provides small quantities of the ions $(H_3O)^+$ and OH^- (Fig. 3.2). Both sorts of positive ions travel towards the cathode, but at the surface of the cathode the $Cu(H_2O)_4{}^{2+}$ ions are discharged in preference to the $(H_3O)^+$ ions (although for

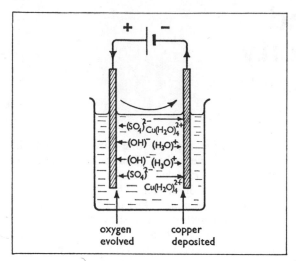

oxygen   copper
evolved   deposited

Fig. 3.2   The electrolysis of copper sulphate solution
with inert electrodes (carbon or platinum)

very large currents some hydrogen may be liberated). A layer of metallic copper is thus deposited on the cathode. Of the two negative ions in solution it is found that OH^- is discharged in preference to SO_4^{2-}; this leads to the evolution of oxygen at the anode, since the hydroxyl groups (OH) react together in pairs according to the reaction—

$$2(OH^-)(aq) \rightarrow H_2O(l) + O(g) + 2e^-$$
(e^- here stands for an electron)

The net result in this case is that the copper is steadily removed from the solution, which therefore becomes paler. At the same time it becomes increasingly acid, since the SO_4^{2-} ions remain in solution, and further dissociation of the water provides the balancing $(H_3O)^+$ ions.

(ii) The electrolysis of copper sulphate solution with copper electrodes (Fig. 3.3).   At the cathode, copper is deposited as in the previous case. But at the anode, instead of OH^- ions being discharged, fresh $Cu(H_2O)_4^{2+}$ ions are formed from the copper of the anode itself. The formation of each ion leaves two surplus electrons behind on the anode; the same charge (carried by two electrons) is imparted to each of the $Cu(H_2O)_4^{2+}$ ions discharged at the cathode. Since the current is the same at the surfaces of both electrodes, the concentration of the electrolyte remains constant; and the net result is

just the transfer of copper from the anode to the cathode. The chief practical importance of this process lies in its ability to purify the copper. The anode is made of a block of impure copper and the cathode is a small sheet of very pure copper; then, as the current passes, pure copper only is taken from the anode and deposited on the cathode. There is a large demand for electrolytically purified copper, since this degree of purity is needed for electrical cables and connectors. Even the smallest trace of impurity can markedly increase the resistivity of the copper.

In modern chemical terminology reactions in which electrons are *removed* from the ions concerned are classified as *oxidations*, while those in which electrons are *added* to ions are classified as *reductions*. Consider for instance the reaction produced by adding metallic zinc to dilute sulphuric acid, with the evolution of hydrogen. Each zinc atom passes into solution as a zinc (II) ion (Zn^{2+}); it imparts the two surplus electrons to two oxonium ions ($H_3O)^+$, converting them into atoms of hydrogen, which combine together in pairs and come out of solution as bubbles of hydrogen. The chemical equation is

$$Zn(s) + \underbrace{2(H_3O^+)(aq) + SO_4^{2-}(aq)}_{\text{sulphuric acid}} \rightarrow$$

$$Zn^{2+}(aq) + H_2(g) + SO_4^{2-}(aq) + 2H_2O(l)$$

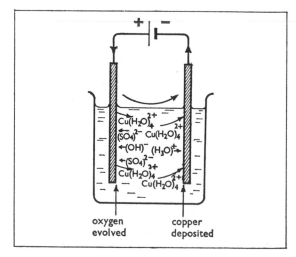

oxygen   copper
evolved   deposited

Fig. 3.3   The electrolysis of copper sulphate solution
with copper electrodes

By the rule given above this constitutes an oxidation of the zinc and a reduction of the hydrogen. This is of course a broadening of the original meanings of the words 'oxidation' and 'reduction'. The student should consult a textbook of chemistry for full explanations of the modern meanings of these terms. It will be seen that in chemical reactions an oxidation is accompanied by a corresponding reduction, since one component of the reaction receives an electron at the expense of another.

In electrolysis the reactions at the electrodes can likewise be classified in this way. At the anode, electrons are removed from the ions and it follows that all anodic reactions are oxidations. Similarly at the cathode, electrons are added to the ions, and therefore all cathodic reactions are reductions. From the chemical point of view the significance of electrolysis is very largely in this—that it enables an oxidation and its corresponding reduction to be conducted at different points in the same solution, so that the products of the actions can be kept apart. In this way many chemical reactions can be brought about which without electrolysis would be impossible. Examples of industrial electrolytic processes may be found in textbooks of chemistry.

3.2   The ionic theory

In our description of electrolytic processes above we have been tacitly assuming an ionic explanation of the phenomena. We shall now consider what the detailed predictions of such a theory must be, and therefore what experimental justification of the theory we can adduce.

The difference between the mass m_a of an atom and the mass m_i of its ion is truly negligible in electrolytic processes. Thus, if we have an atomic species of relative atomic mass A_r, these two masses are both given by

$$m_i = m_a = A_r m_u$$

where m_u is the unified atomic mass constant (p. 5). If the *charge number* of the ion is z_i (i.e. the number of electronic charges it carries), then the actual charge Q_i carried by an ion is given by

$$Q_i = z_i e$$

where e is the electronic charge (1.60×10^{-19} C).

When an electric current I flows for a time t through an electrolytic cell, the total charge Q that passes is given by

$$Q = It \quad \text{(p. 10)}$$

The number N of ions deposited or liberated at one electrode is therefore given by

$$N = \frac{Q}{Q_i} = \frac{It}{z_i e}$$

And the total mass m of this substance deposited or liberated is given by

$$m = N m_a = A_r m_u \frac{It}{z_i e}$$

Re-arranging, we have

$$m = \frac{m_u}{e} \frac{A_r}{z_i} It$$

This expression is a product of three factors:

(*a*) The quantity It, which is equal to the charge Q that passes.

(*b*) The quantity A_r/z_i, which is a constant for any given ionic species; it is a dimensionless quantity, whose value can be derived from chemical evidence alone.

(*c*) The quantity m_u/e, which is a universal constant, the same for all ions. Thus, an ionic theory of the kind we have assumed leads us to make two predictions:

(*i*) The mass of any given substance deposited or liberated in electrolysis is proportional to the total electric charge that passes (i.e. the product of current and time).
(*ii*) The masses of different substances deposited or liberated by the passage of the same quantity of electric charge are in the ratio of the quantities A_r/z_i for the substances.

These two laws were in fact discovered experimentally by Faraday, and were the ground for his first putting forward an ionic theory. The first law is readily tested in any process of electrolysis in which we can relate the quantity of material deposited or liberated to the product of the current and the duration of the electrolysis. The second law was discovered by Faraday by connecting in series a number of different electrolytic cells, using a wide variety of substances. Being in series, the same charge must have passed through all the cells even if the current varied during the experiment. At each electrode the mass of material deposited, liberated or dissolved was measured, and these masses were shown to be proportional to the quantities A_r/z_i for the substances concerned.

There is one other prediction that our ionic theory makes; namely, it enables us to derive from our measurements a figure for the charge per unit mass of the ions we have postulated. This quantity is called the *specific charge* of the ions. Thus

$$\text{specific charge} = \frac{Q_i}{m_i} = \frac{z_i e}{A_r m_u}$$

From our measurements we can write down for any one substance involved in the electrolysis

$$\text{specific charge} = \frac{\text{total charge passed}}{\text{total mass deposited}}$$

For instance from an electrolytic experiment in which hydrogen is liberated we may derive the specific charge of the hydrogen ion (or proton). This gives

$$\frac{e}{m_p} = 9.58 \times 10^7 \text{ C kg}^{-1}$$

Since we have $e = 1.60 \times 10^{-19} \text{ C}$
we can use the measurement to give the mass m_p of the proton.

$$\therefore m_p = \frac{1.60 \times 10^{-19}}{9.58 \times 10^7} \text{ kg}$$

$$= 1.67 \times 10^{-27} \text{ kg}$$

This is in excellent agreement with the deductions of the kinetic theory of gases (p. 2).

Similar measurements can be made of the masses of many different kinds of ions. Many of these we can also measure with a mass spectrometer (p. 251); and the agreement of the two kinds of measurement is excellent. The ionic theory thus takes its place as an essential part of the modern atomic theory of matter.

The mole

Once we have established an atomic theory of matter, an obvious step to take is to measure the amount of a substance in terms of the number of individual particles or entities that it contains. The unit of *amount of substance n* reckoned in this way is called the *mole* (abbreviated to mol).

> The mole is the amount of substance of a system which contains as many elementary entities as there are atoms in 0.012 kg of carbon-12.

If we had a substance whose atomic mass was exactly equal to the unified atomic mass constant m_u, one mole of it would have a mass of 0.001 kg. For, this quantity of it would contain the same number of atoms as there are in 0.012 kg of carbon-12, since

$$m_u = \frac{m_C}{12}$$

Similarly (consulting the table on p. 4) one mole of sodium atoms has a mass of 0.023 kg; one mole of oxygen-16 atoms has a mass of 0.016 kg, etc.

The mass per mole of a collection of *atoms* is called its *molar mass M*; clearly M is proportional to the relative atomic mass A_r of the atoms. Thus

$$M \propto A_r$$

The definition of the mole gives us the constant of proportionality in this, so that we can write

$$M = A_r \times 10^{-3} \text{ kg mol}^{-1}$$

If we have a total mass m of the substance, then the amount of substance n is given by

$$n = \frac{m}{M}$$

E.g. suppose we have 1 kg of oxygen; the amount n of oxygen-16 atoms in this is

$$n = \frac{1}{0.016} \text{ mol} = 61.25 \text{ mol}$$

The mole concept can also be used to state the amount of substance n in a collection of *molecules*. Thus one mole of oxygen-16 molecules (O_2) has a mass of 0.032 kg. And for a species of relative molecular mass M_r the molar mass M is given by

$$M = M_r \times 10^{-3} \text{ kg mol}^{-1}$$

In the same way we can state the amount of substance for *any* specified collection of particles. When using the mole, the elementary entities must therefore be specified, and may be atoms, molecules, ions, electrons or other particles, radicals or any specified group of atoms or particles. A chemical formula may be used to specify the entity concerned; e.g. we can refer without ambiguity to 'one mole of O_2' (whose molar mass is 0.032 kg mol^{-1}). It is not necessary for the entity to be one that has a real separate existence in nature. For instance we could take as our entity half a water molecule $\frac{1}{2}(H_2O)$. Since the relative molecular mass of water is 18, the molar mass of half-water-molecules would be 0.009 kg mol^{-1}.

The Avogadro constant

The number of elementary entities per mole of a substance is an important universal constant known as the *Avogadro constant L*. It is the number of *atoms* per mole of carbon-12; or equally the number per mole of any other atomic species; or the number of *molecules* per mole of any molecular species; or the number of *electronic charges* per mole of electrons (or of any other singly charged particle), etc. If we had an entity whose mass was exactly equal to the unified atomic mass constant m_u, the number of these per mole would be (as always) equal to L. There is thus a connection between the Avogadro constant L and the unified atomic mass constant m_u. For the molar mass of this imaginary entity of mass m_u is 10^{-3} kg mol^{-1}.

$$\therefore Lm_u = 10^{-3} \text{ kg mol}^{-1}$$

$$\therefore L = \frac{10^{-3}}{1.66 \times 10^{-27}} \text{ mol}^{-1}$$

$$= 6.02 \times 10^{23} \text{ mol}^{-1}$$

Similarly for any other atomic species of atomic mass m_a, the molar mass M is given by

$$M = Lm_a = A_r Lm_u$$

If we have an amount of substance n, the total number N of entities in the sample is given by

$$N = nL$$

For example, 1 kg of oxygen, whose amount of atoms was calculated above to be 61.25 mol, contains

$$61.25 \times 6.02 \times 10^{23} \text{ atoms}$$
$$= 3.68 \times 10^{25} \text{ atoms}$$

The Faraday constant

Another important universal constant is the charge per mole of electrons (or of any other singly charged particles, e.g. hydrogen ions). This is known as the *Faraday constant F*. Thus we have

$$F = Le$$

where e is the electronic charge. Substituting for L and e, we obtain

$$F = 9.65 \times 10^4 \text{ C mol}^{-1}$$

When the Faraday constant has been determined, we can write down the specific charge of any given ion. For we have

$$Lm_u = 10^{-3} \text{ kg mol}^{-1}$$

Substituting F/e for L,

$$\therefore \frac{Fm_u}{e} = 10^{-3} \text{ kg mol}^{-1}$$

$$\therefore \frac{e}{m_u} = \frac{F}{10^{-3} \text{ kg mol}^{-1}}$$

For an ion of charge number z_i and relative atomic mass A_r we therefore have

$$\text{specific charge} = \frac{z_i e}{A_r m_u} = \frac{z_i}{A_r} \frac{F}{10^{-3} \text{ kg mol}^{-1}}$$

Substituting the value of F given above

$$\text{specific charge of ion} = 9.65 \times 10^7 \frac{z_i}{A_r} \text{ C kg}^{-1}$$

In an electrolytic process the specific charge is equal to the total charge Q passed through the cell divided by the mass m deposited or liberated.

$$\therefore \text{ specific charge} = \frac{Q}{m}$$

Equating these expressions we have

$$\frac{Q}{m} = 9.65 \times 10^7 \frac{z_i}{A_r} \text{ C kg}^{-1}$$

This result enables us to calculate the charge required to deposit a mass m of any given substance in electrolysis.

3.3 Ionic dissociation

The ionic hypothesis was first introduced by the Swedish chemist Berzelius in 1812. But it took nearly a century of development before it could be presented in a form that met with general acceptance in the scientific world. Following closely on the introduction of the atomic theory by Dalton at the beginning of the nineteenth century Berzelius suggested that the forces holding atoms together in chemical compounds were electrostatic in nature. He supposed that the atoms of metallic elements formed positive particles and non-metallic elements negative ones; and that they were then held together by the electrostatic forces between them. The development of the theory ran into difficulties since, as we now know, other types of force also operate in holding together the atoms in chemical compounds.

In 1833 Faraday investigated the phenomenon of electrolysis. The laws he discovered make an ionic explanation of the process almost inevitable, just as in chemistry the laws of Constant Composition and Multiple Proportions demand an atomic explanation. But the exact nature of the ions and how they came into being could not at that time be clearly explained; and most scientists still remained sceptical. It was Faraday who coined the word *ion* to describe the charged particles that were supposed to be responsible for carrying the current through an electrolyte.

In the 1870's Kohlrausch made extended measurements of the conductivity of electrolytes at varying concentrations. These measurements enabled him to calculate the proportion of molecules ionized in the solutions. It appeared that electrolytes fell into two classes, which Kohlrausch called *strong* and *weak*. The strong electrolytes were completely dissociated into ions under all conditions; such were the solutions of strong acids and alkalis and the salts formed from them. Other electrolytes were only slightly dissociated at large concentrations; but at sufficient dilution even these were completely dissociated. Kohlrausch also used his measurements to give the ionic velocities in the solutions (p. 11). For a potential gradient in the electrolyte of 1 V m^{-1} the velocities obtained were of the order of 10^{-7} m s^{-1}.

It was not until 1887 that a Danish student, Arrhenius, propounded a theory of ionic dissociation that was adequate to explain the phenomena of electrolysis, at least for weak electrolytes. Previous versions of the ionic theory had supposed that the production of ions was initiated by the application of a p.d. between the electrodes, and that the ions did not exist before the flow of current started. But this idea is contradicted by the observation that in many cases of electrolysis current flows no matter how small the applied p.d.; if the p.d. had to supply energy to produce ionic dissociation, we would expect some minimum p.d. to be needed before current could start to flow.

Arrhenius suggested that dissociation into ions starts as soon as an electrolyte goes into solution, and is not dependent on the application of a p.d. Each molecule in solution has a certain chance of dissociating. Since we are concerned in any practical case with enormous numbers of molecules, this means that there will be a steady production of ions at a rate proportional to the number of unionized molecules present. At the same time random movements of the ions in the liquid will bring oppositely charged ions together, when they have a chance of recombining to form uncharged molecules again. The rate of recombination can be taken to be proportional to the numbers of each sort of ion present; the greater the degree of ionization of the substance dissolved, the greater the rate of recombination and the smaller the rate of dissociation (an application of the Chemical Law of Mass Action). A state must soon be reached in which ionization and recombination are proceeding at the same rate. The proportion of molecules ionized then remains constant. Owing to the rapidity of atomic processes this apparent equilibrium is normally reached almost at once.

If some of the ions in the solution are discharged and liberated by electrolysis (or precipitated by some chemical action), for a moment recombination proceeds less rapidly, since there are fewer ions to recombine. The continually proceeding dissociation at once tends to restore the proportion of molecules ionized to its original value. There is thus always a steady supply of ions available to conduct the current, even if as in some cases the proportion of molecules ionized is very small.

Kohlrausch's results indicated that at infinite dilution all electrolytes would be completely dissociated. This is what we should expect on the theory of Arrhenius, since at very low concentrations collisions of ions will be rare events, and the rate of recombination will be negligible.

With strong electrolytes the process is complicated by the tendency of the ions to group together in clusters when they are in strong concentrations; thus the particles in motion become larger with increasing concentration with a consequent decrease in effective mobility. An exact theory was eventually provided on this basis by Debye and Hückel, though this was not until 1923.

Further evidence for ionic dissociation is provided by other properties of electrolytes. Many of their distinctive properties depend only on the presence of a particular ion, and are not affected much by changing the other ions accompanying it. For example, the general properties of acids arise from the $(H_3O)^+$ ions they form in solution; it is these ions that give the sour taste of an acid. Likewise the general properties of solutions of alkalis arise from the hydroxyl ions (OH^-) they contain. For instance in contact with the skin these ions combine with the fats present, producing the thin film of soap that causes the characteristic slippery feeling of alkalis. Other examples are provided by the characteristic colours of solutions containing particular metallic ions; e.g. the blue colour of solutions of the $Cu(H_2O)_4^{2+}$ ion; this colour is not much affected by changing the anion accompanying it (unless this also produces colouring).

Another independent line of evidence for ionic dissociation is provided by the phenomenon of *osmosis* and the changes of boiling and freezing points related with it. Theory shows that the magnitude of the osmotic pressure, the depression of the freezing point of the solution, and the elevation of the boiling point should all depend on the *number of particles* present in the solution. Measurements on electrolytes show that the numbers of particles are much greater than the numbers of molecules of solute; there must therefore be some dissociation of the molecules. The student should consult a textbook of physical chemistry for details of such measurements and of the theory involved.

3.4   The simple cell

If any two different metals are dipped into a tank of any electrolyte, and an electrical connection is made between them, it is found that an e.m.f. acts so as to drive current round the circuit so formed. The resulting electrolysis is accompanied by the usual chemical changes. Such an arrangement is called an *electric cell*. It is in fact a device for producing electrical energy from chemical potential energy; the magnitude of the e.m.f. depends on the energy changes involved at the two electrode surfaces. Many combinations of materials have been used to make electric cells; one arrangement, tradition-

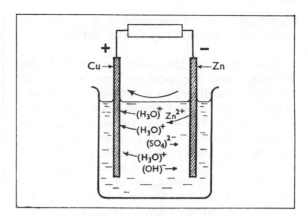

Fig. 3.4   The simple cell

ally known as the *simple cell*, is shown in Fig. 3.4.

It consists of electrodes of zinc and copper in an electrolyte of dilute sulphuric acid. Since zinc reacts to some extent with cold sulphuric acid, it is usual to rub mercury into the zinc electrode forming an amalgam of zinc on the surface. This almost completely prevents the *local action* of the acid on the zinc. When the circuit is completed and a current passes, hydrogen is evolved at the surface of the copper electrode and zinc goes into solution from the other. The current flows (in the conventional sense) from the copper to the zinc outside the cell; and from zinc to copper inside. The copper electrode is thus the *positive pole* or terminal of the cell and the zinc is the *negative pole* or terminal. The total chemical action of the cell is the same as the familiar action of dilute sulphuric acid on zinc—an action that releases a certain amount of energy, some of which, in the operation of the simple cell, is changed into electrical energy.

In terms of the motions of the ions the explanation of the action of the cell is as follows. At the zinc electrode Zn atoms are ionized and pass into solution as Zn^{2+} ions leaving two electrons each on the electrode. For each Zn^{2+} ion formed two $(H_3O)^+$ ions are discharged at the copper electrode, taking two electrons from it. The action continues just as long as electrons can pass from the zinc to the copper through some piece of metal (containing free electrons). If the circuit is broken and the electron flow stopped, charge accumulates on the electrodes, and the electro-static forces prevent any further action.

Note. In a cell the positive electrode is the cathode, since the positive ions (metal and hydrogen) are driven to it. This is the opposite to the examples of electrolysis considered earlier in this chapter. The student should not make the mistake of learning that the cathode is the nega-tive electrode, since this is incorrect for the electrolysis taking place in cells. Rather—*the cathode is the electrode at which the positive ions arrive.* When a current is driven through an electrolyte by some external source of e.m.f. the positive ions are attracted to the negative electrode, which is then the cathode; but in a cell the cathode becomes positive because the posi-tive ions are driven to it by the chemical action.

At the zinc electrode of a simple cell energy is imparted to the circuit, while at the copper elec-trode some of this energy is used in liberating the hydrogen. The rest of the energy is available to drive current round the circuit. Some of it is used in driving the current through the electrolyte of the cell (raising its temperature), and some of it in driving the current round the rest of the cir-cuit. The energy per unit charge (i.e. the electric potential) at various points is shown in the graph in Fig. 3.5. Since we are concerned only with differences of potential, the zinc electrode is here arbitrarily taken as the zero of potential. As with all sources of e.m.f. the p.d. between the terminals of the cell falls when a current is flow-ing on account of the internal resistance; only when the current is zero is the p.d. equal to the e.m.f.

Quite apart from this drop in p.d. because of the internal resistance there is also a decrease in the actual e.m.f. of the cell on account of the effect known as *polarization*. As soon as a cur-rent flows, the electrolysis changes the composi-tion of the cell; the copper electrode is covered with a layer of hydrogen and the concentrations of the ions are altered in the neighbourhood of both electrodes. This gives rise to a *back e.m.f.* acting in opposition to the e.m.f. of the cell and partly cancelling it out. The law known as *Le Chatelier's principle* shows that a decrease in e.m.f. is to be expected when a current flows. (See textbooks of chemistry for an explanation

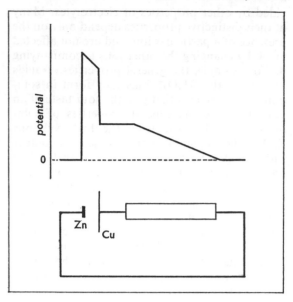

Fig. 3.5   The variation of potential round a circuit containing a simple cell

of this principle.) Unpolarized, the e.m.f. of the simple cell is about 1.1 V, but this may drop to about 0.4 V after a current has been passing for a few seconds. To some extent this defect in the cell as a practical device can be overcome by adding a suitable oxidizing agent to the elec-trolyte. But it cannot be left assembled for any length of time, since the oxidizing agent slowly attacks the electrodes.

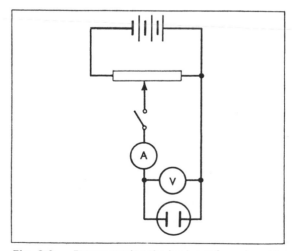

Fig. 3.6   Demonstrating the back e.m.f. produced by polarization

The existence of a back e.m.f. due to polarization may be demonstrated in any electrolysis in which the products change the effective composition of the cell. Fig. 3.6 shows a circuit that may be used to show the effect for the electrolysis of dilute sulphuric acid with platinum electrodes. The potentiometer joined across the 6 V battery enables the p.d. applied to the electrolytic cell to be varied from 0 to 6 V. With the switch closed the p.d. is slowly increased from zero. At first only a very small current flows and no evolution of hydrogen and oxygen can be observed at the electrodes. Only when the p.d. has been raised above about 1.7 V will a larger current start to flow and gas be evolved. Increasing the p.d. above this point causes a steady increase of current, as shown in the graph of Fig. 3.7. We suppose this behaviour to be due to the existence of a back e.m.f. of maximum value 1.7 V arising from the polarization of the cell. If now the switch is opened, this e.m.f. gives a reading on the voltmeter, which at first is 1.7 V but rapidly falls off as the electrodes and electrolyte are restored to their original condition.

From energy considerations we should expect a minimum p.d. to be required for this electrolysis to proceed. The net result is the decomposition of water; to do this a quantity of energy at least equal to the heat of reaction of hydrogen and oxygen must be supplied in electrical form. This energy is 290 kJ mol^{-1} of H_2O. To decompose 1 mole of H_2O requires the passage of 2 mole of electrons through the electrolytic cell. The charge required per mole of H_2O decomposed is therefore $2F$, where

$$F = 9.65 \times 10^4 \text{ C mol}^{-1}$$

The minimum p.d. required to produce continuous electrolysis is equal to the energy that must be supplied per unit charge.

$$\therefore \text{ minimum p.d.} = \frac{290 \times 10^3}{2 \times 9.65 \times 10^4} \text{ J C}^{-1}$$
$$= 1.5 \text{ V}$$

The actual p.d. varies slightly with the type of electrode used; for platinum it is 1.7 V.

Only when the products of electrolysis do not affect the composition of the cell is there no back e.m.f. Such for instance is the electrolysis of

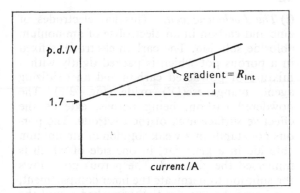

Fig. 3.7   The variation of p.d. with current for an electrolytic cell of dilute sulphuric acid with platinum electrodes

copper sulphate with copper electrodes; in this case electrolysis takes place continuously even for very small applied p.d.'s, and it is found that Ohm's law applies in this case fairly closely. In other cases allowance must be made for the back e.m.f. E. Fig. 3.7 shows that the current I is proportional to the amount by which the p.d. V is in excess of E. Thus we may define the resistance R_{int} of the electrolytic cell by the relation

$$R_{int} = \frac{V - E}{I}$$

This is very similar to the concept of the internal resistance of a battery (p. 19). The difference is that the current is normally being driven through an electrolytic cell *in opposition* to the e.m.f. E, so that V is greater than E; whereas in a battery that is driving a current round a circuit E is greater than V.

3.5   Types of cell

The simple cell is by no means a satisfactory device for making practical batteries. It cannot be left assembled, because of the action of the acid on the zinc electrode. Polarization effects in it are very severe, and the electrolyte is a corrosive substance. A good electric cell should have (*a*) a long shelf life; (*b*) a minimum of polarization effects; (*c*) electrodes and electrolyte that are cheap and safe. We shall describe briefly the composition of a few of the commonest types of cell manufactured at the present time.

(i) The Leclanché cell. This has electrodes of zinc and carbon in an electrolyte of ammonium chloride solution. The carbon electrode is fixed in a porous pot which is packed tightly with a mixture of powdered carbon and an oxidizing agent, manganese(IV) oxide (Fig. 3.8). The powdered carbon, being porous, extends the effective surface area of the electrode. The porous pot stands in a weak solution of ammonium chloride in a glass jar, in one side of which is immersed the zinc rod. The porous pot allows the solution to permeate the inner compartment, but keeps the powder tightly packed round the carbon rod. The e.m.f. is about 1.5 V.

When a current passes, Zn^{2+} ions go into solution at the anode, and (H_3O^+) ions are discharged at the cathode; this causes some polarization, which is gradually removed by the reduction of the manganese(IV) oxide. Subsequently the manganese(IV) oxide is re-formed from the oxygen of the atmosphere dissolved in the solution. The Zn^{2+} ions eventually form with the ammonium chloride an insoluble complex, which is precipitated harmlessly to the bottom of the jar. To maintain the cell, the zinc rod must occasionally be replaced; evaporation must be made up (tap water can be used for this), and some ammonium chloride must be thrown in. The life of the cell is then almost indefinite.

It can deliver a current of about 1 A for a minute or so before polarization becomes excessive, but it quickly recovers afterwards. It has traditionally been used for intermittent use, e.g. for door bells.

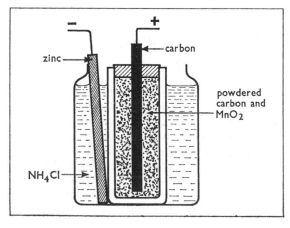

Fig. 3.8   The Leclanché cell

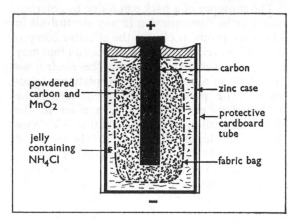

Fig. 3.9   The Leclanché 'dry' cell

(ii) The Leclanché 'dry' cell. Chemically this cell is identical with the Leclanché cell already described; but it is adapted to be portable. The case itself is the zinc electrode (−). The carbon and manganese(IV) oxide are held packed round the central carbon rod by a bag of fabric (Fig. 3.9). The electrolyte is absorbed either in a jelly or in a porous mixture of plaster of Paris and flour; glycerine is often added to it, since this is hygroscopic and helps to maintain the electrolyte in a moist condition. The case is sealed by layers of sawdust and pitch. The cell ceases to function when one of three things happens: (*a*) the electrolyte actually goes dry; (*b*) the zinc perforates so that the moisture runs out; (*c*) the manganese(IV) oxide is entirely used up and permanent polarization occurs.

Its e.m.f. is about 1.5 V, but this falls considerably towards the end of its life.

(iii) The mercury cell. This consists of electrodes of steel and zinc, with an electrolyte of potassium hydroxide. The steel electrode forms the outer case of the cell, which is positive; while the zinc electrode is in the form of an inner cylinder, which is connected to a steel top cap (−). Polarization is largely prevented (provided the current is not excessive) by packing the cathode with a layer of mercury(II) oxide together with powdered graphite to reduce the internal resistance (Fig. 3.10). The result is that the Hg^{2+} ion is the one discharged at the cathode, and this does not cause any polarization.

The cell has an e.m.f. of about 1.4 V, and this remains almost constant throughout its working

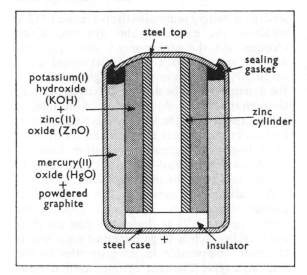

Fig. 3.10   The mercury cell

life, which is very much greater than for a Leclanché dry cell of the same size. A variety of different electrode shapes have been employed. By making the parts in the form of flat discs a very small, compact unit is possible, resembling a button. These cells are used in devices like hearing aids and key-ring lights.

(*iv*) *The Weston cadmium cell.*   This is designed specially for standards purposes. Polarization is eliminated (for very small currents) by having each electrode in a solution of its own ions. The electrodes are mercury (+) and cadmium–

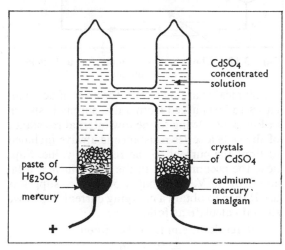

Fig. 3.11   The Weston cadmium cell

mercury amalgam (−), and the electrolyte is a saturated solution of cadmium(II) sulphate. A paste of mercury(I) sulphate and cadmium(II) sulphate solution is placed above the mercury electrode (Fig. 3.11). The e.m.f. is 1.0186 V at a temperature of 293.15 K (20°C), and varies little with temperature. Its internal resistance is deliberately designed to be high (about 500 Ω), and it is usually used in series with a high protective resistance. Its design protects it against polarization provided the current is no more than about 10 μA; but if more than this is passed through it, permanent changes of its e.m.f. can occur.

The Weston cell provides the internationally recognized secondary standard of p.d. (the primary standard is simply the definition). Carefully constructed and used its e.m.f. will remain constant to within 1 part in 10^6 for several years, and to 1 part in 10^5 for very long periods (p. 60).

3.6   Storage cells

In certain types of cell the action is truly reversible. A current driven through it in opposition to its e.m.f. re-forms the original materials of the electrodes, and the cell can therefore be used again and again. It is in fact a device for temporary storage of electrical energy. We shall consider two of the common types of storage cell. In all such cells the chemical actions are of very complicated kinds, and we shall deal with them only in outline.

(*i*) *The lead–acid accumulator.*   The negative electrode is of lead and the positive one of lead(IV) oxide (PbO_2). To increase their effective surface area, in both electrodes the active material is in the form of a spongy paste which is pressed into a grid made of a lead–antimony alloy. The electrolyte is sulphuric acid, initially of relative density 1.21. The initial e.m.f. of the cell is about 2.1 V; this falls rapidly to 2.0 V, at which figure it remains nearly constant until the cell is almost 'discharged'; re-charging should always be begun if the e.m.f. falls as low as 1.8 V.

During discharge the lead in the negative plate and the PbO_2 in the positive plate both change into insoluble lead(II) sulphate ($PbSO_4$). In this process the SO_4^{2-} ions are removed from the solution and the relative density falls. To re-charge the cell current must be driven through it in opposition to its e.m.f.; the chemical actions are then reversed and the relative density of the solution rises again.

It will be seen that the e.m.f. of the cell pro-vides little indication of its state of charge since it remains close to 2.0 V for 90% of the dis-charge period; so the relative density of the acid is normally used to indicate this. The cell should be re-charged when this falls below 1.18.

When re-charging, the e.m.f. of the cell rises quickly to about 2.2 V, and this increases again to nearly 2.7 V when re-charging is complete. At this point the electrolysis consists of the evolu-tion of hydrogen and oxygen, and this 'gassing' at the electrodes is often taken as an indication that charging is finished. Distilled water is added occasionally to make up the loss caused by the decomposition of the water.

A lead–acid accumulator is easily damaged by misuse. The strong acid solution makes the internal resistance of the cell very small—less than 0.01 Ω. This is one of its most useful properties, since it means that the p.d. will be almost constant for all normal currents. But it also means that a short-circuit will give a most damaging current of several hundred amperes.

(ii) The nickel–iron (NiFe) accumulator. The active material in the positive plates is nickel oxide (Ni_2O_3) and in the negative plates finely divided iron. In both plates the material is held in pockets of finely perforated steel. This gives much greater mechanical strength than in the lead–acid cell. The electrolyte is a solution of potassium hydroxide. When discharging, the nickel oxide (Ni_2O_3) is changed into a lower oxide (NiO), and the iron is changed into an iron oxide (FeO); these actions are reversed when the cell is re-charged. The electrolyte appears to take no direct part in the chemical action, and so only a minimum quantity of it need be used.

The e.m.f. of the cell is about 1.2 V, falling to 1.0 V when fully discharged. On re-charging the back e.m.f. is about 1.5 V, rising finally to 1.85 V. This means that it cannot so readily be used for voltage stabilization as a lead–acid cell. For instance in a car battery/dynamo circuit a lead–acid battery maintains the p.d. close to 12 V whatever the e.m.f. of the dynamo, which changes with the engine speed. The equivalent alkaline accumulator (of 10 cells) would allow up to 50% increase in p.d. at high speed when the dynamo would be driving a charging current through the battery. Also the cost of the alkaline cell is greater than the lead–acid type. However, its greater mechanical strength and its freedom from troubles of chemical pollution (such as 'sulphating' in the lead–acid type) are leading to its increasing use for many applications.

An alkaline accumulator using cadmium instead of iron is also often used. This has the same e.m.f. (1.2 V) as the nickel–iron but does not require such a large charging p.d., and is therefore comparable in its properties to the lead–acid type. The cost of these cells is how-ever much higher.

From the point of view of the energy used, the efficiency of the charge–discharge process is about 75% for a lead–acid or a nickel–cadmium accumulator and about 60% for a nickel–iron one. However, the energy stored per unit weight is about 80% greater for alkaline types than for the lead–acid.

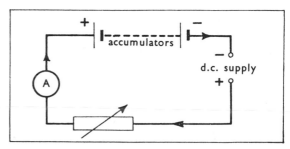

Fig. 3.12   The charging circuit for a battery of accumulators

A typical charging circuit for a set of accumu-lators to be charged from a d.c. supply is shown in Fig. 3.12. Owing to the low internal resistance of the cells a series resistance is always included to limit the current to the required value. Sup-pose the accumulators in the figure have a total e.m.f. of 36 V, and that a 50 V d.c. supply is used; then to obtain a charging current of 5 A we should calculate as follows:

The resultant e.m.f. in the circuit

$$= 50 - 36 \text{ V}$$
$$= 14 \text{ V}$$

$\therefore$ the series resistance required $= \dfrac{14}{5}\ \Omega$

$$= 2.8\ \Omega$$

The total power supplied in this example

= (the p.d. of the supply) × (current)
= 250 W

The power being stored in the batteries

= (the battery e.m.f.) × (current)
= 180 W

The rest of the power (70 W) is lost as heat in the resistance.

3.7 Thermoelectricity

There is a group of phenomena associated with temperature differences between the parts of an electric circuit. We shall consider *two* of these effects.

(i) The Seebeck effect

If a circuit consists of two different metals A and B (Fig. 3.13), an e.m.f. acts in it when the junctions of the metals are maintained at different temperatures. This is known as the *Seebeck effect*. A pair of junctions of this kind is known as a *thermocouple*. The e.m.f. E can be regarded as the difference between two e.m.f.'s V' and V acting in the junctions. (There is also a small additional e.m.f. in the individual metals A and B where there is a temperature gradient along them; but for simplicity we shall ignore this.) Thus

$$E = V' - V$$

A thermocouple may be regarded as a form of *heat engine*. We can compare its functioning with that of a steam turbine. In this, heat energy enters the turbine in the hot steam, which expands (doing work on the rotor) and cools, finally giving up a smaller amount of heat energy to the condenser. The difference is the mechanical energy available for the turbine. Similarly in a thermocouple, when a charge Q flows round the circuit, at the hot junction heat energy $V'Q$ is absorbed; but at the cold junction (where the current is in opposition to the e.m.f. V) heat energy VQ is evolved. The difference,

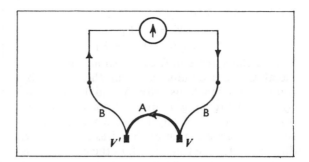

Fig. 3.13   A thermocouple

$(V' - V)Q$, is the electrical energy available in the circuit; it could, in principle, be used to do useful work (e.g. by turning an electric motor). However, in practice the e.m.f. is too small for this, being not more than a few millivolts for a temperature difference between the junctions of 100 K.

Practical arrangements using the Seebeck effect often need to employ more than two metals to make up the circuit. It is important to know what the effect of the additional junctions will be. Consider a circuit made up of three metals, A, B and C (Fig. 3.14). If the circuit is all at the same temperature, there is no possible source of energy that could drive current round it; therefore the algebraic sum of the junction e.m.f.'s (taken in the same sense round the circuit) must be zero.

$$V_1 + V_2 + V_3 = 0$$

When the temperature of the junction A–B is changed, the junction e.m.f. here changes to a new value V_1', but V_2 and V_3 remain the same. The thermoelectric e.m.f. E acting in the circuit is then given by

$$E = V_1' + V_2 + V_3$$

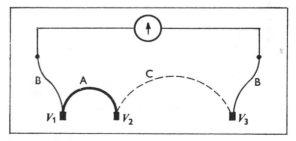

Fig. 3.14   The effect of introducing a third metal into a thermocouple

but from above we have

$$V_2 + V_3 = -V_1$$
$$\therefore E = V_1' - V_1$$

This is the same e.m.f. as we should get if the metal C were eliminated from the circuit by bringing its junctions with A and B into contact. Thus, breaking a circuit at some point and inserting a piece of any metal, whose temperature is the same as that of the wires at the break, does not alter the total e.m.f. in the circuit. This is very useful in practical measurements. It means, for instance, that the existence of a film of solder between the metals A and B in Fig. 3.13 does not alter the total e.m.f. It also means that this circuit could just as well be connected as in Fig. 3.15 without affecting the e.m.f.; in this case the galvanometer takes the place of the metal C in Fig. 3.14, and it makes no difference what metals enter into its construction. It is essential for the whole conducting path inside the galvanometer to be at the same temperature; but provided this is so, we can treat the arrangement as though it formed a simple thermocouple of two metals only, taking the temperature of the second junction of A and B as that of the galvanometer.

Thermoelectric thermometers

When one junction of a thermocouple is kept at a fixed temperature (0°C say) the variation of the

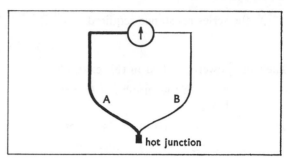

Fig. 3.15   A thermocouple with the cold junction in the galvanometer (at room temperature)

e.m.f. E with the temperature t of the other junction usually follows a *parabolic* law:

$$E = at + bt^2$$

Fig. 3.16 shows the curves for copper–iron and copper–constantan thermocouples. The point at which the curve becomes horizontal is known as the *neutral temperature*; near this point the arrangement is clearly of no use for temperature measurement. But at other temperatures a thermocouple is a useful thermometer for many applications. Such an instrument must always be calibrated at a third fixed point (besides 0°C and 100°C) in order to find the constants a and b. Copper–constantan produces about 10 times the e.m.f. of copper–iron and is the more satisfactory combination for temperature measurement; also its e.m.f./temperature curve is

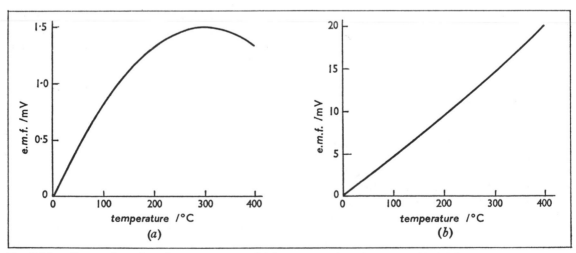

Fig. 3.16   Curves of e.m.f. against temperature: (*a*) for a copper–iron thermocouple; (*b*) for a copper–constantan thermocouple

almost linear over a large range. Another commonly used thermocouple is that employing the nickel–chromium alloys, *alumel* and *chromel*. A combination of platinum and a platinum–rhodium alloy is used for standard temperature measurements between 900 K and 1330 K.

The thermoelectric series

For a given temperature range the direction of the thermoelectric e.m.f. for different combinations of metals may be predicted by arranging the metals in a series; between the temperatures 0°C and 100°C the thermoelectric series for a selection of metals is:

> antimony, nichrome, iron, zinc, copper, gold, silver, lead, aluminium, mercury, platinum–rhodium, platinum, nickel, constantan (eureka), bismuth.

At the cold junction, the e.m.f. acts so as to drive current from the metal earlier in the list towards that later in the list. The series also indicates in a general way the magnitude of the e.m.f.; the farther apart the two metals lie, the greater is the e.m.f. they produce. The largest e.m.f. (with metals) is obtained with an antimony–bismuth thermocouple.

(ii) The Peltier effect

When an electric current is driven round a circuit consisting of two different metals, heat is evolved at one junction and absorbed at the other. This is known as the *Peltier effect*. It is complementary to the Seebeck effect. Thus the energy for the e.m.f. of a thermocouple comes from the heat made available by the difference of temperature between the junctions. When this e.m.f. drives a current round the circuit, the Peltier effect leads to the evolution of heat at the cold junction and the absorption of heat at the hot junction—thus tending to equalize the temperatures. Indeed, by the *second law of thermodynamics* the net transfer of heat must be from the point at higher temperature to the point at lower temperature.

The rate of evolution or absorption of heat is proportional to the current; and the Peltier effect is therefore *reversible*. If the current is driven round the circuit *in opposition* to a See-

beck e.m.f., heat is evolved at the hot junction and absorbed at the cold one—tending to magnify the temperature difference. Starting with the junctions of a thermocouple at the same temperature, the Peltier effect will produce a temperature difference such that the resulting Seebeck e.m.f. acts *in opposition* to the current. (This can be regarded as an example of Le Chatelier's principle.)

The difficulty in demonstrating the small *Peltier effect* lies in separating it from the much larger *Joule effect* (p. 20), whereby heat is evolved in all parts of a circuit in which a current flows. The usual result is that, instead of a rise in temperature at one junction and a fall at the other, we get the temperature rising at *both* junctions—but slightly faster at one than the other. To satisfy ourselves that this is not because of differences of resistance at the junctions, we need to make use of the reversibility of the Peltier effect. Reversing the current in the circuit leaves the Joule effect unaltered ($P \propto I^2$), but reverses the Peltier effect; the temperature difference should then come the other way round.

This may be done with the circuit in Fig. 3.17.

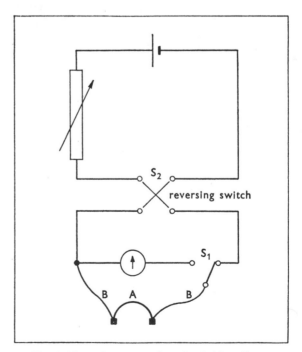

Fig. 3.17   Demonstrating the Peltier effect

First, with the thermocouple at room temperature current is driven through it by the battery. After a short time the two-way switch S_1 is moved over so that the thermocouple is connected to the galvanometer. The deflection observed is a measure of the temperature difference produced between the junctions. When the junctions have once more regained room temperature, the reversing switch S_2 is changed over, and the experiment is repeated with the current in the opposite direction. Now, when the thermocouple is joined to the galvanometer, the deflection is in the opposite direction, indicating that the temperature difference has reversed.

The Peltier effect is normally far too small to produce any significant cooling effect. However, some *semiconductor* materials (p. 23) show a vastly greater Peltier effect than any metal. With a suitably chosen pair of semiconductors a considerable cooling effect may be produced when the current flows the right way through the junction. At present the cost of these materials is high; but no doubt future developments will lead to an efficient refrigerator based on this principle.

4 Circuit Measurements

4.1 Resistance measurement

The direct measurement of a resistance consists in finding values for a steady current I through it and the corresponding p.d. V across it. The resistance R is then given by

$$R = \frac{V}{I}$$

Using an ammeter and voltmeter, either of the arrangements in Fig. 4.1 may be used. In (a) the resistance obtained initially is actually that of the ammeter and unknown coil in series. A subsidiary experiment must be performed with the voltmeter joined across the ammeter (as shown dotted) in order to find the resistance of the latter; the value of R is then the difference of the two results. Likewise in (b) the resistance obtained initially is that of the voltmeter in parallel with R; again a subsidiary experiment must be performed to find the resistance of the voltmeter so that the necessary correction can be applied. In each case several sets of readings are taken with different settings of the rheostat. Only by doing this can we be sure that the currents used are not causing excessive heating of the coil R with consequent change of its resistance.

The accuracy of the method is limited to that of the instruments used. The ordinary ammeter or voltmeter in regular use in a laboratory should not be relied on to an accuracy better than 2%.

If it is an instrument of really good construction it may be calibrated for the occasion (pp. 56, 57). and then its readings might be correct to within 0.5%. But the student should notice that, while the absolute accuracy is probably the same over most of the scale of a dial-reading instrument, the percentage accuracy increases with the deflection of the pointer. For instance a typical ammeter reads from 0 to 1.0 A; and the smallest scale division represents 0.02 A. If we can estimate to one-tenth of the smallest scale division, the absolute accuracy is 0.002 A. Now, at a scale reading of 0.2 A, this gives a percentage accuracy of *reading* the instrument of 1%; but at a scale reading of 1.0 A this becomes 0.2%. Thus for accurate work a dial-reading instrument should never be used at much less than half of full-scale deflection.

Substitution method

As with many scientific measurements, the direct measurement of resistance is not easy to perform accurately. There are, however, a number of satisfactory methods for *comparing* two resistances; one of the simplest of these uses a substitution technique (Fig. 4.2). The reading of the ammeter is first adjusted to a suitable value; •then the unknown coil of resistance R is disconnected and a standard resistance box joined in its place. The resistance of the box is adjusted

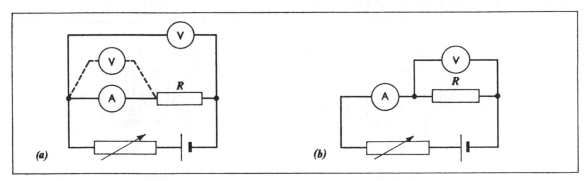

Fig. 4.1   The measurement of resistance with ammeter and voltmeter

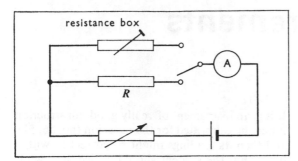

Fig. 4.2   A substitution method

until the ammeter reading is brought back to its former value; the total box resistance is then equal to R. The result does not depend on the correct calibration of the meter, but it does depend on its readings being precisely repeatable. With a good meter used at full-scale deflection an accuracy of 0.2% should be obtained. The method has the special merit that almost any circuit can be adapted for the purpose, and the value of a resistance may therefore be found under the conditions of the circuit in which it is to be used.

Standard resistance boxes are made in a variety of patterns, employing either plugs or switches to bring into use the sets of coils in the box so as to give any desired value of resistance. Fig. 4.3 shows the arrangement used in a *decade box* that switches in steps of 1 Ω from 0 up to 1110 Ω. The resistors are arranged in three banks. The right-hand bank consists of 10 equal resistances of 1 Ω; and the switch is joined up so that any number of these from 0 to 10 can be brought into the circuit. The middle bank consists similarly of 10 resistances of 10 Ω each; and the left-hand bank contains 10 resistances of 100 Ω each. The resistances in such a box are

of the wire-wound type, and they are all non-inductively wound (p. 28) so as to eliminate inductive effects.

4.2   The Wheatstone bridge

For the quick and accurate comparison of resistances the Wheatstone network or 'bridge' is widely used. The arrangement is shown in Fig. 4.4. The four resistances R_1, R_2, R_3, R_4 are joined in a quadrilateral ABCD. A source of e.m.f. is connected across one diagonal AC, and a galvanometer and tapping key across the other BD. One or more of the resistances is adjusted until no deflection of the galvanometer can be detected when the tapping key is pressed. The bridge is then said to be *balanced*. When this condition holds, it may be shown that the relation connecting the four resistances is

$$\frac{R_1}{R_2} = \frac{R_3}{R_4}$$

Thus if R_3 is a known resistance, and the ratio of R_1 to R_2 is known, R_4 may be found. The proof of this result is as follows:

When the bridge is balanced, there is no current through the galvanometer; therefore the same current flows through R_1 and R_2; let this be I_1. Likewise the same current flows through R_3 and R_4; let this be I_2. Also with no current through the galvanometer B and D must be at the same potential.

∴ the p.d. across AB = the p.d. across AD
$$\therefore I_1 R_1 = I_2 R_3$$
Also, the p.d. across BC = the p.d. across DC
$$\therefore I_1 R_2 = I_2 R_4$$

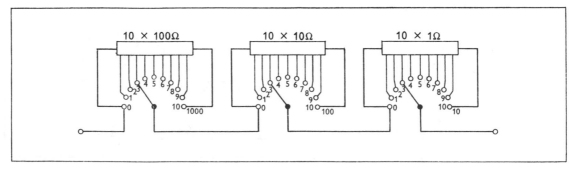

Fig. 4.3   A decade switched resistance box

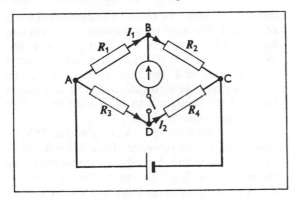

Fig. 4.4   The Wheatstone bridge network

Dividing $\dfrac{R_1}{R_2} = \dfrac{R_3}{R_4}$

The condition for balance is unaffected by inter-changing the positions of the cell and galvano-meter in the circuit. Thus, if the cell is joined across BD and the galvanometer across AC, similar reasoning gives the balance condition as

$$\frac{R_1}{R_3} = \frac{R_2}{R_4}$$

and this may be re-arranged to give as before

$$\frac{R_1}{R_2} = \frac{R_3}{R_4}$$

It does not follow that the bridge is equally sensitive in the two arrangements; but with modern galvanometers there is usually sufficient sensitivity in both arrangements. The accuracy of the method is limited by that of the known resistances of the bridge, and should be better than 0.2%.

A Wheatstone network may be joined up with standard laboratory components. For instance, if we wished to measure a resistance of about 50 Ω, we could connect up the network with the unknown resistance forming R_4, and a standard decade resistance box, like that in Fig. 4.3, as R_3. Then, if we take two fixed standard resistances for R_1 and R_2, choosing values of 100 Ω and 10 Ω respectively, the bridge will be balanced for some value of R_3 in the neighbourhood

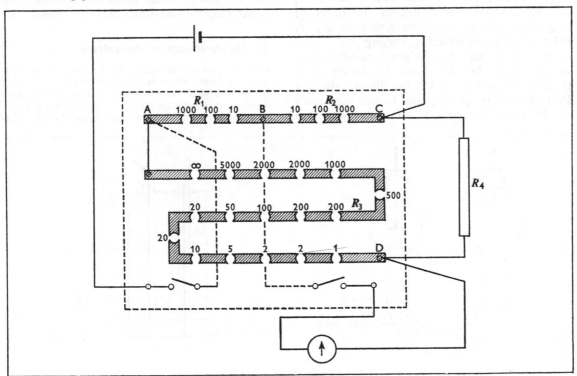

Fig. 4.5   The Post Office box form of Wheatstone bridge with plugged resistances

of 500 Ω; suppose we find the actual value required to be 512 Ω. Then we know that the unknown resistance R_4 is 51.2 Ω.

In a practical measurement it is usual to include in series with the galvanometer a protective resistor to prevent damaging the instrument when preliminary attempts to find the balance condition are being made. The protective resistor can be removed from the circuit when an approximate balance has been found; the maximum sensitivity of the galvanometer is then available for making the final adjustment.

Sometimes it is convenient to have all three of the known resistances, R_1, R_2, R_3, of the bridge already assembled in one box. An arrangement of this sort is traditionally referred to as a *Post office box* (Fig. 4.5). The unknown resistance R_4, the galvanometer and the cell are connected externally to the appropriate terminals of the box. The resistances R_1 and R_2 are arranged to have only three possible values: 10 Ω, 100 Ω and 1000 Ω. But R_3 is adjustable in steps of 1 Ω from 0 to 10 000 Ω. Thus the ratio R_1/R_2 can have any of the values: 100, 10, 1, 0.1, 0.01—chosen according to the order of magnitude of the unknown resistance R_4. A balance is then obtained by varying R_3.

In some designs these values of resistance are selected in the box by means of switches, as in the decade box in Fig. 4.3. But Fig. 4.5 shows a Post Office box using plugs. In this case each of the resistance coils is joined between two adjacent segments of brass strip fixed on the top of the box. The gaps between these segments can be closed by means of well-fitting brass plugs. When the plug is in position the resistance coil is shorted out, and the resistance between the brass segments is zero (ideally). The resistances required are brought into the circuit by pulling out the appropriate plugs.

A factor limiting the accuracy of the Wheatstone bridge is the presence of contact resistances at the points where the resistors are joined in the circuit. It is difficult to ensure that a screwed-down or soldered connection has a resistance of less than 0.002 Ω—and the circuit inevitably includes several such resistances. If all the contacts are clean and firm and all the resistances are greater than 10 Ω, it should be possible to obtain an accuracy of 0.1%—always assuming that the standards are known to this accuracy. When the resistance to be found is less than 1 Ω, the errors due to connection resistances of this sort are likely to be relatively large and the bridge method should be avoided. The potentiometer method given below is the most suitable one for measuring very small resistances.

* Measurements on electrolytes

The condition for the balance of the Wheatstone bridge is not dependent on the steadiness of the e.m.f. of the cell. Indeed, provided the resistances used are free from inductive and capacitive effects, an alternating e.m.f. may be used; the galvanometer is then replaced by a pair of

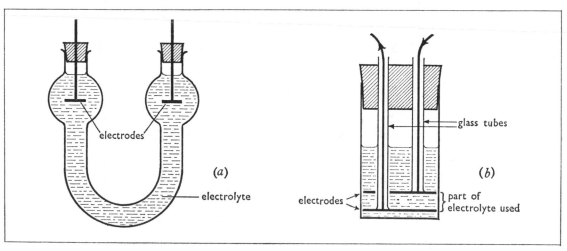

Fig. 4.6   (*a*) U-tube with fixed electrodes for comparing conductivities of electrolytes; (*b*) The Kohlrausch cell for measuring the conductivity of an electrolyte

headphones or some other a.c. detecting device. The best operating frequency for headphones usually seems to be between 500 and 1000 Hz.

An a.c. bridge of this kind can be used to find the resistivity (or the conductivity) of an electrolyte. The use of direct current in this case would lead to polarization (see p. 40) and the introduction of unwanted e.m.f.'s. The use of alternating current prevents this. The conductivities of two electrolytes may be *compared* by measuring the resistances between two electrodes fixed in any suitable vessel, filled first with one liquid and then with the other. A U-tube is often used for this (Fig. 4.6a). For absolute measurements of conductivity the section of electrolyte used must be of known length and cross-sectional area, as in the Kohlrausch cell (Fig. 4.6b). The electrodes are two flat circular plates fitting closely in a cylindrical glass vessel. The leads making connection to the plates are insulated by glass rods so that only the region *between* the plates is used for conduction. The conductivities of electrolytes increase rapidly with temperature. It is therefore important to control the temperature in these experiments with great care, and small currents should be used.

4.3 Potentiometer circuits

A variable resistor connected up as a potentiometer (p. 17) forms the basis of a class of circuits that may be used for comparing potential differences. These circuits can also be adapted for current and resistance measurement. Under the best conditions they enable these measurements to be made with an accuracy of 1 part in 10^6, and this is therefore the only method used for very precise d.c. measurements.

In its simplest form it consists of a uniform resistance wire AB stretched over a scale, through which a steady current is passed from one or more accumulators (Fig. 4.7). The p.d. V to be found is connected at X and Y. Its negative side (at Y) is joined to the negative end A of the wire; and the positive side of the p.d. is joined through a galvanometer and tapping key to the contact maker D. In the loop of the circuit AYXD the p.d. between the points D and A on the wire acts in opposition to the p.d. V under test; if the two p.d.'s are equal, no current flows through the galvanometer. The position of the contact maker D is adjusted to bring about this condition. There is then no deflection of the galvanometer when the tapping key is pressed; and the instrument is said to be *balanced*. The length of wire l between the points A and D is measured.

Let the current through the potentiometer wire

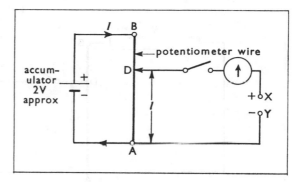

Fig. 4.7   The principle of the potentiometer circuit

be I and let its resistance per unit length be r; then

p.d. V under test = p.d. between D and A

$$= Irl$$

If the wire is uniform and the current through it is steady, I and r are constants, and we may write

$$V \propto l$$

Two p.d.'s V_1 and V_2 may be connected in turn to the wire in this way and the corresponding balance lengths l_1 and l_2 found; then

$$\frac{V_1}{V_2} = \frac{l_1}{l_2}$$

We have assumed that the resistance between the terminal at A and the beginning of the wire at the zero point of the scale is negligible. This may be checked by repeating the measurements with a small resistance in series with the potentiometer wire, which will reduce the total p.d. across it. Both the balance lengths will be changed, but their ratio should be unaltered. This procedure can also be regarded as a check on the uniformity of the potentiometer wire.

If a cell is joined to the points X and Y, no current will flow through it when the potentiometer is balanced, and the p.d. between its terminals is therefore precisely equal to its e.m.f. Thus by using a standard cell—normally the Weston cadmium cell (p. 43)—other p.d.'s may be found by direct comparison with the laboratory standard.

Fig. 4.8 shows the circuit used to compare the e.m.f.'s of two cells. The switch S enables either cell to be connected into the circuit. R_{prot} is a protective resistance in series with the galvanometer to protect it from damage while preliminary

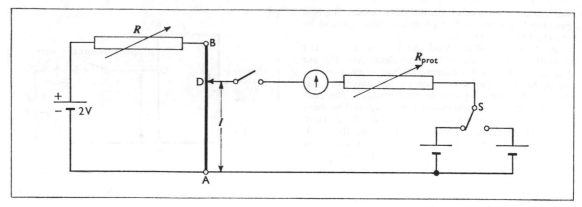

Fig. 4.8   Comparing the e.m.f.'s of two cells with a potentiometer circuit

attempts are being made to find the balance point for the contact maker D. It can then be reduced to zero so as to give the maximum galvanometer sensitivity for the final adjustment. R is a plug resistance box, which is used as described above to obtain different pairs of values for the balance lengths l_1 and l_2, thus giving several independent estimates of the ratio l_1/l_2. Using a potentiometer wire 1 metre long and ordinary laboratory apparatus, the balance lengths can be found to the nearest half-millimetre; and the errors should then be less than 0.1%. Refined methods can greatly improve on this.

The wire AB with its sliding contact D in Figs. 4.7 and 4.8 is in effect a variable resistor connected up as a potentiometer (p. 17). We may if we wish use a simple radio-type potentiometer for this type of circuit. These potentiometers are not designed as precision devices, and even the so-called 'linear' kind are often lacking in uniformity. Also in the wire-wound type the sliding contact arm can make contact only with the wire at a particular point on each turn of the wire round the card that holds it. The resistance therefore changes in a series of small steps in such a potentiometer, rather than continuously. However, if the wire is fine enough so that a large number of turns can be used, this effect is scarcely noticeable. A carefully selected high quality radio-type potentiometer may be used to build a useful compact potentiometer circuit for measurement of p.d.'s; the arrangement is shown in Fig. 4.9. The potentiometer needs to be equipped with a large dial; this could conveniently read

from 0 to 150. We can then arrange easily for the instrument to read p.d.'s in volts directly.

It is calibrated first with a standard cell. The e.m.f. of the Weston standard (p. 43) is 1.019 V; the potentiometer dial is therefore set to 101.9. The instrument is then adjusted with the rheostat R_{cal} until it is balanced at this setting of the potentiometer. The instrument is now calibrated so that each scale division represents 0.01 V, and the whole scale covers the range 0 to 1.5 V. The calibration must of course be repeated each time the instrument is used, since the p.d. of the driving cell is likely to change. Other p.d.'s may now be joined to the points X and Y instead of the standard cell; the potentiometer is quickly balanced, and the p.d. is read off from the dial. As usual, the protective resistor R_{prot} is included in the circuit while preliminary adjustments are being made; this is then reduced to zero to give maximum galvanometer sensitivity for the final adjustment.

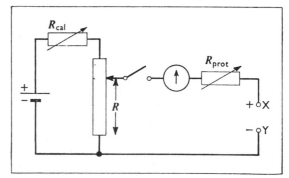

Fig. 4.9   A direct-reading potentiometer circuit made with a simple radio-type potentiometer

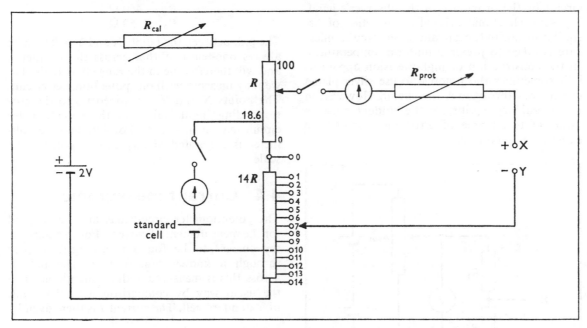

Fig. 4.10   The design of a simple commercial potentiometer apparatus

Fig. 4.10 shows a simplified form of the type of instrument that is available commercially. The potentiometer resistance consists of a continuously variable part R together with a bank of 14 resistances in series, each of which has the same value R as the variable part. A special 15-position switch enables the point Y to be connected to any point in this bank of resistors, while the point X is joined as usual through the galvanometer to the sliding contact on the variable resistance R. The sliding contact moves over a scale marked from 0 to 100. Special internal connections are made for calibration purposes with the standard cell; these are to the point marked '10' on the bank of resistances, and to the point marked '18.6' on the variable one. With the standard cell joined in circuit, R_{cal} is adjusted to bring the galvanometer in series with the cell to read zero. Since the e.m.f. of the standard cell is 1.0186 V, we now know that the p.d. across each section of resistance R is 0.1 V, and each division of the variable resistor represents 0.001 V. The instrument thus provides a precision measurement of p.d. to the nearest millivolt or better.

By extending the bank of resistances or in some other equivalent way, the sensitivity of instruments based on the potentiometer circuit can be increased almost indefinitely. For instance, the e.m.f.'s of thermocouples (p. 46) are generally in the range 0 to 20 mV. To measure temperatures with a thermocouple to the nearest 0.1 K changes of e.m.f. of about 4 μV have to be detected. We then need a potentiometer circuit about 1000 times as sensitive as the one shown in Fig. 4.10. This can be done in principle by increasing considerably the series resistor R_{cal}. If we adjust this until the p.d. across each section R of the potentiometer is 0.0001 V, then each division of the scale represents 1 μV. Other arrangements must now be made of course to effect the calibration in terms of the e.m.f. of the standard cell; and an extremely sensitive galvanometer would have to be employed to detect the proper balance point for the thermocouple.

In the precise measurement of p.d.'s, and particularly if these are very small, the greatest care has to be taken to eliminate stray thermal e.m.f.'s in series with the p.d. under test. For instance, if the potentiometer wire and knife-edge contact maker are made of different materials, the junction will act as part of a thermocouple if its temperature is raised above that of the rest of the circuit by the proximity of the observer's hand. The same effect may arise at any junction of dissimilar metals in the galvanometer circuit, unless care is taken to preserve uniform temperatures. In the construction of high-precision apparatus it is sometimes the practice to use silver-plated conductors throughout the circuit. However, in using ordinary equipment the student must be aware of this source of errors and conduct the experiment accordingly.

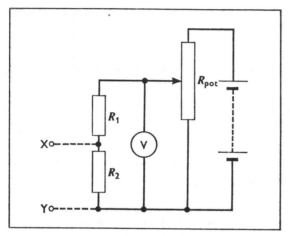

Fig. 4.11   Calibrating a voltmeter with a potentiometer circuit

Calibrating a voltmeter

The potentiometer circuit can be regarded as a sort of voltmeter. But even in its simplest forms it can be much more accurate than an ordinary moving-coil instrument. It may therefore be used to calibrate other kinds of voltmeter, and to measure the errors in their scales. The range of p.d.'s that may be measured with the simple circuits we have described above are from 0 to 1.5 V. If the voltmeter to be calibrated covers a larger range than this, we have to use the arrangement shown in Fig. 4.11. The variable p.d. to deflect the voltmeter is provided by the variable resistor joined across a suitable battery. Across the voltmeter we join a *fixed* potential divider consisting of the resistances R_1 and R_2, by which a known fraction of the p.d. across the voltmeter may be tapped off to be measured by the potentiometer circuit. For instance, suppose our voltmeter has a scale reading from 0 to 15 V, we might then choose values

$$R_1 = 450 \ \Omega$$
$$R_2 = \ \ 50 \ \Omega$$

The p.d. between the points X and Y will then be always one-tenth of that across the voltmeter, and will therefore lie in the range 0 to 1.5 V that we may measure with our potentiometer circuit. The points X and Y are joined up to the corresponding input points of the potentiometer circuit (e.g. as in Fig. 4.10 or 4.9), and the voltmeter is calibrated at any chosen point of its scale.

4.4 Current measurement

The potentiometer circuit may also be adapted for the measurement of current. For this purpose the current to be found is arranged to pass through a known resistance R. The p.d. V across this is measured with a potentiometer in the usual way by comparison with the e.m.f. of a standard cell. The current I is then given by

$$I = \frac{V}{R}$$

The standard resistance R should normally be chosen so that the p.d. V across it is about 1 V, since this is the range for which a simple potentiometer circuit is usually designed.

For precise work resistors to be used with a

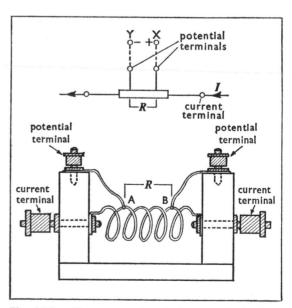

Fig. 4.12   The construction of a four-terminal resistor

potentiometer in this way are made with *four* terminals (Fig. 4.12). Two of these, known as the *current terminals*, are at the ends of the resistance wire, and are used to pass the current into and out of it. The other two, known as the *potential terminals*, are joined by fine wires to two points A and B on the resistance wire; and it is between these points that the specified resistance exists. The p.d. between the potential terminals is measured with the potentiometer and standard cell; this is equal to the p.d. between A and B, since at balance no current passes through the fine wires and potential terminals. With a two-terminal resistor the actual value of the resistance always includes the somewhat uncertain resistance of the contacts through which the current is passing. Even with heavy brass connectors the contact resistances could amount to 0.005 Ω. Only with resistances over 5 Ω could the error on this account be certainly less than 0.1%.

Calibrating an ammeter

The potentiometer circuit is now used to measure the p.d. V across a suitable standard resistance R in series with the ammeter (Fig. 4.13). The current I is then given by

$$I = \frac{V}{R}$$

This can be compared with the ammeter reading, so giving its error at any chosen point of its

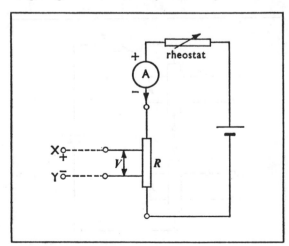

Fig. 4.13   Calibrating an ammeter with a potentiometer circuit

scale. The resistance R is chosen so that the p.d. across it at full-scale deflection of the ammeter is about 1.5 V. For instance, if the ammeter reads from 0 to 0.5 A, the required value of R is given by

$$R = \frac{V}{I} = \frac{1.5}{0.5}\,\Omega$$
$$= 3\,\Omega$$

Electrical calorimetry

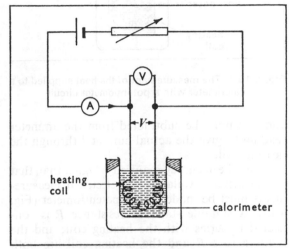

Fig. 4.14   The measurement of the heat supplied to a calorimeter with ammeter and voltmeter

The most satisfactory means of supplying a measured quantity of heat to a calorimeter and its contents is by an electric current. When the accuracy required is no better than 2%, the electrical measurements can be carried out using an ammeter to give the current I through the heating coil and a voltmeter the p.d. V across it. If the current is switched on for a length of time t, then

$$\text{heat supplied} = VIt$$

Fig. 4.14 shows the circuit used. In this circuit a small part of the current recorded by the ammeter passes through the voltmeter instead of the heating coil. If the resistance of the voltmeter is large compared with that of the heating coil, this can be ignored; otherwise a correction must be applied. The current I_v through the voltmeter is given by

$$I_v = \frac{V}{\text{resistance of voltmeter}}$$

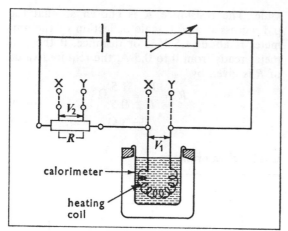

Fig. 4.15   The measurement of the heat supplied to a calorimeter with a potentiometer circuit

eliminating heat losses are matters outside the scope of this book; and the student should refer to textbooks of *Heat* for details about them.)

4.5   Measuring resistance with a potentiometer circuit

The potentiometer circuit provides the most accurate method for the comparison of two resistances. The arrangement used is shown in Fig. 4.16. For precision work both resistors should be of the four-terminal type described above (Fig. 4.12). A suitable current I is passed through the two resistors R_1 and R_2 connected in series. The potentiometer circuit terminals X and Y are joined in turn across each resistor, and the corresponding readings of p.d. are taken. Provided the current I does not change between the two measurements, we then have

$$\frac{R_1}{R_2} = \frac{V_1}{V_2}$$

Only the ratio of the two p.d.'s need be known, so that the potentiometer circuit does not need to be calibrated with a standard cell as with measurements of p.d. and current. It is simplest to choose the current I so that the larger of the two p.d.'s is just under 1.5 V. However, it may well be that the currents must be much smaller than this in order to avoid overheating the coils. In any case the measurements should always be repeated

and this must be subtracted from the ammeter reading to give the actual current I through the heating coil.

When the accuracy required is higher than that of the meters available, the electrical measurements must be made with a potentiometer (Fig. 4.15). A suitable standard resistance R is connected in series with the heating coil; and the p.d.'s across R and the heating coil are both measured with the potentiometer by comparison with the e.m.f. of a standard cell. If these p.d.'s are V_1 and V_2 as shown, then

$$I = \frac{V_2}{R}$$

and the heat supplied in time t is given by

$$\text{heat supplied} = V_1It = \frac{V_1 V_2 t}{R}$$

The measurements are thus carried out by direct reference to the laboratory standards—standard cell and standard resistor. Accuracies approaching 1 part in 10^5 can thus be reached, which exceeds the accuracy of the rest of the measurements in any calorimeter experiment.

The measurement of temperature in such experiments is also often performed by electrical methods. In precise work *resistance thermometers* are usually used, as described below. (The design of calorimeters and the methods used for

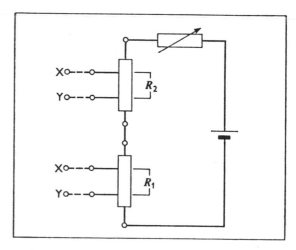

Fig. 4.16   The comparison of resistances with a potentiometer circuit

with various values of the current I through the resistors. Only by doing this can we find out for certain whether R_1 and R_2 are independent of current in the range of currents used. The potentiometer circuit provides the most satisfactory means of measuring very small resistance (less than 1 Ω); the use of the four-terminal technique makes it possible to avoid the errors due to the uncertain connection resistances at the terminals.

The platinum resistance thermometer

Over much of their range resistance thermometers are appreciably more sensitive than any other thermometer; they are also very convenient to use. Precise laboratory measurements of temperature are now largely performed with some kind of resistance thermometer—chiefly the platinum kind. The fine platinum wire whose resistance R_t is to be measured is contained in a hard glass U-tube inside the outer protective sheath. The ends of the fine wire are welded to

thick platinum wires, which in turn are welded to silver leads connecting to the terminals at the top of the thermometer (Fig. 4.17). The fine wire must be of the highest purity; very small traces of impurity can appreciably affect its temperature coefficient of resistance. For this reason the silver leads cannot be joined direct to the fine wire, since the diffusion of silver into its ends would significantly affect its properties. Since the resistance X is to be measured with a potentiometer, four leads are used—two current leads and two potential leads—so that the resistance actually measured is that of the fine platinum wire only. The current leads are used to join the wire in series with a standard resistance R; and the resistances are compared with a potentiometer in the usual way.

$$\therefore \quad \frac{R_t}{R} = \frac{\text{balance length for } R_t}{\text{balance length for } R}$$

In order to measure a temperature, the instrument is first calibrated by finding the resistance of

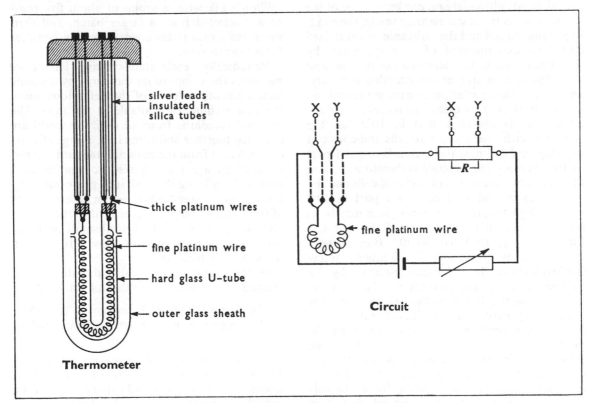

Fig. 4.17   A platinum resistance thermometer

the platinum wire at the *ice-point* (R_0) and at the *steam-point* (R_{100}). Then, if R_t is its resistance at the unknown temperature t, the value of t on the Celsius scale of the platinum resistance thermometer is given by

$$\frac{t}{100} = \frac{R_t - R_0}{R_{100} - R_0}$$

The variation of the resistance of a platinum wire with temperature is not quite linear, so that the value of t is not equal to the standard Celsius temperature except in the immediate neighbourhood of the two fixed points. The deviation from the standard scale is calculated from measurements made at subsidiary fixed points, such as the *sulphur-point* and *oxygen-point*. Details of this technique will be found in textbooks of *Heat*.

4.6   Electrical standards

The accuracy of d.c. measurements in a laboratory depends ultimately on our knowledge of the secondary electrical standards, namely, the e.m.f. of a standard cell and the resistance of a standard coil. The measurement of these quantities by direct reference to the definitions of the volt and the ohm is a matter of considerable difficulty, and is rarely attempted outside the standardizing laboratories of the world. The present accuracy obtainable is about 1 part in 10^5; but the accuracy with which standard cells and coils can be *compared* amongst themselves is about 1 part in 10^6. The present practice is therefore to state the values of the e.m.f.'s of standard cells and the resistances of standard coils to 1 part in 10^6, while recognizing that the values given may be in error in terms of the definition of the volt and the ohm by up to 1 part in 10^5. International uniformity is preserved by frequent intercomparisons of the standards produced by the various standardizing laboratories. Thus, at the present time the values of the volt and the ohm are *fixed by agreement* to within 1 part in 10^6; but future improvements in standardizing techniques may force us in due course to revise these values by anything up to 1 part in 10^5. Indeed there is no certainty that the agreed international values are not drifting by up to this amount from year to year—except that it is

unlikely that all the standard cells and coils in the standardizing laboratories of the world should drift suddenly by the same amount.

Each of the standardizing laboratories preserves its own group of standard cells and standard coils. No single cell or coil could be relied on to remain constant over a period of years. Therefore a group of standards is used, which are regularly compared with each other to check their relative constancy. At the National Physical Laboratory in Great Britain a group of about 20 standard cells is used; these are selected from an even larger batch of cells, the group selected being that which shows the least relative variation over the previous few years. The mean e.m.f. of the group is then assumed to have remained constant since the last international assessment. The composition of the working group is always changing, since it is found that cells develop instability after about 15 years; also it takes 2 or 3 years for a new cell to settle down completely. ('Instability' in this connection means variations of e.m.f. of a few parts in a million!) Likewise, a group of about five resistors is selected from a larger batch, and their mean resistance is assumed to remain constant from year to year.

Periodically, each standardizing laboratory measures the values of its standards by absolute measurements in terms of the definitions, using the best techniques available at the time. The national standards from all over the world are collected together and compared, using cells and coils selected from the national standard groups. Corrections are then applied to the national standards to bring them all into conformity. By this means the standards in use in different parts of the world are kept in agreement to the highest accuracy possible, while their absolute accuracy continually improves with new advances in standardizing techniques.

In the National Physical Laboratory three methods are used to fix the values of standards:

(*i*) *The Lorenz rotating disc apparatus* (p. 115) is used to measure the *resistance* of the standard coils.

(*ii*) *A standard mutual inductance* of known dimensions (p. 114) is used for the direct measurement of inductance and capacitance; with a suitable a.c. bridge it can also be used to measure the *resistance* of the standard coils, as a check on the results of the Lorenz disc experiments.

(*iii*) *The current balance* (p. 10) can be used for the direct measurement of *current*; it can there-fore be used to find the p.d. across an already standardized resistance coil through which this current flows. Using a potentiometric technique this p.d. can be compared with the *e.m.f. of a standard cell*.

5 Magnetic Forces

5.1 Magnets

Magnetic forces arise between charged particles in motion (p. 7). All atoms are therefore inherently magnetic because of the circulating electrons they contain. To a large extent the orbits of the electrons form equal and opposite pairs cancelling out their magnetic effects, but with many atoms there is some residual magnetic effect, caused chiefly by the 'spin' of the electrons. With most substances the effects observed are very small. But with *iron, nickel, cobalt* and a few others interactions between atoms produce intense magnetic properties. These are called *ferromagnetic* materials, and it is from these that magnets may be made.

The properties of ferromagnetic materials are extremely complex and we shall defer the study of them until a later chapter. However, all these properties arise from the behaviour of the circulating systems of charge within the atoms; and we therefore need in this chapter to investigate the nature of the magnetic behaviour of charged particles in motion, i.e. of electric currents.

We give first an elementary account of the behaviour of magnets, since magnets often provide the most convenient sources of magnetic fields; also a small pivoted magnet (or *compass*) is the simplest tool for indicating the direction of a field.

A magnet seems to influence the space around it so that forces act on any other magnet placed nearby. Any region of space in which a magnet would be acted on by forces is called a *magnetic field*. The magnetic properties of a magnet appear to be concentrated in certain parts of it only. Thus if a simple bar magnet is plunged into a bowl of iron tacks, scarcely any adhere to the central parts of the bar, while many tacks cluster round its ends. Likewise the forces on a pivoted magnet when another magnet is brought near appear to act chiefly at its ends. Also the forces on the two ends of the pivoted magnet are in opposite directions. If one end is *attracted* to the magnet brought up to it, the other is *repelled*. When a compass is placed in a uniform magnetic field the forces on it constitute a *couple*—i.e. a pair of equal, opposite and parallel forces (Fig. 5.1). This couple will tend to turn the small pivoted magnet until its *axis* is parallel to the forces; and so the compass (when it comes to rest) indicates the direction of the field. Clearly this will only be an accurate indication in so far as the direction of the field varies very little over the length of the compass. (In a non-uniform field the pivoted magnet not only turns but is pulled in some direction as well.)

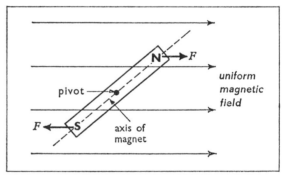

Fig. 5.1   A magnet in a uniform field experiences a couple—a pair of equal, opposite and parallel forces

The apparent concentration of the magnetic properties of a bar magnet at its ends is an illusion. In reality all the material in the magnet contributes to the strength of the magnet. This may be seen if a strip of magnetized material is broken in two (Fig. 5.2). Before the break is made no iron tacks will cling to the central part of the strip, but only to its ends. But the tacks cling to the new ends formed by the break just as much as to the original ends. This shows that the central parts of the magnet were magnetized just as much as the ends, but this was not apparent outside the magnet until the break was made. We can imagine the breaking up of a magnet continued in this way to the limit, until the

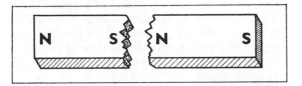

Fig. 5.2 The magnetization of the centre of a magnetized strip is revealed as soon as the strip is broken

specimen is broken down into its constituent atoms. We should then find that the individual atom is magnetized and behaves exactly as though it were a small magnet. This may be demonstrated directly with a narrow beam of atoms in a vacuum. Not only are the atoms rotated by the magnetic field through which they pass, but if the field is intensely non-uniform a resultant force acts as well on each atom and the beam is deflected out of a straight line.

An explanation of this fundamental behaviour of matter occurs to us when we observe that a coil of wire carrying an electric current behaves in the same way. Placed in a magnetic field forces act on it that tend to turn it so that its *axis* is parallel to the field; and in a non-uniform field there is a resultant force acting on it tending to move it as a whole in some direction. The couples and forces can be shown to be of exactly the same nature as those that act on a magnet of the same shape and size as the coil in similar circumstances. Likewise the coil of wire with the current flowing in it produces a magnetic field of exactly the same nature as that of the equivalent magnet. Quite generally it may be shown that the magnetic forces and fields produced by any magnet are identical with those that would arise from electric currents flowing in suitable configurations. The modern picture of the atom with its circulating systems of electrically charged particles leads us to expect that magnetic effects will arise in the individual atom, and we conclude that the magnetic properties of a magnet are simply the summed effect of the magnetic behaviour of the charged particles in the atoms of which it is composed.

The earth itself has a weak magnetic field. This is believed to be caused by electric currents circulating within its core (Fig. 5.3a). The currents are probably generated by convection in the liquid core maintained by radioactive heating of the earth's interior. By analogy with other geographical terms we call the points in which the magnetic axis of the earth meets the surface

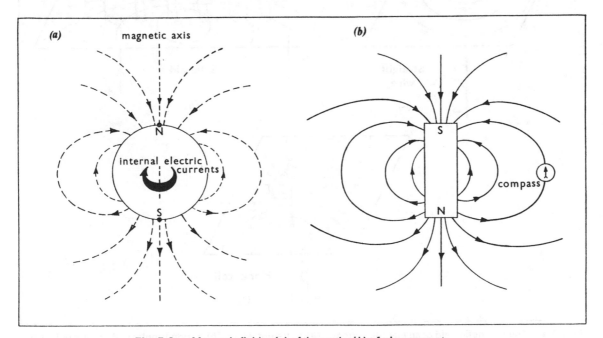

Fig. 5.3   Magnetic fields : (*a*) of the earth ; (*b*) of a bar magnet

the *north* and *south magnetic poles*. In the same way we talk about the *magnetic equator*—the great circle in a plane at right-angles to the magnetic axis; and the *magnetic meridian* at a point—the vertical plane in which the earth's magnetic field acts at the point. A freely suspended magnet (i.e. a compass) therefore tends to align itself so that its magnetic axis is in the magnetic meridian.

This provides us with a simple way of distinguishing and labelling the two ends of a magnet (*poles* they are usually called, by analogy with the geographical terms): the *north pole of a magnet* is defined as the one that is drawn towards the earth's north magnetic pole; and the other end of the magnet is called its *south pole*.

Simple experiments with pairs of magnets show that *like poles repel each other, while unlike poles are attracted together*. (Notice that this implies that in the magnetized core of the earth the northern end behaves like a south-type magnetic pole.)

The magnetic axis of the earth is slowly changing its position. Indeed the direction of magnetization of the rocks of the earth's crust indicate clearly that the earth's magnetic field has even reversed at irregular intervals (of half a million years or so) throughout the earth's history. At present the north magnetic pole is located in northern Canada.

The *direction* of a magnetic field at any point is defined as the direction in which the north

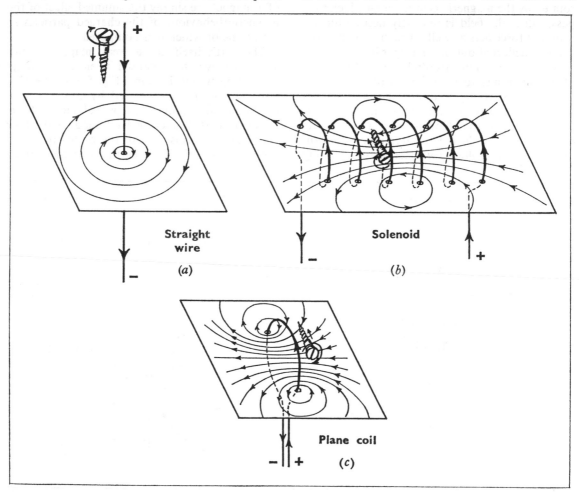

Fig. 5.4   The magnetic field of an electric current: (*a*) a straight wire; (*b*) a long tubular coil or solenoid; (*c*) a plane coil. The direction of the field is given in each case by the right-hand screw rule

pole of a small freely pivoted magnet would point when placed there. Magnetic fields may be described pictorially by means of lines of force. This kind of pattern is most easily revealed by sprinkling iron filings on a piece of card placed in a field. Fig. 5.3*b* shows a sketch of the field of a simple bar magnet derived from an iron filings pattern of this sort. The directions of the arrows on the lines must of course be determined by a compass in the field. Since the north pole of a compass would be repelled from the north pole of a magnet and attracted towards its south pole, the arrows are drawn emerging from the north pole and entering the south pole.

The earth's field (Fig. 5.3*a*) is of course an exception; its lines of force are directed *towards* the north magnetic pole. Near the magnetic equator the earth's field is approximately parallel to the surface—i.e. horizontal. At the magnetic poles the field is vertical—vertically *downwards* at the north magnetic pole. In Great Britain the field is inclined downwards at about 68° to the horizontal. The angle that the field makes with the horizontal at a point is called the *inclination* or the *angle of dip*. The angle between the geographical and magnetic meridians at a point is called the *declination*. The knowledge of these angles at any point on the earth's surface is important for navigational purposes.

The production of a magnetic field by an electric current was discovered by Oersted in 1819, who detected the deflection of a compass placed near a conductor when an electric current was switched on and off in it. A descriptive picture of the fields due to circuits of various shapes may be obtained by observing lines of force in the usual way with iron filings on a card. The diagrams for three cases are shown in Fig. 5.4. Notice in particular that the lines of force near a straight wire are circles concentric with the wire; the field definitely has no component parallel to the wire or radially away from it. The direction of the arrow on the lines of force may be determined as usual with a small compass placed near the wire. This direction may be predicted with the aid of a right-hand screw rule:

The direction of the lines of force round a current is that in which a right-handed screw would be turned in order to advance it in the direction of the current.

The same rule may be applied to describe the direction of the field of a coil. The imaginary right-handed screw may be placed, as shown, at the position of any section of the wire of the coil; this gives the direction of the field near this section, and enables us to predict the sense of the arrows on the lines of force through and near the coil.

In Fig. 5.4*b* with the current in the direction shown the arrows on the lines of force emerge from the left-hand end of the tubular coil of wire (or *solenoid*) and enter its right-hand end. The left-hand end of the solenoid therefore behaves like the north pole of a bar magnet, and the right-hand end like the south pole. Likewise the plane coil in Fig. 5.4*c* produces a field resembling that of a thin disc-shaped magnet with a north pole on the left-hand face and a south pole on the right.

5.2   The force on a current

So far we have considered only the magnetic properties of currents flowing in coils of wire, since it is these that produce effects similar to the behaviour of magnets. We need now to analyse what forces act on single pieces of wire when currents flow in them; and then we need to go further and find what force acts on the single charged particle whose motion contributes to the current.

When a straight length of wire carrying a current is placed at right-angles to a magnetic field, it is found that a force acts on it whose line of action is at right-angles to both the field and the current. The direction in which the force acts is traditionally described by

Fleming's left-hand rule. If the thumb and first two fingers of the left hand are put mutually at right-

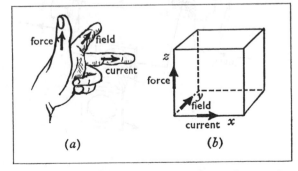

(a)   (b)

Fig. 5.5   Fleming's left-hand rule for the force acting on an electric current in a magnetic field

angles, and the First finger is pointed in the direction of the Field while the seCond finger is in the direction of the Current, then the thumb gives the direction of the force (Fig. 5.5).

This basic magnetic effect may be demonstrated with a simple 'current balance' consisting of a rectangle of copper wire supported on a pair of razor blade knife edges. A magnetic field may be applied locally to one end of the rectangle by means of a suitable U-shaped magnet (Fig. 5.6*a*). When a current is passed through the copper wire via the knife edges the balance is deflected up or down according to the directions of current and field, in agreement with Fleming's

rule. The magnetic field may alternatively be produced by means of a current flowing in a coil placed near the end of the balance (Fig. 5.6*b*), and the same rule is found to be obeyed. In every case the deflecting force is a maximum when the field is arranged to be at right-angles to the current, and it falls to zero if the field is parallel to the current (Fig. 5.6*c*).

The force may be measured approximately by balancing it against the weight of a small rider on the end of the balance. If two or three U-shaped magnets are placed side by side under the end of the balance it may then be shown that the force F is proportional to the length l of

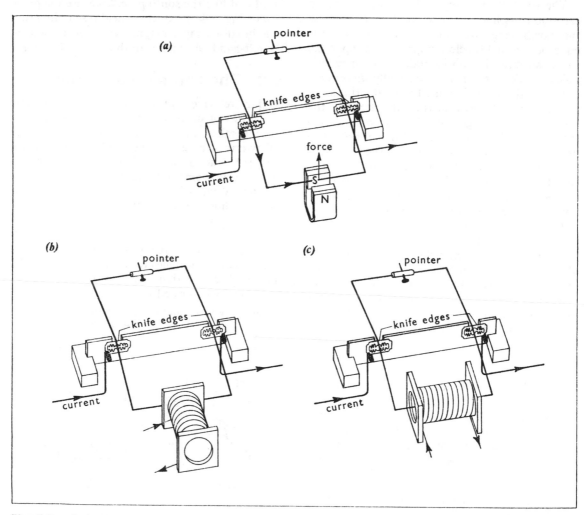

Fig. 5.6   A simple current balance for investigating the magnetic force that acts on an electric current in a magnetic field

wire in the field. Similarly in a given field the force F may be shown to be proportional to the current I flowing in the wires of the balance. The force F also of course depends on the strength of the magnetic field, and may therefore be used as a means of measuring it. The quantity so obtained is called the *flux density* of the field, and is denoted by the symbol B. (The reason for using this name will become apparent later in this chapter, p. 87.)

> The **flux density B** of a magnetic field at a point is the force per unit length that acts on a wire carrying unit current lying at right-angles to the magnetic field.

Thus the force F acting on the wire of length l when placed at right-angles to the field is given by

$$F = BIl$$

If the wire is placed parallel to the field the force drops to zero. In the general case the force can be regarded as due to the *component* of the field *perpendicular* to the current ($B \sin \theta$ in Fig. 5.7); the component parallel to the current produces no force. Thus in Fig. 5.7 the force F is given by

$$F = BIl . \sin \theta$$

The force F is to be measured in *newtons* (N), the length l in *metres* (m), and current I in *amperes* (A). The unit of flux density is called the *tesla* (T), after an American scientist.

Thus $1\,\text{T} = 1\,\text{N A}^{-1}\,\text{m}^{-1}$

Using a simple current balance like that in Fig. 5.6 the flux density B of the various magnets and coil systems used may now be measured directly in teslas.

The couple acting on a coil in a magnetic field

In a uniform field it is clear that opposite sides of a coil will experience equal and opposite forces; thus a system of couples acts on it tending to turn it to a position in which all the forces are in the plane of the coil. Since the magnetic force on a wire is always at right-angles to the field this position must be with the plane of the coil at right-angles to the field—i.e. with its *axis* in line with the field. In a non-uniform field the

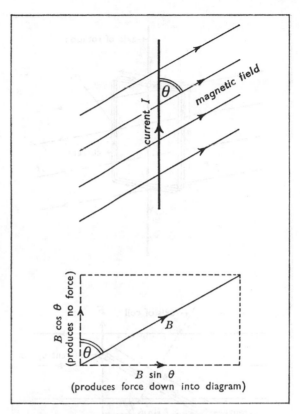

Fig. 5.7   Only the component of the field perpendicular to the current produces any force

pairs of forces on opposite sides of the coil are unequal, and the coil will experience a resultant force acting in some direction as well as a couple tending to rotate it.

In a moving-coil ammeter of the type commonly used in a laboratory part of the current to be measured is passed through a pivoted coil in the field of a permanent magnet. The instrument in effect measures the couple acting on the pivoted coil due to the current flowing in it, and so gives an indication of the current. An ammeter of this kind is therefore a form of current balance using a constant magnetic field. The actual design of meters of various kinds is considered in detail in Chapter 8.

The magnitude of the couple T acting on a coil in a uniform field may be calculated as follows. We shall confine ourselves to the case of a rectangular coil. Consider such a coil, measuring b by h, of N turns carrying a current I free to rotate about an axis in its own plane, this axis

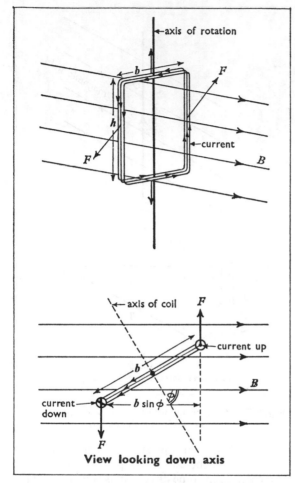

Fig. 5.8   Calculating the moment of the couple acting on a coil in a magnetic field

$$T = Fb \cdot \sin \phi$$
$$= BNIhb \cdot \sin \phi$$
$$\therefore \quad T = BANI \cdot \sin \phi$$

where A is the area of the coil ($= hb$). By an extension of this reasoning the expression may be proved quite generally for a coil of area A and of any shape. The product ANI is called the *electromagnetic moment* of the current-carrying coil, and is given the symbol m. Thus

$$T = mB \cdot \sin \phi$$

where $m = ANI$

5.3   The force on a moving charge

Since an electric current consists of moving charged particles, it is reasonable to suppose that the forces acting on a wire carrying a current are the summed effects of the forces operating on the individual particles in motion. On this assumption we can work out an expression for the force acting on a single charged particle in a magnetic field.

The current I flowing in a conductor is the rate of passage of electric charge past any given point in it. Thus if N electrons each of charge e flow past the point Y in time t (Fig. 5.9), then the current I is given by

$$I = \frac{Ne}{t}$$

If the average drift velocity of the electrons is v, then the length l of the conductor occupied by the N electrons considered is given by

$$l = vt$$

being at right-angles to a field of flux density B (Fig. 5.8). The forces acting on the top and bottom of the coil (through which the axis of rotation passes) are parallel to this axis and produce no turning moment. The force F acting on each of the other two sides is given by

$$F = BNIh$$

These two sides remain at right-angles to the field as the coil rotates so that $\sin \theta = 1$ in all positions. Let the angle between the *axis* of the coil and the field be ϕ. The perpendicular distance between the lines of action of the two forces is $b \cdot \sin \phi$. Therefore the moment T of the couple is given by

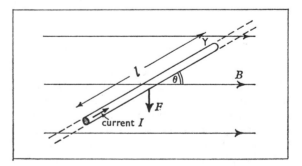

Fig. 5.9   The force on a conductor is the summed effect of the forces acting on the charged particles in it

The force F on this length of the conductor is given by

$$F = BIl \sin \theta$$

$$= B.\frac{Ne}{t}.vt \sin \theta$$

$$= BNev \sin \theta$$

The magnetic force F_m acting on each particle is therefore given by

$$F_m = Bev \sin \theta$$

The force is at right-angles to the field and to the path of the particle in a direction given by Fleming's left-hand rule (not forgetting that the current in the conventional sense is opposite to the motion of an electron).

The effect of this force can be observed directly in the deflection of beams of charged particles in magnetic fields, e.g. electron beams in cathode-ray tubes (p. 228). Since the magnetic force acting on a charged particle is at right-angles to its path, it causes no change of speed but only a change of direction. If the motion is exactly at right-angles to a uniform field, the path is turned into a circle. In the special case of motion along a line of force, the force reduces to zero ($\theta = 0$). In general, with the motion inclined to the field, the path is a *helix* round the lines of force. Thus the effect of a magnetic field on the motions of all charged particles is to 'trap' them in the field, in the sense that their paths are always curved so that they do not move far in directions at right-angles to the field.

An important natural effect of the action of magnetic fields on charged particles occurs in the outer parts of the earth's atmosphere. Charged particles in these regions are caused to spiral back and forth round the lines of force of the earth's field, and so are effectively trapped high in the atmosphere for considerable times. These regions thus contain dense streams of high-speed particles; they are called the *Van Allen radiation belts* after their discoverer. One of the striking results of modern research has been to show that in controlling the large-scale behaviour of matter in the Universe magnetic fields are quite as important as gravitational ones. While gravitational forces predominate in the interactions of large units of matter, such as stars and planets, the movements of charged particles are controlled chiefly by magnetic fields, fields which themselves are produced by currents of the charged particles. It appears that much of the matter in the Universe is in a condition referred to as a *plasma*; this means a gaseous state in which the particles are very largely ionized. A familiar example is an electric arc passing through the air between two conductors. The interiors of the sun and other stars and the gaseous nebulae of interstellar space are all in this condition; and the movements of matter within them are considerably affected by the magnetic fields they contain.

The Hall effect

When a conductor carrying an electric current is placed in a magnetic field, it is found that a p.d. is developed between the sides of the conductor in a direction perpendicular to the field. This is called the *Hall effect*. Thus in Fig. 5.10, if a current is passed through the slab of conductor between the edges P and Q, and a magnetic field is applied at right-angles to the faces of the slab, a p.d. appears between the points R and S on the sides. The Hall p.d. V_H is found to be proportional to the flux density B and to the current density j (i.e. the current per unit area of cross-section); it is also proportional to the breadth b of the slab at right-angles to current and field. Thus

$$V_H \propto Bjb$$

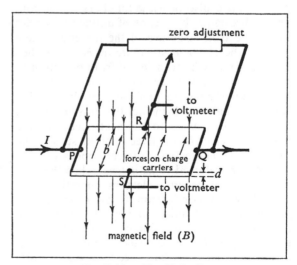

Fig. 5.10   The Hall effect

where $\qquad j = \dfrac{I}{bd}$ (d = thickness)

$$\therefore\ V_\mathrm{H} = R_\mathrm{H} Bjb \qquad\qquad (\mathrm{I})$$

or $\qquad V_\mathrm{H} = \dfrac{R_\mathrm{H} BI}{d}$

where R_H is a constant for the material known as the *Hall Coefficient*.

The effect is readily explained in terms of the forces acting on the moving charge carriers in the conductor. In Fig. 5.10, with the directions of current and field shown, according to Fleming's left-hand rule the forces on the current carriers act across the slab from S towards R, and the concentration of current carriers should increase in this direction. This will manifest itself as a difference of potential between the terminals R and S. The direction of the *force* is the same whichever sign the charge carriers in the material happen to have, so that the sign of the Hall p.d. may be used as a direct test of the sign of the charge carriers in the material. Surprisingly perhaps in many materials conduction is revealed as being by means of *positive* charge carriers, in contradiction to what we might expect if conduction in solids were simply by means of electrons. This paradox is resolved by a more detailed investigation of the manner in which the electrons in a material may transfer charge through it. It is found that as well as the simple migration of free electrons through the material, it is also possible for charge to be conveyed through it by means of unfilled vacancies in the electron structure of the material. These gaps in the electron structure or *holes* (as they are called) may be transferred from one atom to the next through the material; and they behave as though they were positive particles moving through it, although in fact the movement of a hole from one atom to the next occurs by the transfer of electrons between bound sites in the two atoms. These two different modes of conduction are referred to as *n–type* (when the conduction is by *negative* electron migration) and *p–type* (when conduction is by migration of *positive* holes). Semiconducting materials, such as germanium and silicon, are particularly interesting in this respect since the mode of conduction in them may be varied by minute adjustments to the impurities they contain. These

effects are considered in greater detail in the chapter on Electronics (Chapter 14).

The Hall effect thus provides the means of establishing whether the conduction in a material is predominantly *n*– or *p–type*. The Hall coefficient R_H is taken as positive if the p.d. is in the direction caused by positive current carriers— i.e. for *p*-type material—and negative for *n*-type material.

With metals the Hall effect is very small; the coefficient R_H is of the order of 10^{-6} m^3 C^{-1}, and with the electromagnets usually available in schools the Hall p.d. is unlikely to exceed 1 millivolt even with very thin specimens. But in semiconductors the charge carriers move much faster to carry a given current, since there are relatively few of them per unit volume compared with a metal; the magnetic force on a moving charge deflecting it towards the sides of the conductor is therefore much greater, giving a greater Hall p.d.

The chief practical difficulty in demonstrating the effect is in the positioning of the Hall electrodes R and S. It is necessary to ensure that, in the absence of a magnetic field, R and S are at exactly the same potential. Since the Hall p.d. may well be much less than the p.d. applied between P and Q, this adjustment needs to be done with precision. Fig. 5.10 shows one way of doing it. One of the side electrodes R is connected to the slider of a high resistance potential divider joined between the ends of the specimen. Through this electrode a small current can be made to enter or leave the slab, as may be necessary to bring R to the same potential as S. With no magnetic field the slider is positioned to give zero reading on a voltmeter connected between R and S. The Hall p.d. may then be read directly from the meter when a magnetic field is applied.

The measurement of the Hall coefficient enables not only the sign but also the concentration of the charge carriers to be found; and it has therefore provided one of the most powerful techniques for studying the properties of semiconductors. The following approximate analysis shows how this may be done.

When the moving charge carriers migrate across the slab of conductor under the action of the magnetic field, they give rise to an electric field that opposes this lateral movement. The lateral separation of charge continues until the

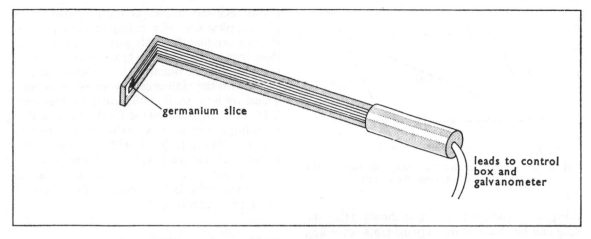

Fig. 5.11   A Hall probe for measuring magnetic fields

magnetic and electrical forces on the charge carriers are equal and opposite. We have already shown (p. 11) that the average velocity of drift v of the charge carriers is given by

$$v = \frac{I}{nAe} = \frac{j}{ne}$$

where $j\ (= I/A)$ is the current density, and n is the number of charge carriers per unit volume.

Therefore the magnetic force F_m on each charge carrier is given by

$$F_m = Bev = \frac{Bej}{ne} = \frac{Bj}{n}$$

and the electrical force F_e is given by

$$F_e = e \times \text{(potential gradient)}\quad \text{(p. 189)}$$
$$= \frac{eV_H}{b}$$

Equating the two forces, we have

$$\frac{eV_H}{b} = \frac{Bj}{n}$$

$$\therefore\ V_H = \frac{1}{ne}.Bjb$$

This expression is in agreement with that obtained experimentally (equation I above); comparing the two equations we see that the Hall coefficient R_H is given by

$$R_H = \frac{1}{ne}$$

Since the electronic charge e is known, this enables the concentration n of charge carriers to be found.

The Hall probe

The Hall effect provides a simple and direct means of comparing magnetic fields. If a constant current is maintained through a suitable slice of germanium, the Hall p.d. generated between its sides is proportional to the component of the magnetic field perpendicular to the slice. For this purpose the germanium slice is usually mounted in the end of a rod through which the necessary leads pass; the balancing potentiometer needed to bring the p.d. to zero in the absence of a magnetic field may be mounted in a separate control box. This arrangement is known as a *Hall probe* (Fig. 5.11). Needless to say a Hall probe is a much more satisfactory instrument for measuring magnetic fields than a current balance such as that described above (Fig. 5.6). The probe gives the component of the field at right-angles to the plane of the germanium slice. To find the direction of the resultant field it is only necessary to turn the probe round until the Hall p.d. is a maximum. To make accurate readings the zero setting of the balancing potentiometer must be checked frequently and the current through the slice must be monitored for constancy.

The vector nature of B

The flux density B of a magnetic field is only completely specified by stating its direction as well as its magnitude. In other words it seems that B must be treated as a *vector*. To establish

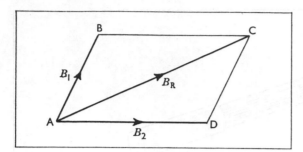

Fig. 5.12   Compounding two magnetic fields using the parallelogram rule for vectors

this point we should be able to show (*a*) that the field can be resolved into components according to the usual rule for resolving vectors; (*b*) that magnetic flux densities can be added together according to the usual rule for the addition of vectors, namely the parallelogram rule. The Hall probe provides a simple means of testing these points.

(*a*) The probe may be set up at right-angles to a magnetic field of flux density *B*, giving a reading V_0 on the millivoltmeter joined across it. When the slice is turned through an angle θ, the reading falls to a value V_θ. If *B* behaves as a vector, its component at an angle θ should be $B.\cos\theta$, and we should find

$$V_\theta = V_0.\cos\theta$$

(*b*) The addition of two magnetic fields according to the parallelogram rule may be demonstrated with the aid of two magnets. First of all each of the magnets is placed in turn on its own marked position on a board, and the direction and magnitude of the flux densities B_1 and B_2 at a point A on the board is measured in each case with the Hall probe. Then with *both* magnets in position the resultant flux density B_R is found. B_R, B_1 and B_2 should then be found to be related according to the usual parallelogram rule. Thus in Fig. 5.12 the sides AB and AD of the parallelogram are drawn parallel to the two flux densities B_1 and B_2 with their lengths in proportion to the magnitudes of the flux densities. The diagonal AC then represents in magnitude and direction the resultant B_R of the two fields.

In an experiment of this sort the earth's field is inevitably superimposed on those of the magnets,

and if necessary allowance must be made for it. The simplest way of avoiding this complication is to ensure that the fields used are much greater than that of the earth, so that the latter may be ignored. But in practice the difficulty scarcely arises with the Hall probes available in schools. These are not usually satisfactory for measuring fields as small as that of the Earth. It can be done by using a very sensitive galvanometer to measure the Hall p.d., but it will probably then be found that the zero setting of the balancing potentiometer is difficult to make precisely. The Hall probe is best reserved for measuring fields greater than 10^{-4} T.

5.4   Induced e.m.f.'s—moving conductors

When a conductor is made to move across a magnetic field, a p.d. is generated between its ends. If we consider the forces acting at the atomic level, such an effect is indeed to be expected. We have seen that a charged particle moving across a magnetic field experiences a force at right-angles to its path and to the field. A moving conductor contains equal numbers of charged protons and electrons, on all of which the magnetic field acts. However, only the electrons are free to move inside the conductor. The result of the motion should therefore be an accumulation of negative charge at one end of the conductor, leaving a surplus of positive charge at the other. Thus, with the directions of the field and the motion indicated in Fig. 5.13

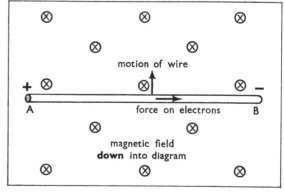

Fig. 5.13   The induced e.m.f. in a conductor arising from the forces acting on the charged particles in it

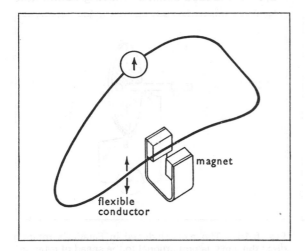

Fig. 5.14   The e.m.f. induced in a conductor that is moved across the lines of force of a magnetic field

the application of Fleming's left-hand rule shows that the force acting on a negative particle moving with the wire will be from A to B; and the end B therefore becomes negative, leaving A positive. It would make no difference in fact whether the conduction in the wire took place by movement of negative or positive particles; if positive particles were free to move, they would be impelled towards the end A, and the same result would occur.

If the moving conductor formed part of a complete circuit, the charge carriers in it would continue in motion as long as the conductor was moving, and a current would flow in the circuit. The moving conductor would in fact behave as a source of e.m.f. in the circuit, the energy being derived from the action of the mechanical forces producing the motion.

This effect is known as *electromagnetic induction*, and the e.m.f.'s generated by such means are called *induced e.m.f.'s*. Electromagnetic induction was discovered by Michael Faraday in a brilliant sequence of experiments—difficult to design in his day because of the insensitive measuring instruments available to him. However, with modern sensitive galvanometers the effect is a very obvious one. Thus in Fig. 5.14 if the flexible conductor is moved up and down between the poles of the permanent magnet the galvanometer joined to its ends registers a small current. The current continues

only while the conductor is moving. The same happens if the magnet is moved while the conductor is held still; relative motion of conductor and source of magnetic field is required to produce the induced e.m.f. Also the motion must 'cut across' the lines of force of the magnetic field. The maximum e.m.f. is produced when the motion is at right-angles to the field, and it falls off to zero for motion in the same direction as the field.

In such experiments allowance has to be made for the relatively slow response of a sensitive galvanometer. When a current through the instrument has ceased, the deflection takes time to fall off to zero. Only if the movements of the conductor are slow in relation to the response time can definite indications of what is happening be obtained. If care is taken in this way, we discover that the e.m.f. induced depends on the speed of the motion, and the results suggest

$$\text{e.m.f.} \propto \text{speed}$$

The simplest way to investigate this is to see how long it takes to move the wire through the field in such a way as to keep the galvanometer deflection constant. If the chosen deflection is then doubled it takes half as long to move the wire the same distance, and so on.

It is easily checked that the direction of the induced e.m.f. is in accordance with the explanation we have given in terms of the forces that must act on the electrons in the conductor. A compass must be used to establish the direction of the magnetic field between the pole faces of the magnet, and we must also ascertain which way the galvanometer deflects for a current passing through it in a given direction.

It is worth examining at this point the parts of a small d.c. dynamo. The moving part (called the *armature*) consists simply of a set of conductors mounted in slots round the iron cylinder (Fig. 5.15) so that when this is turned round they cut across the magnetic field produced by a suitable electromagnet. The wires are connected to the external circuit through sliding contacts at one end so that all those wires that are cutting through the field at any instant are contributing to produce an e.m.f. in the same sense in the circuit. (Armature construction and behaviour are considered in more detail in Chapter 9.)

A simple dynamo of this sort may be used to

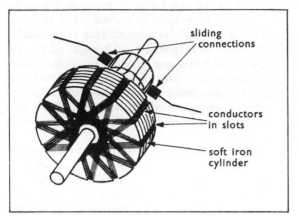

Fig. 5.15   The armature (the moving part) of a d.c. dynamo

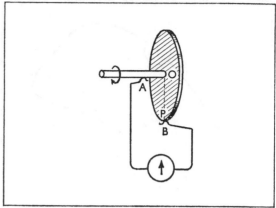

Fig. 5.16   The e.m.f. induced in Faraday's rotating disc; the e.m.f. is proportional to the speed of rotation

test more exactly whether induced e.m.f.'s are indeed proportional to the speed of the moving conductors. For this purpose the magnetic field in which the conductors move must be held constant, and the coils of the electromagnet (called the *field coils*) must therefore be supplied with a constant current. Alternatively, we may use a small dynamo whose field is supplied by a fixed permanent magnet system. A voltmeter is connected to the armature terminals, and arrangements are made to measure the rate at which the armature is rotated. A graph of p.d. measured against speed of rotation will then be found to be a straight line through the origin.

We may also use a simple dynamo to test how the induced e.m.f. depends on the flux density of the magnetic field. Over a range of currents less than the maximum design current the flux density produced by the electromagnet is approximately proportional to the current in the field coils. Using the dynamo at a constant speed of rotation we may then show that

e.m.f. ∝ flux density (*B*)

An even simpler form of dynamo is *Faraday's rotating disc* (Fig. 5.16). It consists of a metal disc that may be rotated at a steady speed in the magnetic field under test. Any given radius (such as OP in Fig. 5.16) cuts through the field at a steady rate; therefore a steady e.m.f. is generated in the circuit formed by the milli-voltmeter and the radius momentarily between the sliding contacts A and B. Again we may demonstrate that the induced e.m.f. is propor-

tional to the speed. If the disc is mounted inside a solenoid supplied with an adjustable current, we may also demonstrate that the e.m.f. is proportional to the flux density *B*.

Induced e.m.f.'s also arise in a coil rotating in a magnetic field. In this case the e.m.f. varies from one position to another, and reverses its direction twice in every revolution; in other words alternating current flows in the circuit of which the coil is a part. A simple way to investigate this behaviour is to use a large rectangular coil in the earth's magnetic field. It may be connected by long flexible leads to a sensitive galvanometer, and may be either held in the hand or mounted so that it can spin about an axis in the plane of the coil (Fig. 5.17). In this case the e.m.f.'s induced in the sides of the coil AB and CD cancel out (by symmetry), and we can concentrate our attention on the e.m.f.'s induced in the other two sides parallel to the axis of rotation (BC and DA). The induced e.m.f. will be a maximum when the plane of the coil is in the magnetic meridian, since at that moment the sides BC and DA are moving directly across the lines of force of the magnetic field and in opposite directions, so that the e.m.f.'s induced in them act in the same sense round the coil. When, however, the plane of the coil is at right-angles to the field the sides BC and DA are moving momentarily in the direction of the field and no e.m.f. is generated in them. If the coil is turned slowly by hand, the way in which the e.m.f. varies may be seen

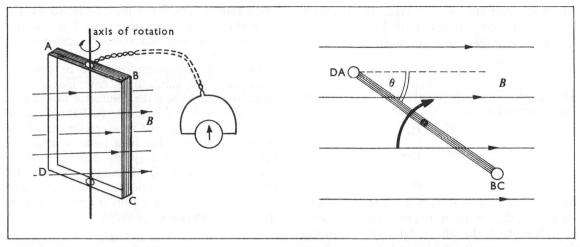

Fig. 5.17   Inducing a slow alternating e.m.f. in a rotating coil

from the deflection of the galvanometer; in fact we are thus generating a very low frequency alternating current.

If sliding connections are provided for the coil (i.e. slip rings, p. 150), the speed of rotation may be increased and the alternating e.m.f. may then be observed by connecting the coil to an oscilloscope. The frequency is equal to the number of revolutions per second of the coil. At steady speeds of rotation a 'sinusoidal' waveform is produced, and the *peak value* of the e.m.f. (p. 28) is proportional to the speed. A simpler mechanical arrangement is however to keep the coil stationary, and to produce the magnetic field by means of a permanent magnet, which can then be rotated without the need for awkward sliding connections (Fig. 5.18). This is essentially the arrangement used in some simple

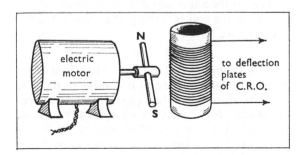

Fig. 5.18   A simple alternator for testing the law of electromagnetic induction; the peak e.m.f. is proportional to the speed

cycle dynamos (which generate a.c.). Connect the output of such a dynamo to an oscilloscope while the jacked-up bicycle is pedalled at a steady rate, and we may then observe the waveform generated. This may not be sinusoidal because of the shape of the magnet used, but it should still be found that the *peak e.m.f.* is proportional to the speed. The peak e.m.f. is given by the height of the oscilloscope trace, while the speed may be measured either by finding the frequency generated or by timing the rate of pedalling.

Using the result deduced above for the force acting on a charged particle in a magnetic field (p. 69), it is quite simple to derive theoretically an expression for the magnitude of the induced e.m.f. E in a moving conductor. The calculation is very similar to that of the Hall p.d. (p. 71). Consider a conductor of length l at right-angles to a magnetic field of flux density B and moving across it at a steady speed v (as in Fig. 5.13). When the conductor is first set in motion we suppose that the magnetic forces cause a migration of electrons in one direction along it, thereby setting up a p.d. between its ends. This migration will continue until the magnetic forces are exactly balanced by the electrical forces tending to restore the electrons to their original uniform distribution. The electrons share the motion of the wire, and the magnetic force F_m acting on one of them is given by

$$F_m = Bev$$

where e = the electronic charge. The electrical force F_e on an electron is given by

$$F_e = e \times \text{(potential gradient)} \quad \text{(p. 189)}$$

$$= e . \frac{E}{l}$$

The electrons reach equilibrium when these forces are balanced.

$$\therefore \ F_e = F_m$$

$$\therefore \ \frac{eE}{l} = Bev$$

$$\therefore \ E = Bvl$$

which is in agreement with the previous experimental results. In the more general case we must remember that the above calculation requires that the value of B used is the *component* B_n of the flux density normal to the plane in which the conductor is moving.

Thus $$B_n = B . \sin \theta$$

where θ is the angle between B and the plane of movement.

An alternative way of describing this result is obtained by observing that the product vl is the area A swept out by the wire per unit time. In calculus notation we can express it

$$E = B_n \frac{dA}{dt}$$

An alternative way of writing this is

$$E = B \frac{dA_n}{dt}$$

where A_n is the area swept out projected at right-angles to the flux density B. For if B makes an angle θ with the plane of movement, we have

$$A_n = A . \sin \theta$$

$$\therefore \ B \frac{dA_n}{dt} = B \frac{d(A . \sin \theta)}{dt} = B . \sin \theta \frac{dA}{dt} = B_n \frac{dA}{dt}$$

We need the result in one of these forms for calculations in which the conductor is not simply a straight wire at right-angles to the direction of motion. We do not need to consider the separate parts of a conductor of irregular shape, provided we can write down the rate at which it sweeps out area. Thus in Fig. 5.13 a kinked wire should give rise to the same e.m.f. as the straight one if the distance between its ends is also l as shown; for if it moves through the field at the same speed, it will sweep out area at the

same rate vl. The point may be tested experimentally with the arrangement of Fig. 5.14 if a deliberately kinked wire is used in place of the 'straight' one.

The same expressions can also be used for calculating the e.m.f. in a case like that of *Faraday's disc* (where the conductor in question is a radius OP that is rotating). Faraday's disc (Fig. 5.16) may be used as a precise test of the result if it is mounted in a uniform magnetic field whose flux density we can measure or calculate. If the disc is of radius r and makes n revolutions per unit time, the rate of sweeping out area by any given radius is $\pi r^2 n$. Therefore the e.m.f. E should be given by

$$E = \pi r^2 n B$$

and this may be checked by measurement. Faraday's disc may be used in this way as a flux density meter, but it is not as convenient as other devices that are available (e.g. the Hall probe), and its use is confined to special applications (p. 115).

We may derive an expression for the e.m.f. induced in a *rotating coil* by considering the areas swept out by its sides. In this case the two sides of the coil BC and DA (Fig. 5.17) cut through the field in opposite directions, and the e.m.f. is proportional to the sum of the rates at which these sides sweep out area at right-angles to the flux density B. This is simply equal to the rate of change of the area A_n of the coil projected at right-angles to B.

$$\therefore \ E = B \frac{dA_n}{dt}$$

Now $$A_n = A . \sin \theta$$

where θ is the angle between the plane of the coil and B. If the coil is turning at a steady angular speed ω, we have

$$\theta = \omega t$$

The maximum (peak) value of e.m.f. is induced when the plane of the coil is parallel to the field ($\theta = 0$), since A_n is then changing at the maximum rate. Fig. 5.19 shows graphically how we may calculate the rate of change of A_n at this moment. The graph of A_n against time t is a sinusoidal one, for we have

$$A_n = A . \sin \omega t$$

But near the origin, where $\theta \, (= \omega t) = 0$, it

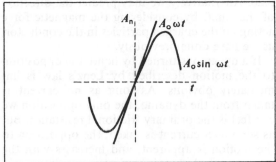

Fig. 5.19   Calculating the peak e.m.f. induced in a rotating coil

approximates to a straight line—which is the tangent to the curve at this point. Now for small values of θ we can write

$$\sin \theta \approx \theta \quad \text{(Appendix, p. 394)}$$

Therefore this tangent is the straight line

$$A_n = A\omega t$$

The maximum rate of change of A_n is the gradient of this line, and so we have

$$\left(\frac{dA_n}{dt}\right)_{max} = A\omega$$

[Using calculus, the same result can of course be obtained more quickly by differentiating:

$$A_n = A . \sin \omega t$$

$$\therefore \frac{dA_n}{dt} = A\omega . \cos \omega t$$

$$\therefore \left(\frac{dA_n}{dt}\right)_{max} = A\omega]$$

If the coil is of N turns, the peak e.m.f. E_0 is therefore given by

$$E_0 = BN\left(\frac{dA_n}{dt}\right)_{max} = BAN\omega$$

The frequency f of the e.m.f. is equal to the number of revolutions of the coil per second; and the angular speed ω is therefore given by

$$\omega = 2\pi f$$

It is worth considering at this point how the e.m.f. induced in a moving conductor is to be detected. In Fig. 5.20 connection is made to the moving conductor AB through the sliding con-

tact with the stationary rails PQ and RS, and a voltmeter inserted in the circuit between the rails would presumably register the e.m.f. But suppose we decided to join the voltmeter directly to the ends of AB and move it along with the conductor; its connecting leads also would then be cutting through the magnetic field, and would be sweeping out area at the same rate as AB. Therefore an equal e.m.f. would be induced in them. We should now have a system consisting of two sources of e.m.f. in parallel and no current would flow in the loop; the voltmeter would therefore register zero. To detect the e.m.f. it is necessary rather for the voltmeter to be kept stationary while the conductor AB is moving (or vice-versa); and the two parts of the circuit must be joined together by flexible or sliding connectors. For an induced e.m.f. to be apparent there must be relative motion of the moving conductor and the measuring instruments. We return to this important matter later.

The calculation of induced e.m.f.'s is best understood by considering some practical examples.

EXAMPLE (i)   Calculate the e.m.f. induced between the wing-tips of an aeroplane of wing span 30 m flying horizontally at 250 m s^{-1}, if the vertical component of the flux density of the earth's magnetic field is 4×10^{-5} T.

The area swept out by the wing per second

$$= 30 \times 250 \text{ m}^2 \text{ s}^{-1}$$

$$\therefore \text{ the e.m.f.} = 4 \times 10^{-5} \times 30 \times 250 \text{ V}$$

$$= 0.3 \text{ V}$$

The student can check that in the northern hemisphere the port (left) wing-tip will be *positive*.

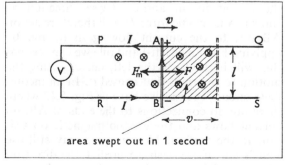

area swept out in 1 second

Fig. 5.20   The e.m.f. induced in a conductor sliding on two rails

EXAMPLE (*ii*) A horizontal gramophone turntable (made of aluminium) of diameter 30 cm rotates at $33\frac{1}{3}$ revolutions per minute in a uniform vertical magnetic field of flux density 1.0×10^{-2} T. Calculate the e.m.f. induced between the centre and rim of the turntable.

A given radius sweeps out an area $\pi \times 0.15^2$ m² per revolution.
The number of revolutions per second

$$= \frac{100}{3 \times 60} \text{ s}^{-1}$$

∴ the rate of sweeping out area by a given radius

$$= \frac{100}{3 \times 60} \times \pi \times 0.15^2 \text{ m}^2 \text{ s}^{-1}$$

∴ the e.m.f.

$$= 1.0 \times 10^{-2} \times \frac{100}{3 \times 60} \times \pi \times 0.15^2 \text{ V}$$

$$= 3.9 \times 10^{-4} \text{ V}$$

If the magnetic field acts vertically *downwards*, with the usual direction of rotation of gramophones the rim will be positive.

Lenz's Law

In any of the examples of electromagnetic induction we have considered so far it will be found that the *direction* of the induced e.m.f. is such that the resulting current *opposes* the movement that caused it. This result is known as *Lenz's Law*. The opposition to the movement occurs only if there is a complete circuit so that a current actually flows. For instance in Fig. 5.20 if the conductor AB is made to slide to the right in contact with the smooth metal rails PQ and RS so that it cuts across a magnetic field, an e.m.f. is induced in it, and a current will flow in the circuit in an anti-clockwise direction as shown. A magnetic force F_m will therefore act on AB due to the current flowing in it, and by applying Fleming's left-hand rule we can see that this will be towards the left, i.e. opposing the motion to the right that caused it. In an incomplete circuit, such as we have in Fig. 5.13 where there are no connections to the ends of AB, no current flows and there is no magnetic opposition to the motion. However, we may still use Lenz's law, if we wish, to predict the direction of the e.m.f. by supposing the circuit to be completed with a suitable conducting path as in Fig.

5.20. Alternatively we may predict the direction of the e.m.f. by considering the magnetic force acting on the charged particles in the conductor, as we have done previously.

If a dynamo is turned by hand, the opposition to the motion described by Lenz's law is immediately obvious. As long as no current is taken from the dynamo, the only opposition we can feel is the ordinary frictional resistance. But as soon as a current is taken, the opposition to the motion is apparent, and increases with the current. Lenz's law is in fact a necessary consequence of the law of conservation of energy. If the e.m.f. acted in the opposite direction to that which opposes the motion the result would be the production of electrical and mechanical energy from nothing; once a dynamo had been started, the current in the armature would act to accelerate the machine; kinetic energy would be created, and at the same time electrical energy could be drawn off for use in the circuit connected to the armature! In fact the magnetic forces brought into being by the current generated must be in the opposite sense so that work has to be done by the applied forces to turn the armature; and this work reappears as the electrical energy in the armature circuit.

We may use reasoning based on the conservation of energy as an alternative way of deriving the expression for the e.m.f. E induced in a moving conductor. Thus in Fig. 5.20 suppose that the conductor AB is being propelled at a steady speed v by a force F along the rails PQ and RS, and let the flux density at right-angles to the plane of motion be B. We must bear in mind that the induced current I produces its own magnetic field, and this modifies the field in which the conductor itself is moving and therefore affects the e.m.f. To simplify the analysis we shall suppose that the effect on the e.m.f. produced in this way is negligible; in other words we assume that the resistance of the circuit is made sufficiently high for the magnetic field produced by the current I to be negligible compared with the applied magnetic field B. Now the opposing magnetic force F_m is given by

$$F_m = BIl \quad \text{(p. 67)}$$

where l is the distance between the rails. If the wire is moving at steady speed, then

$$F = F_m = BIl$$

The rate of doing work

$$= F \times \text{(distance moved per second)}$$
$$= Fv = BIlv$$

The work done reappears as electrical energy in the circuit and is there changed into heat. The total rate of production of heat $= EI$ (p. 14)

$$\therefore EI = BIlv$$
$$\therefore E = Blv$$

in agreement with the previous result. The experimental test of this expression is therefore a confirmation of the law of conservation of energy as applied to the process of conversion of mechanical energy into electrical energy. Conversely a test of the conservation of energy in such a case implies the existence of induced e.m.f.'s to be calculated in the manner we have discussed—always provided we can be sure that there is no other significant form of energy conversion in the given situation.

Frames of reference

Whenever we talk about an e.m.f. induced in a conductor it is important to state what motion the measuring instruments have. In other words we must state in what *frame of reference* the calculations and measurements are being conducted. Consider for instance an aeroplane flying in the earth's magnetic field. It is meaningless to talk of the e.m.f. induced in the wing of a plane or the p.d. between its wing-tips—*unless* we specify also the frame of reference for which the statements are true. For an observer travelling in the plane there is no e.m.f. (provided the field is a uniform one), and there is no p.d. between the wing-tips. Indeed in this frame of reference no evidence can be found that the plane is *moving* through the field. After all, we cannot attach a label to a line of force and watch it drifting past! Only if the field is non-uniform would there be any e.m.f. detectable in a circuit loop fixed in the plane. But another observer making measurements in a frame of reference fixed in the ground analyses the situation rather differently. His observations show that the magnetic field is stationary in *his* frame of reference. He calculates that an e.m.f. must be induced in the wing of the aeroplane as it cuts the flux. To settle the point he contrives to make momentary connection between the wing-tips and his measuring instruments on the ground; and he duly detects the predicted p.d. The observer in the plane agrees that an e.m.f. should be detected in this experiment, but accounts for it in a different way. To him the plane is stationary with respect to the magnetic field, and the voltmeter on the ground with its connecting leads is receding rapidly through the flux; the e.m.f. is therefore induced in the ground-based part of the circuit. Both observers agree, as they must, about the effect—namely the voltmeter reading; but they disagree about the site of the e.m.f. in the circuit. Each believes that the e.m.f. arises in the part of the circuit that is moving with respect to his frame of reference. And a third observer in a train moving with respect to both the others would regard the e.m.f. as arising partly in one section of the circuit and partly in the other.

★ *Relativity*

The above situation is in fact a very simple application of the principle of Relativity. This principle states that the laws of physics must be the same for all observers who are moving at constant velocities with respect to one another. There is no 'preferred observer' whom we can assert to be 'at rest' and whose observations have any sort of priority over the observations of others. Experiments may be conducted with equal validity in any non-accelerating frame of reference, and the same laws will be discovered as in any other similar frame of reference. Two observers may analyse a given situation in terms of forces, velocities, masses, fields, etc., in different ways; but the connections between these quantities, expressed by the laws of physics, must be the same for both. One very important set of laws is that which shows how to relate the values of physical quantities for one observer with those measured by another. This set of laws makes up the Special Theory of Relativity propounded by Einstein in 1905. It is beyond the scope of this book to consider this theory in detail, though some of its results are discussed in other parts of the book (pp. 251 and 345). Generally speaking the analysis of events in the Special Theory of Relativity differs little from the older Newtonian analysis used in elementary mechanics, *except* when the speeds of objects approach that of light. When this happens, almost all the simple laws of mechanics break down. Forces and masses appear to vary as between one frame of reference and another, velocities can no longer be added vectorially, and many of the laws of physics have to be expressed in quite different ways in order that they may take exactly the same form for all non-accelerating observers—as required by Einstein's principle. All these results have been extensively tested by experiment, at any rate for the motions of the elementary particles of matter with which we are concerned at the moment; and the Special Theory of Relativity is generally regarded as one of the corner-stones of modern physics. The older 'classical' forms of physical laws are found in many cases to be approximations to the laws of Relativity appropriate to slow speeds.

As it happens, the laws of electromagnetism are amongst the few that remain unaltered in form in the theory of Relativity. However, the theory does lead us to expect that the analysis of the electromagnetic field into its electric and magnetic components may differ as between one observer and another. Indeed, as we have seen above, this situation arises at quite slow speeds. At such speeds two observers will agree about the resultant force acting on any particle or the resultant e.m.f. in a circuit. But their analyses of the force into its electric and magnetic components or their ideas on the sites at which e.m.f.'s arise in a circuit will in general differ. (At speeds comparable with that of light even the resultant force on a particle will be assessed differently by the two observers.) It is therefore important always to ensure that there is no possible doubt about the frame of reference with respect to which any electromagnetic calculation or measurement is being conducted. What in one frame of reference is simply a magnetic field in another may be a combination of an electric and a magnetic field, and so on.

To see how this works out it is necessary to consider what we mean by the two sorts of field. All electromagnetic effects may be analysed ultimately in terms of the forces acting on charged particles. If the charged particle is at rest (in a given frame of reference) it experiences only the electric force F given by

$$F_e = eE$$

where e is the charge on the particle and E is the potential gradient of the electric field (p. 189). If the particle is moving with velocity v (in the given frame of reference) it experiences in addition the magnetic force F_m given by

$$F_m = Bev$$

where B is the component of the flux density at right-angles to the velocity. The resultant force on the particle is thus the *vector sum* of these two components, which we can express symbolically by

electromagnetic force $= F_e + F_m = e(E + v \times B)$

What we are doing in effect is *defining* the electric force as that which acts on a stationary charge (in our chosen frame of reference) and the magnetic force as the additional component of force that arises when the charge is in motion.

It should now be clear why two observers will disagree about the division of the electromagnetic field into its electric and magnetic components. If a charged particle is at rest in one frame of reference it is moving in the other, and vice-versa. As long as the relative speed of the two observers is small compared with that of light, they will agree about the resultant electromagnetic force, but will apportion it differently between its two components.

Consider for instance the simple situation of a rod cutting the flux of a magnetic field, as in Fig. 5.13. In a frame of reference fixed in the laboratory there is a magnetic flux density B as shown, *and no electric field*. In this frame of reference the charged particles moving as part of the rod are acted on only by the magnetic

force F_m, protons towards A and electrons towards B. If the rod were made of insulating material, the atoms in it would be slightly strained (i.e. *polarized*, p. 193); but there would be no general movement of charge along it, and there would still be virtually no electric field. The situation is different if the rod is of conducting material. A general migration of charge will then take place along it, and this will create an electric field such that the force F_e exactly balances the magnetic force F_m. This electric field then exists also in the region around the wire. For instance if the ends A and B were joined to the plates of a capacitor (p. 161) moving with the rod there would be an electric field in the gap between its plates.

The description in the above paragraph has been for a frame of reference fixed in the laboratory.

Now let us analyse the same situation in a frame of reference fixed in the rod AB and moving with it through the laboratory. If the rod is of insulating material, the *stationary* charged particles in it are observed to be acted on by forces acting parallel to the rod, protons towards A and electrons towards B, and we should find that the material is slightly strained electrically (polarized). We should therefore conclude that we are in a region of electric field created by some outside agency, since there is manifestly no source of electric field in the rod itself. We could also do experiments to establish the existence of a magnetic field by studying the forces on charged particles *moving* in our frame of reference—e.g. we could measure the couple acting on a coil carrying a current. (At slow speeds the flux density B would prove to be negligibly different in the two frames of reference.) If now the rod is of conducting material, the electric field in our frame of reference will cause a migration of charge along it; this will set up an opposing electric field, such that the resultant electric field in the rod is exactly zero. If a capacitor were joined between A and B, there would be no resultant electric field in the gap between its plates.

Both the above descriptions of the electromagnetic field around the rod AB are equally valid, and both are in accordance with the universal laws of electromagnetism, as required by the principle of Relativity.

In this example a pure magnetic field in one frame of reference is interpreted as an electric field (in addition to the magnetic field) in the other. Similarly it can be shown that a pure electric field in the first frame of reference would appear as a magnetic field (in addition to the electric field) in the second. Indeed the very existence of magnetic forces can be explained entirely in this manner. A stationary charged particle near a current-carrying conductor experiences no resultant force, as long as the number of electrons and protons in the conductor remain equal. But if the charged particle is moving with respect to a frame of reference fixed in the conductor, the electric forces on the particle cease to be balanced. According to the theory of Relativity, the force between the moving charged particle and the stationary protons in the conductor is slightly increased as assessed in the given frame of reference; so also is the force between the particle and the electrons in the conductor, *but by a different amount*, since the electrons also are in motion. The apparent difference between the two sets of electric forces is what we interpret as a magnetic force. The detailed analysis of this result is beyond the scope of this book; but enough has been said to show how the principle of Relativity has taken a place as one of the great unifying ideas of modern physics. In fact, by applying this principle the entire theory of electromagnetism may be derived from a single experimental result, namely the law of force for electric charges (p. 203).

5.5   Induced e.m.f.'s— changing currents

An e.m.f. may also be induced in a conductor by changing the magnetic field it is in *without any relative movement of the conductor and the source of the magnetic field*. Thus, if a wire is clamped between the poles of a strong electromagnet and joined to the terminals of a galvanometer, a deflection occurs when the current in the coils of the electromagnet is changing. This effect cannot be explained in terms of the magnetic force acting on a moving charge, since the conductor is stationary in the changing magnetic field; it is an entirely new phenomenon. However, it has many features in common with the simpler form of electromagnetic induction (with moving conductors); and, as we shall see, it is possible to analyse both processes in the same way.

As before, the e.m.f. appears to be proportional to the *rate of change*. If the current in the electromagnet is increased or decreased at a steady rate, the reading of the galvanometer is proportional to the rate at which the change is effected. Indeed the magnitude of the induced e.m.f. seems to be exactly the same whether the change in the magnetic field is brought about by reducing the current in the electromagnet to zero or by moving the conductor out of the field. The point may be tested more exactly by using a pair of solenoids that may be placed one inside the other. This increases the effect by the use of many turns of wire without the complications produced by the presence of iron cores. A variable p.d. may be supplied to the outer solenoid (in series with an ammeter), while the inner one is connected to a sensitive galvanometer (Fig. 5.21). An e.m.f. may now be induced in the latter coil either by moving it slowly out of the outer one or by slowly reducing the current in the outer coil to zero. In both cases the changes may be effected smoothly so as to keep the galvanometer reading nearly constant. It is then found that the same time is taken to reduce

E MP—D

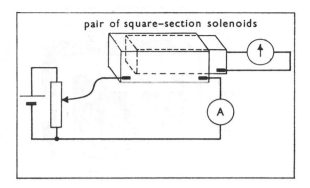

Fig. 5.21   A pair of concentric solenoids; an e.m.f may be induced in the inner one either (*a*) by relative movement of the solenoids; or (*b*) by changing the current in the outer one

the field in the inner solenoid to zero *whichever process is used*. And, as before, the galvanometer reading is doubled by halving the time taken over the change, whichever process is used— showing that the e.m.f. is indeed proportional to the rate of change. *The same e.m.f. is produced by a given rate of change of field no matter how the change is brought about*.

It is found that the directions of the induced e.m.f.'s may again be described by Lenz's law. This predicts that the induced *current* will *oppose* the change that caused it (p. 78). Thus, when the current is reduced to zero in the turns of the outer solenoid, current flows in the *same* sense round the turns of the inner one, so tending to *prevent* the reduction of magnetic field that caused the induced e.m.f. When the current in the outer solenoid is once more increased, the current in the turns of the inner one flows in the opposite sense, tending to *prevent* the magnetic field increasing inside it. There are also mechanical forces between the two solenoids while this is going on, tending to prevent the changes. Such forces will generally be too small to detect, but their existence can be used to help us predict which way induced currents are going to flow in a given case—and this is the function of Lenz's law.

The same pattern of laws may therefore be used to describe the induced e.m.f.'s both in the present case of changing currents in stationary circuits and in the case of relative motion between a circuit and the source of the magnetic

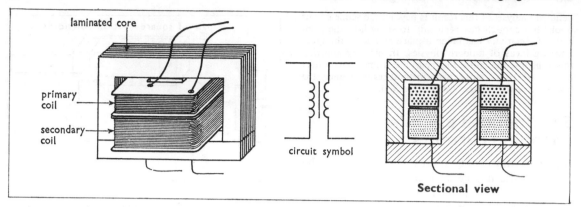

Fig. 5.22   An a.c. transformer

field considered previously. The laws take the following forms:

(*i*) the induced e.m.f. is proportional to the rate of change causing it (Faraday's law).
(*ii*) the induced current opposes the change causing it (Lenz's law).

Although the apparatus on the laboratory bench may seem very similar in the two cases of electromagnetic induction, it is important to realize that the physical reality behind the two cases is very different. In the first case we are dealing with charges moving at steady speeds— a constant current in the coils producing the magnetic field, and a conductor (bearing free charges) moving through that field at a steady speed. The induced e.m.f. is in this case evidence of the force that acts between charged particles in constant relative motion with respect to one another; this is the basic magnetic force between the charged particles of matter. In the second case the conductors are stationary and the current in the coils producing the magnetic field is changing, i.e. the charged particles producing the field are speeding up or slowing down. The induced e.m.f. is now evidence for the existence of a force on a stationary charged particle in the presence of another that is accelerating; the latter particle may be momentarily at rest, but provided it is accelerating this force will operate. The interesting thing is that the induced e.m.f.'s that result from these differing processes should be described by the same laws. There are fundamental reasons why this should be so, as we shall see in due course.

Mutual induction with alternating currents

The a.c. transformer

In the induction of e.m.f.'s between stationary circuits electrical energy is transferred from one circuit to the other through the agency of the magnetic field. The process is sometimes described as *mutual induction*. It can occur only when the current is changing. In d.c. circuits therefore effects due to mutual induction only arise at the moments of switching on or off. But if an *alternating* current is caused to flow in one coil (called the *primary coil*), an alternating e.m.f. of the same frequency is induced in the other coil (called the *secondary coil*). Thus in a.c. circuits energy may be transferred from one circuit to the other through the linking action of the magnetic field.

At high frequencies this process is very efficient even with air-cored coils. But at low frequencies the transfer of energy from one coil to another can only be really effective if the coils are wound on a closed loop of soft iron, which must be laminated to reduce eddy currents (p. 91). This is the basis of the design of the a.c. transformer (Fig. 5.22). One coil, the *primary*, is connected to an a.c. supply, such as the 50 Hz mains. This coil has a large number of turns, and only a small current is needed in it to produce a large alternating field in the core. An e.m.f. is accordingly induced in the other coil, called the *secondary*; the magnitude of this e.m.f. is proportional to the number of turns of the secondary.

Provided the transformer is not overloaded,

$$\frac{\text{secondary p.d.}}{\text{primary p.d.}} = \frac{\text{number of turns in secondary}}{\text{number of turns in primary}}$$

Thus by suitable choice of this *turns ratio* we can produce any desired alternating p.d. at the terminals of the secondary. The a.c. transformer is an exceedingly efficient device. The large ones used in power supply systems handling hundreds of kilowatts are usually more than 99% efficient; and even the smaller ones used in the power supplies of radio sets have very small losses. Thus we can write, approximately

power input to primary

= power output from secondary

∴ primary p.d. × primary current

≈ secondary p.d. × secondary current

Therefore:

$$\frac{\text{secondary current}}{\text{primary current}} = \frac{\text{number of turns in primary}}{\text{number of turns in secondary}}$$

This is the *reciprocal* of the turns ratio. If the p.d. is transformed *up* in a certain ratio, the current is transformed *down* in the same ratio, and vice-versa. The nature of the magnetic interaction between the coils of a transformer is discussed further in Chapter 12 (p. 215).

Transformers are often wound with several secondary coils, electrically insulated from one another. The mains transformer of an audio amplifier is usually made this way. In addition to the main high-tension winding there may be several low-tension windings. These provide power for heaters and indicator lamps. The primary winding has several tapping points near one end. By correct choice of the number of turns in the primary coil the secondary e.m.f.'s can be kept at nearly constant values for a range of possible values of the supply p.d. The heat loss in a coil of a transformer is proportional to its resistance and to the square of the current ($P = I^2R$). To minimize such losses the high-current, low-tension windings are made of thick wire; while the low-current, high-tension coils can well be of relatively fine wire.

We may readily demonstrate the process of mutual induction with a.c. with the pair of air-cored solenoids used previously. The primary

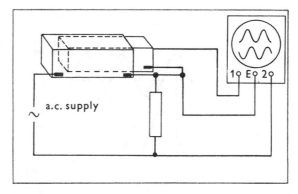

Fig. 5.23   Mutual induction between two solenoids with alternating current

coil is joined to a suitable low-tension a.c. supply, and the galvanometer in the secondary circuit must be replaced with a suitable a.c. meter; a cathode-ray oscilloscope is best since it enables us to see exactly what is happening at different points in the cycle of variation of current. If a double-beam oscilloscope is available we may use one beam to depict the e.m.f. induced in the secondary coil, while the other beam shows the current flowing in the primary. The oscilloscope deflection depends on the *p.d.* between the deflector plates. To show the *current* in the primary coil we must therefore join a (non-inductive) resistance in series with it, and connect the oscilloscope across this (Fig. 5.23). If the current supplied to the primary varies sinusoidally, the e.m.f. in the secondary also varies sinusoidally, but is related to the primary current as shown in Fig. 5.24. At moments such as X the magnetic field is at its maximum value in one direction or the other, but is momentarily unchanging; at this moment the e.m.f. is zero. But at moments such as Y or Z the magnetic field is momentarily zero, but is changing at the maximum rate in one direction or the other. The induced e.m.f. is therefore a maximum or minimum. We can describe this by saying that the induced e.m.f. in the secondary lags a quarter of a cycle behind the variations of current in the primary.

If the frequency of the primary current is varied, while preserving the same sinusoidal waveform and peak current, we expect to find that the peak value of the induced e.m.f. is proportional to the frequency. For, if the primary

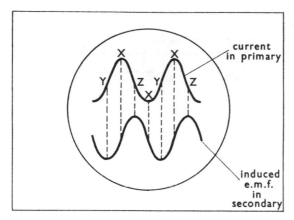

Fig. 5.24   The relationship between the current in the primary and the e.m.f. induced in the secondary for the two solenoids

frequency is doubled, the rate of change of current at every point of the cycle will be twice what it was before, and therefore the induced e.m.f. will also be twice as great at every point of the cycle. Thus

$$\text{peak e.m.f.} \propto \text{frequency}$$

This may be tested with a *signal generator* (an apparatus containing an electronic oscillator, p. 274, producing a sinusoidal output p.d. variable over a wide range of frequencies). At 50 Hz the e.m.f.'s are too small to be detectable (without iron-cored coils). But by using frequencies in the range 1 to 100 kHz, it may be shown that the peak e.m.f. is indeed accurately proportional to the frequency, which is a useful confirmation of Faraday's law of electromagnetic induction.

The a.c. search coil

By using fairly high frequencies (20 kHz or more) we may reduce very considerably the dimensions of the secondary coil in the above apparatus and still get induced e.m.f.'s that can be readily measured with an oscilloscope. At such frequencies a coil of 10 or 50 turns wound on a rod about 15 mm in diameter may be used to explore how the field varies from one part of a solenoid to another. This is called a *search coil*. With it we can show for instance that the field is practically constant over the whole cross-section of a solenoid—a result of great practical importance.

In the next chapter we shall investigate the magnetic fields due to currents flowing in many configurations, and a search coil used with an a.c. field will be one of the most useful tools at our disposal. But for the moment we are concerned with using it to help establish the basic laws of electromagnetic induction.

For instance we expect to find that the induced e.m.f. is proportional to the number of turns N of the search coil; presumably the same e.m.f. is induced in each turn of the coil. By winding a succession of search coils on the same rod we can show that this is indeed so. Again by using rods of different sizes we may show that the induced e.m.f. is proportional to the area of cross-section A of the coils. And by using rods of a variety of cross-sectional shapes (circular, square, triangular, etc.) we can show that the e.m.f. depends only on the area of cross-section and not on the shape.

We may also investigate how the induced e.m.f. depends on the orientation of the search coil. For this purpose we need to produce the magnetic field with an 'open' solenoid of stiff wire, so that the search coil may be inserted through the sides between the turns (Fig. 5.25). The search coil must be of many thousands of turns to make up for the small number of turns in the solenoid. The e.m.f. is a maximum when the plane of the search coil is normal to the field and falls off to zero when it is turned through a right-angle. In general with the plane of the search coil making an angle θ with the axis of the solenoid, we can show that the induced e.m.f. is proportional to $\sin \theta$. Another way of expressing this result is to resolve the flux density B of the magnetic field of the solenoid into two components, $B . \sin \theta$ along the axis of the search coil, and $B . \cos \theta$ at right-angles to it. The result shows that only the component B_n normal to the plane of the search coil is responsible for the induced e.m.f., where $B_n = B . \sin \theta$.

Summarizing these results, the induced e.m.f. E is given by

$$E \propto NA \frac{\mathrm{d}B_n}{\mathrm{d}t}$$

or

$$E = kNA \frac{\mathrm{d}B_n}{\mathrm{d}t}$$

where k is a constant still to be determined.

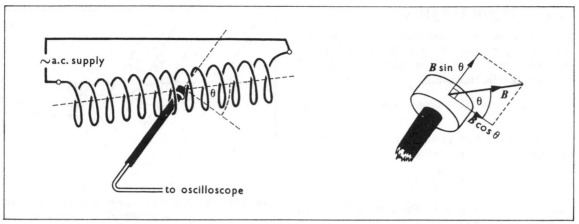

Fig. 5.25   The use of an a.c. search coil to investigate the field in a solenoid

In fact the constant k is exactly 1, and this follows from the experimental result that the same e.m.f. is produced by a given rate of change no matter whether the change is brought about by *moving the coil* or by *varying the field* (p. 81). We could indeed produce the same alternating e.m.f. in our search coil by using a constant field and arranging to spin the coil at the appropriate frequency; this is a perfectly possible experiment with a small coil connected to slip-rings (p. 150). But it is simpler to measure the constant k directly, and settle the point that way.

To do this we have only to measure the peak e.m.f. E_0 induced in the search coil when it is placed with its axis parallel to that of the solenoid. If the alternating flux density in the solenoid is of frequency f and peak value B_0, then we have

$$B = B_0 \sin \omega t$$

where $\omega = 2\pi f$. The maximum e.m.f. is induced at the moments when B is changing at the maximum rate, i.e. when $\omega t = 0$. For small angles we may replace the sine of the angle by the angle itself (cf. Fig. 5.19).

$$\therefore B = B_0 \omega t \quad \text{(for small values of } \omega t)$$

so that the rate of change of B at this moment is given by

$$\left(\frac{dB}{dt}\right)_{max} = B_0 \omega$$

[Using calculus, we may simply differentiate to obtain

$$\frac{dB}{dt} = B_0 \omega \cos \omega t$$

$$\therefore \left(\frac{dB}{dt}\right)_{max} = B_0 \omega]$$

$$\therefore E_0 = k B_0 N A \omega$$

For a coil of known area and number of turns N, we can then find the value of k, provided we know the value of B_0. In the next chapter we shall see how to calculate the flux density in a solenoid from the current flowing in it (p. 104). We can then deduce the value of B_0 from the peak current I_0 in the solenoid. However, at this stage in our development of the subject we can always use a simple current balance to measure the flux density B in the solenoid for a given *steady* current I flowing in it; and then we can calculate the peak flux density B_0 by simple proportions, since we have

$$\frac{B_0}{I_0} = \frac{B}{I}$$

Once we have established that the constant k is unity, we have

$$E_0 = B_0 N A \omega$$

and measurement of the peak e.m.f. E_0 enables the peak flux density B_0 in an unknown alternating field to be found. Mostly we shall seek to use the a.c. search coil to find how a magnetic field varies from one point to another. But we shall sometimes want to use it to find the actual value of the field by the above expression.

In the general case of a coil in a changing field we then have the result:

$$E = N A \frac{dB_n}{dt}$$

5.6 Magnetic flux

We can now bring together the two mathematical expressions we have obtained for the induced e.m.f. in each of the two types of electromagnetic induction.

(a) E.m.f.'s due to charges moving at steady speeds

(i) For each turn of a coil of area A moving in a magnetic field of flux density B we have

$$E = B \frac{dA_n}{dt} \quad \text{(p. 76)}$$

where A_n is the area of the coil projected at right-angles to the field; i.e.

$$A_n = A . \sin \theta$$

where θ is the angle between the plane of the coil and the field. For a coil rotating in a uniform magnetic field this gives

$$E = AB \frac{d (\sin \theta)}{dt}$$

(ii) The same result applies also in the case of the e.m.f. induced in a single conductor moving through a magnetic field; for we then have

$$E = B_n \frac{dA}{dt} \quad \text{(p. 76)}$$

where A is the area swept out by the conductor, and B_n is the component of B normal to the plane of movement. If we bear in mind that the moving conductor must in practice form part of a complete circuit (if we are to detect the e.m.f.), then A is in this case also the area of the circuit, which changes at the rate dA/dt as the conductor moves. So A has the same meaning as before; namely, it is the area of the coil or circuit at right-angles to the appropriate component of flux density B_n ($= B . \sin \theta$).

$$\therefore E = B . \sin \theta \frac{dA}{dt}$$

(b) E.m.f.'s due to accelerating charges

For each turn of a coil of area A in a changing magnetic field of flux density B we have

$$E = A \frac{dB_n}{dt}$$

where B_n ($= B . \sin \theta$) is, as before, the component of the flux density normal to the plane of

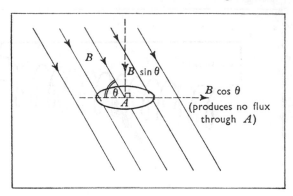

Fig. 5.26   The flux of a magnetic field through a small plane area A

the coil. In this case θ is constant, and we have

$$E = A . \sin \theta \frac{dB}{dt}$$

In each of these cases the induced e.m.f. E is equal to the rate of change of the product (AB_n); for we can write

$$E = \frac{d(AB_n)}{dt} = \frac{d(AB . \sin \theta)}{dt}$$

In case (a) (i)—*the rotating coil*—A and B are constant, and θ changes.

In case (a) (ii)—*the moving conductor*—B and θ are constant, and A changes.

In case (b)—*the coil in a changing field*—A and θ are constant, and B changes.

It is not difficult to devise situations in which two or three of the quantities A, B and θ change. But in every case the e.m.f. E is equal to the rate of change of the product (AB_n).

This product is called the *flux* Φ of the magnetic field through the area A (Fig. 5.26).

The **flux** of a magnetic field through a small plane surface is the product of the area of the surface and the component of the flux density B normal to it.

Thus $\qquad \Phi = AB_n = AB . \sin \theta$

If the field is uniform we can use this definition to give the flux Φ through a large area also (e.g. through the core of a solenoid). If a coil through which the flux passes is of N turns, the *flux linkage* of the field with the coil is N times the flux through any one turn of it.

The unit of magnetic flux is called the *weber* (Wb), after a German scientist.

The **weber** is the flux of a uniform magnetic field of flux density 1 T through a plane surface of area 1 m² placed normal to the flux density.

Thus
$$1 \text{ Wb} = 1 \text{ T m}^2$$

The reason for using the name *flux density* for the quantity B should now be apparent. Thus

flux density = flux per unit area
and $1 \text{ T} = 1 \text{ Wb m}^{-2}$

The word *flux* itself perhaps needs some explanation. In developing the mathematical theory of magnetic (and electric) fields the nineteenth-century scientists were continually making comparisons with the theory of the flow of fluids; the formal mathematics has much in common in the two cases. They found it a helpful aid to the imagination. And so, many of the words used in describing these fields have come down to us from the terms used in the theory of fluids. *Flux* is one of these. We talk about flux 'emerging from' a solenoid, or 'passing through' a coil. But it must be understood that this is only figurative language. The position is a stationary one, and nothing actually flows along the lines of force!

We can now summarize the laws of electromagnetic induction in the following way. The induced e.m.f. is equal to the rate of change of the total magnetic flux linked with a circuit. This is the full statement of Faraday's law of electromagnetic induction. Lenz's law for the directions of the induced e.m.f.'s may be expressed in a formal way in the mathematics by means of a minus sign, since the induced e.m.f.'s always act in such a way as to drive a current that would oppose the flux changes. The laws may therefore be combined in the single statement:

$$E = -\frac{d\Phi}{dt}$$

where $\Phi = AB_n$

This summary of the laws of electromagnetic induction is sometimes known as *Neumann's law*. It reveals incidentally another relationship between the units that it is useful to bear in mind:

$$1 \text{ Wb} = 1 \text{ V s}$$

In some books this is made the basis for a definition of magnetic flux. Flux density is then defined in terms of flux. There are quite a number of different logical paths by which the system of electromagnetic laws, definitions and units may be set out. We cannot claim priority for any particular logical pattern. The set of laws stands or falls together, and with them the definitions of quantities and the units for measuring these. However, in a textbook one has to choose a particular sequence of development related to the experiments that establish the laws. In this book we have chosen to define flux density first, and then arrive at the concept of flux through this; and so we discover the connection between the weber and the volt. It is perfectly possible to approach it all the other way round.

★   The continuity of magnetic flux

The definition of flux we have given so far applies only to small areas, over which the magnetic flux density is virtually uniform. To calculate the flux through a large area like the cross-section of a large coil, which may not be plane and over which the field may well not be uniform, we proceed as follows. Imagine the space inside the coil divided up by a network of lines (Fig. 5.27a) into areas A_1, A_2, A_3, etc., each of which is sufficiently small to be treated as plane and in a uniform field. The total flux through the cross-section is then the summed or integrated total of the amounts through the individual elements of area. Needless to say, with non-uniform fields such calculations can be of extreme difficulty; but the idea behind it is not a difficult one, and this is all that concerns us at the moment.

The question arises, however, whether the e.m.f. induced in such a coil is equal to the rate of change of total flux calculated in this way. That this is so can be seen by the following argument. Suppose the network of lines of Fig. 5.27a were formed by actual conductors surrounding the areas A_1, A_2, A_3, etc. The e.m.f. induced in the circuit loop bounding the area A_1 is the sum (in, say, a clockwise sense) of the e.m.f.'s induced in each of its four sides. The same applies to the area A_2. If now we imagine these two circuit loops joined in series (Fig. 5.27b), the e.m.f. induced in the conductors running along the common boundary cancels out. The e.m.f. induced in the combined loop $A_1 + A_2$ is therefore the sum of the e.m.f.'s induced in A_1 and A_2 separately. This reasoning may be extended step by step to include all the elements of area in the figure. Therefore if the laws apply to the individual elements of the figure they apply also to the whole area, if the flux through it is calculated as we have indicated.

This, however, raises yet another question that is more fundamental. How can we be sure that the total flux through an area will have the same value no matter what network of lines we imagine covering it? For we can imagine any number of surfaces covering the bounding curve of Fig. 5.27a. Like a soap film covering a wire frame, the surface on which the imaginary lines are drawn may be nearly flat (S) or may balloon out (S'), as shown in Fig. 5.27c. Unless the total flux through each of these possible surfaces is the same our method of analysis must break down. Or, to put it another way, the magnetic flux entering the space between the two surfaces S and S' (through S') must be equal to that leaving it (through S). No flux can enter the space without equal flux leaving it at another point. If this is not so, the induced e.m.f. calculated from the flux will not be an exactly defined quantity; it will vary with the choice of surface we make by which to do the calculation. Unless magnetic flux is

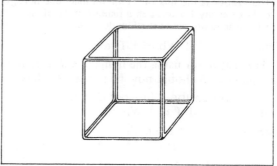

Fig. 5.28   A cube of six square coils for testing the continuity of magnetic flux

continuous in this sense the whole concept of flux is invalid for the purposes of calculating induced e.m.f.'s.

The law of continuity of magnetic flux. The total magnetic flux entering any closed volume is always zero.

The experimental testing of this result is a matter of some importance. We may do it by fashioning a cube out of a set of six square coils, as shown in Fig. 5.28. (Coils completely enclosing any other convenient volume may equally well be used of course.) The coils are connected in series all in the same sense—i.e. in such a way that the e.m.f.'s induced by *increases* in the flux *entering* the cube through them act in the same sense. The resultant e.m.f. (if any) is shown by a galvanometer joined in series with the coils. In general, when the magnetic field in and around the coils is varied, e.m.f.'s are bound to be induced in the individual coils of the set—as may be seen by connecting up any one of the coils alone to the galvanometer. But with the complete set of coils the e.m.f.'s always exactly cancel out. Any increase in the flux entering the cube through one coil is balanced by an exactly equal increase in the flux leaving the cube through the other coils.

The continuity of magnetic flux enables us to see more clearly how the two forms of electromagnetic induction are connected. For instance, in the experiment shown in Fig. 5.21, the same e.m.f. was induced whether a given change of flux was brought about by moving the inner solenoid out of the field or by reducing the current in the outer one to zero. Consider one turn of the inner solenoid (as in Fig. 5.29). This was moved from the position ABCD in the field to a position A'B'C'D' where the field was effectively zero. The side AB of the coil therefore swept out the area ABB'A', and the e.m.f. induced in it was equal to the rate at which it cut through the flux that goes out through this area. The same applies to each of the other sides of the coil. Since the field was effectively zero at the position A'B'C'D', the total flux passing out through the sides of the long tubular channel swept out by the coil was equal to the flux through the initial position ABCD. The total flux cut through by the sides of the coil when it was moved from one position to the other was therefore equal to the change in flux when the

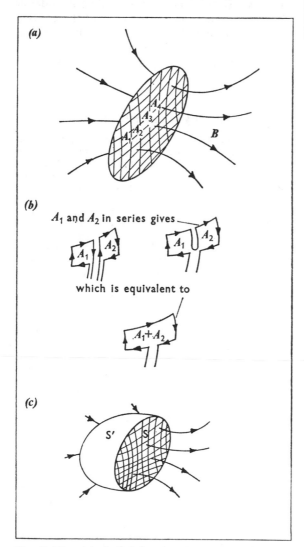

Fig. 5.27   (a) Calculating the flux through an extended area such as the cross-section of a coil; (b) showing that the total induced e.m.f. is equal to the sum of the e.m.f.'s induced in the network of imaginary loops covering the coil; (c) the flux is the same whether calculated through S or through S'

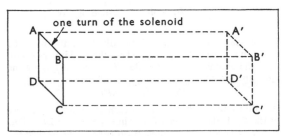

Fig. 5.29   The e.m.f. induced in a turn of a solenoid;
(*a*) by moving it out of the field; (*b*) by reducing the field
in it to zero

field in the position ABCD was reduced to zero without moving the coil. A similar result can be demonstrated for any conceivable process of changing the flux through a coil. All that matters is the change in the total flux linked with the coil; it does not matter by what process this change is brought about.

In the cube experiment of Fig. 5.28 the magnetic fields may be produced in any way whatever—with electric currents or with magnets. In particular, when we push one end of a magnet (say the north pole) into the space inside the cube, there is still no deflection of the galvanometer. Whatever flux comes out of the end of the magnet and passes out through the coils of the cube must enter the cube through the shaft of the magnet. The flux does not 'sprout' from the ends of the magnet, but exists in closed loops—down the inside of the magnet from south to north poles, out of the north pole and back again through the air into the south pole—in fact just like the field of a solenoid. Although we cannot put a probe into the inside of the magnet, we can do experiments to determine the total flux passing through it, and so calculate the average flux density B inside it.

In general the effect of a piece of iron in a magnetic field is to increase the flux. Thus if a bar of iron is placed inside a solenoid that carries a current, the iron becomes partly magnetized by the field—i.e. the individual atoms of iron become to some extent lined up with the field, and add their flux to that of the solenoid. The extra flux through the bar of iron then augments that due to the solenoid itself. The total magnetic field produced by the combination is thus much greater than would be produced by the solenoid alone. This is the principle of the electromagnet. Likewise in an a.c. transformer the primary and secondary coils are wound on a closed loop of iron to ensure that the maximum possible e.m.f.'s are induced in the coils by the alternating currents flowing in them. However, the behaviour of iron in a magnetic field is very complex, and we shall defer detailed consideration of this until a later chapter (Chapter 7).

The measurement of magnetic flux

When the flux linkage Φ of a magnetic field with a closed circuit changes, an e.m.f. E is induced in it and a current flows. Experiments with a search coil thus provide us with the means of measuring *changes* in the flux linked

with it. The simplest technique is that of the a.c. search coil, already considered; with this we measure the *peak* value Φ_0 of an alternating flux. Thus if

$$\Phi = \Phi_0 . \sin \omega t$$

then
$$E = -\frac{d\Phi}{dt} = -\Phi_0 \omega . \cos \omega t$$

and the peak value E_0 of the e.m.f. is given by

$$E_0 = \Phi_0 \omega$$

$$\therefore \Phi_0 = \frac{E_0}{\omega}$$

However, if the flux is necessarily a steady one we must resort to d.c. methods of measurement. One way of doing this is to connect the search coil to a *fluxmeter*. This is a special form of moving-coil galvanometer in which the restoring couple due to the suspension is made very small. Its period of oscillation is thus very long. When the flux linked with the search coil changes, the induced currents cause the coil of the instrument to deflect; since the coil is completely free of mechanical constraints, it moves freely in such a way as to prevent any change in the *total flux through the circuit* (in accordance with Lenz's law). In other words, if the flux linked with the search coil changes, the coil in the instrument moves so that the flux linked with it changes by the same amount (but in the opposite sense in the circuit as a whole). The deflection therefore indicates the change in the flux through the search coil. If this flux was initially zero, the fluxmeter indicates the flux through it when it is moved into some field—say, between the poles of an electromagnet. However, the fluxmeter is essentially a delicate instrument and not very easy to use.

It is possible to use a modern taut suspension mirror galvanometer (p. 137) as a fluxmeter. These instruments are designed so that, when the coil is short-circuited, the currents induced in it if it is set in motion are sufficient to produce heavy damping. Short-circuited like this, the instrument is largely protected against mechanical damage when it is being carried about. With a low resistance joined across the coil, when once the instrument has been deflected, it takes a long time for the spot of light to return to zero. Used with a low resistance search coil it therefore behaves in much the same way as a fluxmeter; and the reading *immediately after* the coil has been placed in a field is an indication of the flux that is then linked with it. However, a moving-coil galvanometer is rather insensitive used in this way as a fluxmeter; and it is better to use it as a ballistic galvanometer as described below.

What we require is an instrument that will *integrate* the e.m.f. E induced in the search coil. For we have

$$E = -\frac{d\Phi}{dt}$$

$$\therefore \int E.dt = \Phi_1 - \Phi_2$$

where Φ_1 and Φ_2 are the initial and final values of the flux linked with the coil.

One way of doing this is to employ an electronic circuit to do the integration. For instance, the integrating amplifier shown in Fig. 14.26 (p. 273) may be used in this way.

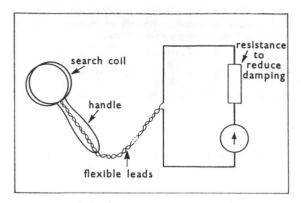

Fig. 5.30   The use of a search coil and ballistic galvanometer to measure flux

With a search coil connected to the input terminals, and a high resistance voltmeter to the output it will behave as a fluxmeter, the voltmeter readings indicating the flux linked with the search coil.

We may also use a sensitive galvanometer to perform this integration for us. When a pulse of current passes through the galvanometer coil an impulse is given to it that sets it swinging. It may be shown that the amplitude of the first oscillation is directly proportional to the total electric charge Q that passed during the pulse, provided the duration of this is small compared with the period of oscillation. Used in this way the instrument is called a *ballistic galvanometer*. The theory of the instrument is discussed in Chapter 8 (p. 146); but the calibration techniques described below provide a test of this result. Thus, if the amplitude of the first 'swing' of the ballistic galvanometer is θ, we find

$$Q \propto \theta$$
or
$$Q = b\theta$$

where b is a constant that is found in the calibration experiment.

If the pulse of current is from a search coil joined to the instrument, the total charge Q that passes is proportional to the change in the flux linked with the coil. For, if the resistance of the complete circuit is R, the current I flowing at any instant is given by

$$I = \frac{E}{R}$$

This current is equal to the rate of flow of charge.

$$\therefore I = \frac{dQ}{dt}$$

$$\therefore \frac{dQ}{dt} = \frac{1}{R}\frac{d\Phi}{dt}$$

Integrating over the time interval during which the change of flux occurs, the left-hand side of the equation gives the total charge Q that passes.

$$\therefore Q = \frac{1}{R} \int d\Phi$$

$$\therefore Q = \frac{\Phi_1 - \Phi_2}{R} = \frac{\text{change of flux}}{R}$$

If the search coil is moved rapidly from the required position in the field to a place where the field is very small, then the charge Q gives the flux Φ_1 that was linking the coil; and we have

$$\Phi_1 = RQ = Rb\theta$$

An alternative technique is to turn the search coil over in the field; the change of flux linkage is then $2\Phi_1$. This needs two people to manage it properly—one to turn the coil, the other to take the reading.

The search coil is joined up to the galvanometer through long, twisted, flexible leads, and is connected in series with a high enough resistance to reduce the damping to a low value (Fig. 5.30), but not so high as to reduce the sensitivity unduly. If we are intending to use the result to give the average flux density B in the coil, its cross-section must be small enough so that the field can be regarded as uniform over the area concerned. The number of turns in the coil must be chosen so that the charge Q is of the right order of magnitude to be measured by the ballistic galvanometer (probably about $1\ \mu C$).

The instrument can be used for *comparing* two fields without any preliminary calibration. For instance the angle of dip ϕ_D of the earth's magnetic field may be found by comparing the horizontal and vertical components B_h and B_v of the field. We then have

$$\tan \phi_D = \frac{B_v}{B_h}$$

For this purpose we could use a large pivoted coil as in Fig. 5.17 (p. 75), but with a high resistance joined in series with the galvanometer. The deflection θ produced by turning the coil through 180° is proportional to the component of flux density normal to the initial position of the plane of the coil. By correct positioning of the coil the deflections corresponding to the horizontal and vertical components may be found. If the ballistic galvanometer is calibrated first, we may use the measurements to give the actual values of B_h and B_v.

If a search coil is made that will fit closely over the cross-section of a magnet, it may be used to investigate the flux in the body of the magnet at different positions. The coil, connected to a ballistic galvanometer, is placed over the magnet in the position required, and is then quickly slipped off and removed to a sufficient distance for the final flux through it to be treated as negligible. The throw of the galvanometer is proportional to the average flux through the coil at the point tested. A maximum reading is, of course, obtained with the coil round the centre of the magnet. By repeating the measurements at a number of positions along the magnet and beyond the ends of it, the 'leakage' of flux from different parts of it may be investigated. With a bar magnet less than half the total flux will be found to emerge from the end faces.

Calibrating a ballistic galvanometer

Fig. 5.31 shows a very simple way of doing this. The galvanometer is joined in series with a battery (of e.m.f. E),

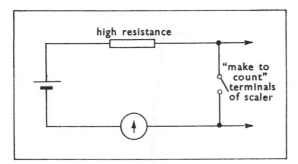

Fig. 5.31   Calibrating a ballistic galvanometer with an electronic timer

a high fixed resistance R and a tapping-key switch. A short pulse of current I can thus be passed through the instrument by pressing the tapping-key for a moment. The duration t of this pulse can be measured with an electronic timer connected to the switch. The charge Q that passes through the instrument is then given by

$$Q = It = \frac{Et}{R}$$

An elegant arrangement for doing the timing is to employ an electronic *scaler* (p. 322). This is a device that counts the electrical pulses fed into it. Many of these instruments are equipped with an internal oscillator producing pulses at a frequency of 1 kHz; they also have an electronic 'gate' so arranged that the gate opens to allow pulses through to the counter circuits when a pair of terminals are joined together. These terminals (marked usually 'make to count') are joined to the two sides of the tapping-key switch. The 1 kHz pulses from the oscillator are then counted only while the tapping-key switch is pressed down; and the reading of the scaler gives the time t directly in milliseconds.

If we plot a graph of galvanometer throw θ against time of contact t a straight line is obtained, establishing the proportionality of throw θ and charge Q that passes, and giving us the calibration constant of the instrument.

Another way of calibrating a ballistic galvanometer is to employ a *standard capacitor*. When a capacitor of capacitance C is charged to a p.d. V, the charge Q stored on the plates is given by

$$Q = CV \quad \text{(p. 165)}$$

The quantity of charge stored in the capacitor when it is charged to a measured p.d. can thus be calculated. When the charged capacitor is then joined to the ballistic galvanometer, the deflection θ produced by the charge Q can be observed, enabling us to calibrate the instrument. The details of this technique are given on p. 162 (Fig. 10.3).

It is important to bear in mind that the calibration obtained by either of these methods is for a circuit of infinite resistance. If the galvanometer is then used in a circuit of sufficiently low resistance to affect the damping

significantly, the sensitivity is altered, and allowance must be made for this. With very high resistance the damping is purely mechanical. But with low resistance it is also electromagnetic, due to induced currents circulating in the circuit by the swing of the coil. If this is significant, we need in each case to derive what the un-damped amplitude of oscillation would be. This is discussed in detail on p. 147; but the following method of analysis is adequate if the damping is small. Suppose the successive turning points of the swinging coil—to right, to left, and then to right again—are $\theta_1, \theta_2, \theta_3$. The interval between the deflections θ_1 and θ_3 is one complete period of oscillation of the coil. The time taken by the coil to reach the deflection θ_1 is a quarter of this. We assume that the damping that occurs in this interval is a quarter of that which occurs during the complete period. The amplitude of oscillation θ_0 that would occur without damping is therefore given by

$$\theta_0 = \theta_1 + \frac{\theta_1 - \theta_3}{4}$$

The ratio θ_0/θ_1 is the ratio in which all the readings should be altered to make allowance for damping.

5.7   Eddy currents

When a conductor moves in a magnetic field, e.m.f.'s are induced in all parts of it that cut the magnetic flux. Previously we have been studying the summed effects of these e.m.f.'s acting in circuits consisting of loops of wire, etc. But if the conductor is a large lump of metal, significant e.m.f.'s can ac. round closed paths inside the lump of metal itself. Although the e.m.f.'s are not usually very large, the resistance of the current paths is so low that large currents may flow. These induced currents circulating inside a piece of metal are known as *eddy currents*. By Lenz's law they must flow in such directions as to *oppose* the motion. They can indeed act as a very effective brake on the motion of a body, the mechanical energy being turned into the electrical energy of the eddy currents, which in turn is converted into heat inside the metal.

The effect of eddy currents may be demonstrated by swinging a pendulum with a thick copper bob between the poles of an electromagnet (Fig. 5.32). When the electromagnet is switched on, the flux through the bob varies rapidly along its path, and considerable eddy currents are generated in it, which produce a very marked braking effect. The eddy currents can be prevented in this case by using a bob with

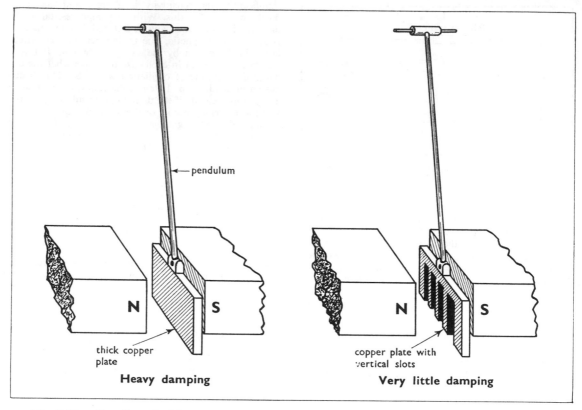

Fig. 5.32   The effect of eddy currents in metal plates swinging between the poles of an electromagnet

a series of slots cut in it so that the currents are now only free to circulate inside the relatively narrow teeth left between the slots. Very little braking effect can then be detected.

A similar demonstration may be performed with a copper cylinder hung from a cotton thread (Fig. 5.33) and set spinning between the poles of the magnet. The eddy current system acts as a very efficient brake in this case also. If the cylinder is made up of a stack of discs insulated from one another, the eddy currents are greatly reduced, since they can now circulate only inside the individual discs and not up and down the whole length of the cylinder. A pile of coins may be used for this purpose; the dirt on the coins acts as sufficient insulation between them, and there is certainly very little braking effect.

Eddy currents arise also in a stationary piece of metal if the magnetic flux through the parts of it is changing. Thus, if a thick brass plate is held just below a suspended magnet performing

torsional oscillations in the earth's field, the movement is seen to be heavily damped on account of the interactions between the magnet and the eddy currents it induces in the plate.

The effect of eddy currents must particularly be allowed for in the design of iron-cored apparatus—motors, dynamos, transformers, etc.—if considerable losses of energy from this cause are to be avoided. It is necessary in these cases to build up the iron parts of the apparatus out of a stack of iron sheets, called *laminations*; these are insulated from one another by a thin layer of paper stuck on one side of each lamination. The loss of energy through eddy currents is approximately proportional to the square of the lamination thickness, so that by suitable design the loss from this cause can be reduced to any desired figure. For low-frequency power apparatus it is usual to choose laminations about 0.5 mm thick.

The eddy current loss is also approximately

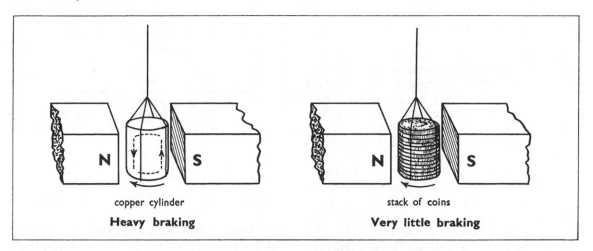

copper cylinder

Heavy braking

stack of coins

Very little braking

Fig. 5.33 The effect of eddy currents in a rotating metal cylinder

proportional to the square of the frequency of oscillation of the magnetic flux. The higher the frequency of an alternating current, the more is the loss from eddy currents in the core of any coil used, and the more the core must be subdivided to keep the eddy current loss down. If thin laminations are not sufficient, a bundle of iron wires may be used or even a core packed with iron dust. But at the frequencies used for radio work even this is not very satisfactory.

Another way of reducing eddy currents is to find magnetic substances that have very large resistivities; such for instance are the iron compounds known as *ferrites*. Many of these substances have such high resistivities that they may be classed as insulators, while their magnetic properties are similar to those of iron. They may be used to make high permeability cores for coils in radio apparatus and introduce very little loss even at high frequencies.

Sometimes eddy current losses are deliberately used as a means of supplying heat to an otherwise inaccessible metal object. For instance in the manufacture of thermionic tubes, cathode-ray tubes, etc., to achieve a permanent vacuum it is necessary to raise the temperature of all parts of the tube as much as possible during evacuation to drive off the 'occluded' gases that might otherwise remain adsorbed on the surfaces. For this purpose the metal parts need to be heated to a much higher temperature than the glass envelope could stand. This is done by

means of eddy currents. A coil carrying a high-frequency current is placed round the tube, thereby generating large eddy currents in the metal parts of it.

According to Lenz's law the effect of eddy currents must be to tend to prevent an alternating magnetic flux from penetrating into a piece of metal. At high frequencies this process is so efficient that we may use a *non-magnetic* metal sheet as a screen for magnetic fields. It is the usual practice in radio apparatus to place aluminium cans round coils and other components that carry high-frequency currents. Eddy currents induced in the cans are sufficient to prevent an alternating magnetic field passing through the can and inducing unwanted e.m.f.'s in other parts of the circuit. (The same cans are also effective in screening the components from stray *electric* fields—p. 200.) However, this magnetic screening would not be effective at low frequencies. A mains transformer working at a frequency of 50 Hz produces an appreciable alternating field in its neighbourhood at this frequency, and this sometimes causes unwanted interference in other parts of a circuit. But nothing would be achieved by surrounding the transformer with a metal screen (of any reasonable thickness), since at this frequency the eddy currents would not be sufficient to give any significant reduction of the field penetrating the screen. The only effective screen in this case would be one made from a soft magnetic material, such as *mumetal* (p. 118).

A similar effect controls the flow of alternating current through any conductor at high frequencies. The lines of force of the magnetic field of a conductor carrying a current are circles concentric with the wire; and inside the conductor the field is of the same form. If the current is a high-frequency alternating one, e.m.f.'s are induced locally in the wire that make the distribution of current over its cross-section non-uniform. The distribution of current must in fact be such as to reduce the alternating flux in the body of the material as much as possible. Thus at high frequencies the current flows mostly in a layer very near the surface. This phenomenon is known as the *skin effect*.

At a frequency of 1 MHz the thickness of the layer through which the current flows is only about 0.05 mm. Theory shows that the thickness of the 'skin' is inversely proportional to the square root of the frequency, so that at the frequencies used for television and radar (10^8 to 10^{10} Hz) it is comparable with the wavelength of visible light! One consequence of the skin effect is that at high frequencies the resistance of a conductor increases very considerably over its d.c. value, since its effective cross-sectional area is then much reduced.

5.8   Inductance

In the process of *mutual induction* e.m.f.'s are induced in a circuit by changes in currents flowing in other circuits. But e.m.f.'s are also induced in a circuit by changes in its own current, since this also produces magnetic flux that links the circuit. This process is known as *self-induction*. From Lenz's law we should expect the e.m.f.'s of self-induction to *oppose* the changes of current that caused them.

When a steady current is flowing in a coil no e.m.f. can arise from self-induction, and the value of the current depends only on the resistance and applied p.d. Thus in d.c. circuits the effects of self-induction are apparent only at switching on or off—while the current is rising to its maximum value or falling again to zero.

At switching on the induced e.m.f. acts in opposition to the applied p.d. so that it delays the

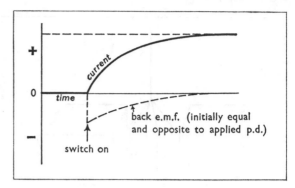

Fig. 5.34   The growth of current in an inductive coil

growth of current. At first the back e.m.f. due to self-induction is nearly equal to the applied p.d. (Fig. 5.34); then as the current approaches its final value, the rate of increase of current becomes less, and the back e.m.f. falls to zero. With a very large iron-cored coil the growth of current may well take several seconds. This may be demonstrated with the circuit of Fig. 5.35.

At switching off the behaviour of an inductive circuit like this is more dramatic. When the switch is opened, the current is obliged to drop almost instantly to zero, and the flux linked with the coil changes extremely rapidly. A very large pulse of e.m.f. is thus induced in it, which acts so as to tend to keep the current going. The peak value of this e.m.f. is often many hundreds of times greater than the original applied p.d., sufficient in Fig. 5.35 to light the neon lamp. If two fingers are placed on the switch terminals, an appreciable shock can be felt even when only a single cell is used in the circuit. (Those with

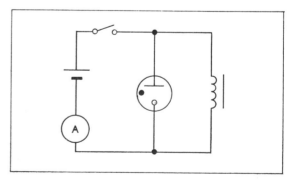

Fig. 5.35   A large e.m.f. is produced on switching off an inductive circuit

weak hearts do well to take this last result on trust!)

This effect poses problems for the designer of switches. If the contacts are separated too slowly at switching off, the e.m.f. due to self-induction may be sufficient to cause an arc between them. The danger is greatest for circuits carrying large currents and containing highly inductive components. In extreme cases the arc can fuse the switch arms altogether. In the domestic light switch it is sufficient to use a spring-loaded switch arm, which results in rapid separation of the contacts. Even in this case trouble sometimes arises when fluorescent lighting is used, since inductive control coils have to be used in such circuits. In some large switch systems jets of compressed air are made to blow out the arc that forms as the contacts separate. With the largest circuit breakers the contacts are mounted in a tank of non-inflammable oil under pressure, which rapidly quenches the arc. Even so, the decomposition and vaporization of the oil produced by the arc leads to the sudden evolution of large volumes of gas, constituting a sizable explosion; and the frame of the tank must be very strong. In the National Grid network the cost of the circuit breakers is a large part of the total capital outlay.

The behaviour of an inductive coil (or *inductor*) with alternating current is quite complex, and is discussed in detail in Chapter 12. But in general terms we can see that the e.m.f.'s due to self-induction must act continually in such a way as to oppose the applied p.d. at every instant. The actual value of the current in such a case depends then not only on the resistance of the circuit, but also on the extent of the self-inductive effect; the larger this is, the smaller the current becomes. This may be demonstrated with a coil of many turns with a removable (laminated) iron core. The coil is joined in series with an electric bulb, and the arrangement is connected to the a.c. mains (Fig. 5.36). With the core removed the self-inductive effect at 50 Hz is fairly small, and the bulb will just be slightly dimmer than usual, due chiefly to the extra resistance of the coil in the circuit. But when the iron core is introduced, the current at once falls to a small value. If it can be arranged so that a closed loop of laminated soft iron is formed through the

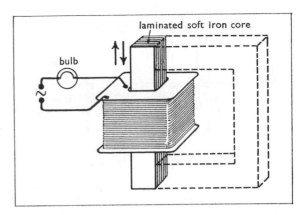

Fig. 5.36   The control of an alternating current with an inductive coil

coil (as suggested by the dotted lines in Fig. 5.36) the effect is even more marked. An inductor used in this way is called a *choke*; an alternating current can be controlled by this means without a great deal of heat loss, such as would occur if a resistor were used instead.

At the moments of switching on and off the behaviour of an a.c. circuit is more complicated than a d.c. one; but the general effects are similar —a slow growth in the magnitude of the alternating current at switching on, and a large pulse of e.m.f. at switching off, with consequent danger of arcing. However, the effects are less violent than in d.c. circuits, and this enables a.c. switches and thermostats to be of lighter construction than is required for direct current.

The induction coil

In the *induction coil* an automatic make-and-break arrangement is included in series with an iron-cored coil so that a rapid succession of large pulses of e.m.f. is produced. Its construction is shown in Fig. 5.37. The core is made up of a bundle of soft iron wires; two coils well insulated from each other are wound on top of this. The *primary* coil is of thick-gauge wire and is designed to carry a large current supplied by a battery; the *secondary* coil has a very large number of turns of fine wire. The turns ratio between primary and secondary may be 100 or more.

When the primary circuit is completed, the current in it starts to grow, magnetizing the iron core; the soft iron armature block A is then

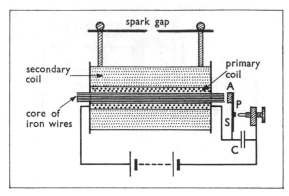

Fig. 5.37   An induction coil

attracted to the core. This breaks the circuit at the contact P, and the core demagnetizes again. The spring strip S now draws the armature back so as to close the contact once more; and so the cycle is repeated indefinitely. Each time the circuit is broken the current falls very rapidly to zero, producing a large pulse of e.m.f. in the primary. But the flux that links with the primary coil links many more times with the secondary, producing in it an even larger pulse of e.m.f. Peak p.d.'s of 50 to 100 kV can be produced by this means.

The contacts of the make-and-break arrangement move apart such a short distance in the induction coil that a special method must be adopted to control the arcing that would otherwise occur between them. For this purpose a capacitor C is joined across them. The property of a capacitor is to store charge on its plates—positive on one plate and negative on the other—when a p.d. is applied between them (p. 162). Thus, for a short time after opening the contacts current continues to flow charging up the capacitor. This gives time for the contacts to move far enough apart for arcing not to occur.

The way in which we measure the self-inductive effect in a coil follows from Neumann's law of electromagnetic induction:

$$E = - \frac{d\Phi}{dt} \quad (\text{p. 87})$$

We shall deal first with the case of coils without cores of iron or other ferromagnetic material, in which the flux linkage Φ of the magnetic field with the coil is always proportional to the current I flowing in it.

$$\therefore \Phi = LI$$

where L is a constant depending on the geometry of the coil.

∴ The induced e.m.f. E is given by

$$E = - \frac{d\Phi}{dt} = -L \frac{dI}{dt}$$

Thus the induced e.m.f. is proportional to the rate of change of current in the coil. The constant L is called the *self-inductance* of the coil—or simply its *inductance*.

Inductance. The e.m.f. E induced in a coil by self-induction is proportional to the rate of change of the current I in the coil;

$$\therefore E \propto - \frac{dI}{dt}$$

or

$$E = -L \frac{dI}{dt}$$

where L is a constant for the coil known as its self-inductance.

The unit of inductance is called the *henry* (H) after an American scientist.

The **henry** is the self-inductance of a conductor in which an e.m.f. of 1 volt is induced when the current in it changes at the rate of 1 ampere per second.

An inductive coil inevitably has resistance R as well as inductance L (unless it is superconducting). If the current in such a coil is I, and this is growing at a rate dI/dt, the p.d. across it must be made up of two components: (*i*) the p.d. required to maintain the current in the resistance—this is equal to IR; (*ii*) a p.d. equal and opposite to the back e.m.f. of self-induction, $-L(dI/dt)$—this is required to maintain the *growth* of current in the inductance (Fig. 5.38). The p.d. V across the inductive coil is therefore given by

$$V = IR + L \frac{dI}{dt}$$

When the current has reached its final steady value, the second term of this expression vanishes, and we have as for any resistance

$$V = IR$$

The first term on the right of the equation is zero at the moment of switching on, when $I = 0$ and before the current has had time to grow appreciably. The growth of current at this moment is

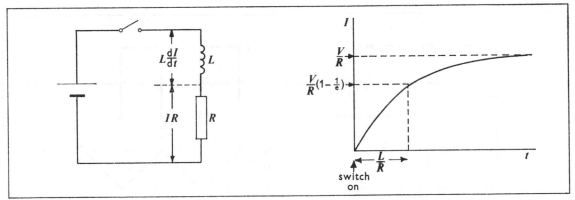

Fig. 5.38   The exponential growth of current in an inductive coil; the time constant is L/R

controlled entirely by the inductance of the coil, and we have

$$V = L \frac{dI}{dt}$$

In other words, at the moment of switching on the back e.m.f. in the coil is exactly equal and opposite to the applied p.d. (see Fig. 5.34). The way this works is best understood by considering a numerical example.

EXAMPLE   A coil of inductance 0.1 H and resistance 4.0 Ω is connected to a 12 V car battery. Calculate (a) the final steady current in the coil; (b) the initial rate of growth of current.

(a) The final current I is given by

$$I = \frac{V}{R} = \frac{12}{4} \text{ A}$$

$$= 3.0 \text{ A}$$

(b) The initial rate of growth of current dI/dt is given by

$$\frac{dI}{dt} = \frac{V}{L} = \frac{12}{0.1} \text{ A s}^{-1}$$

$$= 120 \text{ A s}^{-1}$$

In order to see how the current in an inductive coil grows from zero towards its final steady value we have to find a complete solution of the differential equation

$$V = IR + L \frac{dI}{dt}$$

By direct differentiation the student can check that the required solution is

$$I = \frac{V}{R}(1 - e^{-Rt/L})$$

The way in which I varies with t is shown in Fig. 5.38. The current tends asymptotically to the final value V/R. The quantity L/R that comes in this expression has the dimensions of time; it is known as the time constant τ of the circuit. Thus

$$\tau = \frac{L}{R}$$

It is a measure of the time taken by the current to grow. To be exact, the current grows to within $1/e$ ($= 0.37$) of its final value in the time τ. I.e. when $t = \tau$, the current I is given by

$$I = \frac{V}{R}(1 - 0.37) = 0.63 \frac{V}{R}$$

Another way of thinking of the time constant τ is as the time (L/R) that would be taken by the current to grow to its final value (V/R) if it continued growing at its initial rate (V/L). In fact the rate of growth slows up, so that it only grows to within 1% of its final value in a time of about 5τ. Thus in the example considered above, the time constant τ is given by

$$\tau = \frac{0.1}{4.0} \text{ s}$$

$$= 0.025 \text{ s} = 25 \text{ ms}$$

In this time interval the current grows from zero to the value

$$0.63 \times 3.0 \text{ A} = 1.89 \text{ A}$$

The energy stored in an inductance

Because of the back e.m.f. of self-induction that acts when the current in a coil increases, electrical energy must be supplied in setting up the

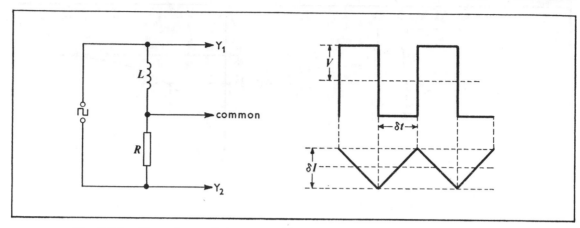

Fig. 5.39   Measuring an inductance with a square-wave generator and oscilloscope

current against this back e.m.f. Thus, multiplying through the differential equation by the
current I, we have

$$VI = I^2R + LI\frac{\mathrm{d}I}{\mathrm{d}t}$$

This can be interpreted to mean:
power supplied = rate of production of heat

   + rate of production of magnetic energy

The magnetic energy referred to is the energy
stored in the magnetic field of the coil; this
grows as the current grows. The total energy
stored in the coil when the current in it is I can
be obtained by integrating the second term on
the right-hand side of the equation. Thus

the energy supplied in time $\delta t = LI\frac{\mathrm{d}I}{\mathrm{d}t}\delta t$

$$= LI\,\delta I$$

where δI is the increase of current that occurs in
time δt.

$$\therefore \text{ total energy supplied} = \int_0^I LI\,\mathrm{d}I$$

$$= \tfrac{1}{2}LI^2$$

Thus in the example considered above, when the
current has grown to its final value of 3.0 A, we
have

energy stored $= \tfrac{1}{2}LI^2 = \tfrac{1}{2} \times 0.1 \times (3.0)^2$ J
$= 0.45$ J

It is this energy that is expended in the spark at
the switch contacts or the flash of the neon bulb
in the experiment of Fig. 5.35.

The measurement of inductance

If the resistance of a coil is negligible we can very
simply measure its inductance by joining it to an
a.c. supply of known frequency f and measuring
the p.d. across it and the current through it with
suitable meters. Thus if

$$I = I_0.\sin \omega t$$

where $\omega = 2\pi f$

then $V = L\frac{\mathrm{d}I}{\mathrm{d}t} = \omega L.\cos \omega t$

The peak p.d. V_0 is then given by

$$V_0 = \omega L I_0$$

$$\therefore \omega L = \frac{V_0}{I_0} = \frac{V_{\text{rms}}}{I_{\text{rms}}}$$

However, when the resistance is not negligible,
this has to be taken into account as well. The
way of analysing this more complicated situation
is discussed in Chapter 12 (p. 219).

Fig. 5.39 shows an elegant method of measuring inductance with a square-wave generator and
an oscilloscope; it illustrates well the principles
discussed above. The square-wave generator
produces an output that changes alternately
between the values $+V$ and $-V$, spending an
interval of time δt at each of these values in
turn. This signal is applied across the inductive
coil and resistance joined in series, and the
p.d.'s across these two components are observed
on a double-beam oscilloscope. When the p.d.
$+V$ is applied, the current starts to grow at the
rate $\mathrm{d}I/\mathrm{d}t$; if the time constant τ of the circuit is
much longer than the interval δt, the growth

proceeds at a constant rate during this interval, and we have

$$\frac{dI}{dt} = \frac{\delta I}{\delta t}$$

The applied p.d. then reverses, and the current falls linearly during the next half-cycle. The waveform observed across the resistance is therefore a 'triangular' one, as shown. By measurements on the oscilloscope screen we can deduce the values of V, δI and δt; and so we have

$$L = \frac{V.\delta t}{\delta I}$$

For the method to be valid the p.d. across the resistor must always be a small part only of the applied p.d. V. The lower trace (Y_2) is therefore considerably more amplified (by the internal amplifier of the instrument) than the upper one; and the frequency of the square-wave oscillator must be high enough for the time interval δt to be very small compared with L/R.

The effect of a ferromagnetic core

When the coil has a ferromagnetic core the result is much more complicated, since the flux linkage Φ is now controlled by the magnetic behaviour of the core, and is not simply proportional to the current I. The same analysis can be applied, and we can define the inductance of the coil in the same way; but we must not expect it to be a precisely measurable quantity. Thus we can write

$$E = -\frac{d\Phi}{dt} = -\frac{d\Phi}{dI}\frac{dI}{dt}$$

$$= -L\frac{dI}{dt} \quad \text{(by definition)}$$

$$\therefore L = \frac{d\Phi}{dI}$$

The relationship between Φ and I is usually of the form shown in Fig. 7.9, p. 123, for a ferromagnetic core, and is by no means of constant gradient. When we refer to the inductance of an iron-cored coil we mean the average gradient $d\Phi/dI$ for particular operating conditions.

Mutual inductance

A similar analysis may be used for the mutual induction between two coils. Thus in the absence of ferromagnetic material the flux Φ_2 linking the second coil due to a current I_1 in the first is given by

$$\Phi_2 = {}_1M_2 I_1$$

where ${}_1M_2$ is a constant depending on the geometry of the coils. Extending the notation in the same way, we have for the other coil

$$\Phi_1 = {}_2M_1 I_2$$

It may be shown that the two constants are equal; i.e.

$${}_1M_2 = {}_2M_1$$

so that we may combine the two equations into

$$\Phi = MI$$

where Φ is the flux linking either coil due to a current I in the other.

Proceeding as before, the e.m.f. E_2 induced by mutual induction between the two coils is given by

$$E_2 = -\frac{d\Phi_2}{dt} = -M\frac{dI_1}{dt}$$

and the constant M is known as the *mutual inductance* between the coils.

The similarity of the defining equations shows that mutual inductance is to be measured in the same units as self-inductance. Indeed the definition of the henry can just as well be framed in terms of the mutual inductance of two coils.

The measurement of mutual inductance may be performed directly in terms of the definition by the method depicted in Fig. 5.21, p. 81. The current in the primary coil may be changed steadily in such a way that the galvanometer joined in the secondary circuit (used here as a millivoltmeter) indicates a steady reading; this reading gives the induced e.m.f. E_2. If the primary current changes by an amount ΔI_1 in time Δt, we have

$$\frac{dI_1}{dt} = \frac{\Delta I_1}{\Delta t}$$

$$\therefore M = \frac{E_2 \Delta t}{\Delta I_1}$$

If the two coils are interchanged in the circuit we can show experimentally that the two mutual inductances are indeed equal:

$${}_1M_2 = {}_2M_1 = M$$

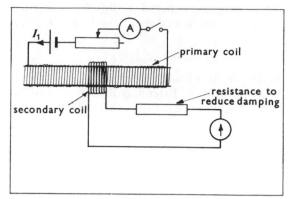

Fig. 5.40   Measuring mutual inductance with a ballistic galvanometer

★   Fig. 5.40 shows a ballistic method of measuring mutual inductance. In this case a high resistance is joined in series with the galvanometer to reduce damping, and we observe the first swing θ of the instrument when a current I_1 flowing in the primary coil is switched off. If the ballistic galvanometer has been calibrated (p. 90), the charge Q that passes through it at this moment is given by

$$Q = b\theta$$

where b is the known constant of the instrument.

Now the flux Φ_2 linking the secondary coil is given by

$$\Phi_2 = MI_1$$

which falls to zero when the switch is opened.

$$\therefore Q = \frac{\Phi_2}{R_2} \quad \text{(p. 90)}$$

where R_2 is the total resistance of the secondary circuit.

$$\therefore Q = b\theta = \frac{MI_1}{R_2}$$

Hence M may be found. Again we may repeat the experiment with primary and secondary coils interchanged, and show that the mutual inductance is the same either way.

In the next chapter (p. 114) we shall see how we may calculate the mutual inductance for a simple arrangement of coils, as in Fig. 5.40; and then it is possible to treat this experiment as a means of calibrating a ballistic galvanometer—which in practice is often simpler than the methods of calibration we have described earlier (p. 91). We also give in the next chapter a simple a.c. method of measuring a mutual inductance in terms of a resistance (or vice-versa)—see Fig. 6.17, p. 115.

The forces between coils

Magnetic forces always act so as to maximize the flux linkage with any coil carrying a current. Suppose the flux linking a coil is Φ; and suppose this would increase to $(\Phi + \delta\Phi)$ if the coil moved a distance δx under the action of the magnetic force F due to the current and field. Then

the work done by the force $= F\delta x$

$$\therefore \text{ the rate of doing work} = F\frac{\delta x}{\delta t}$$

if the change takes place in time δt. The mechanical energy made available in this way is drawn from the electric circuit, for an e.m.f. E arises in it tending to oppose the increase of flux; and, if the current I is to remain constant, the p.d. applied to the coil must increase by an amount δV equal and opposite to E.

$$\therefore \delta V = -E = \frac{\delta\Phi}{\delta t}$$

The power supplied to the coil therefore increases by an amount δP given by

$$\delta P = I\,\delta V = I\frac{\delta\Phi}{\delta t}$$
$$= I\frac{\delta\Phi}{\delta x}\frac{\delta x}{\delta t}$$

This is equal to the rate of producing mechanical energy. Therefore, in the limit as $\delta x \rightarrow 0$, we have

$$F\frac{dx}{dt} = I\frac{d\Phi}{dx}\frac{dx}{dt}$$
$$\therefore F = I\frac{d\Phi}{dx}$$

By similar reasoning we can show that the moment T of the couple acting on a coil in a magnetic field is given by

$$T = I\frac{d\Phi}{d\theta} \quad \text{(cf. p. 68)}$$

where $\theta =$ the angle of rotation.

This result enables us to state a very valuable generalization: *the magnetic forces acting on a circuit tend to move it so as to make the total flux linked with it a maximum;* this includes the flux due to the current in the circuit itself. Consider, for instance, a short solenoid suspended in a magnetic field. The flux linked with it will be a maximum when the field along the axis of the solenoid due to its own current is parallel to the applied field; and this is the position into which the coil tends to turn. If the field is non-uniform, the forces act to move the coil towards the region of maximum flux density. There are also magnetic forces acting on the solenoid tending to make it expand sideways and to contract along its length. All these forces act to increase the total flux of the magnetic field linked with the coil. The student can check these predictions for himself by considering the nature of the field and applying Fleming's left-hand rule (p. 65). But having made the effort he will probably agree that the generalization we have given is a useful way of predicting in many cases the manner in which the magnetic forces will operate.

The above result enables us to write down an expression for the force F acting between two coils in terms of their mutual inductance M. The force (in the x-direction) acting on the second coil due to a current I_1 in the first is given by

$$F = I_2\frac{d\Phi_2}{dx}$$

where I_2 is the current in the second coil, and Φ_2 is the *total* flux linking the coil. Now we have

$$\Phi_2 = L_2 I_2 + M I_1$$

where L_2 is the self-inductance of the second coil; but this first term of the expression for Φ_2 does not change when the coil moves.

$$\therefore \frac{d\Phi_2}{dx} = I_1 \frac{dM}{dx}$$

$$\therefore F = I_1 I_2 \frac{dM}{dx}$$

This is equal and opposite to the force acting on the first coil due to the current in the second. Similarly the moment T of the couple acting to produce rotation about a given axis is given by

$$T = I_1 I_2 \frac{dM}{d\theta}$$

where θ is the angle of rotation about the axis.

These results show that the equality of the two mutual inductances between a pair of coils depends basically on Newton's third law of motion. For if the forces and couples with which the coils act on one another are to form equal and opposite pairs, the symmetry of the above expressions requires

$$_1M_2 = {}_2M_1$$

In the *current balance* used in the standards laboratories, the current I to be measured is caused to flow through both the fixed and suspended coil systems of the instrument (p. 10). The deflecting force F is therefore given by

$$F = I^2 \frac{dM}{dx}$$

The absolute measurement of current with this instrument therefore becomes a matter of making a calculation of mutual inductance between the fixed and moving coil systems, and how this changes with the deflection. Notice that the force F is proportional to the square of the current, so that the sensitivity of the instrument increases with the current.

6 Magnetic Fields

6.1 Measuring flux densities

In this chapter we shall study more closely the patterns of magnetic fields produced by currents flowing in various configurations, and we shall see how the flux densities of these fields may be calculated. We review first the techniques we have already described for investigating magnetic fields quantitatively (iron filings can do no more than give a descriptive picture).

(*i*) *The simple current balance* (Fig. 5.6, p. 66). This is a method for the direct measurement of *B* in terms of the current *I* flowing in the pivoted rectangular loop of wire. It is not of high accuracy, though it may be improved by using a rectangular *coil* of several turns in place of the single loop, thus magnifying the force produced by a given current. The field to be measured needs of course to be constant over the whole of the region occupied by the end of the rectangular loop. In practice such a field is only to be found inside a solenoid, and the use of the simple current balance is therefore confined to this one kind of measurement. Like many fundamental methods of measuring physical quantities, it is best kept in reserve for calibrating the scales of other easier techniques.

The simple current balance used for measuring magnetic field is not to be confused with the type of current balance used in the standards laboratories for absolute measurements of current (p. 10). With the simple current balance we make use of already calibrated ammeters to provide known currents through the balance and the measurements yield values of the magnetic field. The instrument is essentially the same as an ordinary moving-coil ammeter, though in that case a fixed magnetic field is provided by the permanent magnet and the instrument gives the value of the current through the coil.

But in the current balance of the standards laboratories the magnetic field that acts on the moving coils is provided by other fixed coils joined in series, so that the current to be measured flows through both the fixed and moving coil systems. The simple current balance can of course be adapted to this use. We have only to mount the pivoted rectangular coil as though to measure the field inside a solenoid, and then join the solenoid and moving coil in series. Later in this chapter we shall see what the laws are that control the behaviour of such an instrument, and how the measurement of the deflecting force can yield a value of the current in amperes; we can then use the device as a means of calibrating the scale of an ammeter. But for the present we are concerned just to use the simple current balance to measure an unknown magnetic field, making use of an already calibrated ammeter to register the current.

(*ii*) *The Hall probe* (Fig. 5.11, p. 71). This is a simple method for the *comparison* of fields, and is therefore ideal for determining how the field of a circuit varies from one point to another. To find actual values of *B* the probe needs to be calibrated in a known field. Once we have discovered how to calculate the field in a solenoid with a given current flowing in it, this provides the easiest way of producing a known field for calibrating a Hall probe. For the time being we can always measure the field in a solenoid with a simple current balance, and then use this for calibrating the Hall probe.

To make accurate measurements with a Hall probe it is essential to check the zero setting frequently, and to repeat earlier measurements in a series; all semiconductor devices are very sensitive to changes of temperature, and allowance may have to be made for the drift in the calibration. It should not be used for fields less than about 10^{-4} T, otherwise it may be necessary to make awkward corrections to allow for the presence of the earth's field. However, the Hall probe is certainly one of the most convenient methods available for many of the investigations we shall refer to below.

(*iii*) *The a.c. search coil* (Fig. 5.25, p. 85). This provides a measurement of flux density (i.e. of the peak value of the alternating field) through the measurement of the e.m.f.'s induced in the coil. These e.m.f.'s are proportional to the frequency. Therefore we can in principle increase the sensitivity indefinitely by using sufficiently high frequencies. In practice these can well be anything up to 100 kHz. The method is then far more sensitive than any other we have available, and will be preferred for studying the fields of coils, etc., when these are necessarily small. We shall often be able to use this method with precision in peak fields of less than 10^{-6} T.

Only the alternating component of the field is detected with an a.c. search coil; therefore the steady field of the earth or of other equipment nearby does not affect the measurements. However, if care is not taken there may be uncertainties introduced by stray a.c. fields produced by items of equipment in the laboratory other than the coil under test.

(*iv*) *Search coil and ballistic galvanometer* (Fig. 5.30, p. 90). The technique of taking the maximum reading of a swinging galvanometer is a difficult one to manage, and accurate results are possible only if each measurement is repeated several times. This method will therefore be used only where no other is available—i.e. for the measurement of the flux inside a magnet or other solid material. Obviously we cannot insert a probe into the iron of a magnet, but we can wrap a search coil round it, and so find the average flux density inside it (p. 90).

The calibration of the ballistic galvanometer for charge measurement is not easy to do, particularly as it varies with the resistance of the circuit in which the instrument is joined. It is perhaps best treated as a comparison method, to be calibrated for flux measurement in a known magnetic field.

6.2   Solenoids

If we move a Hall probe slowly into a solenoid carrying a steady current, we find that the field increases at first, and then reaches a maximum value which is constant through the whole inner region of the solenoid. Furthermore the field in this inner region is constant over the whole cross-section. The simplest way of producing a uniform magnetic field is therefore to use the central region of a long solenoid. If we place the Hall probe at an end face of the solenoid we shall find that the flux density here is exactly half what it is in the uniform central region. The field at the end is obviously non-uniform, as we can see from the line of force diagram (Fig. 5.4*b*, p. 64). However, if we keep the Hall probe turned so as to register only the component of the flux density parallel to the axis of the solenoid, we shall find that this component is also constant over the whole end face.

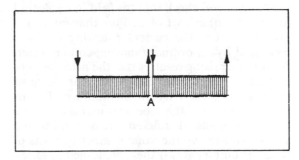

Fig. 6.1   The components of field produced by the two halves of a solenoid

The theoretical explanation of this can be seen by imagining a long solenoid (Fig. 6.1) made up of two parts that meet in the centre at A. The flux density over the cross-section at A is uniform and parallel to the axis. By symmetry, precisely half this flux density must be produced by each half of the complete solenoid (while any components at right-angles to the axis must cancel out). The component of the flux density parallel to the axis at the ends of each part of the solenoid is therefore half what it is in the uniform field regions of either part.

On the outside of the solenoid tube the field is almost zero except near the ends. This may best be tested with an a.c. search coil using a high frequency current in the solenoid. Evidently the fields of the various turns of the solenoid almost completely cancel one another on the outside, and the longer the solenoid the more exactly is this found to be so. If we use a uniformly wound endless solenoid (in the form of a torus, Fig. 6.2),

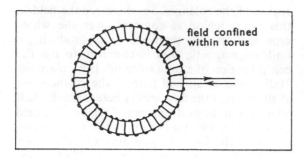

Fig. 6.2   There is no field outside a uniformly wound toroid (an endless solenoid)

we shall find that there is definitely no external field.

We can soon check that the field in a solenoid (or in any other sort of coil for that matter) is proportional to the current I flowing in it—as measured by an ordinary moving-coil ammeter. A more surprising result is that the field does not depend on the cross-sectional area or shape of the solenoid, but only on the number of turns per unit length of the tube. It is useful to have a set of solenoids of different cross-sections and shapes all bearing the same number of turns of wire per unit length. If these are joined in series, so that they all carry the same current, it can be checked with a Hall probe that the flux densities are the same in all of them. The theoretical explanation of this is as follows: consider two identical long solenoids A and B of rectangular cross-section lying side by side (Fig. 6.3). If these are pushed into contact, the currents in their adjacent sides are opposite, and must produce equal and opposite magnetic fields. We could therefore in principle remove these sides, giving

a composite solenoid C of twice the width, but still containing the same magnetic flux density. We can imagine further similar solenoids added in the same way, so as to build up solenoids of any size or shape whatever—and all would have the same flux density inside.

By using (long) solenoids of various lengths and numbers of turns we can show that the flux density for a given current is directly proportional to the number of turns per unit length. An elegant way of doing this is to use an extensible spiral spring made of a suitable non-magnetic material. The current may be fed into and out of this through spring clips, and so the number of turns N in use and length l they are stretched to can be varied. By bending the spring into a curve (i.e. part of a torus), we can show that the field inside is very little different from that of a straight solenoid—at any rate as long as the bend is not too great. The field in a bent solenoid is not strictly uniform over the cross-section, since the same number of turns is now concentrated into a shorter length on the inside of the curve and a greater one on the outside; the number of turns per unit length therefore varies over the cross-section, and the field over the cross-section varies proportionately.

These experiments show that the field of a solenoid is given by

$$B \propto \frac{N}{l}I$$

or we may write this as

$$B = \frac{\mu_0 NI}{l}$$

where μ_0 is a universal constant, called the *magnetic constant*. It is universal in the sense that we can regard it as a property of the Universe. It is the constant that specifies the magnetic force that acts between given electric currents, in the same way that the gravitation constant G specifies the gravitational force that acts between given masses. The above equation shows that the units of the magnetic constant μ_0 are T m A^{-1}. But since we have

$$1 \text{ T} = 1 \text{ N A}^{-1} \text{ m}^{-1} \quad \text{(p. 67)}$$

we can express the units of μ_0 more significantly as N A^{-2}. Its value is fixed as soon as we select the ampere as our unit of current. By using a simple current balance to measure the flux

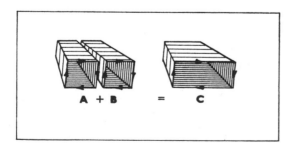

Fig. 6.3   The two square solenoids A and B when put together produce a field identical with that of the rectangular solenoid C

density B inside a long solenoid carrying a current of 1 A (as given by an ordinary moving-coil ammeter), we may find the value of μ_0 implied by our choice of the ampere as the unit of current.

Hence $\mu_0 = 1.26 \times 10^{-6} \text{ N A}^{-2}$

We shall see in due course how this figure may be *calculated from the definition of the ampere* (p. 107).

The field of a sheet of current

If we have a solenoid with a large number of evenly spaced turns, the current flowing in it can be treated as though it were a sheet of current encircling the tube. Thus a solenoid of N evenly spaced turns carrying a current I is equivalent to a current NI flowing in a sheet round the tube. This shows how we may calculate the field produced by such a sheet—a result we shall need later in this chapter. Consider a long solenoid of wide and shallow rectangular cross-section (Fig. 6.4). Near the middle of this we can ignore the contribution of the relatively distant side walls of the tube, and assume that the whole flux density B is produced by two opposite sheets of current lying above and below. Suppose the solenoid has n turns per unit length; i.e. $n = N/l$. Let each turn carry a current I. We define the strength i of the current sheet as the current per unit length of the solenoid—i.e. per unit *width* of the current sheet measured across the line of current flow. Thus

$$i = nI$$

The total flux density inside the solenoid is given by

$$B = \frac{\mu_0 NI}{l} = \mu_0 nI = \mu_0 i$$

The contribution from the top and bottom current sheets of the solenoid is therefore

$$\frac{\mu_0 i}{2} \text{ from each sheet}$$

From considerations of symmetry we must assume that a single current sheet on its own would produce a field $\mu_0 i/2$ on *each* side, the fields on the two sides being in opposite directions (as given by the right-hand screw rule, p. 65). In our flat solenoid the two contributions to the field will add up inside the tube and

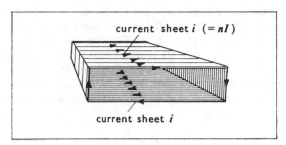

Fig. 6.4   Calculating the field of a current sheet

will exactly cancel outside, where they will be in opposite directions.

Another way of describing this result is to say that the flux density of the magnetic field changes by the amount $\mu_0 i$ from one side of a current sheet to the other. Expressed in this way the result seems to apply equally whether the sheet is flat or curved; for it certainly applies for all solenoids, whatever the shape of the tube round which the current sheet flows.

6.3   Straight wires

We may use the same experimental techniques to investigate the magnetic field produced by a long straight wire, making sure of course that the return leads from the ends of the wire are sufficiently distant from the apparatus not to affect the field significantly. We have already seen that the lines of force of such a field are concentric circles round the wire (Fig. 5.4a, p. 64). It may be confirmed with a Hall probe

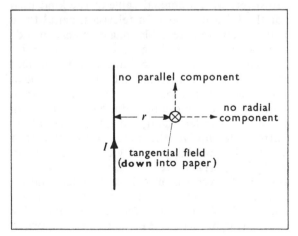

Fig. 6.5   Investigating the field of a straight wire

or an a.c. search coil that the magnetic field has no component parallel to the wire or radially away from it (Fig. 6.5). Turning the probe or search coil so that its plane is parallel to the wire and moving it to various distances from it, we find that the flux density B is *inversely proportional* to the mean distance r from the wire. Thus B is given by

$$B \propto \cdot \frac{I}{r}$$

or

$$B = \frac{kI}{r}$$

where k is a constant whose value may be determined from the actual measurements in this experiment. We thus find

$$k = 2.0 \times 10^{-7} \text{ N A}^{-2}$$

This is not of course the same constant as μ_0 the *magnetic constant* defined in terms of the field of a solenoid; but it is related to it, as we shall see.

Coaxial cables

It is no surprise to discover that practically no field is produced by a pair of conductors close together carrying equal and opposite currents (e.g. like a typical electric light flex). Actually if an a.c. search coil is used with a high frequency supply (say 100 kHz), it may just be possible to detect a small magnetic field very close to the wires of a piece of light flex, where the fields the wires produce do not completely cancel out. However, with a coaxial cable of the kind used for the lead-in wires of a television aerial there is definitely no detectable magnetic field outside the cable. A coaxial cable consists of an insulated inner copper wire that carries the current in one direction, surrounded by an outer tube of fine braided wires that carry the return current. Outside the cable the field due to the current in the outer cylinder exactly cancels the field of the current flowing in the inner conductor. It is worth investigating just how this comes about. For this purpose we can use a tubular conductor of several centimetres diameter, and investigate the field due to a current passing down the length of it just as in the experiments with a long straight wire. We find two important results:

(*i*) The field *outside* the tube is exactly the same as for a single wire running down the centre; i.e.

$$B = \frac{kI}{r}$$

where r is the distance from the centre of the tube, and k is the same constant as above.

(*ii*) The field *inside* the tube is precisely zero (except near its ends where the current flow is irregular).

We can now see what is happening with a coaxial cable. Outside the cable the fields due to the two conductors are exactly equal and opposite. Inside the cable in the space between the two conductors the current in the outer cylinder produces no field at all; and the field here is just that of the inner conductor alone. This enables us to relate the constant k to μ_0 the magnetic constant, using the previous result for the amount by which the magnetic field changes as we pass from one side of a current sheet to the other. The current in the outer conductor is just such a current sheet, not round the cylinder, as with a solenoid, but along its length. The total current I in the sheet is spread out over a *width* (round the circumference) of $2\pi r$, where r is the radius of the outer cylinder. The strength i of the current sheet is therefore given by

$$i = \frac{I}{2\pi r}$$

The flux density changes by the amount $\mu_0 i$ as we pass from one side to the other of such a sheet. The field outside the sheet is zero. Therefore the flux density B on the inside of it a distance r from the centre is given by

$$B = \mu_0 i = \frac{\mu_0 I}{2\pi r}$$

This must be the flux density due to the inner conductor alone at a distance r from the centre. The constant k is therefore given by

$$k = \frac{\mu_0}{2\pi}$$

The useful property of a coaxial cable for technical purposes is that it produces no stray magnetic field in its surroundings. It also incidentally produces no stray electric field, since the outer conductor acts as a complete shield of electric fields (p. 200). With very high frequency currents and p.d.'s this can be an important consideration.

Having derived a value for the constant k, we can now write the full expression for the flux density at a distance r from a *long straight wire* carrying a current I.

$$B = \frac{\mu_0 I}{2\pi r}$$

The magnetic constant μ_0

The modern definition of the ampere (p. 9) envisages an arrangement of two long, straight, parallel conductors. We are now in a position to derive an expression for the forces acting between such a pair of conductors. Suppose they are a distance r apart, and each carry a current I.

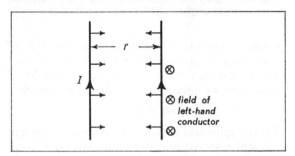

Fig. 6.6   Calculating the force between two parallel conductors

In Fig. 6.6 the field produced by the left-hand conductor at the position of the right-hand one is down into the paper. The flux density B of the field is given by

$$B = \frac{\mu_0 I}{2\pi r}$$

By Fleming's left-hand rule, the force on the right-hand conductor is to the left—i.e. there is an *attraction* between the two conductors. (If the currents are in opposite directions, there is a repulsion between them—Fig. 1.3, p. 7). The force F acting on a length l of one conductor is given by

$$F = BIl = \frac{\mu_0 I^2 l}{2\pi r}$$

In the definition of the ampere the following figures are specified:

$$l = r \ (= 1 \text{ m})$$
$$F = 2 \times 10^{-7} \text{ N}$$
$$I = 1 \text{ A}$$

Substituting, we obtain the value of μ_0 that is implied in the definition of the ampere:

$$\mu_0 = 4\pi \times 10^{-7} \text{ N A}^{-2}$$

It is often convenient to express μ_0 in terms of the unit of inductance, the *henry* (H) (p. 96); and we have the alternative statement:

$$\mu_0 = 4\pi \times 10^{-7} \text{ H m}^{-1}$$

The student can readily check that these units are indeed identical.

6.4 Coils and current elements

In order to calculate the field at some point near a circuit in the general case we need first to have an expression for the field of a *short section of wire* carrying a current, referred to as a *current element*. It should then in principle be possible to divide up any actual circuit into a sufficient number of short sections and find the total field at a given point by a process of summation or integration. Consider a short section of wire at O of length l, carrying a current I; let its distance from the points P and P' at which the flux density B is required be x (Fig. 6.7); and let θ be the angle that this current element makes with the line OP. The problem is to find an expression for B in terms of I, l, x and θ. We shall assume that B is in a direction at right-angles to the

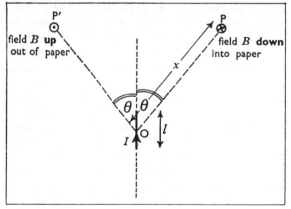

Fig. 6.7   Deriving an expression for the field of a current element

plane of the diagram (Fig. 6.7) in a sense given by the right-hand screw rule. This will at any rate be in agreement with the pattern of field for a long straight wire of which the current element might form part. We are intending to apply the result by adding together the components of field of the successive current elements of a circuit; this implies that B should be proportional to the length l of the element. In any case the matter may be tested by comparing the fields of a set of coils of identical size and shape but with varying numbers of turns; the field can be shown to be proportional to the number of turns, which in this case is a measure of the length of wire l in the coil

$$\therefore \ B \propto l$$

To investigate the dependence of B on the distance x we need to construct a set of flat circular coils of varying radius x, but all containing *the same length of wire l*. The flux densities at the centres of the coils may then be measured with a Hall probe or search coil. All points of any one coil are at the same distance x from its centre, and the line joining each part of the wire to the centre makes the same angle θ with the wire (namely a right-angle in this case). If the same current is passed through each coil in turn, the field at the centre is found to be inversely proportional to the square of the radius.

$$\therefore \ B \propto \frac{1}{x^2}$$

It remains to determine how B depends on the angle θ. We cannot in this case design a complete circuit so that θ is the same for all parts of it, while x and l are kept constant (except for the special case $\theta = $ a right-angle, as at the centre of a flat circular coil); nor of course can we study the field due to a small part of a circuit alone. We have rather to work back to possible expressions involving θ using measurements made on the fields of complete circuits in which θ varies from one point to another. The problem is a very intractable one, and in the early development of the subject several solutions were obtained, each of which appeared in all cases to give correctly the field of a complete circuit. It was eventually shown that there is no unique solution, but rather a range of possible formulae, all of which give the same result for the field of a

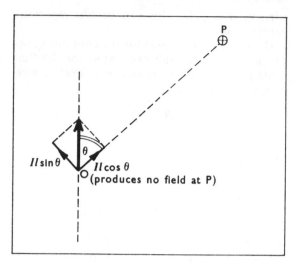

Fig. 6.8   Only the component of the current element at right-angles to OP produces a field at P

complete circuit. It becomes then a matter of convenience which of these expressions we use; and in common with everyone else since Maxwell's day we choose the simplest one, namely

$$B \propto \sin \theta$$

This is equivalent in fact to imagining the current element resolved into two components: $Il.\cos \theta$ along the line OP, and $Il.\sin \theta$ at right-angles to this (Fig. 6.8). We are thus assuming that the component in the line OP produces no field at P, leaving us with a field proportional to the other component $Il.\sin \theta$. But, as we have said, there are other ways of bringing θ into the expression that are equally in agreement when the result is summed for a complete circuit.

We now have the complete expression for the flux density B due to a current element.

$$B \propto \frac{Il.\sin \theta}{x^2}$$

or

$$B = \frac{k'Il.\sin \theta}{x^2}$$

where k' is another constant related to μ_0 the magnetic constant.

The simplest way to calculate the value of k' is to integrate the above expression for all the elements of a long straight wire, and then to compare the result obtained with the previously derived expression. This integration for a long straight wire yields the result

$$B = \frac{2k'I}{r}$$

and from the previous section (p. 107) we have

$$B = \frac{\mu_0 I}{2\pi r}$$

$$\therefore k' = \frac{\mu_0}{4\pi}$$

[This integration is essentially the same as that given below; but we have there assumed the value of k' derived here.]

The flux density B due to a *current element* is therefore given by

$$B = \frac{\mu_0 Il \cdot \sin \theta}{4\pi x^2}$$

and this result may be used to calculate magnetic fields in the general case. It is known as the *Biot-Savart law*.

★ *Deduction of the field of a straight wire from the Biot-Savart law*

Expressing the law in differential form, the component of flux density δB due to a current $I.\delta l$ is given by

$$\delta B = \frac{\mu_0 I \, \delta l \cdot \sin \theta}{4\pi x^2}$$

Consider a point P at a distance r from the wire (Fig. 6.9). The total field B at P is given by

$$B = \int dB = \frac{\mu_0 I}{4\pi} \int \frac{\sin \theta \cdot dl}{x^2}$$

the integration being taken over the whole length of wire. Now, from the figure we have

$$l = l_0 - r.\cot \theta$$
$$\therefore dl = r.\text{cosec}^2 \theta \, d\theta$$

Also $x = r.\text{cosec} \, \theta$

$$\therefore B = \frac{\mu_0 I}{4\pi} \int_{\phi_1}^{\beta} \frac{\sin \theta . r . \text{cosec}^2 \theta . d\theta}{r^2 . \text{cosec}^2 \theta}$$

$$= \frac{\mu_0 I}{4\pi r} \int_{\phi_1}^{\beta} \sin \theta . d\theta$$

$$= \frac{\mu_0 I}{4\pi r} \left[- \cos \theta \right]_{\phi_1}^{\beta}$$

$$= \frac{\mu_0 I}{4\pi r} \{ - \cos \beta + \cos \phi_1 \}$$

Writing this in terms of ϕ_2 we have

$$\phi_2 = 180° - \beta$$
$$\therefore \cos \phi_2 = - \cos \beta$$

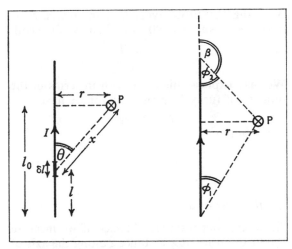

Fig. 6.9   Calculating the field of a straight wire

$$\therefore B = \frac{\mu_0 I}{4\pi r} (\cos \phi_1 + \cos \phi_2)$$

The field of a complete circuit made up of sections of straight wire may then be summed using this expression. For example, the field of a square coil may be found by this means.

A very long straight wire

Using the above result, for a very long wire in the limit both ϕ_1 and ϕ_2 tend to zero. We then have

$$B = \frac{\mu_0 I}{2\pi r}$$

It is useful to know under what conditions a wire may be considered long enough to justify the use of this expression.

Suppose $\phi_1 = \phi_2 = \phi$, and that the angles are small. In this case we may write

$$\cos \phi = 1 - \frac{\phi^2}{2} \quad (\phi \text{ in radians})$$

$$\therefore B = \frac{\mu_0 I}{2\pi r} \left(1 - \frac{\phi^2}{2} \right)$$

The fractional error in taking $B = \mu_0 I / 2\pi r$ is $\phi^2 / 2$.

Now $\phi \approx \tan \phi = \frac{r}{l/2} = \frac{2r}{l}$

where l = total length of wire. If the maximum error allowable is (say) 1%, then we must have

$$\frac{\phi^2}{2} = \frac{2r^2}{l^2} < \frac{1}{100}$$

$$\therefore r < \frac{1}{14} l \quad \text{(approx.)}$$

At the centre of a flat circular coil

Consider a point at the centre of a flat circular coil of radius r containing a total length of wire l.

Applying the Biot-Savart law, in this case we have $x = r$, and $\theta = 90°$ for all parts of the coil.

$$\therefore\ B = \frac{\mu_0 Il}{4\pi r^2}$$

We may express this if we wish in terms of the number of turns N of the coil, for we have

$$l = 2\pi Nr$$

Substituting,

$$B = \frac{\mu_0 NI}{2r}$$

On the axis of a coil

An interesting result is obtained if we measure the flux density at a distance s along the axis of a coil, repeating the measurement for a number of small coils and solenoids, but always at the same distance s chosen so as to be large compared with any of the dimensions of the coils. It is then found that the flux density is proportional to the product (ANI), where A is the area of each of the N turns of a coil. This quantity we have previously called the *electromagnetic moment m* of a coil (p. 68), since it is the quantity that decides the moment T of the couple that acts on it when it is placed in a magnetic field. Evidently the magnetic field produced by a coil also depends on its electromagnetic moment m.

If we now experiment with a relatively large flat circular coil, we find that the flux density B along its axis is given by

$$B \propto \frac{m}{y^3}$$

where m is its electromagnetic moment, and y is the distance from any point of the wire of the coil to the point P (Fig. 6.10).

$$\therefore\ B = \frac{k''m}{y^3}$$

where k'' is yet another constant related to μ_0. This result applies not only for large distances, but also for points close to the coil—provided always they are on its axis. For the special case when the point P is at the centre of the coil, we have

$$y = r$$

and

$$m = \pi r^2 NI$$

Therefore, at the centre,

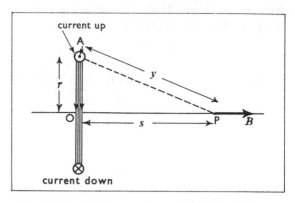

current up

A

current down

Fig. 6.10   Deriving an expression for the field on the axis of a flat circular coil

$$B = \frac{k'' \pi r^2 NI}{r^3}$$

$$= \frac{k'' \pi NI}{r}$$

Comparing this with the previous result:

$$B = \frac{\mu_0 NI}{2r}$$

we have

$$k'' = \frac{\mu_0}{2\pi}$$

Thus the field along the axis of a flat circular coil is given by

$$B = \frac{\mu_0 m}{2\pi y^3}$$

At *large* distances s along the axis of *any* coil or solenoid, we have

$$y \approx s$$

$$\therefore\ B \approx \frac{\mu_0 m}{2\pi s^3}$$

As with all other results for coils or solenoids, it is found that the same result holds also for magnets. Thus at distances large compared with the dimensions of the magnet the flux density is given by the same expression. To test this the electromagnetic moment m of the magnet must be measured by finding the moment T of the couple that acts on it when it is placed with its axis *at right-angles* to a known field B' (p. 68). We then have

$$m = \frac{T}{B'}$$

The flux density B due to the magnet can then be measured at different distances s along its axis;

and we can show that B is given by the same expression as for the equivalent coil:

$$B = \frac{\mu_0 m}{2\pi s^3}$$

6.5   Ampère's law

An alternative way of analysing the magnetic fields of electric currents was devised by Ampère. It provides further insights into the nature of the magnetic field, and is an invaluable tool for calculation in certain cases, particularly in engineering applications.

Ampère's law makes use of a quantity called the *line integral of B round a closed path* in the field. To find this quantity for a closed path such as that shown in Fig. 6.11 we imagine the path divided into short elements such as δl. For each element we then form the product of the length δl and the component of B parallel to it. The line integral round the path is the sum or integral of these products taken right round the path. Thus if the angle between B and the element δl is θ, we can write

$$\text{line integral} = \int B . \cos \theta . dl$$

the integral being taken right round the path. This is usually expressed briefly by the notation

$$\text{line integral} = \oint B . dl$$

Ampère's Law. The line integral of the magnetic flux density round any closed path in a magnetic field divided by the magnetic constant μ_0 is equal to the total current linked with the path.

If the path in question is linked once with a coil of N turns carrying a current I, the law states

$$\frac{1}{\mu_0} \oint B . dl = NI$$

This result may be proved quite generally from the Biot-Savart law; indeed the two laws can be shown to be alternative and equivalent statements. However, the mathematics involved is difficult and beyond the scope of this book. We shall content ourselves with referring to a practical demonstration of the validity of the law. It makes use of a device known as *Rogowski's belt*. This consists of a long solenoid wound uniformly on a flexible former such as a length of plastic tubing. The solenoid is in two

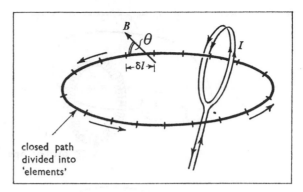

closed path divided into 'elements'

Fig. 6.11   Ampère's law for the line integral of B round a closed path

layers so that the connections can be made to it at the same end. The belt itself is bent into the shape of the 'path' specified in Ampère's law, and the total flux of the field linked with it can be shown to be proportional to the line integral of B round the path. For suppose there are n turns per unit length of the belt, each of area A, then an element of the belt of length δl contains $n\delta l$ turns; and the flux $\delta \Phi$ linked with it is given by

$$\delta \Phi = B . \cos \theta . An \, \delta l$$

The total flux Φ linked with the whole length of the Rogowski belt is therefore given by

$$\Phi = An \int B . \cos \theta . dl$$

If the two ends of the belt are brought together to form the closed 'path' required, we then have

$$\Phi = An \oint B . dl$$

Since A and n are constants of the belt, the flux Φ may be used as a measure of $\oint B . dl$, and we have

$$\frac{1}{\mu_0} \oint B . dl = \frac{\Phi}{\mu_0 An}$$

$$= NI \quad \text{(according to Ampère's law)}$$

The simplest way of using the belt is to supply the circuit whose field is under investigation with alternating current, and to treat the Rogowski belt as an a.c. search coil—using an oscilloscope to measure the peak e.m.f. E_0 induced in the belt.

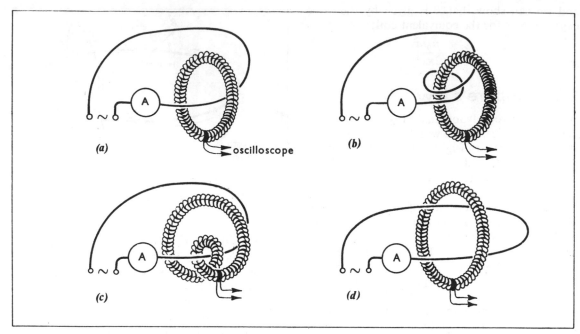

Fig. 6.12   The use of Rogowski's belt to test Ampère's law: (*a*) belt and current linked once; (*b*) and (*c*) belt and current linked twice; (*d*) no linkage at all, and no e.m.f. induced

Without bothering to find the constants of the belt we can soon show that the flux Φ linked with it is independent of the actual path it follows, as long as it is a closed one and is linked with the current N times. Furthermore, if N is varied, we find

$$\Phi \propto N$$

Thus in Fig. 6.12 the height of the oscilloscope trace in (*b*) is twice what it is in (*a*); and in neither case is the induced e.m.f. affected by moving the belt or the wire in any manner whatever—as long as they remain linked as shown. The same e.m.f. is obtained in (*b*) and (*c*), since in both cases the current is linked *twice* with the belt—indeed the arrangement in (*c*) may be changed into that of (*b*) without undoing the belt just by straightening one of the loops in it. In (*d*) the belt and wire are not linked at all, and the e.m.f. is precisely zero.

If the constants A and n of the belt are known, we may use the measurements to confirm Ampère's law quantitatively. Thus if the current I in the circuit under test is given by

$$I = I_0 \sin \omega t$$

Ampère's law gives

$$\frac{\Phi}{\mu_0 An} = NI$$

$$\therefore \Phi = \mu_0 AnNI$$
$$= \mu_0 AnNI_0 \sin \omega t$$

The e.m.f. E induced in the belt is given by

$$E = -\frac{d\Phi}{dt} = -\mu_0 AnNI_0\omega \cos \omega t$$

Therefore the peak e.m.f. E_0 is given by

$$E_0 = \mu_0 AnN\omega I_0$$

Measurement of E_0 and I_0 thus enables us to show that the magnetic constant referred to in Ampère's law is indeed the same as the universal constant μ_0.

Another important point to test is that the flux Φ linked with the belt is not affected by the presence of iron or magnets of any kind. The line integral of B round the closed path depends only on the *actual current* linked with the path. (It would be a different matter if the iron were allowed *inside* the tube of the belt; but as long as the belt itself contains air the presence of iron nearby does not affect the result.)

To illustrate the application of Ampère's law we shall use it to give quick proofs of some of the results we have already considered.

(*i*) *A long straight wire.* Consider a closed circular path of radius r concentric with the wire, which we suppose to carry a current I (Fig. 6.13a). By symmetry the flux density may be assumed to be of constant magnitude B everywhere along this path. Ampère's law then gives

$$\frac{1}{\mu_0} \oint B.dl = \frac{B.2\pi r}{\mu_0} = I$$

$$\therefore B = \frac{\mu_0 I}{2\pi r}$$

(*ii*) *A coaxial cable.* First of all consider the outer cylindrical conductor alone; we can establish the point that the field it produces *inside* the cylinder is zero. The lines of force of the field inside (if any) must by symmetry lie in concentric circles, and the field must be constant at a given radius (Fig. 6.13b). In this case

$$\oint B.dl = 0$$

since the path links no current

$$\therefore B = 0 \quad \text{everywhere inside.}$$

If now the inner conductor of the coaxial cable is considered, the path links this in the same way as for the single long straight wire; and the field

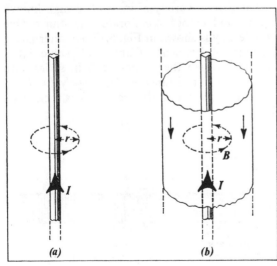

Fig. 6.13   Deducing the field (*a*) of a long straight wire; (*b*) of a coaxial cable

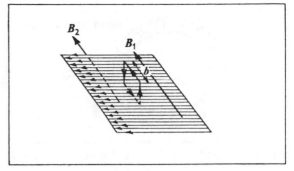

Fig. 6.14   Deducing the field of a current sheet using Ampère's law

between the conductors is therefore given by the same expression:

$$\therefore B = \frac{\mu_0 I}{2\pi r}$$

Outside the outer conductor the field is zero, since any circular path links a total current of zero (I in each direction).

(*iii*) *A current sheet.* Let the current sheet be of strength i (p. 105), and let the flux densities on the two sides of it be B_1 and B_2, as shown in Fig. 6.14. Consider a rectangular path at right-angles to the current flow, and with its top and bottom sides (of length b) parallel to the surface and just above and below it respectively; the ends of the rectangle are thus very short. The current linked with this path is bi.

$$\therefore \frac{1}{\mu_0} \oint B.dl = \frac{1}{\mu_0}(B_1 b - B_2 b) = bi$$

$$\therefore B_1 - B_2 = \mu_0 i$$

showing that the flux density changes by an amount $\mu_0 i$ as we pass through the current sheet of strength i.

(*iv*) *A long solenoid.* Consider a coil of N turns uniformly spaced round the circumference of a torus (Fig. 6.15). By symmetry the flux density is of constant magnitude B at all points on the closed circular path shown, which is of total length l. In this case the current I in the coil links N times with the path. Ampère's law therefore gives

$$\frac{1}{\mu_0} \oint B.dl = Bl = NI$$

$$\therefore B = \frac{NI}{l}$$

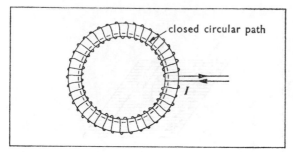

Fig. 6.15   Deducing the field of an endless solenoid

In this case the flux density B varies over the cross-section of the coil, since the length l of a suitable circular path depends on the part of the cross-section considered.

We may pass from this to the case of a long *straight* solenoid by supposing the radius of the torus to be increased indefinitely, while keeping the cross-section of the coil and the number of turns per unit length (N/l) constant. The expression for the field does not change, but in the limit the length l of the path becomes constant for all parts of the cross-section. This shows that the field is constant over the whole cross-section of a long straight solenoid.

We shall see further applications of Ampère's law in the next chapter when we consider the effects of iron in a magnetic system—and this is the main engineering application of the law.

6.6   Calibrating the standards

The ultimate calibration of electrical standards by the standards laboratories depends on finding electromagnetic effects that can be precisely predicted from the geometry of the set-up using only the laws of electromagnetism and the definition of the ampere (the basic electrical standard). Such a measurement must then be translated into a calibration of the practical working standards, namely the standard *ohm* and the standard *cell*, the working standard for the volt (p. 43).

Most of these techniques depend ultimately on a calculation of mutual inductance, which can always be done in principle for a given arrangement of two circuits. The standard mutual inductance is constructed with great care on an accurately machined marble former, so that its

geometry is defined with great precision. However, it is instructive to perform a similar calculation for a simple school laboratory mutual inductance. Suppose this is constructed as in Fig. 6.16. The primary coil consists of a long uniformly wound solenoid of total length l containing N_1 turns of area A. The secondary consists of a short coil of N_2 turns wound round the centre part only of the primary (the part where the field is truly uniform). The flux density B in the core is therefore given by

$$B = \frac{\mu_0 N_1 I_1}{l}$$

where I_1 is the current in the primary. The flux linkage Φ_2 with the secondary is given by

$$\Phi_2 = BAN_2$$
$$= \frac{\mu_0 AN_1 N_2 I_1}{l}$$

Now the mutual inductance M is equal to the flux linkage with the secondary per unit current in the primary; i.e.

$$M = \frac{\Phi_2}{I_1}$$

$$\therefore M = \frac{\mu_0 AN_1 N_2}{l}$$

The area A here is the *mean* area of cross-section of the primary coil within which the flux is confined at the centre.

There are a variety of ways of using a known mutual inductance to calibrate a resistance. One simple way is shown in Fig. 6.17. An alternating current (of frequency f and peak value I_0) is passed through the primary coil in series with the unknown resistance R. The peak value E_0 of the e.m.f. induced in the secondary is then given by

$$E_0 = MI_0 \omega$$

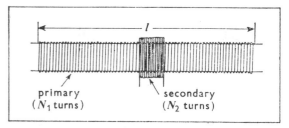

Fig. 6.16   A calculation of mutual inductance

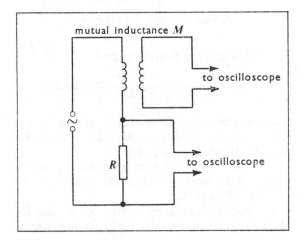

Fig. 6.17   An alternating current method of comparing mutual inductance and resistance

where $\omega = 2\pi f$. The height of the trace on the oscilloscope screen corresponding to this peak e.m.f. is measured. Then the oscilloscope is joined instead across the unknown resistance R, and the height of trace corresponding to the peak p.d. V_0 across this is also measured. Now

$$V_0 = I_0 R$$

The ratio of trace heights is equal to the ratio of p.d.'s, and so we obtain the value of E_0/V_0; we then have

$$\frac{E_0}{V_0} = \frac{M\omega}{R}$$

Thus R is found in terms of M and the frequency f.

A more sophisticated way of standardizing a resistance is that due to Lorenz, in which a Faraday disc is mounted inside a solenoid (Fig. 6.18). The axle of the disc is arranged to coincide with the axis of a long solenoid and the apparatus is turned so that the plane of the disc is exactly in the magnetic meridian. The e.m.f.'s induced in it are then due to the field of the solenoid only. Sliding contacts are made to bear on the rim and axle, as shown; if the distances of these contacts from the central axis are r_1 and r_2, and the disc performs n revolutions per second, then the rate of sweeping out area by the part of a radius of the disc between the two contacts is equal to

$$n\pi(r_1{}^2 - r_2{}^2)$$

The flux density B in the solenoid is given as before by

$$B = \frac{\mu_0 NI}{l}$$

The induced e.m.f. E is equal to the rate of cutting flux by the part of the radius concerned. This e.m.f. is balanced against the p.d. across the small resistance R in series with the solenoid, through which the same current I therefore flows; the adjustment is effected by varying the speed of rotation until the galvanometer gives no deflection.

$$\therefore E = IR = \frac{\mu_0 NIn\pi(r_1{}^2 - r_2{}^2)}{l}$$

$$\therefore R = \frac{\mu_0 Nn\pi(r_1{}^2 - r_2{}^2)}{l}$$

The resistance R is thus found in terms of the dimensions of the apparatus and the speed of rotation n. Again, the calculation amounts to finding the mutual inductance between the solenoid and the disc circuit.

To make the measurement fully satisfactory the simple apparatus in Fig. 6.18 needs many refinements. Friction at the sliding contacts produces heat which leads to thermoelectric e.m.f.'s (p. 45); these are by no means negligible compared with the small induced e.m.f. E. In practice only a small part of the rather large current I can be allowed to pass through the resistance R; a resistance network is therefore used in the circuit instead of the single resistor R. Allowance must also be made for the slight non-uniformity of the magnetic field inside the

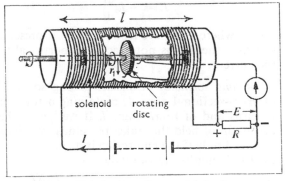

Fig. 6.18   The Lorenz disc method of measuring resistance

solenoid. However, in a suitably refined form this experiment has proved one of the most satisfactory yet devised for measuring the resistance of the coils maintained as standards in the standards laboratories.

To find the e.m.f. of a standard cell as well, we must compare it (by means of a potentiometer, p. 55) with the p.d. across one of the calibrated resistors when a known current is passing through it. A current balance of the type described on p. 10 is used for the measurement of current. The current I to be measured is passed through the fixed and suspended coils in series, and the deflecting force F is given in terms of the mutual inductance between these sets of coils:

$$F = I^2 \frac{dM}{dx} \quad \text{(p. 101)}$$

Again the measurement comes down to a calculation from first principles of the mutual inductance between two sets of coils.

6.7 Calculations of self-inductance

In general the calculation of inductance presents considerable mathematical difficulties. We shall illustrate the technique by dealing with two important examples. The method is to find the total flux linking the coil for a given current I flowing in it; then we make use of the expression

$$\Phi = LI \quad \text{(p. 96)}$$

where L is the inductance. This is equivalent to using the defining formula

$$E = -L \frac{dI}{dt}$$

except where there are iron-cored coils present, in which the flux is not exactly proportional to the current.

(*i*) *A long solenoid.* Consider a long solenoid of cross-sectional area A, containing a total of N turns and of total length l. If we ignore the reduction of field that takes place near the ends (justified only if the end region is a small part of the total length), we can write

$$\text{flux density } B \text{ in the core} = \frac{\mu_0 NI}{l}$$

$$\therefore \text{ flux through the core} = \frac{\mu_0 ANI}{l}$$

$$\therefore \text{ flux linkage } \Phi \text{ with the coil} = \frac{\mu_0 AN^2 I}{l}$$

$$\therefore L = \frac{\Phi}{I} = \frac{\mu_0 AN^2}{l}$$

Two results follow from this; they apply as a matter of fact not only to long solenoids but also to coils of other types as well.

(*a*) For coils of the same dimensions, the inductance is proportional to the square of the number of turns N.

(*b*) For geometrically similar coils, the inductance is proportional to the linear dimensions.

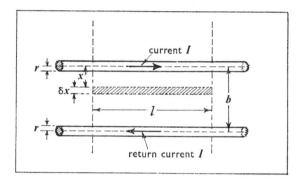

Fig. 6.19   Deducing the inductance per unit length of two parallel wires

(*ii*) *A pair of long, parallel wires.* This is an important practical case, since the transmission lines used for many purposes are of this kind (p. 286). To simplify the analysis we shall suppose that the current flowing in the wires is of sufficiently high frequency to be effectively confined to the surface layers of the wires (p. 94); there is then no magnetic field *inside* the wires. We shall also assume that the separation b of the wires is much greater than their radius r; this means that the current in one wire will not significantly affect the uniform distribution of current over the surface of the other. [When the wires are close together, an effect similar to the skin effect produces a non-uniform distribution of current over their surfaces—the so-called *proximity effect*.]

If the current in the wires is I, the flux density

B produced by *one* wire at a distance x from it is given by

$$B = \frac{\mu_0 I}{2\pi x} \quad \text{(p. 107)}$$

The flux $\delta\Phi$ passing through a rectangular element of width δx and length l in the plane of the wires (Fig. 6.19) due to the current in *one* wire is given by

$$\delta\Phi = \frac{\mu_0 I}{2\pi x} . l \, \delta x$$

Integrating this between the limits r and $(b - r)$ we obtain the flux passing through the gap between the wires in the length l. Each wire produces the same contribution to the flux, so that the total flux Φ linked with the length l of the transmission line is given by

$$\Phi = 2 \times \frac{\mu_0 I l}{2\pi} \int_r^{b-r} \frac{dx}{x}$$

$$\therefore \Phi = \frac{\mu_0 I l}{\pi} \log_e \left(\frac{b - r}{r} \right)$$

$$\approx \frac{\mu_0 I l}{\pi} \log_e \frac{b}{r} \quad \text{(since } r \ll b)$$

The inductance L of this length of cable is therefore given by

$$L = \frac{\Phi}{I} = \frac{\mu_0 l}{\pi} \log_e \frac{b}{r}$$

and the inductance per unit length is

$$\frac{L}{l} = \frac{\mu_0}{\pi} \log_e \frac{b}{r}$$

In almost all practical cases the factor $\log_e (b \, r)$ lies between 3 and 6. The inductance of such a transmission line therefore usually falls between 1.2 and 2.4 $\mu\text{H m}^{-1}$.

7 Magnetic Materials

7.1 Ferromagnetism

When a piece of unmagnetized iron is placed in a magnetic field, it becomes at least temporarily magnetized. One way of doing this is to place the iron near the poles of a magnet. For example, the north pole of the magnet in Fig. 7.1a induces a south pole on the end nearest to it of the bar of iron (and a north pole on the other end). The induced south pole of the bar is then attracted to the north pole of the magnet. It is by this process that a magnet picks up an unmagnetized piece of iron, such as an iron tack.

The induced magnetism of the iron bar produces a magnetic field that is superimposed on that of the magnet, with the result shown in Fig. 7.1a. The lines of force are concentrated together at the ends of the bar, where the field is greatly increased. But the space along the sides is relatively denuded of lines of force, and the resultant field here is very weak. In this way a block of iron can act as an effective shield for magnetic fields. If some object is completely enclosed in iron sheet, it is to some extent screened from the action of external magnetic fields.

Another way of magnetizing a bar of iron is to place it inside a solenoid. When a current passes, it produces a magnetic field along the axis of the coil, and the bar is magnetized accordingly (Fig. 7.1b). The magnetic field of the bar then considerably augments that of the solenoid. An iron core with a magnetizing coil wound around it is called an *electromagnet*; its strength can be adjusted by controlling the current, and it can be used to exert large mechanical forces (Fig. 7.2).

Magnetic materials differ very much in the ease with which they can be magnetized and demagnetized. In the early investigations of magnetism the extremes of magnetic properties were represented by *soft iron* and *hard steel*. Soft iron is very easily magnetized and demagnetized. Hard steels require large magnetic fields to magnetize them, and then retain their magnetization well. Soft iron is therefore a suitable material for the core of an electromagnet, since it requires only a small current to magnetize it. Hard steels are used for making permanent magnets. Modern magnetic materials show much more extreme properties than soft iron and hard steel, but the words 'soft' and 'hard' are still used (in a magnetic sense only) to describe the properties of these new materials. Among the modern 'soft' materials is the alloy known as *mumetal* (74% nickel, 20% iron, 5% copper, 1% manganese); *physically* it is a hard substance, but *magnetically* it is very soft. Even the earth's weak magnetic field is sufficient to reverse the magnetization of a rod of mumetal when it is turned over in the magnetic meridian. Mumetal

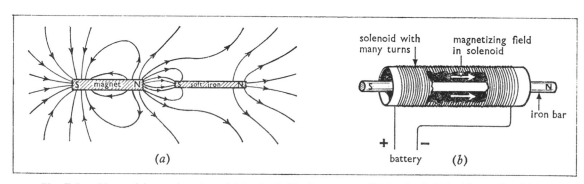

Fig. 7.1   Magnetizing an iron bar: (a) in the field of a magnet; (b) by the field inside a solenoid

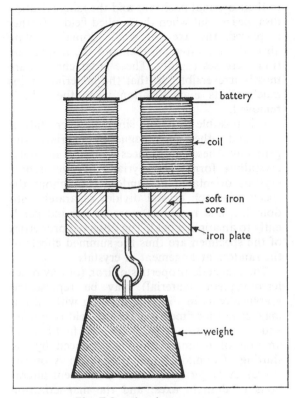

Fig. 7.2   An electromagnet

is one of the best materials to use for making screens to shield pieces of apparatus from stray magnetic fields. Typical of the modern 'hard' magnetic materials are the alloys in the *alnico* range. These are magnetized in manufacture and are then very hard to demagnetize. Compact and powerful magnets can be made with them.

Although the atoms of many elements behave individually as magnetic dipoles, with most materials in bulk only a very small magnetic effect is observed. It appears that the thermal agitation of the atoms is generally sufficient to cause the magnetic axes of the atoms to be arranged almost entirely at random even when strong magnetic fields are applied in an attempt to align them in one direction (p. 68). However, in iron, nickel, cobalt, a few of the rare earths, and some other substances another effect operates. These are called *ferromagnetic* materials. Their characteristics property is that the interactions between neighbouring atoms are such as to align them with their magnetic axes parallel to

one another; and this is so even when no external magnetic fields are applied. Within any one crystal of the substance there is thus a tendency for all the atoms to be lined up in one direction. The actual result is not as simple as this, however, because in a *large* uniformly magnetized crystal strong additional forces would come into play tending to break down the alignment of the atoms. The nature of these forces may be demonstrated if an attempt is made to magnetize a bundle of parallel iron rods—steel knitting needles, for instance. This may be done by bringing the bundle up to one pole of a powerful magnet (Fig. 7.3). The needles are all magnetized in the same sense; and, when they are released, the forces of repulsion between their lower ends cause them to splay out so that their axes are no longer parallel. The same forces operate within a crystal of a ferromagnetic material opposing the natural tendency of the atoms to arrange themselves with their axes parallel. The equilibrium structure of the crystal is one in which it divides up into a series of regions called *domains*; in any one domain all the atoms are aligned with their magnetic axes in one direction, but the direction varies from one domain to another in such a

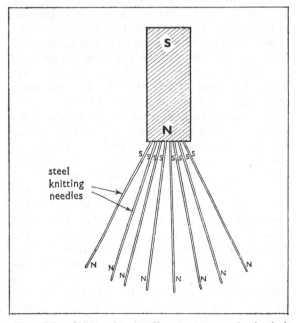

Fig. 7.3   When the needles are magnetized, their lower ends repel one another, and the needles splay out

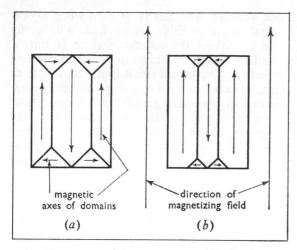

Fig. 7.4   (a) An 'unmagnetized' crystal divides itself into a set of domains whose axes form a pattern of closed loops. (b) In a magnetizing field the domain boundaries move so that more atoms are in line with the field

way that the domain axes form a series of closed loops (Fig. 7.4a). The material is then strained at the boundaries between domains, but this configuration is more stable than a uniformly magnetized crystal would be.

Thermal agitation always tends to disrupt the ordered domain structure. At high enough temperatures the local alignment of the atoms breaks down, and the material ceases to exhibit its characteristic magnetic properties. The temperature at which this happens is known as the *Curie point*. For iron this is 1040 K. But for some ferromagnetic alloys it is below 373 K.

In a freshly formed crystal the magnetic fields of the domains cancel out and virtually no external field can be detected; in this condition we say that the specimen is *unmagnetized*, although on the atomic level there is a high degree of local magnetization. If now it is placed in a magnetic field, couples act on the atoms tending to align their magnetic axes with the field. For small fields the result is a general shifting of the domain boundaries in such a way that the proportion of atoms lined up with the field increases (Fig. 7.4b). The specimen now produces a magnetic field adding to the field that magnetized it. To a large extent this shifting of the domain boundaries is reversible; when the magnetizing field is removed, the boundaries return

to their original positions and the magnetization disappears. But when the applied field is further increased, the axes of whole domains rotate abruptly one after another and the magnetization increases rapidly. These latter changes are mostly irreversible, so that the material retains much of its magnetization when the field is removed.

It is possible to obtain single large crystals of iron and other ferromagnetic materials. But generally these substances solidify in polycrystalline form—i.e. myriads of microscopic crystals orientated at random throughout the specimen. Each crystal divides separately into domains, and responds more or less independently to an applied field. The observed properties of the specimen are thus the summed effects of the random arrangement of crystals.

The magnetic properties of iron (or any other ferromagnetic material) may be represented graphically as in Fig. 7.5. Starting with an unmagnetized specimen a magnetic field is applied and increased slowly from zero. At first a small amount of magnetization is produced by the shifting of domain boundaries (O to A on the curve). A larger field causes permanent alterations of domain axes, and the magnetization then increases more rapidly (A to B on the curve). But beyond a certain strength of magnetic field virtually all the domains are aligned,

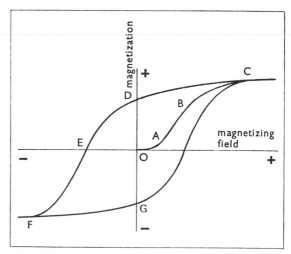

Fig. 7.5   A hysteresis loop showing the way in which the magnetization of a ferromagnetic specimen varies with the magnetizing field

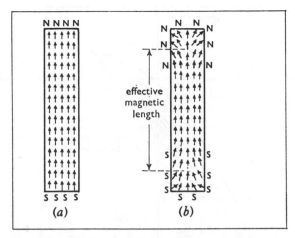

Fig. 7.6   (a) A perfectly hard magnetic material could be made into a uniformly magnetized bar. (b) In practice the magnetization splays out at the ends

and no further increase of magnetization is possible; the magnetization is then said to be *saturated* (C). When the magnetizing field is reduced again, the magnetization does not return to zero, but remains not far below its saturation value (D); an appreciable reverse field has to be applied before it is much reduced again (E). A large enough reverse field changes over the alignment of the domains, and a rapid change of magnetization then occurs until the specimen is saturated in the reverse direction (F). A similar sequence is followed when the field is once more reversed to produce saturation in the original direction (FGC). Thus the magnetization of a specimen does not depend in a simple way on the magnetizing field, but appears to lag behind the changes in the field. This phenomenon is known as *magnetic hysteresis*; and a magnetization curve, like that in Fig. 7.5 obtained by varying the magnetizing field cyclically, is called a *hysteresis loop*. Taking a piece of material through a cycle of magnetization like this requires the expenditure of a certain amount of energy, which appears in the form of heat in the specimen. For a typical soft magnetic material this amounts to about 2×10^{-2} J per kilogram of material; but for a hard magnetic substance the *hysteresis loss* may be thousands of times greater than this.

The magnetization curves obtained for various ferromagnetic materials are all qualitatively similar to Fig. 7.5. Iron alloys do not vary much in

their saturation magnetizations. But the strength of the magnetic field required to reverse the magnetization may vary over a range of 10^4 or more between the softest and hardest materials.

A bar magnet formed from a perfectly hard magnetic material could have uniform magnetization from one end to the other (Fig. 7.6a). But in practice uniform magnetization cannot be achieved because of the demagnetizing action of the forces operating within the magnet itself. The result is that the magnetization is 'splayed out' at the ends of the bar, and the effective *magnetic length* of the magnet is appreciably less than its physical length (Fig. 7.6b). In a bar of soft magnetic material these internal forces are sufficient to break down the alignment of the domains almost completely, and very little magnetization remains once the magnetizing field is removed.

The demagnetizing effect in a magnet is sufficient to produce a gradual loss of magnetization even with modern hard magnetic materials; the

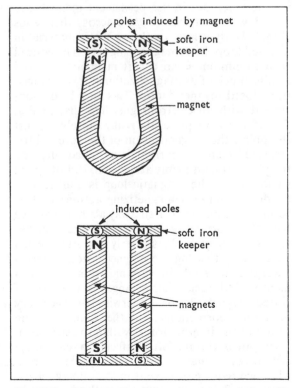

Fig. 7.7   The use of soft iron keepers to reduce the demagnetizing effect in magnets for storage purposes

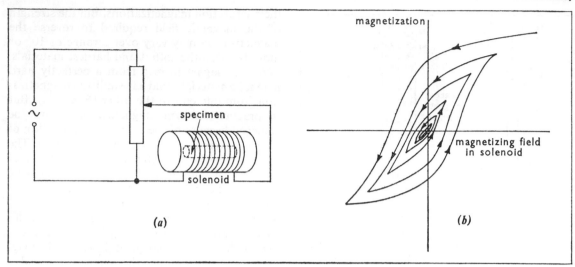

Fig. 7.8 (*a*) An arrangement for demagnetizing an iron specimen. (*b*) As the a.c. field is reduced, the hysteresis loops collapse towards the point where the specimen is completely demagnetized

process is assisted by any mechanical shocks that the magnet receives. In order to avoid demagnetization when not in use permanent magnets are stored with soft iron *keepers* across their ends (Fig. 7.7). By arranging the magnetic material in a closed loop the demagnetizing action is reduced to a minimum. Even a soft magnetic material retains much of its magnetization if it is arranged in a closed magnetic loop. This may be demonstrated with an electromagnet like that in Fig. 7.2. If the iron plate fits really closely against the poles, the magnetization of the core is little reduced when the current is switched off; and the plate remains firmly attached to it. But a very small gap in the magnetic loop is sufficient to produce a large demagnetizing action; and as soon as the plate is prised away a small distance the attraction between core and plate vanishes almost completely. It is clearly important when we are investigating the magnetization of ferromagnetic materials in varying fields to experiment with the materials arranged in the form of closed magnetic loops. Otherwise it is necessary to apply corrections for the demagnetizing action of the internal forces within the specimen.

In spite of this it is often difficult to remove the last traces of magnetization in a piece of iron, when we require to have it in the unmagnetized condition. We then adopt the method shown in Fig. 7.8. The bar to be demagnetized is placed inside a solenoid connected to an a.c. supply, and the current in it is gradually reduced to zero. The bar is thus taken round a series of minor hysteresis loops of steadily decreasing amplitude; and the final result is complete demagnetization.

7.2   Permeability

The effect of filling the core of a coil or solenoid with iron is to increase the flux density B produced by it for a given current. Because of hysteresis the flux Φ in the core is not in general proportional to the magnetizing current I in the coil (Fig. 7.9). The result of this is that electrical components with iron cores, such as transformers and chokes, always introduce some distortion into the waveforms that the circuits are handling. This is very obvious if the input and output waveforms of an a.c. mains transformer are compared, particularly if it is deliberately run for a short time at higher input p.d.'s than it is strictly designed for. If the input current is sinusoidal at a frequency of 50 Hz, the output p.d. is a complex waveform including a strong component of the third harmonic at 150 Hz. In some applications it is essential to have minimum waveform distortion over a wide range of currents and frequencies—for instance in the output transformer of an amplifier. It is then necessary to introduce a narrow air gap in the ferromag-

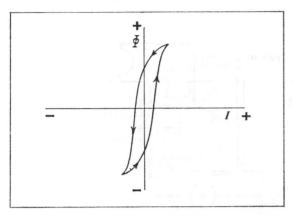

Fig. 7.9   The variation of the flux Φ in a coil with the current in it when it has a ferromagnetic core

magnetizing current I. At best this is an approximation to the real situation; but it is quite a good approximation in many cases, and is easily analysed.

In a closed loop of an ideal magnetic material of this sort the flux density B is increased by a constant factor compared with what it would be if only air filled the coil. This factor is called the *relative permeability* of the material, denoted by the symbol μ_r. If the coil round the specimen ring is in the form of a uniform toroidal solenoid (Fig. 7.11*a*), the flux density B inside the specimen is therefore given by

$$B = \frac{\mu_r \mu_0 N I}{l} \quad \text{(p. 104)}$$

where N is the total number of turns of the solenoid, and l is the total length, i.e. the mean circumference of the toroid.

The product $\mu_r \mu_0$ is called the *absolute permeability* of the material, and is sometimes denoted by the single symbol μ. Thus

$$\mu = \mu_r \mu_0$$

Notice that the relative permeability μ_r is a dimensionless quantity; for a vacuum it is equal to 1, by definition. But the absolute permeability μ is to be measured in the appropriate units (N A^{-2}), and for a vacuum is equal to the magnetic constant μ_0 ($4\pi \times 10^{-7}$ N A^{-2}). The magnetic constant μ_0 is therefore often referred to as the permeability of a vacuum (or of free space).

In the absence of magnetic material in the coil

netic core, as in Fig. 7.10. This introduces some demagnetizing effect, reducing the flux through the core for a given current. But it also has the effect of tilting the curve of flux Φ against primary current I, and provided the core is not taken to saturation the curve now approximates sufficiently to a straight line. The full analysis of such situations, in which we have incomplete magnetic loops, is rather complex and in this book we shall confine our attention almost entirely to coils wound on *closed loops* of ferromagnetic material, so avoiding the complications that arise in analysing the demagnetizing effects in specimens of other shapes.

In this section we shall consider only ideal soft magnetic materials in which we suppose the flux density B to be exactly proportional to the

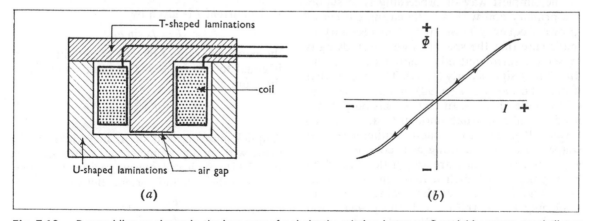

Fig. 7.10   By providing an air gap in the iron core of a choke the relation between Φ and I becomes nearly linear

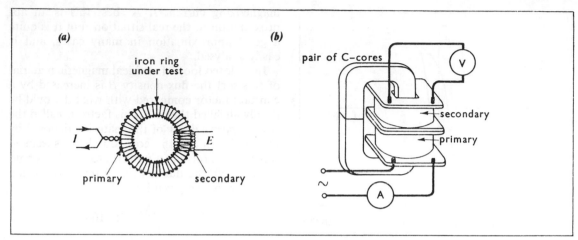

Fig. 7.11   Measuring the relative permeability of ferromagnetic rings: (a) for an iron ring uniformly wound with a primary coil; (b) for a pair of C-cores linking the primary and secondary coils

of Fig. 7.11a the flux density B_0 would be given by

$$B_0 = \frac{\mu_0 N I}{l}$$

which is a quantity we can calculate from the current I, etc. With an iron core we then have

$$B = \mu_r B_0$$

The measurement of the relative permeability μ_r thus consists in finding the flux density B in the ring for a given magnetizing current I. This may be done by winding a secondary coil round part of the torus and investigating the e.m.f. induced in this when the current in the main (primary) coil is changed by a known amount.

The simplest way of proceeding is to supply the primary coil with an alternating current of known frequency f. But it is then necessary to make sure that the specimen we are studying is in such a form that eddy currents are not induced in it significantly (p. 91). This means that it must be *laminated* (p. 92) or be in the form of a bundle of wires. In fact for this purpose we need to make up what amounts to an a.c. transformer. It makes little practical difference what shape the core is, as long as it contains no air gaps; and the results are very little affected by having the coils bunched into convenient units as in Fig. 7.11b, rather than being spread out uniformly round the core. If the primary and secondary coils contain N_1 and N_2 turns re-

spectively, the flux density B in the core is given by

$$B \approx \frac{\mu_r \mu_0 N_1 I}{l}$$

where l is the mean length round the closed core of magnetic material. The flux Φ_2 linked with the secondary coil is given by

$$\Phi_2 = B A N_2$$

The e.m.f. E induced in the secondary is therefore given by

$$E = -\frac{d\Phi_2}{dt} = -A N_2 \frac{dB}{dt}$$

$$= -\frac{\mu_r \mu_0 A N_1 N_2}{l} \frac{dI}{dt}$$

If the current I is given by

$$I = I_0 \sin \omega t$$

the *maximum* rate of change of current (dI/dt) is equal to $I_0 \omega$ (p. 85). The peak e.m.f. E_0 is therefore given by

$$E_0 = \frac{\mu_r \mu_0 A N_1 N_2 \omega I_0}{l}$$

A quick measurement of μ_r may therefore be made with an a.c. ammeter and voltmeter as in Fig. 7.11b. Such instruments register the r.m.s. values of the p.d. and current. But we have

$$\frac{E_0}{I_0} = \frac{E_{rms}}{I_{rms}}$$

and it is only this quotient we need to give the value of μ_r. In this sort of experiment we are in effect treating the secondary coil as an a.c. search coil.

Soft ferromagnetic materials mostly have relative permeabilities of over 10^3, and in some cases values up to 10^5 are found.

7.3 Magnetic polarization

The flux density in a ring-shaped specimen such as that in Fig. 7.11*a* can be regarded as being made up of two components:

(*i*) The part produced directly by the current in the solenoid, which would be the same if the magnetic material were removed; we have called this B_0, the flux density in an air-cored solenoid.

(*ii*) A component of the field produced by the magnetic material; this is referred to as the *magnetic polarization* and is denoted by the symbol J. Thus

$$B = B_0 + J$$

With soft ferromagnetic materials the relative permeability μ_r is so large that B_0 provides a negligible proportion of the total flux density B —as long as the material remains unsaturated. We can then write

$$B \approx J$$

The polarization J reaches a maximum value when the material is magnetically saturated; and, if we want larger values of flux density, the iron core can then contribute nothing further. The saturation polarization of iron and its alloys is about 2 T. For producing fields of 10 T or more it is usual to dispense with the iron core altogether, since it can at best contribute only a small part of the required field. An air-cored solenoid is then used, often at a low temperature at which it will be superconducting; and prodigious pulses of current are passed at the moments when the high field is needed.

For an ideal soft magnetic material the magnetic polarization J is directly proportional to the magnetizing field B_0, as long as the material remains unsaturated. Thus

$$J = \chi_m B_0$$

where χ_m is a constant known as the *magnetic susceptibility* of the material.

Now

$$B = B_0 + J = B_0 + \chi_m B_0 = B_0(1 + \chi_m)$$

But we also have

$$B = \mu_r B_0$$
$$\therefore \mu_r = 1 + \chi_m$$

and for ferromagnetic materials μ_r is so large that we can write

$$\mu_r \approx \chi_m$$

In non-ferromagnetic materials χ_m is very small, and μ_r is then always close to unity. It is in this connection that the concept of susceptibility is chiefly of use. For example, for air

$$\mu_r = 1.000\ 000\ 4,$$

and we prefer to write $\chi_m = 4 \times 10^{-7}$. Like μ_r it is of course a dimensionless quantity.

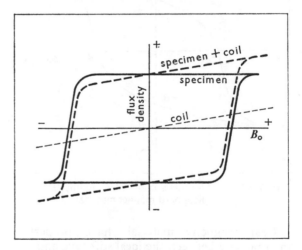

Fig. 7.12   The hysteresis loop of an ideal hard magnetic material

Fig. 7.12 shows the hysteresis loop of a very hard magnetic material, i.e. one whose magnetic polarization J is very little affected by variations of the magnetizing field B_0; J therefore remains constant over a large range of values of B_0. In this case the relative permeability μ_r and susceptibility χ_m have no significance—indeed they both vary over the entire range between $+$ and $-$ infinity in the course of the hysteresis cycle! The value of J must rather be known from the past magnetic treatment the specimen has received; in other words we must know to what point in its hysteresis loop it has been brought.

It is also shown in Fig. 7.12 how the total flux density B varies over the hysteresis cycle. The dashed line shows the component B_0. [This would be a straight line at 45° to the axes if the same scales were used on both axes; but a much larger scale is used for the vertical axis so as to accommodate the larger range of values of B.] The dotted curve then shows the values of $(B_0 + J)$ at every point, i.e. the hysteresis loop of B against B_0. Notice that for the *perfectly hard* magnetic material represented B increases with B_0, because B_0 itself is a component of B.

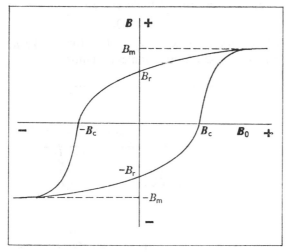

Fig. 7.13   A hysteresis curve of B against B_0 for a typical hard magnet material

Real magnetic materials have properties intermediate between the ideal soft type and the ideal hard. No materials are so soft that their hysteresis loops collapse into straight lines; and none is so hard that J remains constant along the top part of its hysteresis loop. But the properties of any material must be described by the measured curves, and by listing the values of the fields at the most significant points of the hysteresis loop. Both forms of hysteresis curve will be found in technical literature—both the J against B_0 and the B against B_0. Because $B \approx J$, it makes little difference which we use usually. But the electrical engineer generally prefers the curves giving B in terms of B_0, as in Fig. 7.13, since it is only the resultant flux density B that he is concerned with, and it does not matter to him what its components are.

Properties of ferromagnetic materials

(The values obtained depend on the manner of preparation of the specimen, and are therefore only approximate)

Material	$\dfrac{B_m}{T}$	$\dfrac{B_r}{T}$	$\dfrac{H_c}{\text{A m}^{-1}}$	$\dfrac{B_c}{\mu T}$
Iron (99.95% pure)	2.2	1.3	80	100
Cast iron	0.5	0.3	500	630
Mild steel	1.4	0.9	200	250
Silicon iron	1.6	0.8	60	75
Mumetal	0.9	0.6	3	4
Carbon steel	—	1.0	5 000	6 300
Cobalt steel	—	1.0	20 000	25 000
Alnico	—	1.0	60 000	75 000
Nickel	0.7	0.3	400	500
Cobalt	1.5	0.4	800	1 000

The table above gives the values of the important constants that describe the hysteresis loops for a range of soft and hard materials. The first column gives the saturation flux density B_m. The second column gives the flux density B_r that remains after the material has been saturated and the magnetizing field has then been reduced to zero. It is called the *remanence* of the material. It will be noticed that in most materials

$$B_r \approx \tfrac{2}{3} B_m$$

The third column gives the field strength required just to reverse the magnetization of the specimen. This is given in terms of the current per unit length of the solenoid (i.e. the quantity NI/l) that is needed. This quantity is known as the *coercivity* H_c of the material; it is expressed in the table in units of A m^{-1}. The corresponding values of B_0 (denoted by B_c on the curves of Fig. 7.13) are obtained by multiplying H_c by μ_0 ($4\pi \times 10^{-7}$ N A^{-2}); these are given in the last column of the table. It will be noticed that the really big differences between ferromagnetic materials lie in their coercivities, which vary over a range of 20 000 to 1.

We have already seen that the properties of a magnet can be explained entirely in terms of the atomic currents circulating in the individual atoms of which it is made up. The forces and couples that act on a magnet and the field it produces are completely accounted for in this manner. In fact a bar magnet behaves as though it were encircled by a sheet of current, i.e. as though its surface were a solenoid of the same shape and size as the magnet. In the last chapter (p. 105) we found that the flux density B inside a solenoid encircled by a current sheet of strength i is given by

$$B = \mu_0 i$$

Hence we can say that the magnetic polarization J of a bar of material is equivalent to a current sheet of strength J/μ_0 encircling the bar. This quantity is known as the magnetization M of the specimen. Thus

$$M = \frac{J}{\mu_0}$$

The magnetization M is the equivalent current sheet that would produce the same effect as the actual atomic currents circulating in the atoms of the material; M is therefore measured in units of A m^{-1}. If the bar magnet is of length l, the *total* equivalent current flowing in the imaginary sheet encircling it is Ml. The electromagnetic moment m of the magnet may be related to its magnetization M as follows.

The electromagnetic moment m of an ordinary solenoid is given by

$$m = ANI \quad \text{(p. 68)}$$
$$= A \times \text{(total encircling current)}$$

The *electromagnetic moment* m of a bar magnet is therefore given by

$$m = AMl$$

Since the product Al is the *volume* of the magnet, an alternative way of describing the magnetization M is to call it the *electromagnetic moment per unit volume* of the specimen, since we can write

$$M = \frac{m}{Al}$$

[The student is warned that this is an area of the subject in which there has been great confusion in the past. A number of different systems of definitions were in use, and to some extent each author made his own personal selection in the matter. Fortunately the concepts of polarization, magnetization, susceptibility, etc., are not ones that are much needed in a school physics course. But we have given here the currently accepted system of definitions, which can be expected in due course to replace all others.]

Paramagnetism and diamagnetism

All materials exhibit magnetic properties to some extent, although the effects are generally very much smaller than with ferromagnetics. Those substances in which the individual atoms or ions have a net magnetic moment are called *paramagnetics*. The magnetic axes of the atoms or ions tend to align themselves parallel to a magnetic field; but thermal agitation opposes this tendency so that the magnetic susceptibilities of these materials are very small—mostly between 10^{-3} and 10^{-4} at 0°C. The susceptibilities vary inversely with the absolute temperature. At very low temperatures the thermal agitation is weak enough for some paramagnetic materials to exhibit saturation in large fields, just like iron at ordinary temperatures. The only difference between ferromagnetics and paramagnetics is that the latter have no tendency to form a domain structure. At temperatures above their Curie points (p. 120), when the domain structure breaks down, ferromagnetics become paramagnetic, and have susceptibilities of the same order of magnitude as other paramagnetics.

In a magnetic field the forces acting on a paramagnetic specimen are similar to those acting on a piece of soft iron. Thus it is attracted towards the stronger parts of a non-uniform field. This attraction is easily demonstrated with a paramagnetic liquid, such as a concentrated solution of manganese(II) sulphate, contained in a U-tube. If one meniscus is placed between the poles of a powerful electromagnet, as shown in Fig. 7.14, a difference of level of about 15 mm is produced by a field of 0.5 T. Experiments of this sort enable us to find the electromagnetic moment of the individual ions of the solution—in this case the manganese(II) ion.

Those materials in which the atoms or ions have no inherent magnetic moment show an opposite magnetic effect. They are known as *diamagnetics*. In these substances the magnetic moments of the electron orbits are exactly balanced under normal conditions. But when a magnetic field is applied, the motions of the electrons are modified slightly in such a way as to *reduce* the field in the material. This is in fact

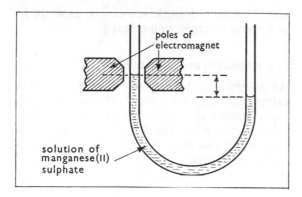

Fig. 7.14   The movement of a paramagnetic liquid in
a magnetic field

an example of the process of *electromagnetic
induction* taking place on an atomic scale. The
susceptibilities of diamagnetic materials are
therefore negative, mostly about -10^{-5}, numerically much less than typical paramagnetic
susceptibilities, and generally independent of
temperature. Presumably the diamagnetic effect
takes place in all materials, including paramagnetics, but is normally swamped by the much
greater paramagnetic effect.

A diamagnetic material is *repelled* from the
stronger parts of a non-uniform field; but the
effect is difficult to demonstrate. If the experiment in Fig. 7.14 is repeated with a diamagnetic
liquid, such as water, there is a just perceptible
fall in the left-hand meniscus when a field of 0.5
T is switched on; but the meniscus needs to be
watched through a low power microscope to see
this.

7.4   The magnetic circuit

We must now turn to the important practical
case of a loop made up of two or more different
materials. An electromagnet is an example, in
which most of the loop is made up of a soft iron
core, but with an air gap which the flux must
traverse. A moving-coil galvanometer provides
another example; here the loop consists of: a
section of permanent magnetic material, soft
iron pole pieces, and two air gaps in which the
sides of the coil move. Arrangements of this
kind in which the magnetic flux is effectively
confined to one or more closed loops are known
as *magnetic circuits*.

Provided the *leakage* of flux from a magnetic
circuit is negligible the total flux through all
cross-sections of it must be the same. Thus the
total flux crossing the air gap of an electromagnet
is equal to the flux passing through its core. If
the cross-sectional area A of the core varies
from one part to another of a magnetic circuit,
the flux density B must vary inversely with A,
since the total flux Φ must be constant throughout. Thus we have

$$\Phi = AB$$

$$\therefore\ B = \frac{\Phi}{A}$$

The calculation of B_0 is now however much
more difficult than for a loop of a single material.
Demagnetizing effects are going to arise, and the
result is that B_0 is not the same at all parts of the
magnetic circuit; it may even be in the opposite
direction to B at some points. We can no longer
calculate B_0 just from the currents linking the
loop; nor can we assume that B_0 is equal to the
value of B in the absence of magnetic material.

However, our experiments with Rogowski's
belt (p. 112) show that Ampère's law still applies
in the presence of magnetic material, as long as
the path along which the line integral of B_0 is
taken is entirely in *air*. We can therefore apply
it to a closed path running in air alongside the
magnetic circuit—or, if we prefer, to a small
imaginary tunnel going down the middle of the
magnetic circuit. For such a path we can still
write

$$\frac{1}{\mu_0} \oint B_0 . \mathrm{dl} = NI$$

where N is the total number of turns of wire
carrying a current I linked with the magnetic
circuit. The quantity on the left-hand side of this
equation is known as the *magnetomotive force*
(or *m.m.f.*) acting in the magnetic circuit, denoted by the symbol F_m. Thus

$$F_\mathrm{m} = \frac{1}{\mu_0} \oint B_0 . \mathrm{dl}$$

Ampère's law states therefore that the magnetomotive force F_m is given by

$$F_\mathrm{m} = NI$$

The unit of magnetomotive force is thus the
ampere (A).

Although we cannot directly calculate the

value of B_0 at any point of the magnetic circuit, we can relate it to the flux Φ in the circuit. For we have

$$B_0 = \frac{B}{\mu_r} = \frac{\Phi}{\mu_r A}$$

where μ_r is the relative permeability of the material in the part of the magnetic circuit considered. If the length of this part of the circuit is l and it is of constant cross-sectional area A, then the contribution to the line integral from the field in this part of the circuit is

$$\frac{1}{\mu_0} \int B_0 \,.\, dl = \frac{\Phi l}{\mu_r \mu_0 A} = \frac{\Phi l}{\mu A}$$

(writing μ for $\mu_r \mu_0$). Now the flux Φ is constant through all parts of the magnetic circuit. Summing the contributions to the line integral from each section of the circuit, we have

$$F_m = \frac{1}{\mu_0} \oint B_0 \,.\, dl = \Phi \sum \left(\frac{l}{\mu A} \right)$$

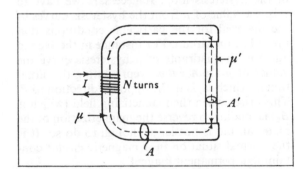

Fig. 7.15   A magnetic circuit of two components

Consider for instance the simple magnetic circuit of two parts shown in Fig. 7.15. It consists of two sections of magnetic material, of lengths round the circuit l and l', cross-sectional areas A and A', and absolute permeabilities μ and μ', as shown. A coil of N turns carrying a current I is wrapped round some part of the circuit; and we assume that the flux Φ through the coil remains entirely confined to the magnetic circuit. In this case we therefore have

$$F_m = \Phi \left(\frac{l}{\mu A} + \frac{l'}{\mu' A'} \right)$$

and Ampère's law gives

$$F_m = NI$$

Knowing the permeabilities and dimensions of the two parts of the circuit we can then calculate the total flux Φ in the circuit; and from this we can obtain the flux density B at any part of it, and hence derive the value of B_0 also.

In a formal way the properties of magnetic circuits prove to be analogous to those of electric circuits. In an *electric* circuit an *electromotive force* is required in some part of the circuit in order to produce an electric current. The effect of the e.m.f. is to produce an electric field, which acts on the charged particles in the circuit to produce the current. The e.m.f. is in fact equal to the line integral of the electric field round the circuit (cf. discussion on p. 297). By analogy, in a *magnetic* circuit we say that a current in the coil wrapped round some part of the circuit produces a *magnetomotive force*, which is given by the line integral of B_0 round the magnetic circuit (divided by μ_0). Ampère's law shows how the m.m.f. is to be calculated from the current in the coil. The two types of circuit obviously differ in that charged particles can move round the electric circuit, whereas nothing actually flows round a magnetic circuit. However, the equations connecting the *flux* in a magnetic circuit with the m.m.f. acting in it are strikingly similar to those connecting the *current* and e.m.f. in an electric circuit; and it is this that makes the analogy a useful one.

Thus for an electric circuit we have

e.m.f. = current $\times$ total resistance

and the resistance R of any section of the circuit of length l and cross-sectional area A is given by

$$R = \frac{\rho l}{A} = \frac{l}{\sigma A} \quad \text{(p. 22)}$$

where ρ is the resistivity and $\sigma \; (= 1/\rho)$ is the conductivity of the material.

$$\therefore \text{ total resistance} = \sum \left(\frac{l}{\sigma A} \right)$$

and we can write

$$\text{e.m.f.} = \text{current} \times \sum \left(\frac{l}{\sigma A} \right)$$

This may be compared with the expression for a magnetic circuit:

$$\text{m.m.f.} = \text{flux} \times \sum \left(\frac{l}{\mu A} \right)$$

The quantity $l/\mu A$ for a part of a magnetic circuit is called the *reluctance* R_m of the piece of magnetic material. Thus

$$R_m = \frac{l}{\mu A}$$

Notice that μ has a function in the magnetic circuit analogous to the conductivity σ in the electric circuit. The permeability μ is seen from this point of view as a measure of the ease with which flux can pass through the material. Thus we have

$$F_m = \Phi R_m$$

The unit of reluctance is therefore the $A\ Wb^{-1}$.

The following example will show how we make use of this method for analysing a simple magnetic circuit.

EXAMPLE   An iron ring of mean perimeter 0.4 m and of cross-sectional area 4×10^{-2} m² is wound with 1500 turns of wire. The relative permeability of the iron may be assumed to remain constant at 2000. Calculate the current required to produce a flux density of 0.3 T in the iron.

How is the result affected if a saw cut 1 mm wide is made in the iron ring?

The total flux Φ required in the ring is given by

$$\Phi = 0.3 \times 4 \times 10^{-4}\ Wb$$
$$= 1.2 \times 10^{-4}\ Wb$$

The reluctance R_m of the ring is given by

$$R_m = \frac{0.4}{2000 \times 4\pi \times 10^{-7} \times 4 \times 10^{-4}}\ A\ Wb^{-1}$$
$$= 4.0 \times 10^5\ A\ Wb^{-1}$$

If the current is I, we have

$$F_m = 1500I = 1.2 \times 10^{-4} \times 4.0 \times 10^5\ A$$
$$= 48\ A$$

$$\therefore I = \frac{48}{1500}\ A = 0.032\ A$$

When the saw cut is made, the reluctance R_m' of the air gap is given by

$$R_m' = \frac{10^{-3}}{4\pi \times 10^{-7} \times 4 \times 10^{-4}}\ A\ Wb^{-1}$$
$$= 20 \times 10^5\ A\ Wb^{-1}$$

The reluctance of the iron ring itself is not significantly reduced.

$$\therefore\ \text{total reluctance} = R_m + R_m'$$
$$= (4 + 20) \times 10^5\ A\ Wb^{-1}$$
$$= 24 \times 10^5\ A\ Wb^{-1}$$

The presence of the saw cut thus increases the reluctance by a factor of 6. To produce the same flux in the iron (and in the gap) therefore requires 6 times the previous current.

$$\therefore\ \text{current} = 0.032 \times 6\ A$$
$$= 0.192\ A$$

This example illustrates well the considerable effect of introducing even a very small air gap into a magnetic circuit.

The usefulness of the concept of the magnetic circuit depends on whether the flux really is constant round the circuit. In practice there is often appreciable flux leakage into the surrounding air (and back again). A magnetic circuit is really analogous to an un-insulated electric circuit immersed in a weakly conducting fluid!

It must also be borne in mind that in a ferromagnetic material the permeability μ is not a constant, but varies from one point to another of the hysteresis loop. If necessary we have to take the value of μ from the hysteresis curves of the material for the particular conditions that prevail in the circuit. For instance in the second and fourth quadrants of a hysteresis curve the values of μ are *negative*, representing the situation in which B_0 is in the opposite direction to B. This arises when the magnetizing field (which is B_0) is tending to reverse the magnetization of the material, but is not large enough to do so. It is the normal situation in a magnetic circuit containing a permanent magnet.

It is instructive to compare the line of force diagrams for the two sorts of field—the B-field, and the B_0-field—in and around a uniformly magnetized bar (Fig. 7.16). Outside the bar the two fields are identical, since $B = B_0$ in air. But inside, the lines of B_0 run in the opposite direction to those of B. The lines of B follow the same pattern as for the equivalent solenoid of the same outline as the magnet. They are always in continuous closed loops (p. 88)—from the north pole of the magnet to the south pole outside it, and back again from south pole to north pole inside (Fig. 7.16a). Turning however to the diagram for B_0 (Fig. 7.16b), there are no actual currents present, and therefore the line integral of B_0 round any closed path in the diagram must be zero. This is only possible if B_0 is in the opposite direction inside the magnet; for we

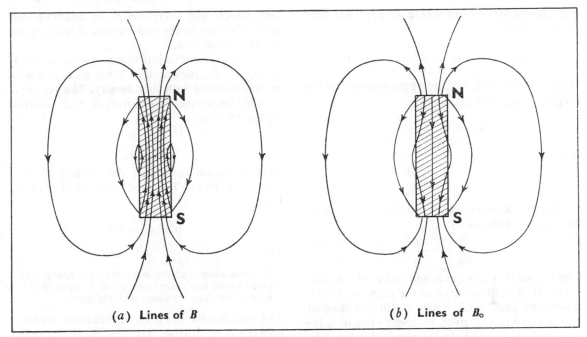

Fig. 7.16   The line of force diagrams of a uniformly magnetized bar: (*a*) the lines of *B*, which form closed loops (*b*) the lines of B₀ that reverse inside the magnet

could choose a path that follows a line of force outside the magnet from one end to the other, and is completed through an imaginary tunnel down the inside of the magnet. Inside the 'tunnel' the field must reverse if the line integral of B_0 is to be zero. Presumably a compass placed in the 'tunnel' would reveal this; but it is in fact revealed in the tendency of a magnet to demagnetize itself with the passage of time. B_0 is the demagnetizing field that produces this effect. Only when the material is formed into a ring do the lines of B and B_0 follow the same closed loops, and there is then no demagnetizing effect. And as we have seen, even a small break in the ring can cause a significant change in the magnetic fields present.

This shows how we may calculate the flux density to be expected in a magnetic circuit that contains a permanent magnet (e.g. the magnetic system of a moving-coil galvanometer). The field is not produced by electric currents, and therefore the m.m.f. in the magnetic circuit is zero. This is possible only if the total reluctance of the circuit is zero also. This comes about through the reluctance of the permanent magnet

being *negative*; B and B_0 are in opposite directions in it, and μ_r (which is equal to B/B_0) is then negative. Fig. 7.17 shows the second (negative) quadrant of the hysteresis curve of the material of the magnet—which is all that we need for this purpose.

Suppose the air gaps in the magnetic circuit have widths totalling x, and that the magnet is of length l. We shall suppose that the reluctance of any soft iron pole pieces is negligible—always

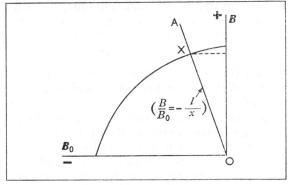

Fig. 7.17   Calculating the flux density in a magnetic circuit containing a permanent magnet

a good approximation where air gaps are concerned. Then

$$\text{total reluctance} = \frac{l}{\mu A} + \frac{x}{\mu_0 A} = 0$$

If the relative permeability of the material of the magnet is μ_r, we then have

$$\mu_r = \frac{\mu}{\mu_0} = -\frac{l}{x}$$

But

$$\mu_r = \frac{B}{B_0}$$

On the hysteresis curve we can therefore draw in the line represented by

$$\frac{B}{B_0} = -\frac{l}{x}$$

This passes through the origin and is of gradient $-l/x$. It meets the curve at the point X, which shows the conditions under which the magnet will operate in the given magnetic circuit, since at this point μ_r will have the required value. The flux density B in the magnet (and in the gaps) can then be read off from the axis. The greater the gap width x, the further the point X moves down the curve and the smaller the flux density B becomes.

7.5 The magnetic field strength H

It is quite possible to continue the analysis of the properties of magnetic materials making use of the two vectors B and B_0 only. However, this obliges us to use a flux density B_0 that lacks the essential property of magnetic flux, namely *continuity*. We have introduced B_0 as a component of the total flux density B. Thus

$$B = B_0 + J \quad \text{(p. 125)}$$

where J (the magnetic polarization) is the component of B due to the magnetization of the material.

As long as we are dealing with a closed loop of a single magnetic material there is no problem. In such a loop J is continuous, and so also therefore is B_0. There are no demagnetizing effects, and B_0 arises entirely from the currents linked with the magnetic circuit. However, in a composite magnetic circuit, J is necessarily dis-

continuous, and therefore B_0 is discontinuous also. The sum of the two vectors B is of course always continuous.

It is convenient to introduce another vector H to replace B_0, measured in other units and with no associations with flux density. The vector H is called the *magnetic field strength*, and is defined by the relation

$$H = \frac{B_0}{\mu_0}$$

This definition is framed deliberately so that Ampère's law can be stated in terms of H in the form

$$\oint H.\mathrm{dl} = NI$$

Thus:

> The line integral of the magnetic field strength H taken round any closed path in air is equal to the total electric current linked with the path.

The unit in which H is to be measured is therefore the A m^{-1}. If we have a solenoid encircled by a uniform current sheet of strength i, then the magnetic field inside it is of strength H given by

$$H = i \quad \text{(p. 105)}$$

The fields due to currents flowing in other configurations can be deduced likewise by dividing the expression for B by μ_0. Thus, using the same symbols as in previous parts of this book, we can write down the expressions for H in the following fields:

Inside a solenoid:

$$H = \frac{NI}{l} = i \quad \text{(p. 104)}$$

Near a long straight wire:

$$H = \frac{I}{2\pi r} \quad \text{(p. 107)}$$

At the centre of a flat circular coil:

$$H = \frac{NI}{2r} \quad \text{(p. 110)}$$

On the axis of a small coil:

$$H = \frac{m}{2\pi s^3} \quad \text{(p. 110)}$$

Inside a piece of magnetic material:

We have $B = \mu_r B_0$
and since $B_0 = \mu_0 H$
we can write $B = \mu_r \mu_0 H = \mu H$

where μ is the absolute permeability ($\mu_r \mu_0$).

It would still not be necessary to introduce another vector to describe the magnetic field if it were not for the irregular behaviour of ferromagnetic materials. If we had no magnetic hysteresis to contend with, μ would be a constant that could be stated precisely for any given material. Knowing μ we could work out the reluctance R_m of any given component of a magnetic circuit; and so we could calculate the flux Φ in the circuit—and all this without bothering with the dubious flux density B_0. As it is, the facts of ferromagnetism force us to use two vectors, since it is only by describing the relation between the two vectors (with a hysteresis curve) that the properties of a ferromagnetic material are specified. We can use B and B_0, as hitherto in this book; or we can use B and H, as described in this section. For the reasons stated, the latter choice is to be preferred; and B and H are the vectors invariably used in electrical engineering.

The student may well be perplexed to find it necessary to have *two* vectors to describe a single physical reality, namely the magnetic field. In point of fact there are *two* independent quantities to be specified: (*i*) the magnetic flux density B that acts on any moving charged particle in the field; (*ii*) the magnetic condition of the material that occupies the field, and itself contributes to it.

But the question should be asked: what exactly do we mean by the magnetic field having 'physical reality'? There have always been those who have asserted that B and H are both figments of our imaginations—Ampère was one of these. The only observed effects, they say, are the interactions of electric currents or of charged particles in motion, and we have no need to invent magnetic fields to occupy the space they move in, let alone suppose the fields to have physical properties apart from the interacting particles. This is a philosophical viewpoint that can be consistently maintained, though it leads to considerable mathematical intricacies. However, the traditional use of the vectors B and H to describe the phenomena of magnetism is also completely consistent; and perhaps this is all that is meant by saying that B and H have physical 'reality'.

A classical illustration

The nineteenth-century scientists who developed the theory of electric and magnetic fields found it a great help to make comparisons with the theory of the flow of fluids. The formal mathematics is virtually identical in the two cases. We shall consider now the obvious features of the behaviour of fluids to show that we should expect to find *two* distinct ways of measuring a magnetic field—or, indeed, any other sort of *vector field*.

Let us consider the flow of water in a river. There are two ways by which the movement of the water at any point might be specified; each is appropriate to a particular sort of calculation.

(*i*) We could draw *lines of flow*, and state the *velocity v* of the water at each point. If we wished to calculate the time taken by a given particle in the water to get from one point to another, it is this approach we should need.

(*ii*) We could state the *mass of water* crossing unit area (placed normal to the flow) per second. If we wished to calculate the total flux of water through a given cross-section of the river, this is the way we should want to measure the flow.

The two methods of measurement are not independent; they are connected by a property of the 'medium'—namely the density of the water ρ. Thus, the mass crossing unit area per second $= \rho v$. This connection is so simple that we should not normally bother to make use of both methods of measurement; but if ρ were not a simple property of the medium, we might have to do so. Similar considerations apply to the analysis of other sorts of vector field; and in this respect magnetic (and electric) fields resemble the 'velocity field' of a fluid. There will always be two ways of measuring such a field, each appropriate to its own type of calculation, and each related to its own class of phenomena:

(*i*) *A line effect*—velocity in a fluid, H in magnetism.

(*ii*) *An area effect*—mass crossing unit area per second in a fluid, B in magnetism.

H is the vector that obeys *circuit* laws like

Ampère's law. B is the vector that obeys *surface* laws, like the law of continuity of magnetic flux.

When the medium has simple properties, the knowledge of a single constant (ρ in fluids, μ in magnetism) enables us to deduce one vector from the other. But when this is not the case, as in ferromagnetism, then both effects must be considered in every situation. Indeed hysteresis effects may well make it impossible to deduce B from H. If both quantities need to be known, they must be found independently by separate methods.

8 Measuring Instruments

8.1 The moving-coil galvanometer

An instrument for detecting or measuring an electric current is called a *galvanometer*. When such an instrument is to be used merely to *detect* currents (as in Wheatstone bridge and potentiometer circuits), it need only be provided with an arbitrary scale marked off in (say) millimetre divisions. Used as a measuring instrument, it may be calibrated to read directly the current in amperes or the p.d. across it in volts, in which case it would be called an *ammeter* or a *voltmeter*. In fact, as we shall see, most ordinary ammeters and voltmeters are simply galvanometers modified by joining suitable resistances in parallel or in series with them. Whether a given instrument is called a galvanometer, an ammeter or a voltmeter depends on the use to which it is to be put and therefore on the manner in which it is connected up in a circuit. The same basic instrument is used in each case. The most common design for modern d.c. instruments is that employing the 'moving-coil' principle.

The pivoted type of instrument

The moving-coil galvanometer consists of a coil wound on a light frame, which is pivoted about an axis in the plane of the coil so that it is free to rotate between the poles of a permanent magnet (Figs. 8.1 and 8.2). The coil is pivoted in jewelled bearings, and its motion is restrained by two hair-springs, one at each end; the hair-springs

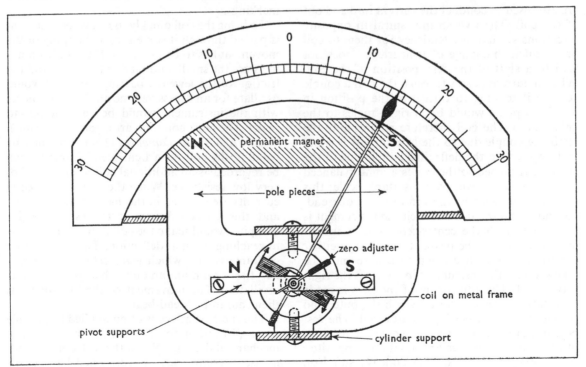

Fig. 8.1   A moving-coil meter (pivoted type)

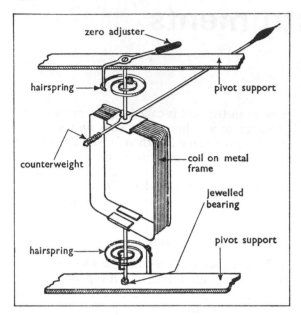

Fig. 8.2 The coil, pivot and hair-spring assembly of a moving-coil meter; the connections to the coil are made through the hair-springs and pivot supports

are also used to conduct the current into and out of the coil. The two springs spiral in opposite directions so that any tendency for them to coil or uncoil with change of temperature does not lead to a shift of the zero position of the coil. When a current passes through the coil a couple acts on it tending to turn it to the position in which its plane would be at right-angles to the field (p. 67). The coil therefore rotates until the restoring couple due to the hair-springs is equal and opposite to the deflecting couple due to the current. Attached to the coil is a light, balanced pointer, which moves over a suitable scale; this may have its zero at the centre or at one end, depending on the use to which the instrument is to be put; with the centre-zero type of instrument current may be passed through it in either direction, but with an end zero care must be taken to join the instrument in a circuit the right way round. The adjustment of the zero is made by rotating a small arm to which the outer end of one of the hair-springs is attached. The rest position of the coil is that in which the tensions of the two hair-springs are equal and opposite.

The magnet is fitted with concave soft iron pole pieces, whose faces are concentric with the

axis of the coil. Also, a soft iron cylinder is fixed centrally between the pole pieces so that it partly fills the space inside the coil. This forms a narrow cylindrical gap in which the sides of the coil can move. The effect of this is to produce what is called a *radial* field; that is, the lines of force in the gaps are along radii to the central axis. Thus, as the coil rotates its plane is always parallel to the field in the gap. In this way the direction of the forces acting on the sides of the coil is in all positions such that the couple is a maximum. Also, by making the width of the gap uniform, the magnetic field in it becomes almost constant in magnitude. The result is that the deflecting couple acting on the coil does not depend on its position in the gap, but only on the current flowing in it. Now, the restoring couple due to the hair-springs is proportional to the deflection. Hence, the use of a radial field makes the deflection proportional to the current, and the instrument then has a uniform scale. This makes it particularly straightforward to calibrate and to read, since the divisions of the scale are evenly spaced, and the fixing of one point on it and the zero decides also all the other markings.

To bring the coil quickly to rest when a current is passed through it some sort of damping of the movement has to be provided. If the only damping were the small amount due to the air and the friction of the jewelled bearings, the coil would oscillate for an excessive time and measurements with the instrument would be very tedious to make. The most common arrangement is known as *electromagnetic damping*. This is provided by winding the coil on a light metal frame; it can be regarded as an additional coil of one 'turn' of very low resistance. When the coil swings, eddy currents are induced in the metal frame (p. 91), and this quickly brings it to rest. Excessive damping would make the coil unnecessarily slow in reaching its final deflection. The ideal design of frame is that which gives *critical damping*— i.e. the minimum amount that just prevents oscillation. The movement of the instrument is then said to be 'dead-beat'.

The same principle is often applied to provide some protection for the galvanometer against mechanical damage. When the instrument is not in use, a copper strip is connected between its terminals, so that the coil and strip form a closed circuit. Any sudden rotation of the coil, such as might result from an accidental blow, then causes currents to be induced in it; this to some extent damps the movement so that it is less likely to lead to damage.

The suspended coil instrument

In very sensitive instruments the slight friction that would occur even with good jewelled bearings would make the movement irregular. It is therefore necessary to dispense altogether with bearings; instead the coil is suspended between two taut phosphor-bronze strips (Fig. 8.3). These are maintained in sufficient tension by the spring strips at top and bottom to keep the coil centred between the pole pieces of the magnet. Connection to the coil is made through the two suspension strips. Instead of a pointer a small plane mirror is fixed to the coil. This is used to cast an image of an illuminated cross-wire on to a scale. Fig. 8.4 shows the optical arrangement used. The bulb illuminates a circular aperture in a metal plate; a fine vertical cross-wire is stretched across this. The short focal length lens

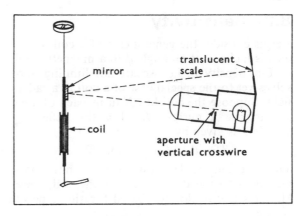

Fig. 8.4   The arrangement of lamp and scale in the box of a mirror galvanometer

casts an image of the aperture and cross-wire on the translucent scale, the beam being reflected from the small mirror on the way. The torsion head carrying the top suspension strip may be rotated to bring the undeflected position of the beam of light to the required point on the scale. The whole galvanometer and optical system is fixed in the one containing box, so that once aligned the optical system needs no further adjustment.

It is not usual in the suspended type of instrument to incorporate any arrangement for producing extra damping; the coil is wound on a non-conducting frame. However, the circuit itself provides a certain amount of electromagnetic damping. When the coil swings, currents are induced in it which damp the movement. If we want the damping to be critical, the effective resistance of the circuit must be adjusted to bring this about. With the coil connected direct to a very low-resistance device, such as a thermocouple, the damping may well be excessive and the movement very sluggish; this can be avoided by connecting a suitable resistance in series with the galvanometer, although this leads to some loss of sensitivity. The value of this resistance is best found by trial and error. On the other hand, in a high-resistance circuit the coil might oscillate excessively; in this case a suitable resistance can be joined in parallel with the instrument to provide a path for the induced currents. Again there is some loss of sensitivity.

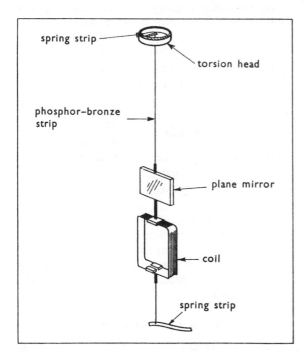

Fig. 8.3   The coil and mirror assembly of a taut-suspension galvanometer

8.2 Sensitivity

Let us consider the general case of a coil whose axis is inclined at an angle ϕ to a magnetic field of flux density B, i.e. we are not limiting ourselves yet to the special case of a coil in a radial field. Suppose the coil has N turns each of area A carrying a current I. Then the deflecting couple T is given by

$$T = BANI \sin \phi \quad \text{(p. 68)}$$

The restoring couple T_k due to the springs is proportional to the deflection θ of the coil, since Hooke's law is closely obeyed for this type of spring.

$$\therefore T_k = k\theta$$

where k is called the *suspension constant*; it is the couple acting per unit deflection. The coil comes to rest in such a position that $T = T_k$.

$$\therefore k\theta = BANI \sin \phi$$
$$\therefore \theta = \frac{BAN \sin \phi}{k} . I$$

In this expression, A, N and k are necessarily constant. To achieve a uniform scale ($\theta \propto I$) we need to ensure that $B \sin \phi$ is constant for all positions of the coil. As we have already shown, this is achieved by the use of a radial field; for in this case $\phi = 90°$, so that $\sin \phi = 1$. Also $B = $ a constant for the uniform gap.

Thus, with a radial field

$$\frac{\theta}{I} = \text{a constant}$$

which is called the *current sensitivity* of the instrument. It is, however, quite common to express the *quality* of the instrument by the reciprocal of the sensitivity I/θ; this would probably be stated in *microamps per division*, and affords a direct indication of the smallest current that could be detected. Thus, for an instrument with a radial field

$$\text{current sensitivity} = \frac{\theta}{I} = \frac{BAN}{k}$$

Sometimes it is more useful to express the sensitivity of the galvanometer in terms of the p.d. V across it. The *voltage sensitivity* is defined as the quantity θ/V. Now, if the resistance of the galvanometer is R, we have

$$I = \frac{V}{R}$$

$$\therefore \theta = \frac{BANV}{kR}$$

$$\therefore \text{voltage sensitivity} = \frac{\theta}{V} = \frac{BAN}{kR}$$

Whether we analyse the behaviour of a galvanometer in terms of its current or voltage sensitivity depends on the application considered. In general we need to take into account the resistance not only of the instrument but of the whole circuit in which it is to be used. The following example shows this:

EXAMPLE   Two moving-coil galvanometers are identical in all respects except that one has a coil of 100 turns of resistance 20 Ω, and the other a coil of 200 turns of resistance 50 Ω. Compare (a) their current sensitivities; (b) their voltage sensitivities; (c) their deflections when joined to a source of constant small e.m.f. of internal resistance 25 Ω.

(a) Now the current sensitivity is proportional to N, since in this case the other factors in the equation are the same for both instruments.

If the current sensitivities are s_1 and s_2, respectively, we have

$$\frac{s_1}{s_2} = \frac{100}{200} = \frac{1}{2}$$

(b) The voltage sensitivities are proportional to N/R; if these are s_1' and s_2' respectively, we have

$$\frac{s_1'}{s_2'} = \frac{100}{200} \times \frac{50}{20} = \frac{5}{4}$$

(c) In this part of the problem we have:

$$\text{deflection} \propto NI$$
$$\text{also} \quad I \propto \frac{1}{\text{total resistance}}$$
$$\therefore \text{deflection} \propto \frac{N}{\text{total resistance}}$$

If the deflections are θ_1 and θ_2 respectively, then

$$\frac{\theta_1}{\theta_2} = \frac{100 (25 + 50)}{200 (25 + 20)} = \frac{5}{6}$$

This shows the general principle to be followed. When the circuit is of low resistance compared with that of the galvanometer, high *voltage* sensitivity is necessary; this usually applies in potentiometer and Wheatstone bridge circuits. But in a circuit of high resistance, the *current* sensitivity must be high.

Measurements on a galvanometer

The measurement of the current or voltage sensitivity of a galvanometer may be carried out with the circuit shown in Fig. 8.5. The resistance

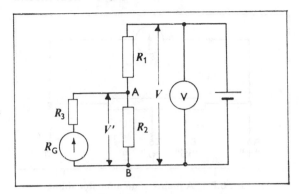

Fig. 8.5   Measuring the sensitivity of a galvanometer

R_3 in series with the galvanometer is made large enough so that (a) the amount of damping is reasonable; (b) the combined resistance (R_3+R_G) is very large compared with the small resistance R_2. We can then assume that the effective resistance between A and B is R_2, with negligible error. R_1 and R_2 form a potential divider so that the p.d. V' between A and B is a calculable fraction of the p.d. V across the accumulator.

$$\therefore \frac{V'}{V} = \frac{R_2}{R_1 + R_2}$$

The current I through the galvanometer is then given by

$$I = \frac{V'}{R_3 + R_G}$$

Measurement of the deflection θ produced by this current gives the current sensitivity; and this quantity divided by the resistance R_G of the galvanometer is the voltage sensitivity. The measurement should be repeated for several points of the scale.

In a practical example we might have

$$\begin{array}{ll} R_1 = 100 \ k\Omega & R_2 = 2 \ \Omega \\ R_3 = 900 \ \Omega & R_G = 100 \ \Omega \\ V = 2 \ V \end{array}$$

$$\therefore \frac{V'}{2} = \frac{2}{100,000} \quad \text{(since 2 is small compared with 100,000)}$$

$$\therefore V' = 4 \times 10^{-5} \ V$$

$$\therefore I = \frac{4 \times 10^{-5}}{1000} \ A = 4 \times 10^{-8} \ A$$

If the deflection θ is 20 divisions,

the current sensitivity

$$= \frac{\theta}{I} = 5 \times 10^8 \ \text{div per amp}$$

$$= 500 \ \text{div per} \ \mu A$$

the voltage sensitivity $= 5$ div per μV

To measure the sensitivity by this method we need first to know the galvanometer resistance R_G. Provided the coil can be clamped, this may readily be found with a Wheatstone bridge arrangement, in which the galvanometer under test forms one arm of the bridge; a second galvanometer is used to obtain the balance. Even with the coil clamped great care must be taken to avoid passing excessive currents; very fine wire is sometimes used in a galvanometer coil, and a small current may overheat it.

When the coil cannot be clamped, the following method due to Lord Kelvin may be used. A Wheatstone type network is connected up as shown in Fig. 8.6, with the galvanometer as one of the four resistances. The resistance in series with the cell must be large enough to give a safe deflection of the galvanometer near the end of its scale. Instead of joining a galvanometer across the middle of the bridge, the points B and D are simply connected together through a tapping key. Now, if the usual bridge balance condition holds,

$$\frac{R_1}{R_2} = \frac{R_3}{R_G}$$

no current flows between B and D even when the tapping key is pressed. Therefore pressing the key can have no effect on the currents in other

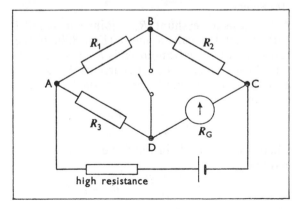

Fig. 8.6   Using a bridge circuit to measure galvanometer resistance

parts of the network; thus, if the bridge is balanced no *change* of the galvanometer deflection occurs when the key is pressed. The other resistances in the bridge are adjusted to bring this about, and so R_G is found.

8.3   D.C. meters

In most modern instruments the basis of the construction is an ordinary moving-coil galvanometer of the pivoted type. This is made as sensitive as possible consistent with reasonable robustness. The required range in amps or volts is then achieved by connecting resistances in parallel or in series with it as described below. A good ammeter or voltmeter is one that disturbs as little as possible the current or p.d. it is required to measure; it must also take as little power as possible. For this reason an ammeter is normally a low-resistance device so that the p.d. across it is small. Likewise, a voltmeter is normally a high-resistance device so that the current taken by it is small.

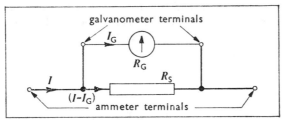

Fig. 8.7   The connection of a shunt to a galvanometer to make an ammeter

Ammeters

A low-resistance shunt R_S is joined in parallel with a suitable galvanometer (Fig. 8.7). If the resistance of the galvanometer is R_G, and the current through it is I_G then the currents through R_S and R_G are inversely proportional to their resistances

$$\therefore \frac{I - I_G}{I_G} = \frac{R_G}{R_S}$$

where I is the total current to be measured. Rearranging,

$$\frac{I}{I_G} = \frac{R_S + R_G}{R_S} = m, \text{ say}$$

The total current I is thus always m times the current I_G registered by the galvanometer. The

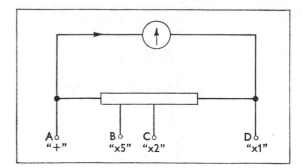

Fig. 8.8   A multi-range ammeter shunt

value of R_S may in this way be chosen so as to make the range of the instrument any desired multiple of that of the unshunted galvanometer. The best type of instrument to make into an ammeter is one that has *high voltage sensitivity*, since this will require only a small p.d. across it to drive it to full-scale deflection, and the resulting ammeter is then of low total resistance.

If the same instrument is to be used on a number of different ranges, it is possible to provide it with several separate shunts; one of these is attached to the galvanometer terminals for each range. But a better method is a universal shunt of the type shown in Fig. 8.8. The resistance R_S is first adjusted to provide the lowest required range of the ammeter; connection for this range is made to the terminals A and D. Further tappings are then made at points along R_S, as shown, to give the higher ranges. For instance, if the resistance of the part of the shunt between A and B is $R_S/5$, it may be shown that the ammeter range obtained by making connection to A and B is 5 times the basic range. Similarly, if the point C is halfway along AD, the range obtained by using the terminals A and C is 2 times the basic range. Such a multi-range instrument would have the terminals A, B, C, D marked '$+$, $\times 5$, $\times 2$, $\times 1$', in that order. The ammeter is always connected so that the current enters at the '$+$' terminal (this is then the positive side of the meter); the other terminal is selected according to the range required.

Voltmeters

A high resistance R_S is joined in series with a suitable galvanometer (Fig. 8.9). If the resistance

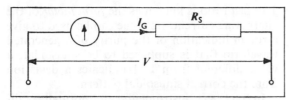

Fig. 8.9 The use of a series resistance to make a galvanometer into a voltmeter

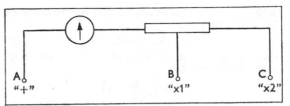

Fig. 8.10 The arrangement of resistors in a multi-range voltmeter

of the galvanometer is R_G and the current through it I_G, then the p.d. V across the combination is given by

$$V = (R_S + R_G)I_G$$

Thus the p.d. V is proportional to the current I_G registered by the galvanometer, whose scale may therefore be calibrated to read p.d.'s directly. The series resistance R_S is chosen to make the instrument operate over the desired range. The best type of galvanometer to make into a voltmeter is one that requires only a small current to drive it to full-scale deflection—i.e. one having *high current sensitivity*.

Voltmeters are often specified by their resistance per volt at full-scale deflection. Thus, if I_0 is the current and V_0 the p.d. that gives full-scale deflection, then

$$\text{resistance per volt} = \frac{R_S + R_G}{V_0} = \frac{1}{I_0}$$

This quantity is therefore a constant for a given meter, the same no matter what series resistance is used with it. A voltmeter should have a resistance of at least 1000 Ω per volt.

The arrangement of resistors for a multi-range voltmeter is shown in Fig. 8.10. The lowest range of the instrument is obtained by making connection to terminal A (the '+' terminal) and terminal B (marked '×1'). By connecting to A and C (marked '×2') the p.d. to be found is given by twice the scale reading; to achieve this the resistance between A and C must be twice that between A and B.

★ *The ohmmeter*
In a multi-range combined ammeter and voltmeter it is usual to make provision also for the instrument to measure resistance directly. The circuit used is shown in Fig. 8.11a; the dry battery is incorporated inside the case of the meter. The variable resistance R_1 is of such a value that the combined resistance of the galvanometer and R_1 is small compared with the series resistance R_2. The ohmmeter is used as follows. First, the external terminals of the meter, A and B, are short-circuited, and the resistance R_1 is varied to produce full-scale deflection. Then the resistance R to be measured is joined to A and B and the current falls. As shown below, the use of this procedure makes the latter current almost independent of the e.m.f. E of the battery. Thus the scale may be calibrated to read the resistance R directly. This type of scale runs 'backwards'; i.e. the resistance is zero for maximum current and infinite for zero current (Fig. 8.11b).

Let the full-scale current flowing through the meter

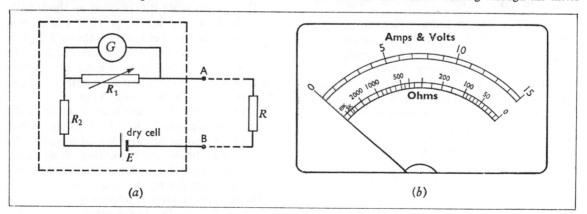

Fig. 8.11 (a) The ohmmeter circuit; (b) the 'ohms' scale of a universal meter

and its shunt be I_0, and let this fall to I when the resistance R is in circuit. Then,

$$E = I_0 R_0 = I(R_0 + R)$$

where R_0 is the total internal resistance of the meter circuit. Rearranging,

$$R = R_0\left(\frac{I_0}{I} - 1\right)$$

Since the resistance of the galvanometer and shunt is small compared with R_2, R_0 is almost unaffected by the small changes in the value of R_1 that occur when the meter is adjusted; and it may therefore be treated as a constant. Knowing R_0 and the currents, the values of R may be calculated in terms of the current readings, and the corresponding points of the scale marked accordingly. Notice that when

$$I = \tfrac{1}{2}I_0 \qquad R = R_0$$

i.e. the reading is R_0 for half-scale deflection.

In addition to the moving-coil instruments considered above, nearly all the instruments designed for a.c. work may also be used for d.c. measurements. Moving-iron meters in particular are much cheaper and more robust than moving-coil instruments, and are used to some extent with direct currents when no great sensitivity or precision is demanded.

8.4   A.C. meters

We have seen in an earlier chapter that there are three ways used for expressing the magnitude of an alternating current or p.d. (p. 29). Thus, we may measure:

(a) the peak value;
(b) the mean (or half-cycle average) value;
or (c) the r.m.s. value.

It is important to know which aspect of an alternating quantity a given meter measures. As long as we are dealing only with sinusoidal waveforms, it makes no difference whether an instrument has its scale marked to give peak, mean or r.m.s. values, since in this case the three quantities bear known ratios to one another (p. 30):

$$\frac{\text{r.m.s. value}}{\text{peak value}} = \frac{1}{\sqrt{2}} = 0.707$$

$$\frac{\text{mean value}}{\text{peak value}} = \frac{2}{\pi} = 0.637$$

$$\frac{\text{r.m.s. value}}{\text{mean value}} = \frac{\pi}{2\sqrt{2}} = 1.11$$

This last ratio is known as the *form factor* of the alternating current. Its value can be used to give a rough indication of the purity of a generated waveform that is supposed to be sinusoidal; a value different from 1.11 indicates a deviation from the correct sinusoidal pattern.

In fact a.c. instruments are normally calibrated to read r.m.s. values, on the assumption that the form factor will be 1.11. However, when using an instrument whose deflection depends on mean values, we must bear in mind that with a non-sinusoidal waveform the reading is not truly the r.m.s. value, but rather a quantity 1.11 times the mean value.

The deflecting force of an a.c. instrument is a rapidly pulsating quantity; the pointer reaches a steady deflection because its inertia prevents it following the oscillations. The reading therefore depends on the mean value of the deflecting force. For an instrument to indicate r.m.s. values the deflecting force must depend on the *square* of the current through it (or the square of the p.d. across it in the case of a voltmeter). With such an instrument the same deflection will be obtained with a direct current I or with an alternating current whose r.m.s. value is I. It may be calibrated using direct current; and then the readings obtained with alternating current will be r.m.s. values.

(i) *Moving-iron instruments.*   The deflection is produced by the forces acting on a small piece of soft iron, which is magnetized by the current flowing in a coil. Two types in common use are shown in Fig. 8.12. In the *attraction type* (a) a small soft iron plate P is pivoted on the spindle through O that carries the pointer. The current to be measured passes through the coil C; this partly magnetizes the plate P causing it to be attracted into the coil where the magnetic field is a maximum. This is the same for both directions of current in C. In the *repulsion type* (b) two soft iron rods A and B are mounted parallel to the axis of the solenoid S. A is fixed to the framework of the instrument and B is attached to the pointer. Whichever way the current flows in the solenoid, A and B are magnetized in the same sense and therefore repel one another.

The restoring couple is sometimes provided by hair-springs, as in (b), in the same way as in the moving-coil instrument. But a cheaper arrange-

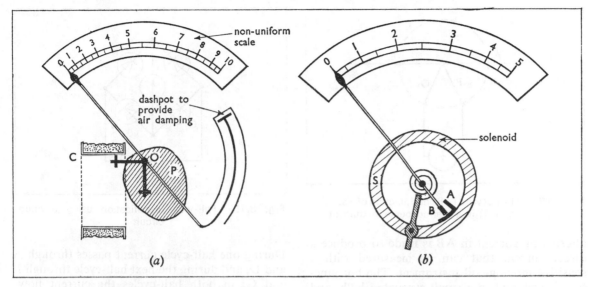

Fig. 8.12   Moving-iron meters: (*a*) attraction type; (*b*) repulsion type

ment is to use *gravity control*, as in (*a*). In this case the moving parts are not balanced about their axis of rotation, and their weight is used to provide the restoring couple. When a current flows in the solenoid, the pointer comes to rest in the position in which the moment of the restoring force is equal and opposite to the moment of the mean deflecting force. With gravity control the zero is adjusted by levelling. Neither method of control gives a uniform scale, since the deflecting force is not proportional to the current. However, in some applications this is even an advantage, since we may arrange the scale so that the part of it to be used is highly expanded, making the instrument more sensitive in the required range.

These instruments are usually provided with some form of pneumatic damping, as shown in Fig. 8.12*a*. Attached to the moving parts is a light aluminium piston or vane, which moves inside a curved 'cylinder', known as a *dashpot*.

The readings of moving-iron meters ideally indicate r.m.s. values, since in both types of instrument the deflecting force depends on the *square* of the current through the solenoid. This may be seen as follows:

(*a*) In the attracted type the force acting on the moving piece of iron depends on the product of the magnetization induced in it and the strength of the field which acts on it. With an 'ideal' mag-

netic material the magnetization would be proportional to the magnetizing field. Thus the deflecting force in a given position is proportional to the square of the field strength, and therefore depends on the square of the magnetizing current.

(*b*) In the repulsion type the deflecting force is proportional to the product of the magnetizations of the two soft iron rods. The magnetization of each of these is 'ideally' proportional to the magnetizing field, and so also to the current in the solenoid. Again, the deflecting force depends on the square of the current. In practice these simple results do not hold exactly on account of hysteresis effects in the iron, which make the magnetization not truly proportional to the current.

(*ii*) *Thermocouple instruments.* In these instruments the deflection depends on the heating effect of an electric current. This is proportional to the *square* of the current ($P = I^2R$), and the readings therefore indicate r.m.s. values. The current to be measured passes through the fine wire AB (Fig. 8.13), raising its temperature. Attached to its centre are the ends of two other fine wires P and Q of dissimilar metals; this forms one junction of a thermocouple, the other junction being at the temperature of the surroundings (p. 46). Thus, by its heating effect the

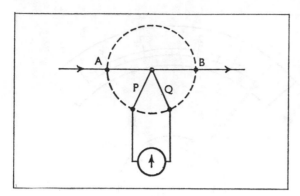

Fig. 8.13   The hot wire (AB) and junction of dissimilar metals (P and Q) in a thermocouple instrument

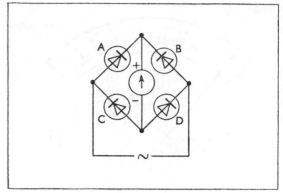

Fig. 8.15   Full-wave rectification using a bridge circuit

alternating current in AB is made to produce a direct current that can be measured with a sensitive moving-coil instrument. The hot junction is enclosed in a small evacuated bulb, and so is not affected by convection currents. It is particularly suitable for high-frequency measurements, since the short wire AB has very low impedance.

(iii) Diode instruments.   A diode is a device that allows current to pass through it in one direction only; in the reverse direction it has a very high resistance (p. 258). Connected to an alternating supply a diode passes a pulsating but unidirectional current which will give a steady deflection on a moving-coil instrument joined in series with it (Fig. 8.14). Current flows on the positive half-cycles only, and the deflection is proportional to the *half-cycle average* or *mean value* of the current. A more satisfactory arrangement is to use four diodes joined in a bridge circuit (Fig. 8.15). This gives *full-wave* rectification.

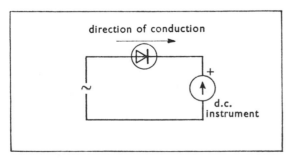

direction of conduction

d.c. instrument

Fig. 8.14   Half-wave rectification using a semiconductor diode

During one half-cycle current passes through A and D, and during the next half-cycle through B and C; in both half-cycles the current flows through the meter in the same direction, and the reading is again proportional to the half-cycle average or mean current.

A diode conducts very little for *small* p.d.'s in the forward direction; and it is therefore usual to make use of a step-up transformer to ensure that the applied p.d. is large enough to give efficient rectification. This in turn means that a given instrument can be calibrated correctly for only a limited range of frequencies.

Being built round a moving-coil galvanometer a rectifier instrument is more sensitive than other kinds of a.c. meter. This arrangement is used in universal multi-range instruments, that incorporate a.c. as well as d.c. ranges. All the shunts, series resistances, diodes and transformers are then included in the one case with the meter, and the connections for the various ranges are provided by switches. The a.c. ranges are calibrated to read r.m.s. values, but this calibration applies only to *sinusoidal* waveforms, and then only in a limited frequency range.

(iv) Dynamometer instruments.   The dynamometer consists of a coil pivoted about an axis in its own plane and free to rotate in the field of a pair of fixed coils (Fig. 8.16). The coil is pivoted in jewelled bearings and its motion is controlled by two hair-springs, just as in the moving-coil d.c. instrument. No iron or other magnetic material is used, however, and the field produced by the fixed coils is therefore strictly propor-

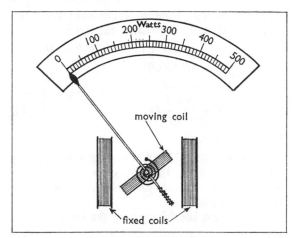

Fig. 8.16   The arrangement of fixed and moving coils
in a dynamometer

tional to the current in them. The absence of magnetic material also makes the field much smaller than in the d.c. instrument; and the dynamometer is therefore rather insensitive. It is also liable to disturbance by stray magnetic fields from other apparatus, though this source of error can be avoided by enclosing it in a mumetal shield (p. 118).

Used as an *ammeter* the fixed and moving coils are joined in series, both being of low resistance. The deflecting couple is obviously in the same direction whichever way the current flows. It is in fact proportional to the *current* in the moving coil and to the *field* of the fixed coils; both of these are proportional to the current I to be measured.

$$\therefore \text{ deflecting couple} \propto I^2$$

and the instrument measures r.m.s. values. The scale obeys a *square law*, expanding towards its upper end.

It is worth noting that the *current balance* (p. 10) is merely a form of dynamometer, in which the geometry is such that the forces between the coils can be calculated from first principles. It is normally used for measuring steady currents. But it can equally be employed to make an absolute measurement of the r.m.s. value of an alternating current.

The commonest use of a dynamometer is as a *wattmeter*. For this purpose the moving coil is joined in series with a high resistance *across* the supply, and the fixed coils are joined in series

with the load (Fig. 8.17). Thus the fixed coils carry the same current I as the load, while the moving coil carries a current proportional to the p.d. V of the supply. The deflecting couple is proportional to the product of these two currents.

$$\therefore \text{ deflecting couple} \propto VI$$

Now, $VI =$ instantaneous power supplied

$\therefore$ mean deflecting couple

$$\propto \text{ mean power supplied}$$

This result applies even if V and I are not in phase, as happens in an inductive or capacitive circuit (p. 213). In this case the product of the r.m.s. values of p.d. and current does *not* give the mean power; but the wattmeter reading still does. Since the deflecting couple is proportional to the power, the instrument has an approximately uniform scale when used as a wattmeter.

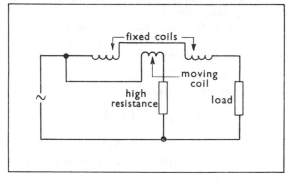

Fig. 8.17   The connection of a dynamometer as a
wattmeter

Other types of a.c. measuring instrument are described in later chapters. They are briefly listed here for the sake of completeness.

(*v*) *The electrostatic voltmeter* (p. 177). This may be used as an a.c. instrument indicating r.m.s. values. It is only really suitable for rather large p.d.'s; but it has the merit of taking negligible current. The same applies to the *gold-leaf electroscope* (p. 178), which can be regarded as a form of electrostatic voltmeter.

(*vi*) *The diode peak voltmeter* (p. 263). This kind of instrument records *peak values* of the applied p.d., whatever the waveform may be. It can be adapted for use at almost any frequency.

(*vii*) *The cathode-ray oscilloscope* (p. 231). With a good modern tube the deflection of the spot of light is very nearly proportional to the p.d. between the deflector plates. With an alternating p.d. joined to the *Y*-plates the height of the trace is therefore proportional to the *peak p.d.* This instrument has the obvious merit that it enables the waveform to be studied at the same time as its magnitude is being measured.

8.5   The ballistic galvanometer

When a current passes for a very short time through the coil of a galvanometer, it gives it an impulse that sets it swinging. Provided the duration of the pulse of current is short compared with the period of oscillation of the coil, the amplitude of oscillation does not depend on the manner in which the current varies during the pulse; it is simply proportional to the total charge Q that passed.

Using the symbols adopted on p. 68, the couple T_I acting on the coil when a current I flows in it is given by

$$T_I = BANI$$

∴ the angular impulse

$$= \int T_I \, dt = BAN \int I \, dt$$

$$= BANQ$$

This is equal to the angular momentum imparted to the coil.

$$\therefore BANQ = J\omega$$

where J = the moment of inertia of the suspended system; and ω = the angular velocity produced.

It is assumed here that the deflection is still negligible by the time the pulse of current has finished, so that the action of the restoring couple can be ignored during the impulse.

∴ the kinetic energy gained $= \frac{1}{2}J\omega^2$

$$= \frac{(BANQ)^2}{2J}$$

When the coil eventually comes to rest momentarily, this kinetic energy has been changed into elastic potential energy in twisting the suspension fibres. The couple T_k due to the fibre is given by

$$T_k = -k\theta$$

where k is the suspension constant (p. 138). The work done in twisting the fibre

$$= \text{(average couple)} \times \theta = \frac{1}{2}k\theta^2$$

$$\therefore \frac{1}{2}k\theta^2 = \frac{(BANQ)^2}{2J}$$

$$\therefore \theta = \frac{BAN}{\sqrt{kJ}}.Q \qquad (1)$$

i.e. $\theta \propto Q$

The constants in this expression are not usually known. However, they may be expressed in terms of other readily measurable properties of the instrument.

The current sensitivity s of the instrument is given by

$$s = \frac{BAN}{k} \quad \text{(p. 138)}$$

Also, it is shown below that the period of oscillation T of the coil is given by

$$T = 2\pi \sqrt{\frac{J}{k}}$$

Equation (1) above can then be written

$$\theta = \frac{BAN}{k}\sqrt{\frac{k}{J}}.Q = s.\frac{2\pi}{T}.Q$$

$$\therefore \theta = \frac{2\pi s}{T}.Q \qquad (2)$$

The period T may be found by timing a number of complete oscillations with a stop-watch, and the sensitivity s can be measured by the method given on p. 139. However, this analysis has ignored one important point: the damping of the movement of the coil is never really negligible, with the result that the actual deflection θ_1 obtained at the end of the first quarter-swing is always rather less than the theoretical value θ that we have calculated. We need therefore to consider in more detail the equation of motion of the coil.

Once the impulse due to the pulse of current has finished, there are two couples acting on the coil:

(*i*) the restoring couple T_k, proportional to θ.

(*ii*) the damping couple T_λ, which is due to air resistance and to the effects of any currents induced in the coil by its movement in the magnetic field; this couple is proportional to the angular speed $d\theta/dt$.

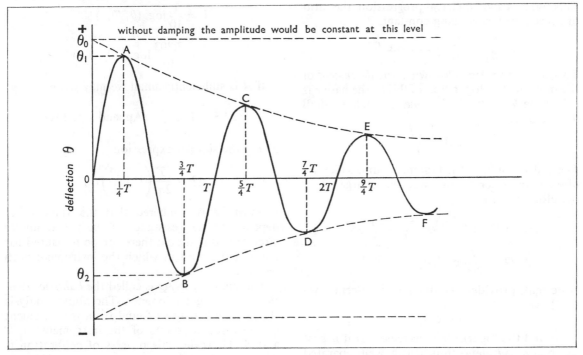

Fig. 8.18   The motion of a galvanometer coil after the passage of a charge through it—an oscillatory motion with exponentially decreasing amplitude

In order to get the signs right we must then write:

$$T_k = -k\theta;$$

and

$$T_\lambda = -\lambda \frac{d\theta}{dt}$$

where λ is the *damping constant*.

The equation of motion then is

$$-k\theta - \lambda \frac{d\theta}{dt} = J \frac{d^2\theta}{dt^2}$$

$$\therefore J \frac{d^2\theta}{dt^2} + \lambda \frac{d\theta}{dt} + k\theta = 0$$

It may be verified by direct differentiation that a solution of this equation is

$$\theta = \theta_0 \, e^{-\lambda t/2J} \sin\left\{ \sqrt{\left(p^2 - \left(\frac{\lambda}{2J}\right)^2\right)}\, t \right\}$$

where $p^2 = k/J$.

In the present case we deliberately keep the damping very small, so that we can ignore the second term under the square-root sign in comparison with the first.

$$\therefore \theta = \theta_0 \, e^{-} \qquad \sin pt$$

The variation of θ with t is shown in Fig. 8.18. It

is an oscillatory motion with exponentially decreasing amplitude. From the oscillatory factor ($\sin pt$) we get the period of oscillation T since

$$p = \frac{2\pi}{T}$$

$$\therefore T = \frac{2\pi}{p} = 2\pi \sqrt{\frac{J}{k}}$$

If there were no damping ($\lambda = 0$), the amplitude of oscillation would be θ_0, as shown by the dotted line. This is the quantity that is really given in equation (2) above:

$$\therefore \theta_0 = \frac{2\pi s}{T} \cdot Q$$

To allow for the exponential damping factor, we proceed as follows. Let the numerical values of the deflections at successive turning points (A, B, C, . . . in the graph) be $\theta_1, \theta_2, \theta_3, \ldots$, respectively. The values of t at these points are

$$\tfrac{1}{4}T, \tfrac{3}{4}T, \tfrac{5}{4}T, \ldots \text{etc.}$$

$$\therefore \; \theta_1 = \theta_0 \, e^{-\lambda T/8J}$$
$$\theta_2 = \theta_0 \, e^{-3\lambda T/8J}$$
$$\theta_3 = \theta_0 \, e^{-5\lambda T/8J} \quad \text{etc.}$$

This series is a geometrical progression, the ratio of successive terms being constant.

$$\therefore \frac{\theta_1}{\theta_2} = \frac{\theta_2}{\theta_3} = \ldots = e^{\lambda T/4J}$$

This common ratio is known as the *decrement* of the motion. The logarithm of this to the base e is called the *logarithmic decrement*, which we shall denote by Λ.

$$\therefore \Lambda = \frac{\lambda T}{4J}$$

The deflection θ_1 at the first maximum, which we observe in order to measure the charge Q, is therefore given by

$$\theta_1 = \theta_0 e^{-\Lambda/2} = \frac{2\pi s}{T} e^{-\Lambda/2} . Q$$

$$\therefore Q = \frac{T}{2\pi s} e^{\Lambda/2} . \theta_1$$

Note that, provided the damping is constant we still have

$$Q \propto \theta_1$$

To find the logarithmic decrement Λ it is best to observe the deflection at two well-separated turning points. For instance, we might record the 1st and 11th, θ_1 and θ_{11}.

$$\therefore \frac{\theta_1}{\theta_{11}} = \left(\frac{\theta_1}{\theta_2}\right)^{10} = (e^\Lambda)^{10} = e^{10\Lambda}$$

$$\therefore \Lambda = \frac{1}{10} \log_e (\theta_1/\theta_{11})$$

$$= \frac{2.303}{10} \log_{10} (\theta_1/\theta_{11})$$

If Λ is sufficiently small we may write

$$e^{\Lambda/2} = 1 + \frac{\Lambda}{2} \quad \text{(Appendix p. 394)}$$

giving the simpler expression

$$Q = \frac{T}{2\pi s}\left(1 + \frac{\Lambda}{2}\right).\theta_1$$

It must be remembered that the value of Λ depends on the resistance of the galvanometer circuit; and it should therefore be measured for the actual circuit in which the instrument is to be used.

The quantity θ_1/Q is called the *ballistic sensitivity* of the galvanometer. The above analysis shows how it may be found from measurement of the three constants of the instrument—s, T and Λ. However, this process of calibration is rather a tedious one; and, having satisfied ourselves on the theoretical point that the charge passed is indeed proportional to the first 'throw', it is usually simpler to use one of the direct methods of calibration discussed on p. 91.

9 Generators and Motors

9.1 A.C. generators

Consider a flat coil rotating about an axis in its own plane at right-angles to a magnetic field (Fig. 9.1). The flux linkage of the magnetic field with the coil alternates in the manner shown in the graph. In positions (i) and (iii) the plane of the coil is normal to the field, and the flux linkage reaches its maximum value, but is momentarily unchanging; thus the rate of change of flux is zero. At these positions the e.m.f. is therefore zero. In positions (ii) and (iv) the flux linkage is momentarily zero, but it can be seen from the graph that it is changing at the maximum rate—decreasing in position (ii) and increasing in position (iv). At these positions (with the coil parallel to the field) the e.m.f. therefore reaches its maximum magnitude. By Lenz's law the e.m.f. acts so that the current driven by it opposes the changes of flux; it is therefore positive at (ii) and negative at (iv). Thus an alternating e.m.f.

is generated in the coil, whose frequency is equal to the number of revolutions per second. (An expression for the actual value of the e.m.f. was deduced on p. 77.)

Fig. 9.2 shows the principle of a possible form of simple a.c. generator or *alternator*. The coil rotates between the poles of an electromagnet, the current for which comes from a separate d.c. supply. The connections to the coil must be made through sliding contacts, and the figure shows diagrammatically the system that is used in such a case. The ends of the coil are joined to two insulated copper rings (known as *slip-rings*) on the rotating shaft. The connection is then made through fixed graphite blocks (called *brushes*), which are pressed onto the slip-rings by springs.

In practice, however, it is more satisfactory to keep the coil fixed and to rotate the magnetic field in relation to it. Fig. 9.3 shows the arrangement used in diagrammatic form; the coil is

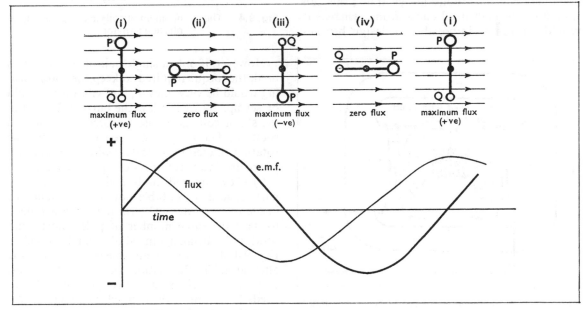

Fig. 9.1   The variation of flux through a rotating coil, and the e.m.f. induced in it

149

to carry currents of this magnitude. In small alternators (e.g. cycle dynamos) a rotating *permanent* magnet is used; and no slip-rings are then needed at all.

To increase the flux through the coils they are wound on soft iron cores, which are laminated to reduce eddy currents. The rotating electromagnet is called the *rotor*, while the fixed system of coils with their core is called the *stator*.

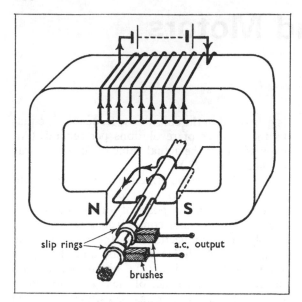

Fig. 9.2   Generating an alternating current by rotating a coil in a magnetic field

wound in two parts in series, between which the electromagnet rotates. The advantage is that the slip-rings and brushes are now required to carry only the relatively small current needed to magnetize the rotating electromagnet. In a large alternator the current generated in the fixed coils may run into thousands of amps; this can now flow through fixed connections, whereas it would be almost impossible to design brushes

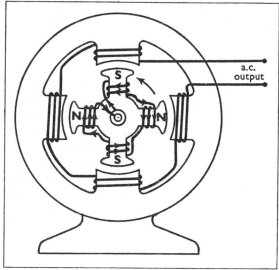

Fig. 9.4   The arrangement of coils in a 4-pole, single-phase alternator

However, if only a single pair of stator coils is used, the torque required to keep the rotor turning would be very uneven, and vibrations would be set up. By Lenz's law the current flowing in the stator coils gives rise to forces *opposing* the rotation. The torque required to maintain the rotation therefore pulsates with the current. There are two ways of reducing this effect.

(*i*) The stator coil can be split up into 4, 6 or 8 parts spaced on suitable pole pieces round the inside of the stator. The rotor is then arranged to have the same number of poles, north and south poles being arranged alternately round it. Fig. 9.4 shows the arrangement of a four-pole alternator. The fluctuations of torque still occur, but are less serious; also the speed of rotation needs to be only half as much in order to produce the same frequency of alternating current.

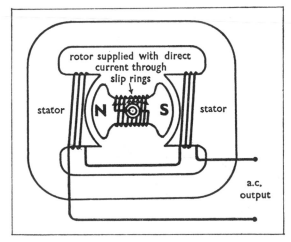

Fig. 9.3   A simplified a.c. generator or alternator (2-pole, single-phase type)

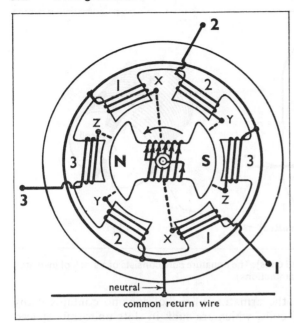

Fig. 9.5   The arrangement of coils in a 2-pole, 3-phase alternator

in the three sets of coils. Each set of coils is connected to a separate distribution line; but the three lines share a common return wire. This is known as a *three-phase system*. Fig. 9.5 shows the arrangement of a two-pole, three-phase alternator. The simpler system employing a single distribution circuit connected to a single set of stator coils is called a *single-phase system*.

The public electricity supply employs a three-phase system, since this makes for greater efficiency not only in generation but also in distribution; this is discussed further in Chapter 12 (p. 218).

9.2   D.C. generators

When a coil rotates in a magnetic field, an alternating e.m.f. is generated in it. To produce a current flowing in one direction only, connection to the coil must be made through some form of rotary switch, which reverses the connections twice in every revolution.

Fig. 9.6*a* shows in diagrammatic form the method that might be used for a single coil. On the same shaft as the coil are mounted two insulated semi-cylindrical plates, one connected to each end of the coil. Contact with these is made by two carbon brushes, which bear on the plates on opposite sides of the shaft; the arrangement is called a *commutator*. It will be noticed that the positions of the brushes are such that the

(*ii*) The same rotor can be used to generate alternating current in *three separate sets* of stator coils spaced on pole pieces round the inside of the stator. The reaction between any one set of stator coils and the rotor is still a pulsating one, but their combined reactions are constant, since the peaks of current occur at different moments

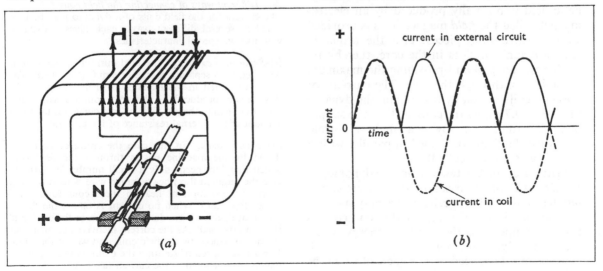

Fig. 9.6   Generating direct current with a rotating coil and commutator

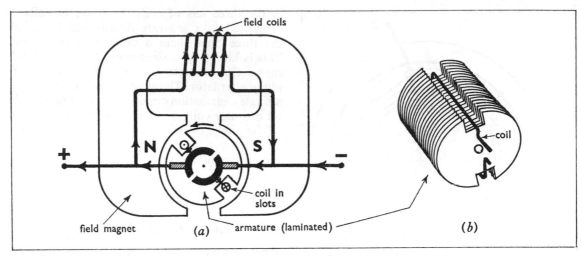

Fig. 9.7   A simplified d.c. dynamo (with shunt-wound field coils). The armature core consists of a stack of insulated iron discs (or laminations)

taps of the commutator come under them at the moments when the plane of the coil is at right-angles to the field; in this position the e.m.f. is momentarily zero, but is just about to reverse. In Fig. 9.6b the dotted curve shows the alternating current in the coil, while the full line gives the current in the external circuit. This current is unidirectional, but pulsating.

To increase the flux the coils are wound on a soft iron core, which is laminated to reduce eddy currents (p. 92). The rotating coils and their core are referred to as the *armature*. The magnetic field is normally produced by an electromagnet called the *field* magnet. This is provided with concave pole pieces; also the armature coils are sunk in slots in the core, thereby reducing the air gap and providing the maximum possible flux through the core. The current for the field coils is usually taken from the dynamo itself. Fig. 9.7 shows the construction of a simplified d.c. dynamo employing a single-coil armature with the field coil joined in parallel with it (or *in shunt*, as it is called).

When a current is taken from the dynamo, by Lenz's law a torque arises in the armature *opposing* the rotation. The mechanical energy supplied to maintain the rotation against the action of this torque reappears as the electrical energy generated in the dynamo circuit.

The e.m.f. produced by a single coil would be by no means steady, as we have seen. In practice

the armature of a d.c. dynamo contains many coils, which are sunk in slots uniformly spaced round the armature core (see Fig. 5.15, p. 74). They are connected to a commutator of many segments in such a way that at any given instant about two-thirds of the coil system is contributing to the e.m.f. The design of d.c. machines is a specialized topic of no little complexity; the student who wants to pursue it further should consult textbooks of electrical engineering.

★   *Different ways of connecting the field coils*
The current for the field coils of a dynamo may be provided in several different ways; each gives a dynamo with its own special characteristics.

(*i*) *Separately excited.*   The field current in this case is provided by a separate source of e.m.f. (Fig. 9.8a). The e.m.f. generated in the machine is proportional to the flux density produced by the field coil. As with other sources of e.m.f., the p.d. falls when a current is taken on account of the internal resistance of the armature.

(*ii*) *Series-wound.*   In this case the entire current produced by the machine passes through the field coil, which therefore needs only a small number of turns to give the required flux; heavy gauge wire is used to keep down the evolution of heat in these coils to a small value (Fig. 9.8b). When no current is taken, the flux is only that due to the residual magnetism of the core, and the e.m.f. is small. As the current increases, so also does the flux produced by the field coils and with it the e.m.f. For a given speed of rotation the e.m.f. is thus approximately proportional to the current taken, until the point is reached at which the magnetic core is saturated.

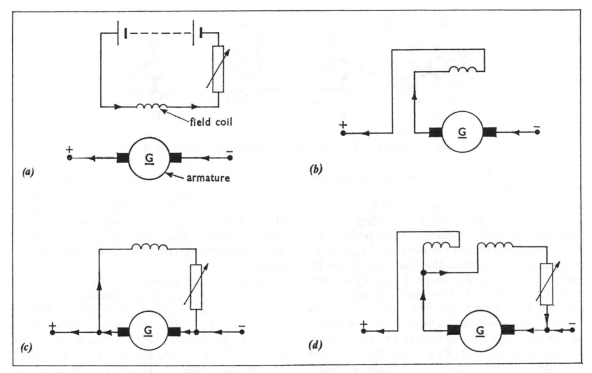

Fig. 9.8 Different ways of connecting the field coils: (*a*) separately excited; (*b*) series wound; (*c*) shunt wound; (*d*) a compound dynamo, in which the field magnet is equipped with both series and shunt windings

Owing to the unsteadiness of its p.d. the series-wound dynamo does not find many practical applications.

(*iii*) *Shunt-wound.* In this case the field coil is joined in *parallel* with the armature, usually with a rheostat in series to control the field current (Fig. 9.8*c*). To avoid using a large current in the field coil, this is made up of a large number of turns of relatively fine wire. For a given speed and setting of the rheostat, the field current (and with it the flux density) is proportional to the p.d. across the armature. The maximum e.m.f. is therefore obtained when no current is being taken by the external circuit. When the current taken increases, the p.d. drops on account of the internal resistance of the armature, and also because of the resulting drop in field current. The fall of p.d. with current is therefore greater than with a separately excited dynamo, but is still small enough to make this the most useful sort of dynamo for many purposes. The e.m.f. can readily be controlled with the field rheostat; in some designs this adjustment is made automatically so as to keep the p.d. nearly constant.

(*iv*) *Compound.* In this case the field magnet carries two windings: a series winding of a few turns of heavy gauge wire, and a shunt winding of many turns of fine wire (Fig. 9.8*d*). When no current is being taken, the flux is provided entirely by the shunt windings. However, as the current increases, the series windings cause an increase of flux, and with it of the e.m.f. By careful choice of the

numbers of turns in the two windings it can be arranged for this rise in e.m.f. exactly to balance the internal drop of p.d., so that the p.d. at the output terminals remains nearly constant for all currents. The machine is then said to be *level compounded.*

9.3 D.C. motors

A coil carrying a current in a magnetic field experiences a couple that tends to turn it with its plane at right-angles to the field—i.e. so that the flux through it is a maximum. To produce continuous rotation a commutator must be used to reverse the connections to the coil whenever it passes through the equilibrium position. Fig. 9.9 shows the action of a commutator for a single-coil motor. The ends of the coil are connected to two insulated semi-cylindrical plates on the motor shaft; contact with these is made through carbon *brushes* in the same way as for a d.c. generator. Fig. 9.9 gives a sectional view of the coil and commutator in four different positions. In position A the couple acts on the coil so as to turn it to position B. In this position the gaps

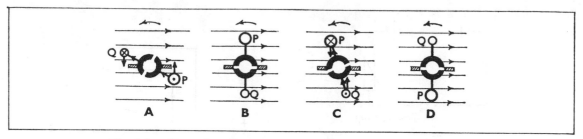

Fig. 9.9   The action of a commutator in producing continuous rotation

in the commutator come opposite the brushes. The momentum of the coil must be sufficient to carry it through to position C, when the current in it is reversed; the couple then acts to produce a further half revolution to position D. To give the maximum torque over as much as possible of the rotation the flux is increased by sinking the coil in slots in a soft iron cylindrical core (laminated to reduce eddy currents); and concave pole pieces are fitted to the electromagnet that produces the field. Fig. 9.10 shows the construction of a simplified d.c. motor with a single-coil armature. In practice the armature is made with many coils sunk in slots spaced round the core and connected to a commutator of many segments. The coils are connected so that at any given moment at least two-thirds of them contribute to the torque. In fact in all respects the construction of a d.c. motor is identical with that of the equivalent d.c. dynamo (compare Fig. 9.10 with Fig. 9.7). Any d.c. machine may be used equally as a dynamo or a motor. The difference between the two is only a question of function. When mechanical power is being supplied to the machine, and is being changed into electrical power, we call it a *dynamo*; and when electrical power is being converted into mechanical power, we call it a *motor*.

Closer thought shows that the d.c. machine must always behave *simultaneously* both as a dynamo and as a motor. Any current flowing in the armature gives rise to a torque. Thus, when current is taken from a dynamo, a torque arises, as in a motor; by Lenz's law this must *oppose* the rotation. Likewise, whenever the armature of a motor rotates, an e.m.f. is induced in it, just as in a dynamo; by Lenz's law the e.m.f. must act *in opposition* to the applied p.d. The fact is that a rotating coil carrying a current in a magnetic field both generates an e.m.f. and experiences a couple. The difference between its functions as a dynamo and a motor is in the *direction of the current*.

AS A DYNAMO: the current is in the direction of the e.m.f.—which is of course responsible for driving the current through the armature and round the circuit. The product of e.m.f. and current is equal to the electrical power generated, which in turn is equal to the mechanical power *input* needed to maintain the rotation against the opposing electromagnetic torque.

AS A MOTOR: the current is driven through the armature by the applied p.d. in opposition to the induced e.m.f. The product of induced e.m.f. and current is now equal to the mechanical power *output*.

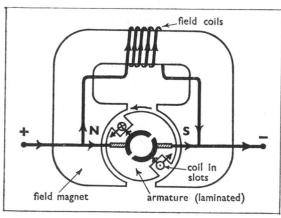

Fig. 9.10   A simplified d.c. motor (with shunt-wound field coils)

This can be compared with the similar behaviour of a reversible electric cell (such as a lead–acid accumulator). Connected to an electric circuit containing no other source of e.m.f. it drives current round the circuit, and so converts chemical energy into electrical energy. But if another source of e.m.f. greater than that of the accumulator is connected so as to drive current through it in the reverse direction, then the result is the conversion of electrical energy into chemical; and the accumulator is restored to its 'charged' condition (p. 44). An accumulator is a device for the mutual conversion of electrical and *chemical* energy; whereas the d.c. machine is for the mutual conversion of electrical and *mechanical* energy. But the action of the two devices in an electric circuit is essentially the same. A practical example may help to make this clear.

Consider an electric delivery van worked from a 100 V battery, climbing over the brow of a hill. As it ascends the hill, it travels fairly slowly; the back e.m.f. generated in its motor is proportional to the speed, and in this case is less than the battery e.m.f.; suppose it is 95 V. The difference between the two e.m.f.'s, namely 5 V, drives a large current through the low-resistance armature, and provides the torque to keep the van moving up the hill. The net result is the conversion of the chemical energy of the battery into potential energy of the van. As the van approaches the brow of the hill, the torque required decreases, and the vehicle accelerates. The increased speed raises the back e.m.f. generated in the armature to a figure very near the battery e.m.f.; it might be 99 V. The difference between the two, now 1 V only, drives a current one-fifth as great through the motor; but this is sufficient to maintain the speed on the smaller gradient and to supply the small amount of power needed to overcome friction. As the van starts to descend the other side of the hill, it speeds up further, until the induced e.m.f. is equal to the battery e.m.f. At this point the loss of potential energy is just sufficiently rapid to make up for the frictional loss; no current flows, and the motor is 'idling'. When the speed increases still further, the e.m.f. becomes more than 100 V, and the d.c. machine, now behaving as a dynamo, drives current in the reverse direction through the battery. Much of the potential energy acquired by the van as it ascended the hill is now being changed back into chemical

energy. The current flowing in the machine now produces a torque which opposes the motion and therefore acts as a brake.

If the e.m.f. of a dynamo is E, the current through the armature I, and the p.d. across it V, then

$$E - V = IR$$

where R = the internal resistance of the armature. Multiplying this equation by I and rearranging, we have

$$EI = VI + I^2R$$

this can be interpreted as:

power generated = electrical power output
                 + rate of heating armature

Likewise with a motor, using the same symbols

$$V - E = IR$$

since in this case $V > E$, and the current flows in opposition to the back e.m.f. E. Multiplying by I and rearranging,

$$VI = EI + I^2R$$

and this means:

power supplied = mechanical power output
                + rate of heating armature

To achieve high efficiency in a d.c. machine the armature resistance R clearly needs to be as low as possible.

Starting a motor

At the moment of starting the e.m.f. E is zero, and it increases only slowly to its final value as the motor gathers speed. If no other resistance were included in the circuit, the starting current I' would therefore be given by

$$V = I'R$$

In a large motor with low armature resistance I' could reach a large and damaging value, unless steps were taken to limit it. For instance, a motor operating on a 200 V d.c. supply might have an armature resistance of 0.05 Ω. Then

$$I' = \frac{200}{0.05} = 4000 \text{ A}!$$

When starting it is therefore necessary to join a suitable resistance in series with the armature. Suppose in the present example we must limit

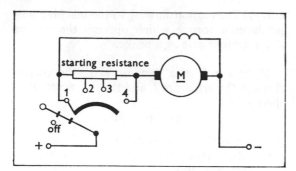

Fig. 9.11   The starting resistance and switches used with a large electric motor

the current to 20 A. The total resistance required must then be

$$\frac{200}{20}\Omega = 10\,\Omega$$

As the motor speed increases the back e.m.f. rises from zero and the starting resistance is progressively switched out of the circuit; near full speed the back e.m.f. alone is sufficient to limit the current to a safe value, and the starting resistance is switched out of the circuit altogether. Fig. 9.11 shows the arrangement used for this purpose. With small d.c. motors (< 0.2 kW) the resistance of the armature is usually sufficient without the aid of any starting resistance.

Reversing an electric motor

The direction of rotation of a d.c. motor is not altered by reversing the polarity of the supply; this would merely reverse both the armature and field currents, leaving the torque still in the same direction. To reverse the rotation we need rather to reverse the connections to one only of the coils—say, the field coil.

However, very small motors, such as those used in toy electric trains, are often made with *permanent magnets*. Reversing may then be effected by simply changing the connections at the d.c. supply.

9.4 The properties of d.c. motors

The field coils of d.c. motors may be connected in the same variety of ways found in dynamos (Fig. 9.8).

Series-wound, shunt-wound and compound motors each have their own characteristic properties, making them suitable for particular applications. We shall consider these properties under the following headings:

(*i*) *The steadiness of the speed under changing loads.*   The speed of a d.c. motor settles down at such a value that the back e.m.f. is a little less than the applied p.d. If the load increases, the armature current must rise to provide the extra torque. This is brought about by a drop in speed, which in turn causes a drop in the back e.m.f., thereby allowing the current to rise. However, the drop in e.m.f. required is usually very small. For example, in a motor working on a 100 V supply the back e.m.f. might be 99 V. This has only to drop to 98 V to double the armature current.

Now for a given motor the back e.m.f. E is proportional to the magnetic flux in the core Φ and to the speed ω.

$$E \propto \Phi\omega$$

With a *shunt-wound motor* the flux Φ is constant, since the field current does not vary. (The same applies with a *permanent magnet* motor.)

$$\therefore E \propto \omega \quad \text{in this case}$$

The small drop in E that must accompany an increase of load is brought about by a proportionately small drop in ω. Thus, provided it is not overloaded, the speed of a shunt-wound motor varies little under changing loads.

However, with a *series-wound motor* the field current is the same as the armature current. Since the changes in E are small we can regard it as approximately constant.

$$\therefore \Phi\omega = \text{constant (approximately)}$$

Provided the field magnet is not saturated, we also have approximately

$$\Phi \propto I$$

where I is the current.

$$\therefore \omega \propto \frac{1}{I}$$

Thus, if the current needs to double to cope with an increase of load, the speed drops to nearly half. It is clear that the speed of a series-wound motor is very unsteady. Furthermore, if the load is suddenly disconnected, as might happen by the accidental breaking of a coupling belt, the current required would drop to a low value and the motor might race dangerously. It is therefore necessary to provide protective devices to guard against this sort of mishap.

(*ii*) *The ease with which the speed can be controlled.* In the *shunt-wound motor* the speed is very easily controlled by altering the field rheostat. A decrease in the field current causes a fall in the flux Φ; this leads to an increase in speed in order to return the back e.m.f. E to its former value, just less than the applied p.d. V. Since the field current is not large, the controlling rheostat can be of quite light construction.

In the *series-wound motor* speed control is not so simple. It is possible to use a variable resistance in *parallel* with the field coil, so that some of the total current is by-passed. However, such a resistance must be designed to carry a substantial fraction of the total current.

Both types of motor may be controlled by means of rheostats in series with the armature. But this is usually practicable only with low-power motors taking small currents.

(*iii*) *The magnitude of the starting torque.* With all electric motors the starting current can be allowed to be greater than that used under steady running conditions. Since the torque is proportional to the current I, this means that all d.c. motors produce their best torque under starting conditions. This is a very useful property. But the torque T is also proportional to the magnetic flux Φ.

$$\therefore T \propto \Phi I$$

With a *shunt-wound motor*, the flux Φ is constant for a given setting of the field rheostat.

$$\therefore T \propto I \quad \text{in this case}$$

With the *series-wound motor*, the flux Φ depends on the total current, and provided the magnet is not saturated, we have approximately

$$\Phi \propto I$$
$$\therefore T \propto I^2 \text{ in this case}$$

Suppose the current allowable at starting is k times the normal current used at full speed. The starting torque of a *shunt-wound motor* would then also be k *times* the steady-running torque—which is good. However, the starting torque of an equivalent *series-wound motor* using the same current would be k^2 *times* its steady-running

torque—which is much better than with the shunt-wound model.

As might be expected, the compound motor combines the properties of the shunt-wound and series-wound types; its exact behaviour depends on the proportion of the field provided by each of the windings. Thus, by making it predominantly series-wound it can have the good starting characteristics of this type of motor, while the small shunt winding suffices to prevent it from racing dangerously if the load is removed. This type of motor is regularly used for traction purposes, e.g. for electric trains and trolley-buses. Series-wound motors are suitable for heavy-duty work where initial torque is all important—e.g. for cranes and winches—or for high-speed, direct-coupled motors, such as fans and grind wheels. Shunt-wound motors are used wherever control of speed is essential; e.g. for lathes and record players. (However, many record players employ synchronous a.c. motors in which the speed is locked to the frequency of the supply.)

It is interesting to compare the properties of an electric motor with those of an internal combustion engine. The electric motor runs satisfactorily at all speeds; it gives its best torque at starting, and is highly efficient at top speed. The engine of an ordinary car is only reasonably efficient within a limited range of engine speeds, and this efficiency is much less than that of the electric motor. Also at low speeds the torque falls to zero and the engine 'stalls'. A diesel engine is even more limited in its performance, though it is more efficient at its best running speed. Thus cars and buses require gear-boxes to match a limited range of engine speeds to a much larger range of possible travelling speeds. No gear-boxes would be needed if electric motors could be used. However, the means of storing electrical energy (e.g. accumulators) are extremely bulky. At present, electrically operated vehicles have proved satisfactory only as delivery vans and milk floats—except where a permanent supply network can be installed, as with trolley-buses and electric trains. If an efficient means could be found of converting the energy of the primary fuels, such as oil, directly into electrical energy, the electric motor might really come into its own for driving vehicles. This is attempted in the diesel-electric engines, which are in use to some extent on the railways. Here a diesel motor running at its optimum speed drives a d.c. dynamo, which then supplies power for one or more electric motors that drive the train. But it is questionable whether this multiple system is any more efficient than other modern engines, such as the gas turbine.

9.5   A.C. motors

Since a.c. power supplies are now widely used, the vast majority of motors encountered in industrial and domestic use are a.c. kinds. The subject is a vast one, and we can do no more here than indicate the principles behind some of the more important types.

(*i*) *The series-wound, commutator motor.*   If the direction of the current through both the armature and the field windings of an ordinary d.c. motor is reversed, the torque remains in the same direction. Thus, in principle, a d.c. motor may be used with alternating current. However, the shunt-wound type cannot be used in this way, because self-inductive effects in the field coils would make the current in them lag behind the p.d. (p. 212); the result would be that the peaks of magnetic field would occur at different moments from the peaks of armature current, and the torque developed would be very small. But in the series-wound type the magnetic field and the armature current are necessarily in phase. Provided the core of the field magnet is laminated, the same motor may be adapted to run on alternating or direct current. Like the d.c. version, its a.c. counterpart is essentially a high-speed motor, whose speed is rather variable under changing loads.

(*ii*) *The three-phase synchronous motor.*   This type of motor is in principle a three-phase alternator operated as a motor instead of as a generator. The three-phase supply is connected to the three sets of stator coils (Fig. 9.5, p. 151). It may be shown that if the magnitudes of the currents in the three coils are equal, then the resultant magnetic field they produce is of constant magnitude and rotates at a steady speed. With the stator in Fig. 9.5, the field would rotate

once for each cycle of alternation of the current —i.e. 50 times a second with the usual a.c. supply. A direct current from an auxiliary supply is passed through the rotor coils, as usual. If now the rotor can be brought to rotate at exactly the same speed as the field, then the magnetic forces will hold it 'locked in' to this speed, provided the load is not too great. The motor then necessarily remains in synchronism with the supply. The problem is to get its speed up to that of the field in the first place. This is sometimes done with an auxiliary starter motor, and sometimes the one rotor is made to contain also windings of the type used in an induction motor (see below). These provide the starting torque and cease to function when the speed approaches synchronism.

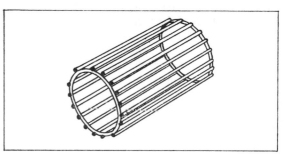

Fig. 9.12   The network of conductors in the rotor of a squirrel-cage induction motor

(*iii*) *The three-phase induction motor.*   In this type of motor a three-phase stator is used just as in the synchronous motor or in the alternator. Thus, the magnetic field in and round the rotor is a rotating one of constant magnitude. The construction of the rotor is shown in Fig. 9.12. It consists of a number of copper bars welded at both ends onto copper rings that encircle the ends of the rotor. The arrangement is referred to as a *squirrel-cage*. The bars are embedded as usual in a laminated iron core. When the rotating field sweeps past the bars, large currents are induced that circulate internally round the system. They could be described as a system of controlled eddy currents. The action of the magnetic field on these currents must be to speed up the rotor towards the point at which there is no relative motion of rotor and field. The motor can never in fact reach synchronism, since if there were no relative motion no currents would be induced in the rotor, and there would be no torque. The rotor therefore 'slips' behind the field; the greater the load, the more the rotor speed falls below that of the field.

(iv) The single-phase induction motor. This type of motor uses a squirrel-cage rotor of similar construction to that of the three-phase machine. A single-phase stator is used that gives an approximation to a rotating field. First, it may be noted that a simple alternating (non-rotating) field can be regarded as the resultant of two fields of equal amplitude rotating at the same speed in opposite directions. Fig. 9.13 shows successive positions of the two component fields; their resultant is clearly an alternating quantity with always the same line of action. With the rotor at rest, there can be no resultant torque. But if the rotor can be set moving in one direction, then it can be shown that the torque on it due to the 'component' of the field that rotates that way grows at the expense of the torque due to the other component. The starting torque is

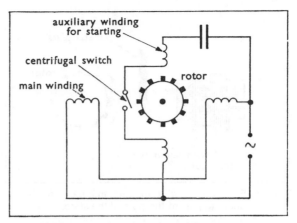

Fig. 9.14   The use of phase splitting to produce a rotating field to start a single-phase induction motor

provided by using an auxiliary set of coils arranged to give a field at right-angles to that of the main stator coils (Fig. 9.14). By including a capacitor in series with the auxiliary windings the current in these is made to lead approximately 90° on the current in the main windings (p. 211). This system is known as *phase splitting*. We have now in effect a two-phase stator, which produces a rotating field—though not of constant magnitude as in the three-phase version. As soon as the rotor has reached sufficient speed to maintain its motion in the right direction, a centrifugal switch operates to disconnect the auxiliary coil. The efficiency of this motor is not as high as that of most other types, but it is adequate for low-power applications.

(v) The shaded pole induction motor.   Another way of producing a rotating magnetic field is to cover part of the iron core of the stator with a conducting loop or plate. This is called *shading* the pole of the stator. The action can be demonstrated with a squirrel-cage rotor and an iron-cored coil arranged as in Fig. 9.15. Before the metal plate is brought up the currents induced in the rotor cannot possibly cause it to turn in either direction. The metal (aluminium) plate is now brought up so that it partly covers the end of the iron core; and at once the rotor starts to turn, the side nearest the stator moving towards the metal plate as shown.

The effect is caused by the eddy currents circulating in the plate. These currents lag about a quarter of a cycle behind the oscillations of

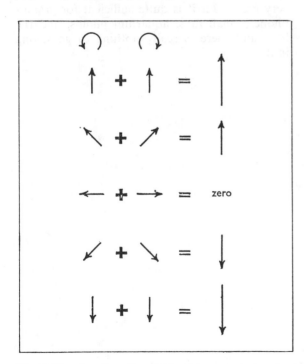

Fig. 9.13   An alternating field can be regarded as the sum of two fields of equal amplitude rotating at the same speed in opposite directions

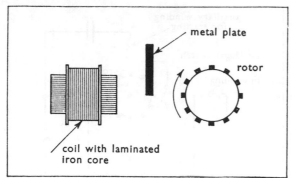

Fig. 9.15   Producing a moving magnetic field by shading part of the pole of the stator

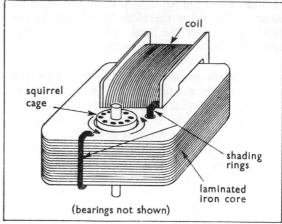

Fig. 9.16   A shaded pole induction motor

flux in the core of the stator. The magnetic field produced by the eddy currents is then added to that of the coil, and the result is that in front of the plate the peaks of magnetic field occur *later* than in front of the unshaded part of the pole. It is not strictly a rotating field that is produced; but there is a local progression of the alternating field towards the shaded part of the pole. This is enough to give the rotation observed.

This principle is employed in the very simple shaded pole induction motor shown in Fig. 9.16. The rotor is the usual squirrel-cage with a lamin-ated iron core. The laminated stator core is shaded by two (or more) welded copper loops. These cause the alternating flux to be delayed in the parts of the core they enclose, and the rotor turns towards these as shown. The efficiency of this little motor is not high, nor is its torque very great. But it is quite sufficient for gramo-phones, small fans, aquarium pumps, and the like; and there is really nothing to go wrong in it.

10 Capacitors

10.1 Electric charge

An arrangement of two conductors in close proximity, but insulated from one another, is called a *capacitor*. A capacitor may take many different forms—two sets of interleaved sheets, two insulated sheets of metal foil rolled up into a 'cigar', a pair of concentric tubes, etc. These different forms are described later in this chapter (p. 172). But for the present we consider only the simplest arrangement, which consists of just a pair of parallel metal plates (Fig. 10.1); the symbol of a capacitor is based on this arrangement, whatever the actual form of construction in a particular case. A series circuit containing a capacitor is therefore an incomplete one, and no continuous current can flow in it. However, when the capacitor is first connected to the source of e.m.f. in the 'circuit' it is found that there is a momentary flow of current. To demonstrate this a very sensitive galvanometer must be used, and it must be stood on an insulating slab so as to avoid confusion from small leakage currents through the insulation of the instrument (which is not designed to be used with high p.d.'s). A very high resistance should be included in the circuit as a protection in case of accidental touching of the plates together.

A galvanometer used like this to detect a momentary pulse of current is called a *ballistic galvanometer* (pp. 90, 146). It has the useful property that the first swing of the coil after the pulse of current has passed through it is proportional to the total charge passed during the pulse. Thus when the capacitor is first connected by joining the flying lead P to the point A on the high tension supply (Fig. 10.1), the galvanometer swings as a small pulse of current flows through it, and the maximum deflection in the first swing is a measure of the charge that has passed. If now the galvanometer is allowed to come to rest, and the lead P is taken out and joined to the earthed point B, the galvanometer deflects in the opposite direction; and it is clear that the same charge as before now flows in the opposite direction in the capacitor leads, discharging the capacitor. This may be repeated with different values of the p.d. V, and it is found that the charge Q that passes during charging or discharging is proportional to V.

$$Q \propto V$$

For a given value of V the experiment may be repeated with different values of the separation d of the two plates; and it is then found that

$$Q \propto \frac{1}{d}$$

and the maximum value of Q is obtained when the plates are close together. By moving the plates sideways we can also vary the area of overlap A at constant separation; and we find that

$$Q \propto A$$

The insulator between the plates also affects the charge that flows onto and off the plates in this experiment. If a slab of any good insulating material (such as polythene) is fixed between

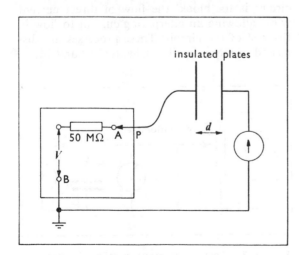

Fig. 10.1 Demonstrating the momentary flow of current in an incomplete capacitor 'circuit'

161

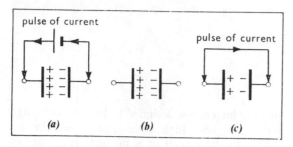

Fig. 10.2   The flow of charge onto capacitor plates: (*a*) charging; (*b*) isolated, and charge stored; (*c*) discharging

the plates, the charge Q is found to be considerably greater than with air between them.

When a capacitor is connected to a source of e.m.f. as in the above experiment there is thus a momentary flow of charge in the circuit as shown in Fig. 10.2*a*. This leaves a surplus of positive charge on one plate of the capacitor and a surplus of negative charge on the other, as shown. The capacitor is then said to be *charged*, and provided the insulation between the plates is good enough this charge ($+Q$ on one plate and $-Q$ on the other) will remain isolated upon them even when the source of e.m.f. is disconnected (Fig. 10.2*b*). When the plates are subsequently connected together (Fig. 10.2*c*) a pulse of current flows in the circuit, carrying a charge Q from one plate to the other and so discharging the capacitor once more. The simple arrangement of two parallel plates in Fig. 10.1 cannot store very much charge; the insulation strength of the air gap sets a limit to the p.d. that may be used, and the charge Q is certainly less than $1\mu C$. (It may be measured directly if the ballistic galvanometer has been previously calibrated, p. 90.) However, the tests above show what must be done to design a capacitor capable of storing larger quantities of charge. We must increase the area of overlap A as much as possible, and decrease the separation d of the plates consistent with preserving the necessary strength of insulation between them for the p.d. to be used; the use of a suitable insulator (or *dielectric*, as it is usually called in this connection) can also increase the charge stored.

With a suitable commercial capacitor we can then experiment using less sensitive instruments and lower p.d.'s. We can test the proportionality

of Q and V with the arrangement shown in Fig. 10.3, and measure just how much charge is stored at a given p.d. With the two-way switch on contact X the capacitor is charged to the p.d. V indicated by the voltmeter. The switch is then moved to the contact Y, discharging the capacitor through the galvanometer; the maximum deflection on the first swing is noted. The measurement may be repeated with a deliberate delay between opening the contact X and touching it on Y so as to test whether there is any significant leakage of charge through the insulation.

With some older forms of capacitor (e.g. those designed with waxed paper insulation between the plates) it is found in such an experiment that not all the charge is removed when the capacitor is discharged through the galvanometer. It seems that the insulation is in a strained condition under the electric forces between the plates, and this strain is not entirely removed when the capacitor is first discharged. The result is that a second smaller discharge can be obtained after a short interval, and sometimes several more discharges after that.

If an alternating supply is joined to the plates of a capacitor, charge flows onto and off the plates as the p.d. between them oscillates; and considerable alternating current may flow in the connecting wires. This can well be sufficient to light a large electric bulb in series (Fig. 10.4). One of the functions of a capacitor in an electric circuit is to 'block' the flow of direct current while allowing an alternating current to flow in the rest of the circuit. These processes are discussed in more detail in Chapters 12 and 14.

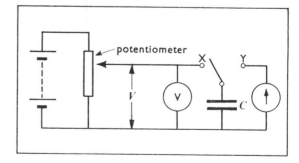

Fig. 10.3   Testing the proportionality of charge and p.d. for a commercial capacitor

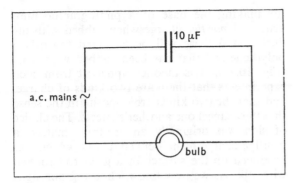

Fig. 10.4   The flow of alternating current 'through' a capacitor

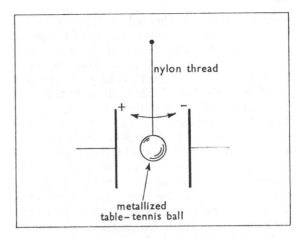

Fig. 10.6   A metallized table-tennis ball between charged capacitor plates

Electric fields

In the space between the plates of a charged capacitor strong electric forces operate. This is obvious enough when the insulation breaks down. A solid insulator will have holes punched in it, and chemical breakdown of the material usually occurs. The presence of these forces can be demonstrated by placing a flexible test strip in the gap between two plates charged to a high p.d. (Fig. 10.5); almost any piece of thin foil on an insulating handle may be used, but a strip of thin metallized plastic is usually best. If the test strip is touched on the positive plate it picks up from this a small positive charge, and is then seen to be bent towards the negative plate, so revealing the electric force that acts on any positive charge in the gap. Similarly if the test strip is touched on the negative plate, picking up a sample of negative charge, it is deflected in the opposite direction, showing the force that acts on a negative charge in the gap. These forces are simply the electrical forces that always operate

in an electric circuit, driving the positive charge in one direction and the negative charge in the other. In a metallic circuit there are free electrons able to move through the conductors, and an electric current (i.e. a flow of charge) is produced. But in an insulator (such as air) there are no free charged particles, and no charge flows; but the forces still act on any charged particle introduced into the space, even if it is not free to move. Any space in which charged particles are acted on by forces is called an *electric field*. There is thus an electric field between the plates of a charged capacitor, as evidenced by the deflection of the test strips above. The field is produced by the charges on the surfaces of the plates. The behaviour of the test strip shows that *like charges repel* one another, while *unlike charges attract* one another. Thus a test strip with a positive charge on it is repelled from the positive charge on one of the plates and attracted to the negative charge on the other. We may take two test strips and charge them both by touching them on one plate or the other, and then show the forces acting between them even when they are far removed from the capacitor plates. With like charges on the strips they bend away from one another, and with unlike charges they bend together.

The nature of an electric current is revealed in a striking way if a light metallized sphere is suspended on an insulating thread between the plates of a capacitor (Fig. 10.6). As long as the

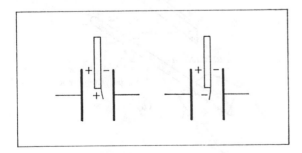

Fig. 10.5   Detecting the electric field between capacitor plates with a flexible test strip

sphere remains uncharged it experiences no force. But as soon as it touches one of the plates (say, the positive one) it acquires from it a positive charge and is impelled towards the other plate; here it delivers up this charge, receiving a negative one instead. It then returns to the positive plate, and so the cycle is repeated indefinitely. A galvanometer joined in series with the capacitor and the high tension supply registers a steady deflection. Presumably the current flows in a series of pulses, but the galvanometer response is too slow to do more than show the average current. If the plates are disconnected from the supply leaving the sphere bouncing back and forth, it discharges the capacitor just as leaky insulation would do, and the rate of discharge slowly falls off as the p.d. between the plates decreases.

Frictional electricity

Historically, electric charges were first produced by friction. When insulating bodies are rubbed together they are found to become charged, i.e. forces of attraction and repulsion are observed to act between them. Polythene, for instance, rubbed with a woollen duster acquires a *negative* charge; and two polythene rods so treated will be found to repel one another (Fig. 10.7a). Perspex and cellulose acetate (the substance used

for making the base of a photographic film) acquire a *positive* charge when rubbed with the duster, and these will be attracted towards a polythene rod that has been rubbed with wool (Fig. 10.7b). It is at once apparent from such experiments that there are two kinds of charge, and that the two kinds are opposite in the sense that they cancel one another's effects. The choice of sign was originally an arbitrary matter, a positive charge being defined as that which was produced on the surface of a glass rod rubbed with silk. As we now know, choosing our signs in this way leads to the electron having a negative charge and the proton a positive one. We suppose that when glass is rubbed with silk electrons are transferred from the glass surface to the silk, thereby leaving a deficit of electrons on the glass, i.e. a positive charge.

But we can now assume the signs of electric charges to be given by the sign of the battery terminal in the circuit in which they are produced. Current is taken to flow in the positive direction out of the positive terminal of the battery (or other source of e.m.f.); and a capacitor plate joined to this terminal therefore acquires a positive charge. We can test the sign of the frictional charges on polythene rods, etc., by introducing them into the space between charged capacitor plates (as in Fig. 10.1); we must be careful to rub them *very lightly* for this purpose,

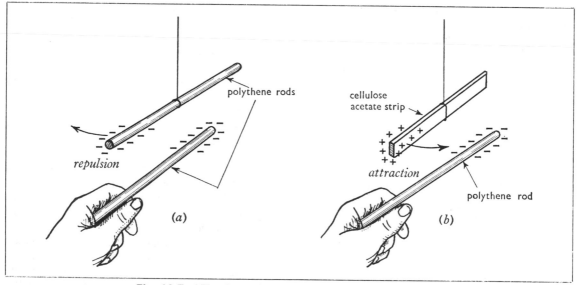

Fig. 10.7   The forces between frictionally produced charges

as otherwise the charges they carry can be large enough to disturb the distribution of charge on the plates considerably and so give anomalous results. Alternatively we can bring up near the frictionally charged piece of insulator a test strip that has been charged by contact with one of the capacitor plates.

Sometimes we find that unwanted frictional charges appear on parts of our apparatus producing disturbing effects. An insulating support may easily become charged through inadvertent contact with a coat sleeve, and then the distribution of electric charge in the apparatus is disturbed by the frictional charge, giving spurious results. Discharging of charged insulators is easily effected by using a flame (e.g. a match or small bunsen burner flame). The flame contains large numbers of ions of both signs, and when it is placed in an electric field some of these are drawn out of the flame and pass quickly and invisibly through the air towards the sources of the field. Thus a negatively charged polythene rod held a few centimetres from a flame attracts positive ions out of it, and is discharged completely in about a second.

If a small flame is held near the capacitor plates in Fig. 10.1 the current carried across the gap by the ions can be quite large, and is registered by the galvanometer. By the same token, if we are using any apparatus to measure very small movements of electric charge, we do well to make sure that there are no bunsen burners (or other sources of ionization) on the other end of the bench.

Until the advent of modern plastics with their excellent insulating properties the difficulties attending experiments on frictional charges were enormous. With most insulators the normal humidity of a temperate climate deposits a conducting layer of moisture upon them, and the materials have to be kept specially dry and clean if any effects are to be observed at all. However, many plastics have water repellent properties, and with these the experiments are relatively straightforward. Indeed the difficulty is often rather how to keep them free from frictional charges when they are not wanted.

Normally frictional charges are observed only on good insulators, since with a conductor any charge at once leaks away to the earth. But if a conductor is carefully insulated from the earth and rubbed with an insulator, charges are separated just as with a pair of insulators. Frictional processes in the exhaust of an aeroplane can lead to the separation of large charges; accidents have been caused by the sudden discharge that occurs as the plane touches down. This is now prevented by adding substances to the rubber of the tyres to make them slightly conducting; the charge on the plane then leaks away gently during the first moments of contact with the ground.

10.2   Capacitance

We have seen that when a potential difference V is connected between the plates of a capacitor, a charge Q flows onto the plates charging them up; i.e. a charge $+Q$ appears on one plate, and a charge $-Q$ on the other. Also we have found that

$$Q \propto V$$

We can therefore write $Q = CV$

where C is a constant which is called the *capacitance* of the capacitor.

Capacitance. The charge Q on one plate of a capacitor is proportional to the p.d. V between the plates;

$$\therefore Q \propto V$$

or $\qquad Q = CV$

where C is a constant for the capacitor known as its capacitance.

The unit of capacitance implied by this definition is therefore the $C V^{-1}$; but it is convenient to have a special name for this much used unit, and it is called the *farad* (F) after Michael Faraday.

The **farad** is the capacitance of a capacitor on the plates of which there are charges of $+$ and -1 coulomb when a p.d. of 1 volt exists between them.

The farad turns out to be an extremely large quantity, and capacitances are usually expressed in μF (10^{-6} F) or even pF (10^{-12} F).

When a *steady* current flows into a capacitor, the charge on the plates increases at a steady rate, and the p.d. across them therefore grows at a steady rate also. The current I is equal to the rate of flow of charge.

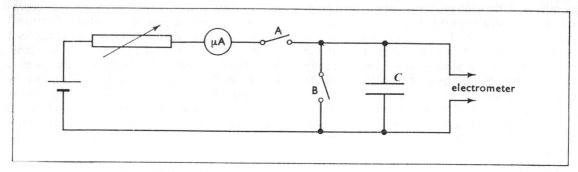

Fig. 10.8   Charging a capacitor with a constant charging current

$$\therefore I = \frac{\mathrm{d}Q}{\mathrm{d}t}$$

$$= \frac{\mathrm{d}(CV)}{\mathrm{d}t}$$

$$= C\frac{\mathrm{d}V}{\mathrm{d}t} \quad \text{(since } C \text{ is constant)}$$

If then the current I is kept *steady* while the p.d. grows from zero to V in time t, we have

$$I = C.\frac{V}{t}$$

This behaviour can be observed with the arrangement shown in Fig. 10.8. The voltmeter used to measure the p.d. V needs to be a very high resistance one that draws truly negligible current from the capacitor; a cathode-ray oscilloscope is suitable (used with 'd.c.' connection to the plates), or else an electrometer (p. 175). First both switches are closed, and the current is adjusted to a convenient value with the rheostat. Then the switch B is opened, and a stop-clock is started at the same moment. The current that is being registered by the microammeter then starts flowing into the capacitor, charging it up and causing the c.r.o. or electrometer to deflect. As the p.d. across the capacitor grows the current tends to fall. Continual adjustments are therefore made to the rheostat to keep the current constant. After a suitable time interval t the switch A is opened and the p.d. V is read. A graph of V against t is found to be a straight line, confirming the above analysis; and from the graph the capacitance C may be calculated. This method of measurement avoids the difficulties of handling a ballistic galvanometer, C being found from the readings of ordinary electrical meters and a clock.

Energy

The existence of a potential difference between the plates of a capacitor implies that the charge stored there carries electric potential energy. The quantity of energy may be calculated from the definition of p.d. (p. 12).

Suppose the p.d. between the plates of a capacitor is V. If now a small quantity of charge δQ is allowed to flow from the positive to the negative plate, the energy converted from electrical to other forms (chiefly heat in the connecting wire) is $V.\delta Q$. This movement of charge partly discharges the capacitor, and the p.d. between the capacitor plates falls slightly. It can be seen in Fig. 10.9 that this loss of energy is represented by the area of the shaded strip under the graph of V against Q. If further quantities of charge δQ are allowed to flow from one plate to the other, further quantities of

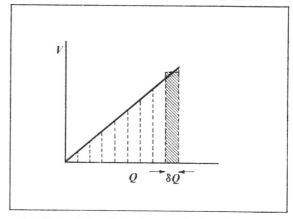

Fig. 10.9   The relation between p.d. and charge for a capacitor; the energy stored is given by the area under the line

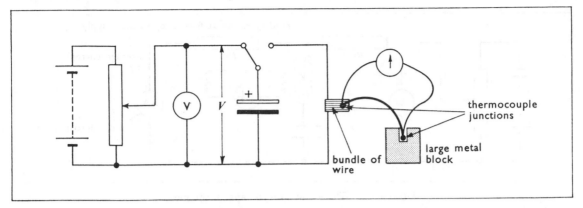

Fig. 10.10   Demonstrating that the energy stored is proportional to V^2 by discharging through a bundle of resistance wire

energy $V.\delta Q$ are released which decrease steadily as the p.d. V falls. The total energy E stored in the capacitor is represented by the total area under the graph of V against Q.

$$\therefore E = \tfrac{1}{2}QV$$

Since $Q = CV$, we may substitute for Q or V in this expression to obtain the three alternative forms:

$$E = \tfrac{1}{2}QV = \tfrac{1}{2}CV^2 = \tfrac{1}{2}\frac{Q^2}{C}$$

In calculus notation we can express this reasoning in terms of an integration.
Thus

$$E = \int_0^Q V.\mathrm{d}Q$$

But

$$V = \frac{Q}{C}$$

$$\therefore E = \int_0^Q \frac{Q}{C}.\mathrm{d}Q = \tfrac{1}{2}\frac{Q^2}{C}$$

For a given capacitance C the energy stored is proportional to the square of the p.d. This result may be tested experimentally with the arrangement shown in Fig. 10.10. A large electrolytic capacitor (e.g. 10 000 μF) is charged to a p.d. V, and then discharged through a length of resistance wire wound into a small bundle. The heat delivered is sufficient to raise the temperature of this 'calorimeter' by about 0.2 K, and this may be registered by a thermocouple joined to a sensitive galvanometer. One junction

of the thermocouple is inserted into the bundle of wire, and the other into a hole in a large block of metal to act as a heat sink whose temperature will be constant for the few minutes taken by the measurements. After each discharge a few seconds must be allowed for the reading to become steady; but the temperature rise is so small that very little subsequent cooling of the bundle of wire will occur. The reading of the galvanometer is a measure of the temperature rise and so of the heat delivered; and this will be found to be proportional to V^2. Thus, if the p.d. V is doubled the change in deflection of the galvanometer is four times as much.

The vibrating reed capacitance meter

In this instrument the capacitance C under test is charged to the p.d. V of the supply, and is then discharged through a microammeter. But this sequence of connections is performed many times a second by means of a metal strip (or reed) maintained vibrating at a known frequency (Fig. 10.11). One way of doing this is to employ a device known as a *reed-switch*. In this kind of switch the contacts to be joined are carried on the ends of two steel strips fixed in the ends of a glass tube (Fig. 10.12a). When the strips are magnetized by current flowing in a coil round the tube, the contacts are drawn rapidly together. The action is extremely rapid, allowing the switch to be closed and opened 400 times a second or more. For the purpose of this experiment a change-over version of the switch is employed (Fig. 10.12b) in which the moving reed

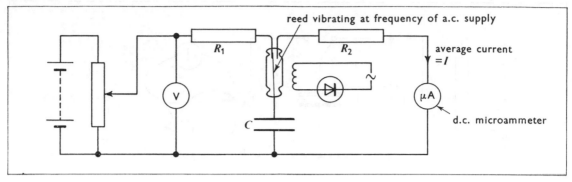

Fig. 10.11   The vibrating reed capacitance meter

springs back in the unmagnetized condition on another (non-magnetic) contact. The reed switch is mounted inside a coil which is joined to an a.c. supply of known frequency through a semiconductor diode (p. 258). The switch then closes in the conducting half-cycle of the diode and springs back in the other half-cycle, when the diode is non-conducting. The number of discharges per second is equal to the frequency of the a.c. supply.

If the pulses of current through the microammeter follow one another at high enough frequency the meter shows a steady deflection which records the *average* current I passing through it.

Thus $I =$ charge passing per second

The charge Q stored in the capacitor at a p.d. V is given by

$$Q = CV$$
$$\therefore I = fCV$$

where $f =$ frequency of vibration of the reed. The capacitance is thus obtained in terms of the

two meter readings and the frequency of the a.c. supply. Since the current I is directly proportional to C for given values of f and V, the scale of the microammeter can be calibrated to read capacitance directly. The scale range may be changed by choosing suitable values of f and V.

The resistances R_1 and R_2 are included in the charging and discharging circuits (Fig. 10.11) to avoid excessive pulses of current as the contacts close. It is interesting to observe the effect of varying these resistances. Up to a certain value they have negligible effect on the average discharge current through the microammeter. However, if R_2 is increased beyond a certain point only partial discharge of the capacitor takes place during the time that the switch contacts are closed, and the microammeter reading falls off. Likewise if the other resistance R_1 is made too large, the p.d. across the capacitor never rises to the full p.d. available. A cathode-ray oscilloscope may be joined across the capacitor to show the way in which the p.d. varies during the charging and discharging cycle; and the falling off of the recorded current may be correlated with the change in waveform seen on the oscilloscope screen. However, the quick test that the two resistances are small enough not to affect the result is to see whether they may be changed (say, halved) without altering the microammeter reading.

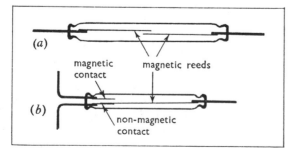

Fig. 10.12   Reed switches: (*a*) a simple switch; (*b*) a change-over switch

The vibrating reed capacitance meter may also be used to test the relationship for the energy E stored in a capacitor: $E = \frac{1}{2}CV^2$

If a small bulb is substituted for the micro-

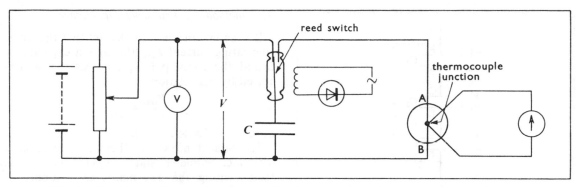

Fig. 10.13   Testing the relation $E = \frac{1}{2}CV^2$ using a reed-switch and thermocouple instrument

ammeter in Fig. 10.11, the brightness of the bulb is proportional to the average rate at which energy is being dissipated in it. For a given frequency f and capacitance C this will be found to depend on the *square* of the p.d. (whereas the microammeter reading was *directly* proportional to p.d.). The current taken by even a small bulb is liable to overload the reed-switch, and a better arrangement is to use in place of the bulb a thermocouple meter of the kind used for measuring high frequency alternating currents (p. 144). The current from the discharging capacitor is passed through the fine resistance wire AB (Fig. 10.13). This quickly reaches a steady temperature depending on the rate of supply of energy from the capacitor and reed-switch. One junction of a thermocouple is in good thermal contact with the centre of AB; the galvanometer reading then depends on the temperature of the wire, which in turn depends on the rate of supply of energy to it. The simplest way of testing the result is to change the p.d. V and then to restore the galvanometer reading to its former value by altering either the capacitance C or frequency of discharge f. Thus, if the p.d. is halved, either C or f would need to be increased four times to restore the galvanometer reading.

In order to predict the behaviour of systems of capacitors and other components we need to be able to calculate the effective capacitance of various combinations of capacitors.

(i) Capacitors in parallel

In this arrangement (Fig. 10.14) the p.d. V is the same across the capacitors C_1, C_2 and C_3,

but the charges on them will be different. The combined effective capacitance C_{tot} is given by

$$Q = C_{\text{tot}}V$$

where Q is the total charge on the capacitors, i.e. the sum of their separate charges.

Now
$$Q = C_1 V + C_2 V + C_3 V$$
$$= V(C_1 + C_2 + C_3)$$
$$\therefore C_{\text{tot}} = C_1 + C_2 + C_3$$

(ii) Capacitors in series

Some caution is necessary before we pronounce a set of capacitors to be in series. Let us consider what happens when the capacitors in Fig. 10.15 are charged by applying a p.d. V between the points A and F. This will cause a charge $+Q$ to flow onto the plate A and $-Q$ therefore onto the other plate B. The plates B and C and the connecting link between them form a single insulated conductor, whose total charge must remain zero throughout the process. Therefore if a charge $-Q$ appears on plate B a charge $+Q$ must appear on plate C of the next capacitor, and so on through the sequence of capacitors. Thus, with capacitors in series the *same charge* is stored in each capacitor, $+Q$ on one plate and $-Q$ on the other.

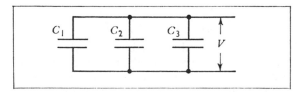

Fig. 10.14   Capacitors in parallel

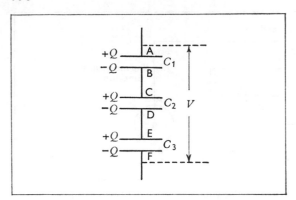

Fig. 10.15   Capacitors in series

But it is quite possible to charge these capacitors in a different way. We might for instance join separate sources of e.m.f. of any magnitudes whatever across each capacitor individually; when the capacitors are charged, these sources of e.m.f. could then be removed so that the circuit diagram *looks* the same as in the first case. The p.d. across the combination is still the sum of the separate p.d.'s, but in general the charges on the capacitors will be *different*, and we cannot say they are in series, since this process of charging has involved also making temporary connections to the intermediate points of the chain.

However, in the case in which the same charge Q is stored in each capacitor, the combined effective capacitance C_{tot} is given by

$$Q = C_{tot}V$$

where V is the total p.d. across the combination.

Now
$$V = \frac{Q}{C_1} + \frac{Q}{C_2} + \frac{Q}{C_3}$$
$$= Q\left(\frac{1}{C_1} + \frac{1}{C_2} + \frac{1}{C_3}\right)$$

But
$$\frac{V}{Q} = \frac{1}{C_{tot}}$$
$$\therefore \frac{1}{C_{tot}} = \frac{1}{C_1} + \frac{1}{C_2} + \frac{1}{C_3}$$

The above results may be tested very simply with a vibrating reed capacitance meter, measuring the capacitances of a set of capacitors individually and then trying them in various combinations in parallel and in series.

A.C. methods of measuring capacitance

It is shown in the next chapter that the r.m.s. alternating current I_{rms} 'through' a capacitance C and the r.m.s. p.d. V_{rms} across it are connected by the relation

$$\frac{V_{rms}}{I_{rms}} = \frac{1}{\omega C}$$

where $\omega = 2\pi \times$ frequency.

The capacitance C may therefore be measured with a.c. ammeter and voltmeter as in Fig. 10.16*a*, using the a.c. mains as a supply of known frequency. The resistance of the ammeter to some extent affects the current; to avoid excessive error it is necessary for this to be small compared with $1/\omega C$.

Another arrangement is that shown in Fig. 10.16*b*. A known non-inductive resistance R is joined in series with the capacitor, and a cathode-ray oscilloscope is used as a high-impedance a.c. voltmeter (p. 232) to measure in turn the magnitude of the alternating p.d.'s V_1 and V_2 across the resistor and capacitor, respectively. Then

$$V_1 = IR$$

and
$$V_2 = I\frac{1}{\omega C}$$
$$\therefore \frac{V_1}{V_2} = \omega CR$$

enabling C to be calculated.

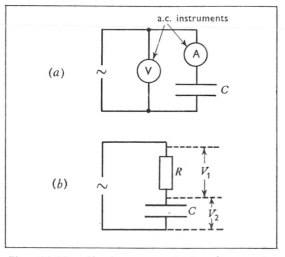

Fig. 10.16   Simple a.c. methods of measuring capacitance

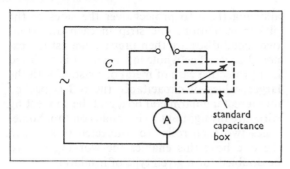

Fig. 10.17 An a.c. substitution method

To make a satisfactory measurement, V_1 and V_2 must be of the same order of magnitude, and the values of R and of the frequency must be chosen accordingly.

There are many ways of devising substitution techniques for measuring capacitances, if a set of standard capacitors can be used. Capacitance 'boxes' are available which are designed so that any chosen value may be obtained by manipulating switches or plugs. One arrangement using the a.c. mains and an ammeter is shown in Fig. 10.17. In this case we do not depend on the calibration of the ammeter, nor does its resistance affect the result; nor does the frequency of the supply need to be known, as long as it is constant. The current reading is taken first with the unknown capacitance in circuit. The capacitance box is then substituted for this and its capacitance varied until the ammeter reading is brought back to its original value; and so the capacitances can be equated.

A bridge method

Many types of a.c. bridge circuits have been devised for measuring capacitance (and other quantities). One of the simplest, known as the De Sauty bridge, is shown in Fig. 10.18. It resembles the Wheatstone bridge, except that in two arms of the bridge the resistors are replaced by capacitors. C_1 is the unknown capacitance, and C_2 is a known fixed capacitance; R_1 is a known fixed resistance, and R_2 is an adjustable resistance box. A suitable alternating p.d. is connected across the diagonal of the bridge AC; and the sound in the headphones joined between B and D is adjusted to zero by varying the resistance R_2. When the bridge is balanced, no current flows through the headphones, and the current I_1 is the same 'through' both capacitors; likewise the current I_2 is the same through both resistors. Also at balance

$$\text{p.d. across } C_1 = \text{p.d. across } R_1$$

$$\therefore I_1 \frac{1}{\omega C_1} = I_2 R_1$$

Also p.d. across C_2 = p.d. across R_2

$$\therefore I_1 \frac{1}{\omega C_2} = I_2 R_2$$

Dividing these equations

$$\therefore \frac{C_2}{C_1} = \frac{R_1}{R_2}$$

The balance condition does not depend on the frequency of the supply. With this bridge it is therefore possible to use a mixture of frequencies such as would be produced in a circuit containing a simple make-and-break buzzer run from a battery. This is not the case with most types of a.c. bridge circuits; in general, the balance condition depends also on the frequency, and a source producing a pure sinusoidal waveform must be used.

In practice it is often impossible to reduce the sound in the headphones actually to zero. This is because of the stray capacitances between the experimenter (who is earthed) and the headphones and other parts of the apparatus. But for frequencies up to 1000 Hz the effect of the stray capacitances may usually be reduced to manageable proportions by earthing a suitable point in

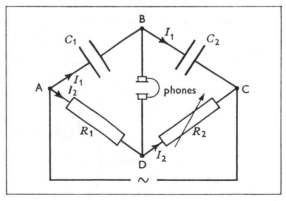

Fig. 10.18 The De Sauty bridge for comparing capacitances

the buzzer or other a.c. supply arrangement. The correct balance then produces a clear *minimum* of sound in the headphones, even thought it is not strictly zero.

10.3   The construction of capacitors

A capacitor consists of two conducting sheets insulated a short distance apart by a suitable dielectric. To produce large values of capacitance it is necessary (*a*) to use sheets of large total area; (*b*) to use very thin layers of dielectric; (*c*) to choose a dielectric of high permittivity (p. 192). Breakdown of the insulation occurs in any given dielectric for electric fields above a certain critical value and there is therefore a maximum p.d. that can safely be applied across a given thickness of it. The thickness of the layer of dielectric is thus decided by the p.d. at which the capacitor is to operate. The choice of dielectric material generally depends on the frequency range of the alternating currents that will be flowing in the capacitor and on the stability of capacitance required.

Plastic capacitors

The electrodes consist of two long strips of metal foil. These are interleaved with two similar strips of plastic, polystyrene or polythene. (At one time paper impregnated with paraffin wax was used, but this type of capacitor was not suitable for high frequencies.) The composite strip is then rolled up into a tight cylinder (Fig. 10.19). Connection is made to the two strips of foil by

allowing them to project over the sides of the plastic insulation, one strip in each direction; two metal discs are then pressed against the exposed edges. The whole arrangement is enclosed in a protective case of metal or plastic. With the larger values of capacitance the roll is packed into a rectangular metal box, and the connecting wires are brought out to terminals on top. Sometimes two such rolls are connected in series in the one box; this enables the working p.d. to be doubled for a given type of insulation.

The insulation resistances of these dielectrics are very high (often more than $10^{10}\ \Omega$), and they are suitable for use at all frequencies. The values of capacitance made in this design cover the range from about $10^{-3}\ \mu F$ to $10\ \mu F$.

Mica capacitors

Mica is a mineral that has the useful property that it can be split into very thin uniform sheets —a thickness of about 0.05 mm is common in capacitors. The capacitor is built up of alternate sheets of metal foil and mica (Fig. 10.20). The odd-numbered sheets are taken out beyond the dielectric at one end and the even-numbered sheets at the other. The two groups of sheets form the two electrodes and are joined to connecting wires at each end of the stack. The whole capacitor is then bound tightly in a hermetically sealed plastic case.

In an alternative form of construction the metal foils are replaced by very thin layers of silvering on the surfaces of the mica sheets themselves. The result is a more compact unit; also in this way all air gaps are eliminated between

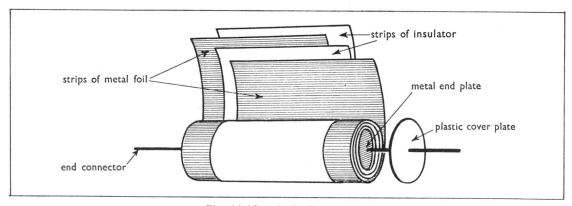

Fig. 10.19   A plastic capacitor

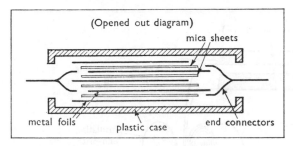

Fig. 10.20   A mica capacitor

the electrodes and the mica, which makes for greater stability of capacitance.

Mica has a relative permittivity of about 6 and produces very low losses for all frequencies up to about 300 MHz. The values of capacitance generally available cover the range from 5 pF up to 0.01 μF. The stability of capacitance is better than with the plastic type. With careful design variations can be kept below 1%—sufficient for laboratory standards; special capacitors are made for this purpose with values up to about 1 μF.

Ceramic capacitors

There is a wide range of ceramic materials that can be used as low-loss dielectrics at all frequencies. The basis of the 'mix' is often steatite (talc), which has a relative permittivity of about 8. But by the addition of other materials, notably titania, the relative permittivity may be increased to 100 or more; and with barium titanate and related compounds relative permittivities up to 5000 have been realized! These enable very compact capacitors to be made; but the stability is not good with the highest values of relative

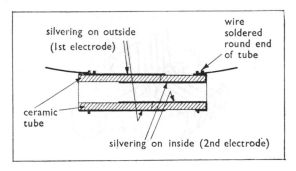

Fig. 10.21   A ceramic capacitor

permittivity. A useful property of some of these materials is that they have a *negative* temperature coefficient of relative permittivity. Most forms of capacitor show an increase of capacitance with temperature. But by combining a suitable ceramic capacitor in parallel with another type the resulting capacitance can be made almost independent of temperature.

A common form of construction is shown in Fig. 10.21. The purified material is pressed into a suitable shape, usually a small cylindrical tube. It is fired at a high enough temperature for vitrification to occur; this is necessary to make the ceramic non-porous. The cylinder is then silvered over parts of its inside and outside surfaces, as shown, thus forming the electrodes. Connection is made to these by means of wire rings soldered round the silvered layers at the ends of the tube. The whole tube is then given a coat of protective lacquer or is mounted inside another ceramic tube as a container.

Electrolytic capacitors

When an electric current is passed through a solution of ammonium borate using aluminium electrodes, a very thin film of oxide forms on the anode; this is an example of the process known as *anodic oxidation* or *anodizing*. The thickness of the film depends on the p.d. used and on the time of deposition. In the electrolytic capacitor the oxide layer is the dielectric; its thickness may be no more than 10^{-7} m, but it shows an insulation strength of about 10^9 V m^{-1}—vastly greater than any other available dielectric. An extremely compact unit of high capacitance can thus be produced. The two 'plates' of the capacitor are the *anode* and the *electrolyte* (to which connection is made via the cathode). Electrolytic capacitors come into their own particularly in low-voltage applications since the thickness of the dielectric layer can be reduced almost indefinitely. The physical size of a 6 V, 1000 μF capacitor may well be no more than that of a 350 V, 1 μF, plastic type. However, in order to maintain the electrolytic deposit it is necessary for a small 'leakage' current to flow continually through the capacitor in the right direction. This limits its use to circumstances in which an alternating p.d. is superimposed on a direct polarizing p.d. Care must always be taken to

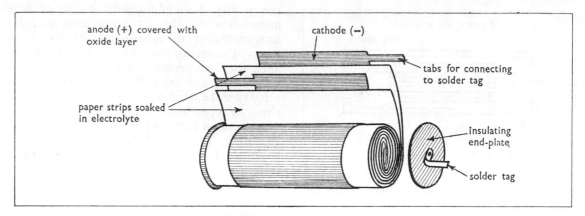

anode (+) covered with
oxide layer

cathode (−)

tabs for connecting
to solder tag

paper strips soaked
in electrolyte

insulating
end-plate

solder tag

Fig. 10.22   An electrolytic capacitor

connect an electrolytic capacitor so that the polarizing current flows through it in the right direction; one terminal is normally coloured to indicate that this must be kept positive with respect to the other. The stability is very poor, since the thickness of the dielectric film depends on the average p.d. at which the capacitor is working. However, these drawbacks are of no consequence in many applications, such as the storage and smoothing capacitors in rectifier circuits (p. 262). Their use is limited to low frequencies—up to about 10^4 Hz or 50 kHz in special cases.

A typical design is shown in Fig. 10.22. Two long strips of aluminium are interleaved with two strips of paper soaked in the electrolyte. The composite strip is rolled up into a cylinder. Tags are attached to the sides of the foil by which connection is made to the terminals of the capacitor. The whole arrangement is then sealed in a leak-proof container. Sometimes the case is made of aluminium, and is used as one terminal.

Even more compact capacitors can be produced by using *tantalum* foil instead of aluminium. In this case the polarizing current need be only a few micro-amps, so that its electrical properties compare favourably with those of a plastic capacitor.

Air capacitors

Capacitors of this type have almost perfect electrical properties at all frequencies. Such losses as do occur arise chiefly in the insulation used in supporting the plates. However, air suffers from the disadvantage of having very low insulation strength—about 1/20 of the insulation strength of mica or plastic. The spacing of the plates therefore has to be relatively large; also the plates themselves have to be thick enough to be mechanically rigid. Air capacitors are therefore extremely bulky, and *fixed* capacitance types are used only for laboratory standards.

However, air insulation provides an obvious means of constructing a *variable* capacitor (Fig. 10.23). It consists of two sets of interleaved parallel plates (or vanes). One set is fixed to the frame, the other to the central shaft. Rotation of this varies the area of the interleaved parts of the vanes, and therefore changes the capacitance. Miniature variable capacitors are also made with sheets of a plastic dielectric held loosely between the plates; this enables the

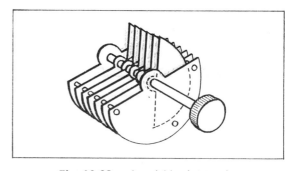

Fig. 10.23   A variable air capacitor

plates to be thinner and closer together than with air insulation alone.

10.4   Electrometers

Many experiments in this department of the subject require an instrument for measuring p.d.'s with an exceptionally high resistance. An ordinary moving-coil voltmeter is often quite inadequate because of the current it must draw. Even an instrument giving full-scale deflection for a current of only 1 μA would discharge a 1 μF capacitor within a few seconds; and often we shall be using capacitances much less than this. Sometimes we can use a cathode-ray oscilloscope as a high resistance voltmeter, but even this is found to take a significant leakage current when we are dealing with very small quantities of electric charge. We therefore need voltmeters operating on quite different principles that will draw currents of 10^{-12} A or less. Such instruments are called *electrometers*.

The most useful kind of electrometer for many purposes makes use of electronic components (transistors and thermionic tubes) to produce a high-gain *d.c. amplifier* or *measuring amplifier*, as it is often called (p. 272). The p.d. at the input terminals is amplified, and the output is connected direct to a moving-coil meter, which may be calibrated to read the values of p.d. at the input (Fig. 10.24a). The instrument is thus primarily a very high resistance voltmeter, but it may also be adapted to measure a very small current I by passing this through a known high resistance R joined across the input terminals (Fig. 10.24b), since we have

$$I = \frac{V}{R}$$

By this means currents of 10^{-12} A or even 10^{-13} A may be measured without difficulty.

Used in this way the electrometer amplifier enables us to measure the resistivities of many materials that are classed as insulators. If a p.d. of 100 V or 1000 V is applied to opposite faces of a thin slab of insulator, a current of a magnitude that may be measured with the electrometer will flow, and so the resistivity can be calculated (p. 23). Polythene and nylon have resistivities of 10^{14} Ω m or more; but poly-(tetrafluoroethene) or PTFE is at least an order

of magnitude better as an insulator, and must be used in electrical apparatus when the highest insulating qualities are demanded.

At low potentials and in the absence of any particular ionizing agency, the insulating property of air and other gases is almost perfect. However, many types of radiation are able to disrupt gas molecules into separate ions; the conduction of current through the gas can then go on as long as the radiation continues. These ionizing radiations are: (a) The shorter wavelength electromagnetic radiations, ultra-violet

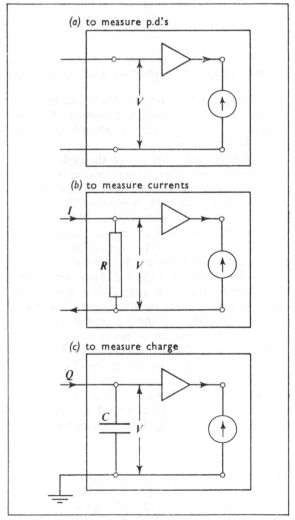

Fig. 10.24   A d.c. amplifier used as an electrometer: (a) to measure p.d.; (b) to measure current; (c) to measure charge

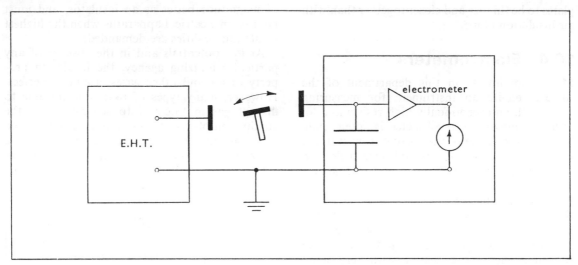

Fig. 10.25   'Spooning' equal quantities of charge from a high-voltage supply into an electrometer

rays, X-rays and γ-rays; (*b*) the rapid charged particles arising in nuclear processes, e.g. the α-particles and β-particles produced in radio-active disintegrations.

If ionizing radiation passes through the air between two electrodes, a small *ionization current* will flow when a suitable p.d. is maintained between them. An arrangement of this sort is called an *ionization chamber* (p. 314). At low p.d.'s a certain amount of recombination of the ions takes place before they are collected by the electrodes; under these conditions the ionization current increases with the p.d. However, if the p.d. exceeds a certain value, *all* the ions are collected before they can recombine; the current is then independent of the p.d. and is said to be *saturated*. Measurement of the saturation current enables the rate of production of ions to be found.

By joining a known capacitance C across the input terminals of an electrometer the instrument may also be adapted to give a direct reading of a quantity of charge Q delivered to the input terminals (Fig. 10.24*c*), since we then have

$$Q = CV$$

The capacitor slowly discharges through the input of the amplifier, but this leakage current is so small that the p.d. falls to a negligible extent over a period of some minutes while the measurement is being made.

Used in this way an electrometer enables us to measure small quantities of charge collected on another insulated conductor or stored on the plates of a small capacitor. For instance a small metal disc on an insulating handle (usually called a *proof-plane*) may be used to collect a sample of the charge on some other conductor, and then this may be measured by touching the proof-plane on one terminal of the capacitor across the electrometer input. In practice almost the *entire* charge on the proof-plane is given up to the capacitor when this is done; this may be tested by discharging the capacitor, and then seeing if there is any further charge on the proof-plane when it is touched a second time onto the input terminal. It is instructive to use an arrangement of this sort to take successive equal quantities of charge from a plate maintained at a high potential and to 'spoon' them into the electrometer capacitor (Fig. 10.25). It will be found that the reading of the instrument goes up in equal steps, confirming that the p.d. across the capacitor is directly proportional to the charge stored in it.

The same thing may be done to measure the capacitance C of a small capacitor. The capacitor is charged up to a suitable p.d. V (say 100 V), and then it is joined across the input capacitor of the electrometer, and so gives up almost its entire charge Q to it. Then we have

$$C = \frac{Q}{V}$$

Again we can test whether the capacitor does effectively give up its entire charge by seeing whether there is any further charge in it when the electrometer is discharged and the small capacitor joined to it a second time.

If the capacitance C is so small that there is insufficient deflection of the instrument to provide an accurate measurement, we can repeat the transfer of charge a number of times, recharging the capacitor each time from the source of e.m.f. (V), and then delivering the charge to the electrometer. In this way the total charge collected can be made large enough to give a reasonable meter reading.

If the capacitance C to be measured is comparable with the electrometer capacitance, we must use only low p.d.'s (one dry cell, say), and then we must make allowance for the incomplete transfer of charge that will occur when C is joined in parallel with the electrometer capacitance (see Question 19, p. 371).

Before the advent of efficient electronic amplifiers the only way of designing electrometers was to make use of the force that acts between two charged conducting surfaces. When a potential difference is produced between two conductors, the force acts in such a way as to draw them together, since the charges on the two conductors are of opposite sign. The force F is proportional to *each* of the charges carried on the conductors, and is therefore proportional to the *square* of the p.d. V between them.

$$\therefore F \propto V^2$$

One consequence of this behaviour is that the sensitivity of an electrometer of this kind increases with the p.d. These instruments are therefore insensitive for small p.d.'s, but are capable of measuring quite small *changes* in a large p.d.

The force F acts as an attraction between the electrodes whichever way round the p.d. is connected; electrometers are therefore suitable for measuring alternating p.d.'s as well as steady ones. Since the deflection depends on V^2, their readings indicate r.m.s. values (p. 142).

An instrument of robust design is the *electrostatic voltmeter* shown in Fig. 10.26. This consists of two sets of parallel plates (or vanes) interleaved with one another, rather like the

electrode system of a variable capacitor. One set of vanes is fixed to the frame of the instrument; the other (sometimes only a single vane) is pivoted on an axle between jewelled bearings, and carries a light pointer. The motion is controlled by two hair-springs; these fix the zero position with the vanes fully opened out. When a p.d. is applied, the electrostatic forces draw together the movable and fixed vanes. The pointer comes to rest at the position in which the deflecting couple due to the electrostatic forces is equal and opposite to the restoring couple due to the hair-springs.

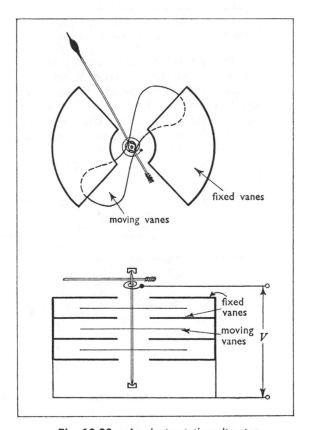

moving vanes

fixed vanes

fixed vanes

moving vanes

V

Fig. 10.26   An electrostatic voltmeter

The instrument with which most of the early investigations in electrostatics were made is the *gold-leaf electroscope* (Fig. 10.27). It consists of a rectangular piece of fine gold-leaf attached at its top edge to the flattened side of a brass rod. This is fixed in an insulating plug in the top of a conducting box with glass windows. A small metal

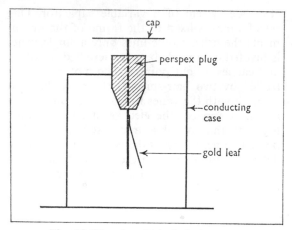

Fig. 10.27   A gold-leaf electroscope

The gold-leaf electroscope suffers from the defect of being rather fragile. To increase its robustness and at the same time achieve maximum sensitivity in the measurement of charge it is necessary to reduce the scale of the apparatus as much as possible and to make the moving part very light. This is the idea behind the design of the *fibre electrometer* (Fig. 10.28). A very fine metal-plated quartz fibre forms the moving part; this is attached at both ends to a wire support which is held by an insulating plug in the metal tube that forms the case of the instrument. The fibre is deflected away from its support when a p.d. is applied between it and the case. The restoring force is provided by the elasticity of the fibre, rather than by gravity as with the gold-leaf electroscope. The tube incorporates a low power microscope with a graduated scale in the focal plane of the eyepiece to enable the position of the image of the fibre to be registered. The scale usually covers a range of p.d.'s between 100 V and 200 V. But the capacitance of the electrometer is only about 1 pF, so that a flow of charge of 10^{-10} C onto or off the electrodes may be sufficient to take the fibre right across the scale. A particular use

disc, called the *cap* of the electroscope, is usually mounted at the top of the brass rod. When a p.d. is produced between the cap and the case, the electrical forces draw the gold-leaf out at an angle to the vertical, as shown. The instrument is mostly used for qualitative observations; but, if necessary, it can be calibrated against another voltmeter so that the angle of deflection indicates the p.d.

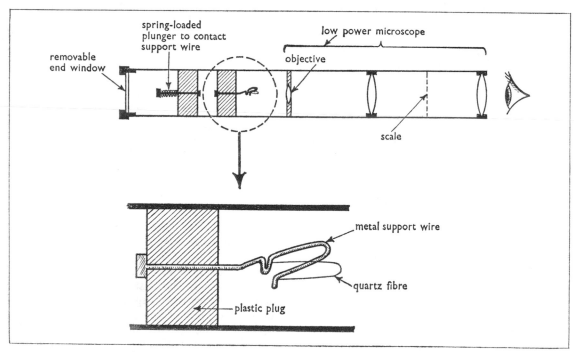

Fig. 10.28   A quartz fibre electrometer or radiation dosimeter

of this type of electrometer is as a *radiation dosimeter*, for monitoring the total dose of dangerous radiation received by people working with radioactive materials (p. 316).

10.5 Charging and discharging

In general the charging and discharging of a capacitor is an oscillatory process. This happens because of the inevitable self-inductance of the circuit in which the capacitor is joined. Even when the wire of the circuit is not formed into a coil, the magnetic flux produced by the current in it links the circuit and gives rise to small self-inductive effects. Thus, if a capacitor is discharged by joining a short length of wire between its plates, the circuit diagram of the arrangement should be as shown in Fig. 10.29, where L represents the small self-inductance of the wire and R its resistance. The variation of the p.d. across the capacitor during the discharge process is shown in Fig. 10.30. At the point A the switch is closed, but the self-inductance delays the growth of current and the discharge only gradually gathers momentum. At B the capacitor is fully discharged, but the rate of discharge—i.e. the current—is a maximum; the

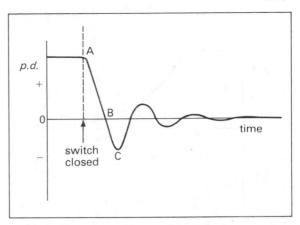

Fig. 10.30   The oscillatory discharge of a capacitor

effect of self-induction is now to maintain the current, thus charging up the capacitor in the reverse sense. As the p.d. grows in the reverse direction, the current slowly decreases until the point C when the p.d. across the capacitor is again momentarily steady. This process repeats itself many times over, giving rise to the oscillations shown in the figure. These are of decreasing amplitude since at each pulse of current some of the energy originally stored in the capacitor is turned into heat in the resistance R. Another cause of energy loss from such a circuit is the emission of electromagnetic radiation; it is found that radio waves are emitted from any circuit in which a rapidly changing current flows (p. 293). The discharge of a capacitor will usually be found to make a 'crackle' in a nearby radio set. However, the loss of energy from this cause is almost always much less than the dissipation of heat in the resistance of the circuit.

Increasing the resistance R of such a circuit leads to heavier damping of the oscillations, and for values of R above a certain amount no oscillation occurs at all; the p.d. then falls asymptotically towards zero without ever reversing (Fig. 10.31).

Similar effects occur when a capacitor is charged by connecting it to a source of e.m.f. If the resistance of the circuit is low, the p.d. oscillates before settling down to its final value. But in a high-resistance circuit the p.d. rises without oscillation towards its final value equal to the e.m.f. of the source.

The loss of energy as heat, etc. in the charging

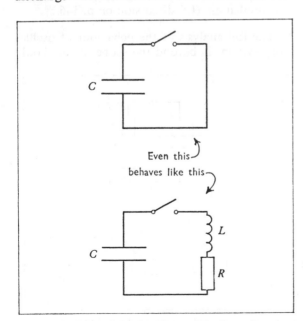

Fig. 10.29   The discharging of a capacitor

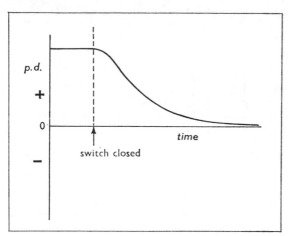

Fig. 10.31 The non-oscillatory discharge of a capacitor in series with a large resistance

process is the same as in the discharge of the capacitor. The source of e.m.f. must in fact supply twice as much energy as is eventually stored in the capacitor. Thus, if the e.m.f. is E and the charge stored Q, then

the energy stored in the capacitor $= \frac{1}{2}QE$

and the energy supplied by the source of e.m.f.

$$= \text{charge} \times \text{e.m.f.} = QE$$

$\therefore$ the energy lost as heat, etc.

$$= QE - \frac{1}{2}QE = \frac{1}{2}QE$$

This may be compared with the similar behaviour of a spring to which a weight is suddenly attached. The weight proceeds to oscillate about the position in which it will finally come to rest; the energy of oscillation is gradually dissipated as heat by frictional effects. If the weight is W and the final extension it produces x, then it may be shown that in the final rest position

the energy stored in the spring $= \frac{1}{2}Wx$

and the potential energy lost by the weight

$$= Wx$$

$\therefore$ the energy lost as heat $= Wx - \frac{1}{2}Wx = \frac{1}{2}Wx$

Again, the weight must supply twice as much energy as is eventually stored in the spring, exactly half of it being lost as heat as the oscillation is brought to a stop. If the frictional forces are much larger (suppose the weight is hanging in treacle), then the weight will move to its final position without oscillation; but it will still be true that exactly half the potential energy given up by the weight will be dissipated as heat by friction. A detailed discussion shows that the behaviour of oscillating mechanical systems is closely analogous to that of oscillating electric circuits. The inductance L of the electric circuit occupies the same place in the equations as mass (inertia) in the mechanical system; the capacitance C fulfils the same function as the stretching of the spring (or whatever else is responsible for the storage of potential energy); the resistance R corresponds to the frictional forces that damp the oscillation. (Cf. discussion on p. 146.)

The full analysis of the behaviour of oscillatory systems is beyond the scope of this book.

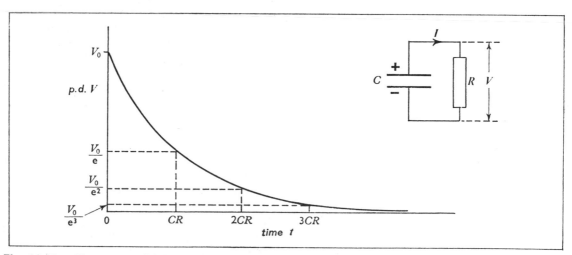

Fig. 10.32 The exponential decay of p.d. across a capacitor when the inductance of the circuit is negligible

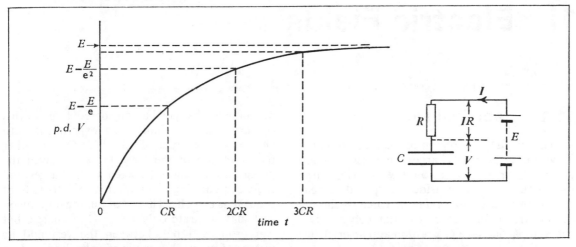

Fig. 10.33   The exponential rise of the p.d. across a capacitor charged through a high resistance

But we shall consider here the special case of the discharge of a capacitor through a *very large resistance*. The charge Q on the capacitor is given by

$$Q = CV$$

where V is the p.d. at any instant across the capacitor (Fig. 10.32). The *discharge* current I is therefore given by

$$I = -\frac{dQ}{dt} = -C\frac{dV}{dt}$$

Also $V = RI$

$$\therefore V = -CR\frac{dV}{dt}$$

Re-arranging and integrating,

$$\frac{1}{CR}\int_0^t dt = -\int_{V_0}^V \frac{dV}{V}$$

where $V_0 =$ the initial p.d.

$$\therefore \frac{t}{CR} = -\log_e\left(\frac{V}{V_0}\right)$$

$$\therefore V = V_0\, e^{-t/CR}$$

Thus the p.d. across the capacitor decays accord-ing to an *exponential law* (Fig. 10.32). The effective time taken by the decay process is measured by the product CR.

Thus, when $t = CR$, $V = \dfrac{V_0}{e} = 0.37V_0$

and when $t = 2CR$, $V = \dfrac{V_0}{e^2} = 0.14V_0$, etc.

The product CR has the dimensions of time; it is called the *time constant* of the circuit, and is measured in *seconds*.

A similar result is obtained when a capacitor is being *charged* through a large resistance by a source of e.m.f. E (Fig. 10.33). The differential equation in this case is

$$E = V + CR\frac{dV}{dt}$$

and the solution is

$$V = E(1 - e^{-t/CR})$$

The p.d. therefore rises exponentially (with the same time constant CR) towards its final value equal to the e.m.f. E of the battery.

11   Electric Fields

11.1   Electrical forces

An *electric field* is a region of space surrounding a system of electric charges; in such a region electrical forces will act on any electric charge placed there. Leaving aside the special nuclear forces, there are *three* types of force (including the electrical force) that can act between two charged particles—e.g. a proton and an electron:

(1) An *electrical* force;
(2) a *magnetic* force, if the particles are in motion (p. 7);
(3) a *gravitational* force.

All three forces are of the type that varies with distance according to an inverse square law—i.e. doubling the distance apart of the two particles reduces the force between them to a quarter as much, etc. But the three types of force are of very different orders of magnitude. The electrical force is incomparably greater than the gravitational force—the ratio is about 2×10^{39} for a proton and an electron. The magnetic force depends on the speeds of the particles, but is always less than the electrical force—much less, provided the speeds do not approach that of light. (In the limiting case of two charged particles travelling in parallel paths at the speed of light the magnetic and electrical forces would be equal.)

The magnitude of the electrical force is such that it is impossible to disturb the balance of protons and electrons in a piece of matter to any appreciable extent at all. For instance if we take a piece of copper wire 1 mm in diameter and remove from it 1 in 10^{13} of the electrons it contains, the electric field produced by the surplus positive charge left on it will be sufficient to disrupt and ionize the molecules of air in contact with the wire, thereby discharging it again. With a thicker wire the proportion that could be removed would be even less.

For this reason in experimental electrostatics we handle very small electric charges; a few microcoulombs will be considered large in this context, and we shall often experiment with charges of the order of 10^{-12} C! On the other hand because of the large forces involved the potential differences between such charges and their surroundings will often run into tens of thousands of volts. The act of pulling off a nylon vest may separate only 10^{-10} C of charge, but can produce a p.d. between the vest and its wearer of many thousands of volts, leading to a breakdown of the insulation of the intervening air and the familiar crackling sound! Even in a lightning discharge the quantity of charge concerned is only about 20 C; this much flows through a 250 V, 1 kW electric fire in five seconds.

There is thus a very big difference between the orders of magnitude of the charges and p.d.'s that we handle in electrostatics and in current electricity. This makes experimental work in the two departments of the subject appear very different, and the student must be prepared for some surprises. For instance many substances that we regard as excellent insulators at the low p.d.'s of current electricity will be treated as excellent conductors in electrostatics; a wooden ruler may have a resistance of 10^8 Ω; but this is sufficiently low to conduct away a charge of 10^{-10} C at a p.d. of 10^4 V in about 1 μs. In fact most ordinary insulators are quite inadequate for work in electrostatics on account of the film of moisture they carry on their surfaces except under especially dry conditions; glass, wood, ebonite, paper, the human body—are usually 'excellent' conductors. Without special drying the only adequate insulators are water-repellent substances, such as paraffin wax, nylon and many modern plastics. In most experiments we can assume that the bench, the walls, the experimenter and the earth are in good electrical contact. The only electric fields that exist will be between these objects and others that have been specially insulated and charged.

Electric fields may be described by means of

lines of force in much the same way as magnetic fields, the lines indicating the direction of the forces that would act on any charged particle placed in the field. The arrow on a line of force is always drawn to show the direction of the force that would act on a *positive* charged particle. In a line of force diagram the lines are therefore drawn starting from surfaces carrying positive charge and ending on those carrying negative charge. The lines thus give the directions in which current would flow (in the conventional sense) under the action of the field if a conducting substance occupied that region of space.

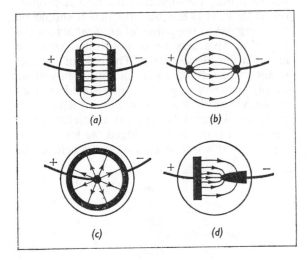

Fig. 11.1   Line of force diagrams of the electric fields of various arrangements of electrodes

It is possible to plot the lines of force of an electric field in a rather similar way to that used with magnetic fields (p. 65). It is found that a small particle or fibre tends to set itself with its longest axis parallel to a line of force. Particles of semolina are very convenient for the purpose. This material will float on a layer of tetrachloromethane (carbon tetrachloride) in a small dish; the evaporation of this rather volatile liquid can then be inhibited by pouring a layer of castor oil on top of it, and the particles float at the interface of the two liquids. Pairs of electrodes of various shapes can be inserted in the liquid and connected to a source of a few thousand volts. The particles of semolina immediately reveal the pattern of lines of force. A similar demon-

stration may be performed with small clippings of nylon or other such fibre.

Fig. 11.1 shows the patterns of lines of force found for a variety of arrangements of electrodes with the arrows inserted according to the usual rule. The simplest field is that between a pair of parallel plates (11.1*a*). Near the edges of the plates the field is non-uniform, varying in direction and magnitude from point to point; but in the central region between the plates the lines of force form a pattern of parallel straight lines across the gap between the plates; and the field is a *uniform* one. By this we mean that the force acting on a small charge placed in the field is the same in magnitude and direction at all points. This may be demonstrated roughly with the aid of a flexible test strip as described in the last chapter (p. 163). If the test strip is charged by touching it on one of the plates, and is then moved about to different positions between them, the amount it bends under the action of the electrical forces is clearly the same at all points—except near the edge of the gap between the plates, where the strength of the field falls off and is obviously no longer uniform. A more precise test of the uniformity of the field between parallel plates is provided by observing the movements of charged oil drops as in Millikan's experiment (p. 189). The force acting on such a drop is found to be precisely the same in all positions in the central region between the plates.

An electric field is the immediate cause of the flow of electric charge that constitutes an electric current. When an isolated conductor is placed in an electric field there is a momentary flow of charge within it so that part of it acquires a net positive charge and part a negative one. This can be shown with the aid of two small metal plates on insulating handles (e.g. proof-planes, p. 176), and an electrometer (p. 175). The proof-planes are introduced into the field, carefully avoiding contact with the charged electrodes; they are then brought into contact with one another face to face (Fig. 11.2). The action of the field separates opposite charges on the proof-planes as shown. They are now separated while still in the field and carefully withdrawn. The charges remain insulated on them; and, using the electrometer, they may be shown to be equal in magnitude and of opposite sign. The magnitude of the charge separated provides a

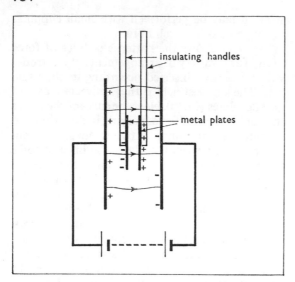

Fig. 11.2 The separation of charge in a conductor placed in an electric field

rough measure of the strength of the field. By repeating this at a number of points between a pair of parallel plates we may again demonstrate the uniformity of such a field. The pair of proof-planes may also be used to show that there is a strong electric field in the air near the terminals of a high-voltage supply unit. It is this electric field that causes charge to move in any conductor connected to the terminals and leads to a continuous current if there is a complete circuit. A source of e.m.f. of any kind is simply a device that uses an external supply of energy to produce a separation of charge within it. The unbalanced distribution of charge produces an electric field that acts on free electric charges in the circuit causing the flow of current.

We measure an electric field by the force that acts on unit electric charge placed in it. The unit of electric field strength is therefore the newton per coulomb ($N\ C^{-1}$). In fact the presence of an isolated charge of as much as 1 C in a laboratory would produce such vast electric forces that it would completely alter the electric field it was introduced to measure. It is assumed here and elsewhere that a test charge introduced to measure a field is small enough to produce negligible disturbance of the field. In the situations described in this chapter such charges will usually be of the order of 10^{-12} C or less. The force per

unit charge is known as the *strength of the field* and is denoted by E.

> The **strength of an electric field** E at a point is the force per unit charge that acts on a small positive electric charge placed there.

Thus the force F_e acting on a charge Q in the electric field is given by

$$F_e = QE$$

Uniform fields

We need to know how the intensity of the field between parallel plates depends on the p.d. V between the plates and their separation d. It is not surprising perhaps that the field is proportional to V. It is less obvious that it should be proportional to the *potential gradient* across the gap V/d. This may be demonstrated with the aid of a flexible test strip. The pair of parallel plates is connected to a suitable high-voltage supply and the charged test strip is held as usual between them. If V and d are now varied in such a way as to keep V/d constant, the deflection of the test strip remains unaltered (as long as the separation d is not so great that the field in the

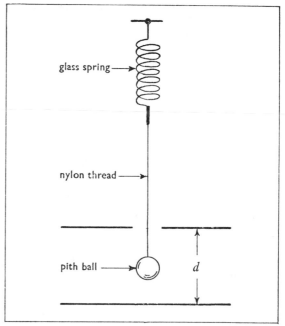

Fig. 11.3 Investigating the field between capacitor plates with a small charged object hanging on a glass spring

gap can no longer be considered uniform). Thus if the p.d. V and separation d are both doubled, the deflection of the test strip stays the same.

The test strip provides only a rough and ready indication of the strength of the electric field E. A better way is to use the arrangement shown in Fig. 11.3. A light metallized pith ball is suspended by a nylon thread from a fine spring (e.g. a glass spring). The two plates are arranged horizontally and connected to a source of p.d. V; the thread passes through a hole in the top one, so that the pith ball hangs freely in the space between the plates. When the pith ball is charged by touching it momentarily on the top plate, the electrical forces acting on it produce a slight extension of the spring. This is most easily measured by illuminating the space with a small source of light to one side and observing the shadow cast on a screen. The force is proportional to the extension produced in the spring. By this means it can be shown that the electric field is directly proportional to V and inversely proportional to d.

Thus we have

$$E \propto \frac{V}{d}$$

11.2 Electric potential

We need at this point to generalize our concept of electric potential. Hitherto we have used the term only to describe the electrical situation in a *circuit*. Current flows (in the conventional sense) from points at high potential in a circuit to points at low potential, and the strength of the current depends on the potential difference between these points (p. 12). But the same idea is useful for describing an electric field in an insulating medium such as the air. The electrical forces still act on any charged particle that happens to be present, and if it is a positive charge they act to drive it in the direction of the arrow along a line of force. Just as with an electric circuit we can say that the potential decreases as we pass along a line of force from the positive electrode towards the negative one.

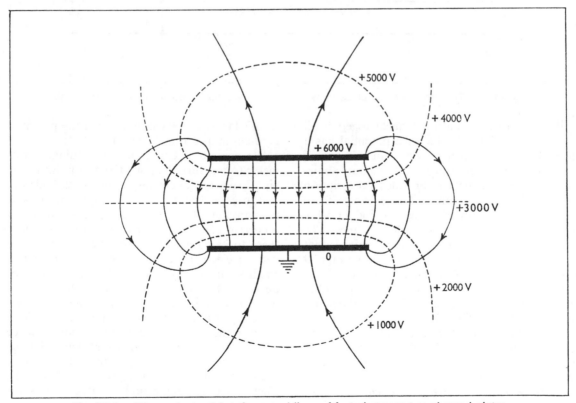

Fig. 11.4   The equipotential surfaces and lines of force between two charged plates

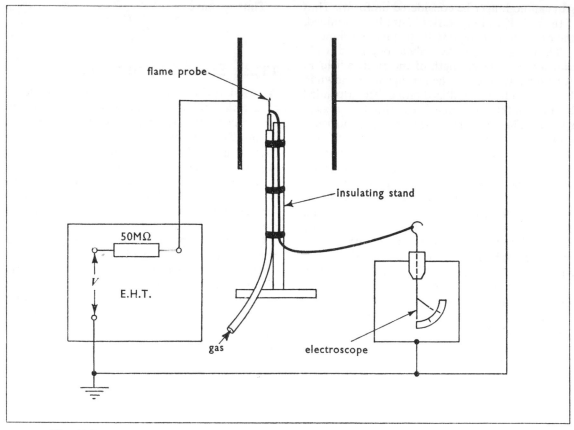

Fig. 11.5   The use of a flame-probe to measure the distribution of potential in an electric field

We may in imagination connect together all points in the field at which the potential has a given value; the surface formed in this way is called an *equipotential surface*. A diagram showing equipotential surfaces provides a description of the field complementary to the line of force picture we have already used. In Fig. 11.4 the equipotentials are shown by dotted lines. The direction of the line of force at any point is that in which the potential changes most rapidly with distance; this means that the lines of force are always at right-angles to the equipotential surfaces. In the central region between the plates in Fig. 11.4 the field is uniform, i.e. the equipotentials are planes parallel to the plates, and the potential decreases uniformly through the air gap from one plate to the other. As we have seen, the strength of the electric field E is directly proportional to the *potential gradient* across the gap. (Compare the discussion on p. 24, where

Ohm's law is expressed in terms of the potential gradient in a resistance wire.)

We cannot of course use an ordinary voltmeter to measure the potential at a point in the air. Even an electrometer or a cathode-ray oscilloscope draws a small leakage current, and in an insulating medium there is no way in which a current can flow to whatever probe we place there to connect to the instrument. Even if we had an electrometer that drew zero current, a small electric charge needs to flow into it to produce a deflection. It is true that such an instrument does indicate a reading when a probe joined to it is put at a point in an electric field, because the electric field causes a movement of charge between probe and electrometer. But the electric field would be altered to some extent by the charge induced on the probe itself, and the electrometer reading would not necessarily show the true potential in the absence of the probe.

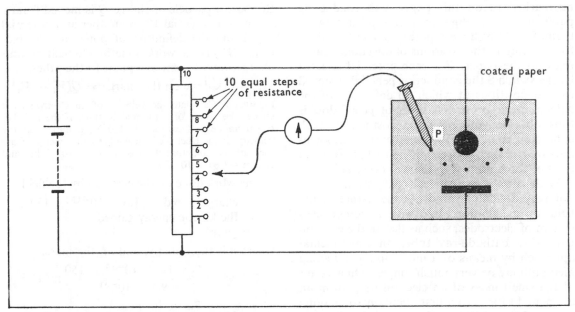

Fig. 11.6   Studying the distribution of potential on paper coated with a uniform resistance layer

We may get round this difficulty by the simple expedient of incorporating a source of ionization in the probe. The easiest way of doing this is to make the probe out of a hypodermic syringe on the end of which a minute gas flame is maintained. The flame produces a copious supply of ions which will move out into the surrounding air if there is any charge at all on the probe. When there is no charge on the probe, we know that the electric field nearby is undisturbed by it, and the distribution of potential in the field must be the same as though the probe were not there. In other words an electrometer joined to a flame-probe shows the potential at the position of the probe without any disturbance due to the measuring instrument. A gold-leaf electroscope is a suitable kind of instrument for the purpose; it may be calibrated beforehand so that the deflections give actual p.d.'s between probe and earth.

We may use the flame-probe and electroscope to study the distribution of potential in any field, such as for instance the field between the parallel plates in Fig. 11.5. One plate is joined to the case of the electroscope and earthed, and the other is joined to a suitable high-voltage source (not more than 1000 V for most gold-leaf electroscopes). The flame-probe may then be moved about to measure the potential at any required point in the field, and the shapes of the equipotential surfaces may be discovered both in the space between the plates and outside the gap.

If we wish to investigate the pattern of equipotential surfaces round a complicated system of electrodes it is usually easier, however, to make a *scale model* of the electrode system and to immerse this in a weakly conducting liquid. The p.d.'s between the electrodes can then be scaled down also, and the potential at any point can be measured with a high resistance voltmeter joined between the earthed electrode and a small metal probe placed at the point. The current flows along the lines of force in the liquid; and the potentials are the same at all points as though an insulator filled the space. Polarization in the electrolyte (p. 41) can create difficulties, but this can be avoided by using a low-frequency a.c. supply to the electrodes, and measuring the alternating p.d.'s between probe and earth with a suitable a.c. instrument (e.g. a cathode-ray oscilloscope).

In many cases it will be sufficient to study the nature of the field in two dimensions only, and we can then use a similar but much simpler procedure with paper coated with a uniform resistance layer. The positions of the electrodes can be marked in on this paper with conducting paint, and a suitable low p.d. applied between them. Fig. 11.6 shows a potentiometer technique for marking in the positions of the equipotential lines on the paper. For each potential selected by the switch S the conducting pencil P is moved to find positions at which no deflection of the galvanometer occurs. A series of pencil dots is thus obtained outlining the position of each equipotential line. From a map of the equipotentials we can then proceed, if we wish, to draw in lines of force normal to them at every point.

With simple geometrical configurations it is not very difficult to find out the nature of the field by calculation, but with a complicated system of electrodes, such as that in the electron gun of a cathode-ray tube, an experimental approach by means of a model in a conducting medium may be very much simpler. On account of the small mass of an electron its path in an evacuated tube is to a first approximation along a line of electric force, though this is of course modified by the action of any magnetic fields present.

Having generalized the concept of potential so that we can apply it to insulators through which electric fields pass as well as to conducting circuits, we need also to generalize the definition of potential (p. 12) so that we can proceed to analyse how the variation of potential from one point to another can be related to the strength of the electric field.

> The **electric potential** at a point is the work done by the forces of the electric field per unit positive charge moved from that point to a place at zero potential.

For most practical purposes the earth is taken as having zero potential. In the laboratory the walls, the floor, the bench top and the experimenter himself are all normally sufficiently good conductors to be regarded as being at a common zero of potential. In theoretical calculations on isolated systems it is usually convenient to take a point at an infinite distance from the system as the zero of potential.

A knowledge of the potentials at two points enables us to calculate the work done by the electrical forces in moving a charged particle from one point to the other, without having to work out the field strength at intermediate points.

Suppose a particle of charge Q is moved from a point at a potential V_1 to another at a potential V_2. From the definition of potential, the p.d. $(V_1 - V_2)$ is the work done by the field on unit charge moved from one point to the other.

$$\therefore \text{ work done on the charge} = Q(V_1 - V_2)$$

EXAMPLE   Calculate the velocity of an electron as it strikes the anode of a thermionic tube, if the p.d. between anode and cathode is 150 V; assume that the velocity of the electron is negligible as it emerges from the cathode, and that its mass is 9.1×10^{-31} kg, and charge -1.60×10^{-19} C.

The work done on the electron by the field

$$= \text{charge} \times \text{p.d.} = 1.6 \times 10^{-19} \times 150 \text{ J}$$
$$= \text{the kinetic energy gained}$$
$$= \tfrac{1}{2}m_e v^2$$

where v is the velocity, and m_e the mass.

$$\therefore v^2 = \frac{2 \times 1.6 \times 10^{-19} \times 150}{9.1 \times 10^{-31}} \text{ m}^2 \text{ s}^{-2}$$
$$\therefore v = 7.3 \times 10^6 \text{ m s}^{-1}$$

The electron-volt

The energy acquired by a charged particle accelerated by an electric field in a vacuum depends only on its charge and the p.d. through which it falls. A proton, having a larger mass than an electron, would acquire a smaller velocity with the same potential difference, but would gain the same kinetic energy. In any case in many atomic and nuclear calculations we are not much concerned to know the velocity; only the energy is important. It is therefore convenient in such applications to define a new unit of energy specially adapted for this purpose; it is called *the electron-volt* (eV).

> The **electron-volt** is the quantity of energy gained by an electron in accelerating through a p.d. of 1 volt.

The electronic charge $= 1.6 \times 10^{-19}$ C

$$\therefore 1 \text{ eV} = 1.6 \times 10^{-19} \text{ J}$$

Some idea of the magnitude of the electron-volt may be gained if we say that the energy of vibration of an atom of a solid at room temperature (say, an atom on the page you are reading) is about 0.03 eV ($= 4.8 \times 10^{-21}$ J).

EXAMPLE   Calculate the energy in eV and in J of an α-particle (helium nucleus) accelerated through a p.d. of 4×10^6 V.

(The charge on an α-particle $= 2e$, where $-e$ is the charge of an electron.)

The kinetic energy gained

$$= 2 \times 4 \times 10^6 \text{ eV}$$
$$= 8 \text{ MeV}$$
$$= 8 \times 10^6 \times 1.6 \times 10^{-19} \text{ J}$$
$$= 1.28 \times 10^{-12} \text{ J}$$

The relation between field strength and potential may be derived as follows.

Consider two points A and B a short distance δx apart in an electric field; let E be the component of the strength of the field in the direction $\overrightarrow{AB}$ (Fig. 11.7). If a charge Q moves from A to B, the work done by the electrical force F_e is given by

$$\text{work done} = F_e \cdot \delta x$$
$$= QE \times \delta x$$

If the potential difference between B and A $= \delta V$, we have

$$\delta V = - \frac{\text{work done by field}}{Q} = -E\,\delta x$$

(The potential *decreases* in the direction of the field; hence the minus sign.)

In the limit, as $\delta x \rightarrow 0$,

$$E = - \frac{\mathrm{d}V}{\mathrm{d}x} = -(\text{potential gradient})$$

This enables us to calculate the field given the potential or vice versa. The result is of course consistent with what we obtained experimentally with the test strip or with the glass spring apparatus described above (p. 184). Because of this relationship it is common to measure electric fields in volts per metre (V m^{-1}) rather than in N C^{-1}. The student can readily show that the two units are identical (cf. discussion on p. 104).

In the case of the uniform field between parallel plates, the potential decreases uniformly as we pass across the gap from one plate to the other. We therefore have in this case

$$E = \frac{\text{p.d.}}{\text{distance apart}}$$

As an example of the use of this result, with the glass spring apparatus described above (Fig. 11.3) we could if we wish actually measure the charge Q carried by the pith ball. For this purpose the glass spring must be calibrated by

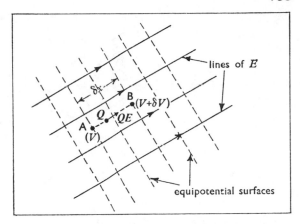

Fig. 11.7   Deducing the connection between the electric field strength E and the potential gradient

hanging known small weights on it so that its extensions indicate the actual force F_e (in newtons) acting on the pith ball in the electric field E. Then we have

$$E = \frac{F_e}{Q}$$
$$\therefore \ Q = \frac{F_e}{E} = \frac{F_e d}{V} \quad \text{(p. 184)}$$

where $d =$ the separation of the plates and $V =$ the p.d. between them.

The result is of no great significance, except that it is instructive to find the small magnitude of the charge Q involved (probably less than 10^{-12} C). But the same principle is used in the very important experiment devised first by Millikan for measuring the charge carried by small oil drops, which we now describe.

11.3 The charge of the electron

Millikan's experiment

The principle of this experiment is to measure the electrical force acting on a charged oil drop in a known electric field, and so to find the charge it carries. The drops used are of microscopic size and the charges they carry are many orders of magnitude smaller than the smallest charge that could be measured by any other means.

Millikan found that the charges on such oil drops are always integral multiples of a certain

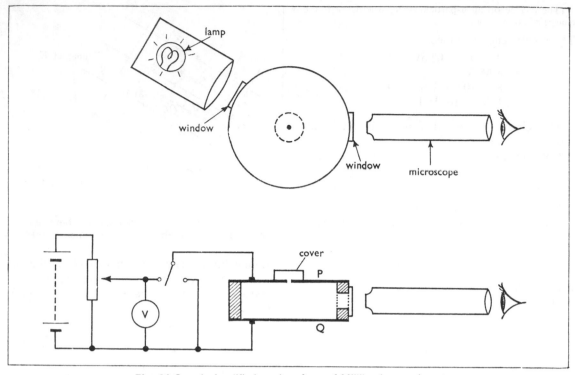

Fig. 11.8   A simplified modern form of Millikan's experiment

smallest quantity of charge e; i.e. the charges obtained were

$$\pm e, \quad \pm 2e, \quad \pm 3e, \ldots \quad \text{and so on}$$

The charge was never $\frac{1}{2}e$, $1\frac{1}{2}e$ or any other fractional multiple. The experiment demonstrates conclusively the atomic nature of electric charge. Millikan assumed that e was the magnitude of the charge of one electron; and this is in agreement with the results of other methods of measuring the electronic charge. Millikan's experiment has not proved eventually to be the most precise method available. But other methods do no more than measure the *average* charge of electrons. It is the special achievement of Millikan's experiment that it demonstrates that all electrons (and presumably protons too) carry the *same* charge.

A simplified form of Millikan's apparatus is shown in Fig. 11.8. The electric field is produced by a steady p.d. applied between two horizontal plates P and Q; these are held exactly parallel by an accurately made spacer ring of insulating material. The ring has two windows let into it, as

shown. The top plate has a small hole in the centre through which oil drops are allowed to fall from a spray. Frictional effects in the nozzle of the spray result in at least some of the oil drops being charged. A beam of light is concentrated into the space between the plates through one window; and a low-power microscope at the other is used to observe the drops by means of the light scattered from them. They are seen as sharp points of light against a relatively dark background. The microscope has a graduated scale in its eyepiece, by means of which distances in the object plane (where the drops are located) can be measured.

When the field is switched on between the plates, the motion of some of the slowly falling drops is reversed because of the electrical forces acting on them; a suitable drop is selected, and by switching off and on alternately it may be held near the centre of the space until all other drops have landed on one plate or the other. Measurements are now conducted on this single oil drop. The oil must be of the type used in vacuum apparatus; this has a very low vapour pressure so that evaporation of the drop is slow, and its weight remains practically constant for a considerable period.

First, with the electric field switched off (and the plates connected together) the drop is allowed to fall under the action of gravity. Using a stop-watch the time is measured for the spot of light to move between two selected divisions of the eyepiece scale; hence the velocity of the drop is obtained. The drop is so small that it reaches its terminal velocity almost at once; the viscous force due to the air is then exactly equal to the weight. According to Stokes's law the force F acting on a sphere of radius r moving with velocity v through a medium of viscosity η is given by

$$F = 6\pi r \eta v$$

Now, the force of gravity on the drop $= \frac{4}{3}\pi r^3 \rho g$ where $\rho =$ the density of the oil.

The upthrust of the air on the drop

$$= \text{the weight of air displaced}$$
$$= \tfrac{4}{3}\pi r^3 \rho_A g$$

where $\rho_A =$ the density of the air. The apparent weight W of the drop is then given by

$$W = \tfrac{4}{3}\pi r^3 g(\rho - \rho_A)$$

If the terminal velocity $= v_1$, we have

$$W = \tfrac{4}{3}\pi r^3 g(\rho - \rho_A) = 6\pi r \eta v_1$$

$$\therefore r^2 = \frac{9\eta v_1}{2g(\rho - \rho_A)}$$

knowing r, we may then also calculate W.

The field is now switched on in such a direction as to oppose the force of gravity on the drop. A first approximation to the charge Q on the drop may be obtained by adjusting the p.d. between the plates until the drop is held exactly stationary. If the electric field strength is E, we then have

$$F_e = QE = W = \tfrac{4}{3}\pi r^3 g(\rho - \rho_A)$$

Now, $E = \dfrac{V}{d}$

where $V =$ the p.d., and $d =$ the separation of the plates.

$$\therefore Q = \frac{4\pi d r^3 g(\rho - \rho_A)}{3V}$$

However, this adjustment is not easy to make, and it is usually better to use a rather larger field and then measure the speed v_2 as the particle ascends between the plates again. Then

$$QE - W = 6\pi r \eta v_2 = W.\frac{v_2}{v_1}$$

$$\therefore \quad QE = \frac{V}{d}.Q = W.\frac{v_1 + v_2}{v_1}$$

$$\therefore \quad Q = \frac{Wd}{Vv_1}(v_1 + v_2)$$

The same drop is timed again and again as it falls and rises between the plates, and accurate values of v_1 and v_2 are obtained. The charge on the drop can then be changed by holding a radioactive source nearby. This slightly ionizes the air between the plates. The drop will soon collide with one or more ions and its charge is then changed, and with it the value of v_2. The measurements are repeated with many different quantities of charge on the drop—and again with other drops of differing sizes. It is found that, within the limits of error expected, the quantities of charge measured are always integral multiples of the smallest charge that can be obtained on a drop. This smallest quantity is taken to be the charge of the electron.

The accuracy of the result depends on the knowledge of the viscosity of air, which is difficult to measure with high accuracy. Also it can be shown that Stokes's law is not strictly applicable when the diameter of the sphere is comparable with the mean free path of the gas molecules.

The best values of e, the electronic charge, have been obtained indirectly from measurements of atomic quantities which theory shows to depend on e. One example has been given on p. 37. According to the ionic theory,

$$e = \frac{F}{L}$$

where $F =$ the Faraday constant, $L =$ the Avogadro constant. F may be found very precisely from electrolytic experiments. L is best found by X-ray diffraction experiments on suitable crystals (e.g. calcite); these give a very precise measurement of the spacing of the atoms in the crystal. From this we can work out the number of atoms per unit volume and hence L, the number of atoms per mole (p. 306). The presently accepted value is

$$e = 1.602 \times 10^{-19} \text{ C}$$

11.4   Permittivity

We have shown that the strength E of the electric field between the plates of a capacitor is equal to the potential gradient across the gap. Thus

$$E = \frac{V}{d}$$

where V = the p.d. between the plates, and d = their separation. We need now to consider how the intensity E is related to the charges on the surfaces of the plates that produce the field.

In the last chapter (p. 161) we showed how the charge Q carried on the plates depends on their area and separation. Thus we found that the capacitance C of a pair of plates is directly proportional to the area A of overlap of the plates and inversely proportional to d.

$$\therefore \ C \propto \frac{A}{d}$$

But, by definition $\quad C = \dfrac{Q}{V}$ (p. 165)

$$\therefore \ \frac{Q}{V} \propto \frac{A}{d}$$

Re-arranging, we have

$$\frac{Q}{A} \propto \frac{V}{d}$$

The quantity Q/A is the charge per unit area on each plate, and is referred to as the *surface density of charge* on the plates; it is denoted by the symbol σ (Greek letter 'sigma'), and is measured in C m^{-2}. On the other side of the proportionality above, V/d is equal to the electric field strength E. Therefore we can write this

$$\sigma \propto E$$

Thus the electric field strength E is directly proportional to the charge density on the plates σ. It is not affected by the separation d of the plates as long as the charge upon them remains constant. We may demonstrate this result with the flexible test strip used previously (p. 163). The plates are connected for a moment to a suit-

able high-tension source and are then *isolated* from it. If the insulation is good enough, the charge on the plates then remains constant. The test strip is charged and held as usual in the gap to indicate variations in field strength. When the separation d of the plates is now varied, the deflection of the test strip stays the same, as long as d is not so great that the field can no longer be considered uniform in the gap.

In this experiment it follows that the p.d. between the plates must *increase* with the separation if the potential gradient is to remain constant. If d is doubled, the p.d. must double also if E is to remain unchanged. This may be shown directly by joining the plates to the terminals of a gold-leaf electroscope. When the plates are charged and isolated, the p.d. between is then seen to increase as they are moved apart; and it falls when they are brought together again. [If the plates are kept *connected* to a constant high-voltage supply, the p.d. between them stays constant; and the field *decreases* as they are moved apart, since in this case the charge on the plates decreases.]

Since $\sigma \propto E$ we can therefore write for the field between parallel plates

$$\sigma = \varepsilon E$$

where ε (Greek letter 'epsilon') is a property of the insulator in the gap between the plates known as its *permittivity*. The permittivity of a vacuum is denoted by ε_0; it is also called the *electric constant*. Like μ_0 it can be regarded as a property of the Universe (p. 104); ε_0 is the constant that specifies the electric force that acts between those particles that carry electric charge, while μ_0 specifies the magnetic force, i.e. the force that arises additionally when those particles are in motion (p. 7). In magnetism we found that most substances through which the field acts have little effect on its strength; only with the few ferromagnetic materials was any considerable effect produced. In electric fields the position is quite different. All insulators produce a measurable effect on an electric field; this is small in gases, but in solids and liquids the field is usually *reduced* by a factor of between 2 and 10, and in a few materials by very much more than this.

In atomic terms this may be explained as follows. In an insulator the charges are *bound* in the sense that they cannot move away from the site in the crystal lattice to which they belong; but they are capable of small amounts of movement under the action of an electric field, the positive charges (atomic nuclei) in one direction and the negative ones (electrons) in the other. Suppose we have a slab of insulator between two parallel plates (Fig. 11.9a). When a p.d. is applied between them, the electric field acts on the bound charges in the insulator slightly displacing them. There is thus a small electric current in the insulator, called a *displacement current*, which flows momentarily while the electric field is changing; as soon as the field is steady, no further displacement of the charges takes place. The effect of this is to produce two layers of bound charges on the surfaces of the insulator; the field of these partly cancels out the electric field in the slab. The bound charge on the surface of the insulator is always less than that on the neighbouring metal plate, so that the field in the insulator is never completely cancelled. When the charges in an insulator are displaced in this way, we say that it is *polarized*.

Although a steady current cannot pass through an insulator, an electric field can; it is for this reason that an insulator is often referred to as a *dielectric*. By contrast, a piece of metal acts as a screen to an electric field. In a metal some of the charges (electrons) are completely free to move; under the action of an electric field they will continue in motion until they have taken up positions that reduce the field inside the metal everywhere to zero. Only while a current is flowing is there any electric field inside a conductor. When a sheet of metal replaces the slab of insulator between the plates of the capacitor (Fig. 11.9b), the charges induced on the surfaces of the sheet are *equal* to those on the neighbouring plates of the capacitor, thus reducing the field inside the sheet to zero.

The effect of an insulating medium in the gap is thus always to *reduce* the field compared with its value in a vacuum; or, to put it the other way round, a greater density of charge σ is required on the plates to produce a given field strength E in the gap. The permittivity ε of all insulating materials is therefore greater than ε_0. The ratio $\varepsilon/\varepsilon_0$ is called the *relative permittivity* of the material and is denoted by the symbol ε_r. (An older name for ε_r is the *dielectric constant* of the material.) To distinguish ε from ε_r we call the former the *absolute permittivity* of the material. The relative permittivity ε_r is a dimensionless quantity, and is equal to 1 for a vacuum, by definition. The absolute permittivity has the dimensions of

$$\left[\frac{\text{charge per unit area}}{\text{potential gradient}}\right]$$

and must be found for a given medium by experiment. The units of ε are therefore the $C\ V^{-1}\ m^{-1}$, but any dimensionally identical combination of

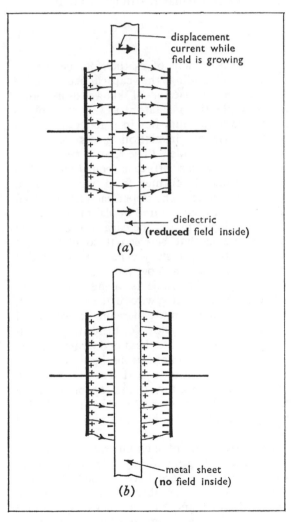

displacement current while field is growing

dielectric
(**reduced** field inside)

(a)

metal sheet
(**no** field inside)

(b)

Fig. 11.9   (*a*) The polarization of a dielectric, leading to a *reduction* of the field inside it. (*b*) The separation of charges in a metal, producing *zero* field inside it

units may of course be used. A simpler combination is the F m⁻¹, and this is usually used nowadays, since it is more obviously related to the way by which ε is measured.

To measure ε_0 we need to find what charge density σ is required on a pair of parallel plates to produce a given potential gradient (V/d) between them. Thus

$$\sigma = \frac{Q}{A} = \varepsilon_0 \frac{V}{d}$$

Since Q/V is equal to the capacitance C of the pair of plates, we may rearrange this to read

$$C = \frac{\varepsilon_0 A}{d} \qquad (I)$$

The measurement of ε_0 therefore becomes a question of finding the capacitance C of a parallel-plate capacitor of known dimensions. In principle the insulator between the plates should be a vacuum, but in practice it is sufficient to use air insulation unless very high accuracy is demanded. (The relative permittivity of air at the earth's surface is about 1.0005.) The capacitor may be formed from two horizontal rectangular plates. The upper one rests on a set of small accurately ground insulating spacers on top of the lower plate. The capacitance we need to measure is that between these plates. But inevitably there are stray capacitances included in the arrangement—between the unearthed plate and earthed objects round about, and between the connecting wires of the arrangement. The measurements must therefore be conducted in such a way as to allow for these stray capacitances and to eliminate them in the calculations. Also there are complications because of the region of non-uniform field that exists at the edges of the capacitor. If the gap between the plates is small, however, this effect is negligible at the accuracies to be reached in a school laboratory.

The capacitance of such an arrangement is of the order of 100 pF, and any suitable method of measuring capacitance may be used. The simplest is probably the vibrating reed method (p. 167). The p.d. V and the frequency of vibration f should be the maximum that the design of the apparatus allows. Even then the current I may be only a few microamps, and a sensitive mirror

galvanometer must be used. The total capacitance C is then given by

$$I = fCV \quad \text{(p. 168)}$$

$$\therefore \; C = \frac{I}{fV}$$

But this includes the stray capacitance C_0 as well as the capacitance between the two plates that is given by equation (I) above. These two capacitances are in parallel, and we therefore write

$$C = \frac{\varepsilon_0 A}{d} + C_0$$

To eliminate the stray capacitance C_0 we repeat the measurements for a range of values of gap width d. A graph is then plotted of total capacitance C against $1/d$, as in Fig. 11.10. The intercept on the axis of C gives the value of C_0 and the gradient is equal to $\varepsilon_0 A$. The precision of the straight line obtained shows that the capacitance is indeed inversely proportional to d, and at the same time shows that the edge effects are not significant to the accuracy obtained.

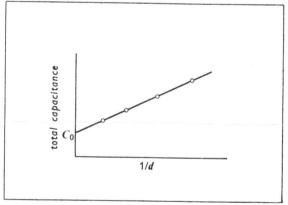

Fig. 11.10   The variation of capacitance with reciprocal of plate separation

By displacing one plate sideways it is also possible to investigate how the capacitance depends on the area A of overlap of the two plates, providing a further check on the theory of the parallel-plate capacitor.

The result obtained from such experiments is

$$\varepsilon_0 = 8.84 \times 10^{-12} \; \text{F m}^{-1}$$

The smallness of ε_0 is an indication of how small an electric charge is required to produce a large electric field. In other words it shows how enormous are the electric forces that act between the charged particles of matter (p. 183).

A leaky capacitor

Every insulator is to some extent imperfect and allows a small leakage current to pass. We have already seen how the lines of current flow in a conducting medium follow the lines of force of the electric field; in a parallel-plate capacitor this is straight across the gap between the plates. The area of cross-section of the slab of material in a capacitor (treated as a resistor) is A, while the 'length' of the conducting path is d. The resistance R of the slab of insulator is therefore given by

$$R = \frac{\rho l}{A} \quad \text{(p. 22)}$$

In the previous chapter (p. 181) we showed that the time taken by a capacitor to discharge is measured by the *time constant CR* of the capacitor and resistor combination. (In the time CR the p.d. across the capacitor falls to 0.37 of its initial value.) The time constant τ of the leaky capacitor is therefore given by

$$\tau = CR = \frac{\varepsilon_r \varepsilon_0 A}{d} \cdot \frac{\rho d}{A}$$

$$\therefore \ \tau = \varepsilon_r \varepsilon_0 \rho$$

The time constant of a leaky capacitor thus depends on the properties of the insulator between its plates, and is independent of its dimensions. The best modern insulators have time constants of thousands of hours.

Electric flux density

When a slab of insulator is introduced into an electric field between the plates of a capacitor, the material polarizes by the movement of the bound charges within it; and the effect is that the electric field is reduced by the factor ε_r compared with its value in the adjoining air gap (Fig. 11.9a, p. 193). Thus

$$E_{ins} = \frac{E_0}{\varepsilon_r}$$

where E_{ins} = field strength inside the insulator, and E_0 = field strength in the air gap. Now these field strengths are related to the charge density σ on the plates by

$$E_0 = \frac{\sigma}{\varepsilon_0}$$

$$E_{ins} = \frac{\sigma}{\varepsilon_r \varepsilon_0}$$

$$\therefore \ \varepsilon_0 E_0 = \varepsilon_r \varepsilon_0 E_{ins} = \sigma$$

Thus the quantity εE (where $\varepsilon = \varepsilon_r \varepsilon_0$) remains the same as we pass from one medium to another across the gap between the capacitor plates; furthermore it is equal in both media to the charge density σ on the surface of the plates. It is referred to as the *flux density* of the electric field.

> The **flux density** of an electric field at a point is the product of the absolute permittivity ε of the medium and the electric field strength E at that point.

Since the electric flux density is equal to the charge density on which it arises, it will be expressed in the same units as σ, namely $C \ m^{-2}$.

We have considered here only the case of the uniform electric field between parallel plates. But the result can be shown to apply quite generally. The electric flux density (εE) at the surface of a conductor is equal to the charge density on which it arises.

$$\varepsilon E = \sigma$$

In the general case of non-uniform fields the strength of the field varies rapidly as we move away from the surface, and the above result holds only for the field immediately in contact with the surface.

The field between parallel plates is in fact produced by the combined action of the charges of opposite sign on both the plates. It is worth analysing what must be the field produced by a *single* sheet of charge. If the charge is a positive one the field must be directed *away* from the charged surface in both directions; and we would expect to find the same flux density on both sides of it. This provides a consistent description of the pattern of electric fields inside and outside a parallel-plate capacitor (Fig. 11.11). For consider the field close to the surface of the

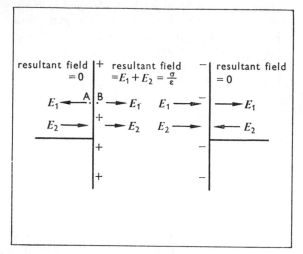

Fig. 11.11 The combined action of the sheets of charge on capacitor plates, producing zero field beyond the plates, and a field of strength σ/ε in between them

left-hand plate. It can be regarded as the sum of two components:

(*i*) The field E_1 due to the sheet of charge of density σ on the left-hand plate. This in isolation would amount to a field directed away from the surface layer of charge in both directions;

(*ii*) The field E_2 due to the opposite sheet of charge on the other plate. This component cannot change direction in the small distance between the two points A and B close to the left-hand plate and on either side of it. E_2 must therefore act to the right through this surface at both points.

At A, to the left of the left-hand sheet of charge, the resultant field strength is zero (both inside the material of the capacitor plate and in the space beyond). We therefore have

$$E_1 + E_2 = 0$$

But between the two capacitor plates the two components of field act in the same direction, and we have

$$E_1 + E_2 = \frac{\sigma}{\varepsilon}$$

$$\therefore\ E_1 = E_2 = \tfrac{1}{2}\frac{\sigma}{\varepsilon}$$

In other words precisely half the field strength between the plates is produced by each sheet of charge. We have considered here only the case of the uniform field between parallel plates. But this result can in fact be generalized to apply to the field at the surface of any charged conductor. In non-uniform fields E varies rapidly as we move away from a charged surface. But we find that the field can always be regarded as the sum of two components:

(*i*) The field E_1 due to the local charge on the surface, directed away from the surface in both directions;

(*ii*) The field E_2 due to the charge on the rest of the conductor and on other charged objects nearby; this cannot change direction in passing a short distance from one side of the surface to the other.

Inside the material of the conductor the resultant field is zero (if no currents are flowing in it). Again we find that just half the field at the surface is produced by the local surface charges, and half by the more distant charges.

$$\therefore\ E_1 = E_2 = \tfrac{1}{2}\frac{\sigma}{\varepsilon}$$

The force on a charged surface

Consider a small element of area δA of the surface of a conductor carrying a charge density σ. To calculate the force F on the element we need to find the product of the charge $\sigma.\delta A$ that it carries and the field strength E_2 due to the *other* charges (i.e. those not on the element δA itself); the field E_1 due to the charge on the element cannot contribute to this force.

$$\therefore\ F = \sigma.\delta A.E_2 = \frac{\sigma^2.\delta A}{2\varepsilon}$$

$$\therefore\ \text{force per unit area} \left(\frac{F}{\delta A}\right) = \frac{\sigma^2}{2\varepsilon}$$

Whatever the sign of σ, this force acts *outwards* normal to the surface. In some respects it resembles a hydrostatic pressure acting inside the surface.

When a potential difference is produced between two conductors, the surface forces act in such a way as to draw them together. This is the nature of the deflecting force in many electrometers and electroscopes. In these the deflecting force is proportional to σ^2; the charge on the moving part is proportional to the p.d. V between the electrodes. Therefore, as we have already seen (p. 177)

$$F \propto V^2$$

The same force acts as an attraction between the plates of any charged capacitor. With parallel plates in air, if the area of overlap of the plates is A, the total force F acting over the whole surface area of one plate is given by

$$F = \frac{\sigma^2 A}{2\varepsilon_0}$$

If the p.d. between the plates is V and their separation d, then we have

$$\sigma = \varepsilon_0 E = \frac{\varepsilon_0 V}{d}$$

$$\therefore F = \frac{A}{2\varepsilon_0}\left(\frac{\varepsilon_0 V}{d}\right)^2 = \frac{\varepsilon_0 A V^2}{2d^2}$$

By measuring the force F and the dimensions of the apparatus it is thus possible in principle to make an absolute measurement of the p.d. V. This is done in the *attracted disc electrometer*, originally designed by Lord Kelvin. But the instrument is difficult to use satisfactorily, and has gone out of use for the absolute determination of the volt.

The above result throws an interesting light on the way in which energy is stored in a capacitor. We can imagine the plates of the capacitor moved slowly apart, the plates being isolated so that the charge density σ upon them remains constant; the field intensity E between them and the force F attracting them together are then kept constant also. The work done against the force of attraction as the plates are moved apart from a small separation to the final value d is therefore given by

$$\text{work done} = Fd = \frac{\sigma^2 A d}{2\varepsilon_0}$$

and we can regard this as energy stored in the extra volume of electric field created between the capacitor plates by the movement. This is of course consistent with the expression previously obtained for the energy stored in a capacitor (p. 167); for we found

$$\text{energy stored} = \tfrac{1}{2}CV^2 = \tfrac{1}{2}\frac{Q^2}{C}$$

Now $Q = \sigma A$; and $C = \varepsilon_0 A/d$

$$\therefore \tfrac{1}{2}\frac{Q^2}{C} = \frac{(\sigma A)^2}{2\varepsilon_0 A/d} = \frac{\sigma^2 A d}{2\varepsilon_0}$$

But $Ad = $ total volume created between the capacitor plates

$$\therefore \text{ energy per unit volume} = \frac{\sigma^2}{2\varepsilon_0}$$

$$= \frac{\varepsilon_0 E^2}{2}$$

since $\sigma = \varepsilon_0 E$.

This shows us that there is limit to the energy that can be stored in a capacitor of given volume, since for any insulator there is a maximum field strength E that the material can sustain without breakdown. For most insulators used in the construction of capacitors the insulation strength (i.e. the maximum electric field before breakdown) is between 10^7 V m^{-1} and 10^8 V m^{-1}, while the relative permittivity lies between 3 and 8. The maximum energy that may be stored is therefore of the order of 10^5 J m^{-3}. In an electrolytic capacitor the maximum field strength in the dielectric is about 10^9 V m^{-1}, and the maximum energy that can be stored is two orders of magnitude greater than for any other kind of capacitor—hence the small physical size of such capacitors. In air, the maximum electric field is about 10^6 V m^{-1}, and the maximum energy that may be stored is about 4 J m^{-3}.

Electric flux

The *flux* of an electric field through a surface can be defined in the same manner as the equivalent quantity for magnetic fields (p. 86).

> The **flux** of an electric field through a small plane surface is the product of the area of the surface and the component of the flux density normal to it.

Thus the flux Ψ of an electric field of strength E through a small plane surface of area A to which the field is inclined at an angle θ is given by

$$\Psi = \varepsilon EA.\sin\theta \quad \text{(Fig. 11.12a)}$$

If the field is uniform, we can use this definition to give the flux through a large area also. Thus, the flux Ψ crossing the gap between two capacitor plates is given by

$$\Psi = \varepsilon EA = \sigma A$$

where A = area of overlap of plates (ignoring edge effects). Now σ is the charge density on the plates, and σA is the total charge Q on one plate.

$$\therefore \Psi = Q$$

In other words the total flux crossing the gap between the plates is equal to the charge on one plate. In fact it can be shown that electric flux possesses the same property in every case. Namely, a body carrying a charge Q is found to be the source of an equal quantity of electric flux Q; this flux passes through the space around the charged body and ends on an equal and opposite charge on another body (or bodies) nearby. In the next section we shall describe experiments by which this result may be tested. But we shall not consider the general theoretical proof in this book.

To calculate electric flux in the general case, of non-uniform fields through large areas, we

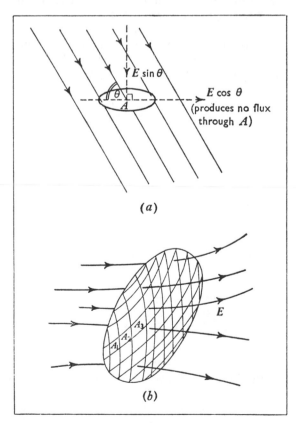

(a)

(b)

Fig. 11.12   Calculating the electric flux through (a) a small plane area A; (b) through a large area

have to resort to a process of summation or integration. Thus, to calculate the total flux through the surface depicted in Fig. 11.12b we imagine it divided up by a network of lines into small areas A_1, A_2, A_3, etc., each of which is small enough to be treated as plane. The total flux is then found by summation of the flux through the separate elements of area.

It is useful at this point to compare and contrast the properties of magnetic flux Φ (p. 86) and electric flux Ψ. Magnetic flux is produced by electric currents (whether currents in circuits or electron motions at the atomic level in magnetic materials). Magnetic flux always exists in 'closed loops'—or, as we say, magnetic flux is always continuous; whatever flux enters some part of a closed surface must emerge from it at some other part of the surface. It is also possible to produce electric flux in closed loops. This happens in fact in the familiar process of electromagnetic induction. When the magnetic flux through a circuit changes, an electric field is produced that acts in a closed loop—as evidenced by the induced e.m.f. in the circuit. Electric flux of this kind is always continuous also. However, this sort of electric flux arises only in the presence of *changing* magnetic fields. In static situations electric fields are produced by electric charges only; and each coulomb of positive electric charge is the source of one coulomb of electric flux, which spreads out into the space around the charge and ends somewhere else on an equal negative charge. In a closed space *that contains no electric charge* electric flux is always continuous; whatever flux enters the space must emerge from it at some other point, and the total flux out of the surface is zero. But if the closed space contains any charge, then the total electric flux leaving the space is equal to the quantity of charge contained. This result is known as *Gauss's Theorem*, and can be made the basis of the formal development of electric field theory. A detailed consideration of this theorem is beyond the scope of this book, though we shall have occasion to make use of Gauss's insight as an aid to understanding and analysing the nature of electric fields.

★   In studying the properties of magnetic materials we found it convenient to introduce another magnetic field vector H in addition to the magnetic flux density B with which most of the theory of magnetic fields was handled (p. 132). The connection between the two vectors was

$$B = \mu H = \mu_r \mu_0 H$$

where μ = the absolute permeability of the material. As long as we are dealing with straightforward materials for which μ is constant, the second vector H is redundant, since we have $B \propto H$ in all cases. However, with the important class of materials known as *ferromagnetics*, of which iron is an example, we find that μ does not have a single value, and the relation between B and H has to be described by a hysteresis loop (p. 126). The second vector H, called the magnetic field strength, is then a help in analysing the properties of these materials.

A similar situation exists in describing electric fields. In this case the vector that arises naturally in describing electric fields is the field strength E. This does not have the character of a flux density (like the magnetic vector B). But we have found that the product εE *does* have the nature of a flux density and is 'continuous' in the space between the positive and negative charges which are the sources of the field. As long as the insulating materials with which we deal have simple properties, i.e. if the permittivity ε is a *constant* of the material, there is no point in introducing a second electric vector to replace the product εE. There is, however, a small class of materials with irregular dielectric properties rather similar in a formal way to the magnetic properties of ferromagnetic materials. These are known as *ferroelectrics* (though they have nothing to do with iron!). The materials used in ceramic capacitors (e.g. barium titanate) are mostly of this kind (p. 173). In these materials the permittivity ε does not have a single value, and the relation between the electric flux density and the field strength E has again to be described by a hysteresis loop; i.e. the polarization of the material is not just a function of the electric field strength E in it, but depends also on the past history of electric fields in which the specimen has been placed. It is convenient then to introduce a second electric field vector to denote the electric flux density, and the symbol D is usually used. Thus

$$D = \varepsilon E = \varepsilon_r \varepsilon_0 E$$

If a ferroelectric material is introduced into the gap between a pair of capacitor plates, the charge Q on the plates is no longer proportional to the p.d. V between them. The flux density D is, as always, equal to the charge density σ on the plates, and is therefore proportional to Q. The electric field strength E is equal to the potential gradient across the gap, and is therefore proportional to V. The hysteresis loop (of D plotted against E) can thus be revealed by investigating the relationship between Q and V as the applied p.d. is taken through a suitable cycle of variation. Needless to say, the capacitance C has no clearly defined value in such a capacitor —a factor that needs to be taken into account in using some ceramic capacitors.

11.5   Fields near conductors

In the absence of any source of e.m.f. producing a continuous supply of electrical energy, the charges in a system of conductors reach equilibrium under the action of the electric field in a very short time (less than 10^{-12} s in most cases). When this has happened the field inside the material of a conductor must be everywhere zero. If not, the free charges in it will move under the action of the field until the distribution of charge is such as to make this so. Only if a current is flowing can the field inside a conductor be non-zero. It follows that under static conditions the potential inside the material of a conductor is everywhere the same. In particular, the surface of the conductor is an equipotential surface. Only when a current is flowing can the potential vary through the material, decreasing in the direction of current flow.

By similar reasoning we can see that the field at the surface of a conductor (under static conditions) is everywhere normal to the surface. If it were not, it would have a component parallel to the surface and would set in motion the free charges in the surface. It follows that the equipotential surfaces very close to the conductor are everywhere *parallel* to the surface.

A hollow conductor

The charge on a hollow conductor is entirely on its *outside* surface. There is no charge at all on the inside surface of the hollow space. Strictly speaking this applies only for a *completely* closed-in hollow space, but in practice it is sufficiently clearly demonstrated by using a deep metal can on an insulating support. Faraday did his original experiments on this subject with an ice-pail, which happened to be to hand at the time. So the cocoa tins, etc., used in repeating this experiment are traditionally referred to as 'ice-pails'! The ice-pail is given a charge by raising it to a high potential. We now make a test of the charge on different parts of it with a proof-plane (Fig. 11.13). When placed on a conducting surface the proof-plane becomes for a moment electrically part of the surface, and collects a sample of the charge (if any) upon it. The charge may be tested by touching the proof-plane on the terminal of an electrometer. It is found that no charge can be collected from the *inside* of the ice-pail, provided the proof-plane is not touched too near the lip of it. The charge is found to reside entirely on the outside. [A word of warning here for the experimenter: if there is any frictional charge on the handle of the proof-plane, this will give rise to spurious electric fields inside the ice-pail when the proof-plane is introduced into it, and this may cause a flow of charge onto the disc of the proof-plane whenever it touches another metal object. Great care must therefore be taken to keep the handle free of frictional charges. If necessary it may be discharged by holding it for a few seconds near a small flame (p. 165).]

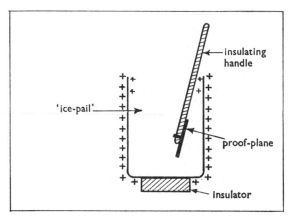

Fig. 11.13   No charge can be collected from the inside surface of a hollow charged conductor

A useful experimental technique may be derived from this result: if we wish to communicate the *entire* charge on a proof-plane (or any other metal object) to an electrometer, we may do this by attaching a hollow metal can to the terminal of the electrometer and touching the proof-plane to the *inside* surface of the latter; touching it on the outside surface would cause only part of the charge to be given up. In many cases the capacitance of the electrometer is so much greater than that of the proof-plane that virtually the entire charge is given up to the instrument in any case (p. 176). But sometimes this is not so, and then the ice-pail technique may be used to deliver up the entire charge for sure.

If there is no charge on the inside surface of a hollow space in a conductor, it follows that the electric field must be zero everywhere inside the hollow. An electric field can arise only from a charge, and if there is no charge inside there is no place where electric flux could start or end. The point may be tested with a gold-leaf electro-

scope. In general if the electroscope is placed in an electric field, a difference of potential is created between cap and case, and the leaf deflects. For instance a frictionally charged plastic rod produces a large electric field all around it, and causes the leaf of the electroscope to deflect when the rod is brought anywhere near the cap. A gold-leaf electroscope and cap is thus a sensitive indicator of the presence of an electric field. But if the electroscope is placed in a closed metal box, no deflection of the leaf can be produced in any circumstances. The box must have windows of metal gauze through which the instrument can be observed. Large electric fields may be produced near the box; or it may be stood on an insulating slab and raised to a high potential. In no case will any deflection be observed. There cannot therefore be any electric field inside the hollow space.

The result is of practical importance in the design of high-voltage apparatus. A hollow metal compartment may be raised to a high potential, and the apparatus and observers inside it are not in any way affected by the large fields outside. The ability of a closed metal screen to shield apparatus inside it from stray electric fields outside (and vice-versa) is also made use of in high-frequency radio equipment. Many of the components—coils, thermionic tubes, etc.—are mounted inside metal cans; without these the fields between different parts of the circuit would make its behaviour quite unpredictable. At high frequencies the cans also provide shielding from stray magnetic fields (p. 93).

Induced charges

If the picture of the electric field that we have been developing is consistent, we expect to find the electric flux that starts on a given quantity of charge ending on an equal and opposite charge at some other place. Thus an insulated electric charge should always cause an equal and opposite charge to flow onto its surroundings. The latter charges are referred to as *induced charges*; and we expect induced and inducing charges always to be equal in magnitude.

To test this result experimentally an ice-pail is connected to a gold-leaf electroscope; this is

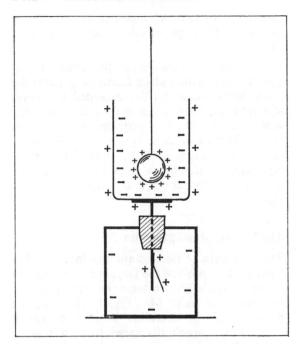

Fig. 11.14   Demonstrating that the induced charge is
equal and opposite to the inducing charge

often done by standing it on top of the cap. A
charged metal object is then lowered on an
insulating handle inside the ice-pail without
touching it. This induces an opposite charge on
the inside surface, and leaves a charge of the
same sign as the inducing charge on the outside
surface: a proportion of this charge is shared
with the electroscope. The flow of charge raises
the potential of the electroscope and a deflection
is observed. If the charged object is moved about
inside the ice-pail (Fig. 11.14), no change is
observed in the electroscope deflection, showing
that the induced charge is not affected by the
position of the inducing charge, provided it is
well inside the ice-pail. Now the charged object
is *touched* on the inside of the ice-pail; this is the
crucial test. We know from the previous experiment (Fig. 11.13) that the total charge inside the
ice-pail is now zero, and careful observation
shows that the deflection of the electroscope has
not changed at the moment of making contact.
It follows that the induced and inducing charges
exactly cancelled at the moment of contact, and
that therefore they were equal and opposite.

The same experiment may of course be performed with any kind of electrometer. An electronic one (p. 175) is in many ways more satisfactory because of its greater sensitivity. But
for this very reason greater care has to be taken
to eliminate frictional charges on the insulation,
which can have a very disturbing effect.

Charge density and surface curvature

On a charged conductor of irregular shape the
charge density varies from point to point of its
surface. A simple test with a proof-plane and
electrometer shows that the charge density is
always greatest at the most highly curved convex
parts of the surface. Thus the charge is found to
concentrate chiefly near corners and points,
where the curvature of the surface is greatest.
Samples of charge may be taken from various
parts of the charged conductor with the proof-
plane and then transferred to the electrometer.
It will be found that the deflection of the electrometer is greatest when the proof-plane has
been touched on the most highly curved parts of
the surface, less on any flat parts, and least on
any concave part (almost zero in any deep
concavity). The charge density also depends on
the distribution of other conductors and charges
in the neighbourhood. An earthed conductor
(the hand) placed nearby greatly increases the
charge density on the part of the surface nearest
to it; and a charged object held close to the
conductor may even produce a change of sign of
the charge at some part of the surface.

A consequence of this feature of the electric
fields near conductors is that it is possible,
without using very high potentials, to produce
a very large field locally near a sharp point or
fine wire. The field strength E at the surface of the conductor is proportional to the
charge density σ. Near a sharp point or fine wire
it may even be sufficient to ionize the air there.
Those ions which have the same sign of charge
as the conductor are violently repelled from it,
giving rise to an appreciable electric 'wind'. If a
lighted candle is held near a sharp needle raised
to a high potential, the wind can be sufficient to
blow the flame horizontal. It carries away the
charge on the needle into the surrounding air;
this is described as a *point discharge* or *corona
discharge*.

Any object placed in the way of the 'wind' may

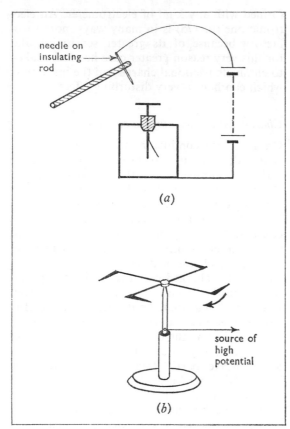

(a)

(b)

Fig. 11.15   (a) The corona discharge from a sharp
needle. (b) A windmill driven round by the electric wind
from the points

collect some of the charge on its surface. This
may be shown by holding a sharp needle connected to a high-tension source near the cap of
an electroscope (Fig. 11.15a); a rapidly increasing deflection will be observed, as the charge
blows onto the cap. The existence of the electric
wind may also be demonstrated with the electric
windmill shown in Fig. 11.15b. When this is
raised to a high potential, the electric wind
directed away from the points is sufficient to
produce rapid rotation. Corona discharge has to
be taken into account in the design of high-
tension equipment of all sorts:

(a) All parts of apparatus at a high potential
must be gently curved; all sharp corners or pro-
truding ends of wire must be eliminated. The
high-tension parts of a television set will be seen
to be designed like this. With equipment working

at millions of volts all high-potential parts are
covered with large metal domes of uniform
curvature.

(b) There is a maximum p.d. that can economi-
cally be used with a given diameter of cable for
power transmission. If this is exceeded, the power
loss into the air through corona discharge becomes significant. This provides an additional
reason for making large power cables for use in
air of aluminium (p. 23), since for a given
resistance an aluminium cable has a larger diameter than a copper one and so has a less curved
surface.

The Van de Graaff generator

This is a form of electrostatic machine that has
been used to provide the large p.d.'s needed for
atomic particle accelerators; the largest models
give p.d.'s up to 10 MV. Fig. 11.16 shows the
principle of the machine. Positive charge is
sprayed by corona discharge from a row of
points P on to a moving belt of insulating
material—usually rubberized silk. This is car-

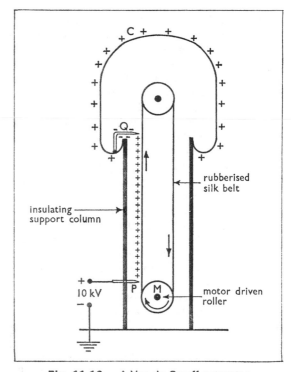

Fig. 11.16   A Van de Graaff generator

ried up into the hollow conductor C and induces a negative charge on the inside of it. This negative charge is sprayed by the comb Q onto the belt, neutralizing the positive charge. The net result is the continuous transfer of positive charge from the points P to the belt and so to the upper conductor. A considerable tension is set up in the belt due to the repulsion between the charge on the upper conductor and the charge being carried up on the belt. The energy is supplied in the form of work done against this repulsion by the motor that drives the lower roller M. The maximum potential reached is decided chiefly by the quality of the insulation round the machine. For the highest p.d.'s the entire apparatus must be enclosed in an outer shell in which the gas pressure is raised.

In the simplified type of Van de Graaff machine often used in schools the row of points P is earthed, and the action depends on the frictional charge developed on the inside of the belt by its contact with one roller (which is made of insulating material). This charge induces an opposite charge on the row of points P, and this is then sprayed onto the outside of the belt and carried up to the hollow conductor C.

11.6   Spherical and cylindrical fields

An isolated charged particle

The electric field of an isolated particle must have spherical symmetry. We assume that the total electric flux Ψ emerging from the particle is equal to the charge Q that it carries. Imagine a spherical surface of radius r concentric with the particle, and suppose that the strength of the field at the distance r (in air) from the centre is E. Then the flux Ψ passing through the spherical surface is given by

$$\Psi = \varepsilon_0 E . 4\pi r^2 = Q$$

$$\therefore E = \frac{Q}{4\pi\varepsilon_0 r^2}$$

If a second particle with charge Q' is placed at the distance r from the first one, the force F acting between them is given by

$$F = Q'E$$

$$\therefore F = \frac{QQ'}{4\pi\varepsilon_0 r^2}$$

The force between the two particles thus varies inversely with the square of their distance apart —a result known as the *inverse square law* for electric charges. We have derived the law theoretically from the assumption of *continuity of electric flux* from the charged particle, and it can be seen that the two results stand or fall together; one implies the other. With a different law of force between charged particles we should find, perhaps, the electric flux from a charged particle decreasing with distance from it; the result would be that the induced charges on neighbouring conductors would depend on their distance away, and quite certainly the induced charge would not always be equal and opposite to the charge inducing it. The ice-pail experiment described above (Fig. 11.14) would therefore yield a different result. In fact if any other law than an inverse square law operated, the method of describing an electric field in terms of an electric flux would be inconsistent. We might find fields inside hollow closed conductors, or charges on their inside surfaces. We might find regions of positive charge from which no flux emerged, or perhaps flux emerging from a region of positive charge and petering out in empty space rather than on an equal and opposite negative charge. Direct tests of the inverse square law are not easy or satisfactory, but indirect tests that rely on some result that would not hold on any other law of force are plentiful enough.

However, it is a good thing to attempt a check of so important a result by direct measurement. This may be done by measuring the force F between two charged metallized pith balls. One of these is fixed on an insulating support, while the other is suspended from a pair of long nylon threads (Fig. 11.17a). If the pith balls carry charges of the *same* sign, the repulsion between them deflects the suspended pith ball; the sideways displacement of this is best measured by observing its shadow cast on a screen by a small light source. If the displacement is small compared with the length of the supporting threads, the force F is proportional to the *horizontal* movement s of the pith ball (Fig. 11.17b). For, taking moments about the centre of rotation O of the pith ball, we have

$$mg.s = Fl.\cos\theta$$

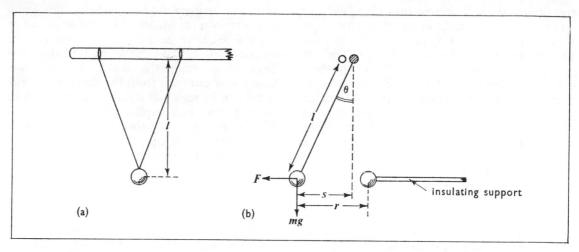

Fig. 11.17   Testing the inverse square law for the force between small charged particles

where $m =$ the mass of the pith ball, and $l =$ the length of the support system. If θ is small, $\cos \theta \approx 1$, and we have

$$F \propto s$$

It is therefore necessary to show that the horizontal movement s is inversely proportional to the square of the distance apart r of the pith balls; and both these distances may be found from the positions of the shadows on the screen. Leakage of charge from the pith balls produces uncertainties, but with care it is possible to confirm the inverse square law to an accuracy of about 10%. An experiment of this kind was first performed by the French scientist Coulomb, after whom the unit of electric charge was most appropriately named. The indirect tests referred to above have enabled the law to be confirmed to an accuracy of better than 1 part in 10^9.

If the two pith balls in the above experiment are charged in the same manner so that we can assume their charges to be equal, we can use the results actually to measure this charge. It is likely to be of the order of 10^{-11} C. The coulomb is a very large quantity of charge which it would be quite impossible to isolate on its own. It may be instructive to work out the force that would act between two isolated charges each of 1 C placed 1 km apart in a vacuum.

Now 1 km $= 10^3$ m.

$$\therefore F = \frac{1 \times 1}{4\pi \times 8.84 \times 10^{-12} \times (10^3)^2} \text{ N}$$
$$= 9 \times 10^3 \text{ N}$$

This is approximately equal to the gravitational weight of 1 tonne of matter at the earth's surface!

Electric potential near a point charge

The potential V at a distance r (in air) from a charged particle Q may be calculated as follows. The connection between the field strength E and potential V is:

$$E = -\frac{\mathrm{d}V}{\mathrm{d}r} \quad \text{(p. 189)}$$

Integrating between the limits r and ∞ and taking the potentials as zero at infinity, we have

$$V = \int_0^V \mathrm{d}V = -\int_\infty^r E.\mathrm{d}r$$
$$= -\int_\infty^r \frac{Q.\mathrm{d}r}{4\pi\varepsilon_0 r^2} = \frac{Q}{4\pi\varepsilon_0 r}$$

Thus the potential varies inversely with the distance r from a point charge. Using this result we may arrive at the potential due to any system of charges by a process of summation. The calculation of potential is almost always easier than the calculation of the field strength E in such cases. The reason is that the field strength E is a *vector*, while the potential V is a *scalar*. The addition of the field strength due to a number of charges can be a tedious process involving the parallelogram rule; but the potentials due to the charges may be compounded by simple addition.

A conducting sphere

Again the field has spherical symmetry, and the total flux arising from it is equal to the charge Q it carries. *Outside the sphere* the field is exactly the same as though the charge Q carried on its surface were concentrated at its centre. The field strength E at a distance r from the centre of the sphere (in air) is therefore given by

$$E = \frac{Q}{4\pi\varepsilon_0 r^2}$$

and the potential V is given by

$$V = \frac{Q}{4\pi\varepsilon_0 r}$$

If the radius of sphere is r_s, then at its surface the field strength E_s and potential V_s are given by

$$E_s = \frac{Q}{4\pi\varepsilon_0 r_s{}^2}$$

$$V_s = \frac{Q}{4\pi\varepsilon_0 r_s}$$

The potential has the same value as this throughout the interior of the sphere, where the field (which is equal to the potential gradient) is zero.

The variation of potential V with distance r from the centre of a sphere may be investigated with the aid of a flame-probe (p. 186). The sphere (e.g. a metallized football) needs to be suspended on insulating threads in the centre of a room, well away from walls, ceiling, floor, bench top, etc. It is joined by a fine wire (producing negligible field itself) to one terminal of a high-tension source, the other terminal of this being earthed. The flame-probe is then joined by a long, well-insulated lead to an electrometer by which the potential of the probe is indicated—a calibrated gold-leaf electroscope is the most convenient type for this purpose. The potential may be measured reliably with the electroscope and flame-probe up to about 2 m from the sphere, provided the experimenter keeps his own person well away from the probe and is very careful about the insulation of the wires going from probe to electrometer. In principle we could test the result graphically in either of two ways: (*a*) by plotting V against $1/r$, or (*b*) by plotting r against $1/V$; and we could expect to obtain straight lines in both cases. However, it is preferable to choose the latter method of displaying the results, since any uncertainty in locating the centre of the sphere on the scale that gives the distances r will not affect the nature of the graph beyond displacing the whole straight line sideways.

The capacitance of an isolated sphere

A conducting sphere carrying a charge $+Q$ necessarily induces a charge $-Q$ on its surroundings. We have assumed above that this induced charge $-Q$ is at a great distance, and have ignored any effect due to it. But its existence is always implied. When we talk about an isolated object like this as a capacitor, we are implying that the other 'plate' of the capacitor is another conductor completely surrounding the sphere and at a very great distance from it. The potential V of the isolated sphere is therefore the p.d. across the capacitor formed by the sphere and the imaginary conductor at infinity that forms the other electrode of the system. The capacitance C (in air) of the sphere is therefore given by

$$C = \frac{Q}{V} = 4\pi\varepsilon_0 r_s$$

Concentric spheres

A charge $+Q$ on the inner sphere induces a charge $-Q$ on the inside of the outer one, which we take to be at zero potential (Fig. 11.18). The electric flux passing radially across the gap

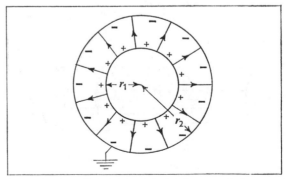

Fig. 11.18   Calculating the field strength between concentric spheres

between the spheres is also Q. The field in this gap is therefore the same as that due to a point charge $+Q$ at the centre of the system. Except for the region between the spheres the field is zero. The p.d. V is therefore given by

$$V = \int_{r_2}^{r_1} - E\,dr = - \int_{r_2}^{r_1} \frac{Q}{4\pi\varepsilon r^2}\cdot dr$$

$$= \frac{Q}{4\pi\varepsilon}\left(\frac{1}{r_1} - \frac{1}{r_2}\right)$$

where r_1 and r_2 are the radii of the spheres. The capacitance C is therefore given by

$$C = \frac{Q}{V} = \frac{4\pi\varepsilon}{(1/r_1 - 1/r_2)} = \frac{4\pi\varepsilon r_1 r_2}{r_2 - r_1}$$

If $r_2 \gg r_1$, this reduces to the expression for the capacitance of an isolated sphere.

Concentric cylinders

This is an important practical case, since a coaxial cable is in effect a capacitor of this kind. If the cylinders are very long, end effects can be ignored, and the field must be everywhere radial (by symmetry). Imagine a closed cylindrical surface S, of length l and radius r, between the two conductors and concentric with them (Fig. 11.19). The flux through S is $\varepsilon E.2\pi rl$, where E is the strength of the field at a distance r from the centre. If the charge on the inner conductor is q per unit length, then by Gauss's theorem we have

$$\varepsilon E.2\pi rl = ql$$

$$\therefore E = \frac{q}{2\pi\varepsilon r}$$

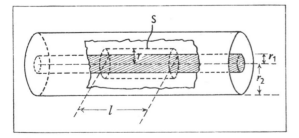

Fig. 11.19　A capacitor consisting of two concentric cyclinders (e.g. a coaxial cable)

But E is also the potential gradient

$$\therefore E = - \frac{dV}{dr}$$

$$\therefore -\frac{dV}{dr} = \frac{q}{2\pi\varepsilon r} \qquad (1)$$

(The minus sign implies that with *positive* charge on the inner conductor the potential *falls* with increasing r.) Integrating, the p.d. V between the two conductors is given by

$$V = \frac{q}{2\pi\varepsilon} \int_{r_1}^{r_2} \frac{dr}{r} = \frac{q}{2\pi\varepsilon} \log_e \frac{r_2}{r_1}$$

where r_1 and r_2 are the radii of the surfaces of the two conductors. The capacitance C of the length l of the arrangement is given by

$$C = \frac{ql}{V} = \frac{2\pi\varepsilon l}{\log_e (r_2/r_1)}$$

$\therefore$ the capacitance per unit length

$$= \frac{C}{l} = \frac{2\pi\varepsilon}{\log_e (r_2/r_1)}$$

A typical polythene-insulated coaxial cable has a capacitance of about 75 pF per metre.

A pair of long, parallel wires

This also is an important practical case, since the transmission lines used for power distribution and for telephones are often of this kind. To simplify the analysis we shall suppose that the radius r of the conductors is small compared with their separation b. This enables us to assume that the charges $+q$ and $-q$ on unit length of the wires is uniformly distributed on their surfaces, so that we may use equation (1) above to calculate the potential due to the charges carried by each conductor. To preserve the symmetry of the arrangement we shall take the zero of potential as that of the centre plane between the conductors (Fig. 11.20).

Let x_1 and x_2 be the distances of any point from the centres of the positive and negative conductors, respectively. Then the potential V_1 at the surface of the positive conductor *due to its own charge* is given by

$$V_1 = \int_{b/2}^{r} \frac{dV_1}{dx_1}\cdot dx_1 = -\frac{q}{2\pi\varepsilon_0} \int_{b/2}^{r} \frac{dx_1}{x_1}$$

$$= \frac{q}{2\pi\varepsilon_0} \log_e \frac{b}{2r}$$

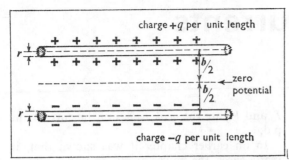

Fig. 11.20   Calculating the capacitance per unit length of a pair of parallel wires

The potential V_2 at the same point *due to the charge on the other conductor* is given similarly by

$$V_2 = \int_{b/2}^{b-r} \frac{\mathrm{d}V_2}{\mathrm{d}x_2} . \mathrm{d}x_2 = -\frac{-q}{2\pi\varepsilon_0} \log_e \frac{b-r}{b/2}$$

$$\approx \frac{q}{2\pi\varepsilon_0} \log_e 2 \quad (\text{since } r \ll b)$$

The total potential at this point is then

$$V_1 + V_2 = \frac{q}{2\pi\varepsilon_0} \left(\log_e \frac{b}{2r} + \log_e 2 \right)$$

$$= \frac{q}{2\pi\varepsilon_0} \log_e \frac{b}{r}$$

The potential at the surface of the negative conductor lies the same amount below zero. The p.d. V between them is therefore given by

$$V = 2(V_1 + V_2) = \frac{q}{\pi\varepsilon_0} \log_e \frac{b}{r}$$

The capacitance C of a length l of the transmission line is given by

$$C = \frac{ql}{V} = \frac{\pi\varepsilon_0 l}{\log_e (b/r)}$$

∴ the capacitance per unit length is

$$\frac{C}{l} = \frac{\pi\varepsilon_0}{\log_e (b/r)}$$

The factor $\log_e (b/r)$ lies in almost all cases between 3 and 6. The capacitance of a parallel-wire transmission line therefore usually falls between 4 and 8 pF m^{-1}.

12 Alternating Currents

12.1 Describing alternating quantities

Alternating currents and p.d.'s may have many different waveforms—saw-toothed, square-wave, sinusoidal, and many irregular forms; some examples are shown in Fig. 12.1. Each of these waveforms has its own special uses in electrical technology. But the simplest form, in terms of which all others may be analysed, is the *sinusoidal waveform* (Fig. 12.1c), expressed by

$$I = I_0 \sin \omega t \quad \text{(p. 28)}$$
$$\text{and} \qquad V = V_0 \sin \omega t$$

where $\omega = 2\pi \times$ frequency. If the alternating current is being generated by a simple two-pole alternator (Fig. 9.3, p. 150), then ω is the angular velocity of the rotor in radians per second;

I_0 and V_0 are the *peak values* of the current and p.d.

In an earlier chapter it was shown that, in many cases, the power produced in a.c. circuits may be calculated in the same way as for d.c. circuits (section 2.5, p. 28). This applies when the electric and magnetic fields of the circuit have negligible effect on the currents in it; that is, when capacitive and self-inductive effects are negligible and the currents are controlled almost entirely by the resistances of the circuit. In this case the current I in a resistance R and the p.d. V across it are proportional to one another at every instant, and we have

$$\frac{V}{I} = R$$

The current and p.d. are therefore *in phase*—i.e.

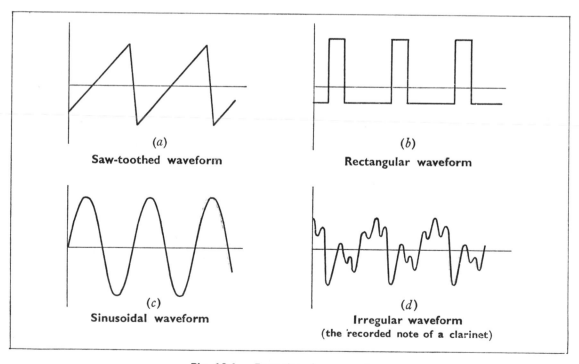

(a)
Saw-toothed waveform

(b)
Rectangular waveform

(c)
Sinusoidal waveform

(d)
Irregular waveform
(the recorded note of a clarinet)

Fig. 12.1   Examples of a.c. waveforms

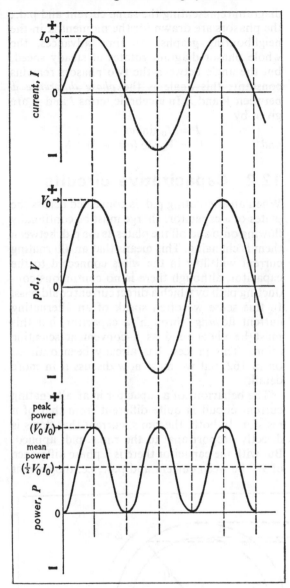

Fig. 12.2 Parallel graphs of current, p.d. and power dissipated, for a resistor

the peaks of current and p.d. occur at the same moments. The power P dissipated in the resistor *at any instant* is given by the three alternative formulae

$$P = VI = I^2R = \frac{V^2}{R}$$

and the *mean power* $\bar{P}$ by

$$\bar{P} = V_{rms}I_{rms} = I_{rms}^2R = \frac{V_{rms}^2}{R}$$

where I_{rms} and V_{rms} are the r.m.s. values of current and p.d. In the special case of a sinusoidal waveform it was shown that

$$I = \frac{I_0}{\sqrt{2}} \quad \text{and} \quad V = \frac{V_0}{\sqrt{2}} \quad \text{(p. 30)}$$

In terms of the peak values I_0 and V_0 the mean power $\bar{P}$ is then given by

$$\bar{P} = \tfrac{1}{2}V_0I_0 = \tfrac{1}{2}I_0^2R = \tfrac{1}{2}(V_0^2/R)$$

The relationship between these quantities is displayed in the parallel graphs of current, p.d. and power in Fig. 12.2.

But in general the behaviour of alternating current circuits is not as simple as this. When capacitive or inductive effects are appreciable, the currents are no longer controlled by the resistances only, nor are the currents and p.d.'s necessarily in phase—i.e. the peaks of current and p.d. may occur at different moments. The details of these processes are discussed later in the chapter; for the moment we are concerned with the way of describing these *phase differences*, as they are called.

Phasor diagrams

Any sinusoidal alternating quantity may be represented by the following geometrical model (Fig. 12.3):

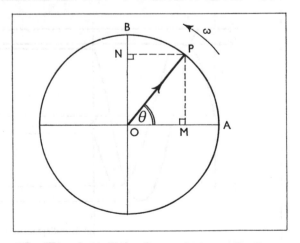

Fig. 12.3 The geometrical model of a rotating line, or phasor, to represent an alternating quantity

OP is a line (or *phasor*, as it is called), which we imagine to be rotating at a steady angular speed ω about the point O. If OP makes an angle θ with the line OA, then

$$\theta = \omega t$$

where t is the time measured from the point where $\theta = 0$. Then the projections of OP on OB and OA respectively are

ON $= $ OP sin ωt and OM $= $ OP cos ωt

We say that OP sin ωt and OP cos ωt are the *components* of the phasor OP in the directions OB and OA. Now, if the *length* of the phasor OP is made proportional to the peak value of the alternating quantity to be represented, then the component of OP in a suitably chosen direction gives its instantaneous value.

The angle θ in this model is called the *phase* of the current or p.d. Thus, the full description of a sinusoidal alternating quantity requires the specification of

(*i*) its frequency
(*ii*) its peak value
(*iii*) its phase

The phasor model provides a simple means of showing the relations between these quantities for two or more alternating currents or p.d.'s.

Fig. 12.4 shows the graphs of the current I through a piece of apparatus and the p.d. V across it; these are of the same frequency but are not in phase. Alongside is shown the phasor diagram representing the same current and p.d.; the phasors are drawn for the moment X in the neighbouring graphs. As time advances, the whole phasor diagram rotates at steady speed, but the angle between the two phasors remains constant; this angle is the *phase difference* ϕ between V and I. In algebraic terms I and V are given by

$$I = I_0 \sin \omega t$$

and

$$V = V_0 \sin (\omega t + \phi)$$

12.2 Capacitative circuits

When an alternating p.d. is connected across the plates of a capacitor, charge must be continually flowing onto and off the plates as the p.d. between them is changing. This means that an alternating current will flow in the wires connected to the capacitor, although there is no continuous conducting path by which a direct current could pass. In this sense we often speak of an alternating current flowing 'through' a capacitor, but this must be understood as a convenient scientific idiom. This process has already been outlined on p. 162, but we must now discuss it in more detail.

The behaviour of a capacitor in an alternating current circuit is quite different from that of a resistor. In both, the r.m.s. current that flows is directly proportional to the r.m.s. p.d. applied. But with the capacitor there is a phase difference of 90° between current and p.d.; i.e. the current

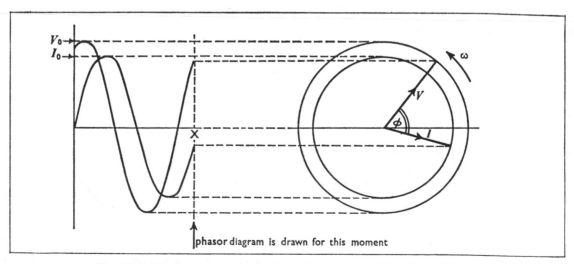

phasor diagram is drawn for this moment

Fig. 12.4 Curves of current and p.d., and the corresponding phasor diagram

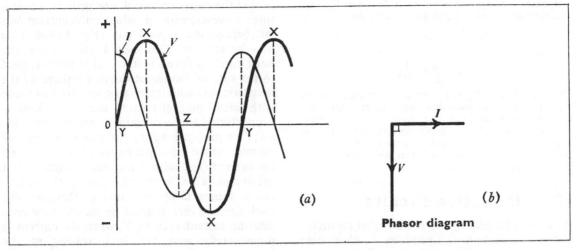

Fig. 12.5   Curves of current and p.d. for a capacitor, and the corresponding phasor diagram

reaches its maximum value at the instants when the applied p.d. is momentarily zero (Fig. 12.5*a*). This is often described by saying that the current and p.d. are *in quadrature*.

This behaviour can be explained by the following reasoning. The charge on the plates of a capacitor is directly proportional to the p.d. between them. At the moments marked X in Fig. 12.5*a* the applied p.d. is a maximum, but is momentarily unchanging; therefore the charge on the plates is momentarily constant, and the current flowing onto them (which is the rate of flow of charge) is zero. Now consider an instant such as Y, when the applied p.d. is momentarily zero, but is increasing at the maximum rate; the rate of flow of charge onto the plates (i.e. the current) is then also a maximum in the *positive* sense. Likewise, at an instant such as Z, the applied p.d. is again zero, but is now decreasing at the maximum rate; the current is again a maximum but in the *negative* sense. Thus the current and p.d. are 90° out of phase as shown. It will be noticed that each peak of current occurs *less far along the time axis* (i.e. earlier) than the corresponding peak of p.d. Thus the peaks of current are 90° ahead of the peaks of p.d. This is described by saying that '*the current leads 90° on the p.d.*' Fig. 12.5*b* shows the phasor diagram representing this situation.

If we increase the frequency *f* of the supply while keeping the applied p.d. constant, the same charge as before has to flow onto and off the plates, but it now has to do this in a proportionately shorter time. The current *I* is therefore proportional to the frequency.

$$I_{rms} \propto f$$

Increasing the capacitance *C*, while keeping the frequency and the applied p.d. constant, causes a larger charge to flow onto and off the plates in the same time, and so again increases the current. Thus

$$I_{rms} \propto C$$

★ Using calculus methods, the reasoning of this section can be put rather more briefly and effectively. Let the applied p.d. *V* be given by

$$V = V_0 \sin \omega t$$

where $\omega = 2\pi \times$ (frequency). Then the charge *Q* on the capacitor plates is given by

$$Q = CV = CV_0 \sin \omega t$$

Now

the current $I = \dfrac{\mathrm{d}Q}{\mathrm{d}t} = \omega C V_0 \cos \omega t$

$$= \omega C V_0 \sin (\omega t + \pi/2)$$

The current therefore leads 90° on the p.d., as represented in the curves of Fig. 12.5*a*. The expression shows also that the current is proportional to both the frequency and the capacitance. The peak current I_0 is given by

$$I_0 = \omega C V_0$$

$$\therefore \frac{V_0}{I_0} = \frac{1}{\omega C}$$

The same relationship also holds between V_{rms} and I_{rms}, the r.m.s. values of p.d. and current, since

$$V = \frac{V_0}{\sqrt{2}} \quad \text{and} \quad I = \frac{I_0}{\sqrt{2}}$$

$$\therefore \frac{V_{rms}}{I_{rms}} = \frac{1}{\omega C}$$

The quantity $(1/\omega C)$ is known as the *reactance* of the capacitor. It has the same dimensions as *resistance* and is therefore measured in Ω. But the two concepts should not be confused. In a resistance the current and p.d. are in phase, and heat is evolved. In a reactance the current and p.d. are in quadrature, and there is no net dissipation of energy (p. 214).

12.3 Inductive circuits

When an alternating current flows in an inductive coil (or *inductor*), it produces an alternating flux linkage with it. By the law of electromagnetic induction this gives rise to e.m.f.'s in the coil, and according to Lenz's law these must act in opposition to the applied p.d. The way in which the back e.m.f.'s can limit the current has already been discussed in general terms on p. 95; but we must now deal with the matter in greater detail. For the present we shall confine our attention to the case in which the self-inductance of the coil is large and its resistance small, so that the latter has negligible effect on the current.

The behaviour of an inductor connected to an alternating supply resembles that of a capacitor in that the current and p.d. are again in quadrature; however, with an inductor the current *lags* 90° behind the applied p.d. (Fig. 12.6a). This may be seen as follows. By Faraday's law, the back e.m.f. induced in the coil is proportional to the rate of change of the flux linkage of the magnetic field with it; for air-cored coils or those with 'ideal' magnetic cores, the flux linkage is proportional to the current. Therefore the back e.m.f. is proportional to the rate of change of current. At the moments marked X in Fig. 12.6a the current (and with it the flux linkage) is at a maximum value (in one direction or the other), but is momentarily unchanging; therefore the back e.m.f. is zero at these moments. Now consider an instant such as Y, when the current is momentarily zero but is increasing at the maximum rate; the back e.m.f. is therefore a maximum, but in the *negative* sense, since by Lenz's law it must act so as to tend to prevent the change of current. Likewise, at an instant such as Z, the current is again zero, but is now decreasing at the maximum rate; the back e.m.f. is therefore a maximum in the *positive* sense. Now, assuming the resistance of the coil to be small, the applied p.d. must at every instant be almost equal and opposite to the back e.m.f. to maintain the alternating current. It can be seen from Fig. 12.6a that the current and p.d. are 90° out of phase. Also the peaks of current occur further along the time axis than the cor-

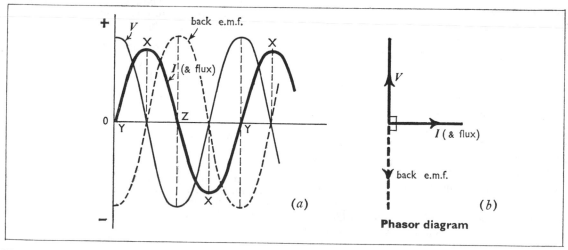

Fig. 12.6 Curves of current and p.d. for an inductor, and the corresponding phasor diagram (the flux is in phase with the current)

responding peaks of p.d., and thus occur *after* them. This is described by saying that '*the current lags 90° on the p.d.*' This situation is represented by the phasor diagram in Fig. 12.6*b*.

In this case an increase in the frequency *f* of the supply allows a shorter time for the changes of current to take place. But if the applied p.d. is to remain constant, then the back e.m.f. must also be constant, and therefore so also must the rate of change of current. The current is therefore inversely proportional to the frequency.

$$I_{rms} \propto \frac{1}{f}$$

If the self-inductance *L* of the inductor is increased, this again leads to an increase in back e.m.f. unless the current decreases in proportion. Therefore for constant applied p.d. the current varies inversely with the self-inductance.

$$I_{rms} \propto \frac{1}{L}$$

★ Using calculus methods we may again greatly shorten this reasoning. Let the current flowing in the inductor be given by

$$I = I_0 \sin \omega t$$

where $\omega = 2\pi \times$ (frequency). Then by Neumann's law (p. 87), the back e.m.f. *E* is given by

$$E = -\frac{d\Phi}{dt}$$

where $\Phi = $ the flux linkage of the magnetic field with the inductor.
Now

$$\Phi = LI = LI_0 \sin \omega t$$
$$\therefore E = -\omega L I_0 \cos \omega t = -\omega L I_0 \sin (\omega t + \pi/2)$$

The applied p.d. *V* is given by

$$V = -E = +\omega L I_0 \sin (\omega t + \pi/2)$$

The current and p.d. are thus 90° out of phase, and the p.d. is leading (or the current lagging), as represented in the curves of Fig. 12.6*a*. The expression also shows that the current is inversely proportional to both the frequency and the self-inductance. The peak p.d. V_0 is given by

$$V_0 = \omega L I_0$$
$$\therefore \frac{V_0}{I_0} = \frac{V_{rms}}{I_{rms}} = \omega L$$

The quantity ωL is known as the (inductive) *reactance* of the coil; like capacitive reactance it is measured in Ω.

12.4   Power calculations

Consider the general case of a circuit in which sinusoidal alternating current flows; suppose

there is a phase difference ϕ between the current and p.d. Then we can write

$$I = I_0 \sin \omega t$$
and $$V = V_0 \sin (\omega t + \phi)$$

The instantaneous power supplied *P* is given by

$$P = VI = V_0 I_0 \sin \omega t \sin (\omega t + \phi)$$

The mean value of this is no longer $\frac{1}{2} V_0 I_0$, as with a purely resistive circuit, in which current and p.d. are in phase. There is a well known trigonometrical identity, which in its general form runs

$$\sin A \sin B = \tfrac{1}{2}[\cos (B - A) - \cos (B + A)]$$

Applying this to the above expression, we have

$$P = \tfrac{1}{2} V_0 I_0 [\cos \phi - \cos (2\omega t + \phi)]$$

The first term in the bracket is constant, while the second term is an oscillating quantity whose mean value taken over a whole cycle must be zero. Therefore $\bar{P}$, the mean power supplied, is given by

$$\bar{P} = \tfrac{1}{2} V_0 I_0 \cos \phi = V_{rms} I_{rms} \cos \phi$$

The product $V_{rms} I_{rms}$ is sometimes called the *apparent power*. It is equal to the true power $\bar{P}$ in the special case of *V* and *I* being in phase ($\phi = 0$); $\cos \phi$ is known as the power factor of the piece of apparatus. Thus

$$power\ factor = \frac{true\ power}{apparent\ power}$$

This expression is taken as the definition of power factor in all cases, including those of non-sinusoidal currents and p.d.'s. It is only in the special case of sinusoidal waveforms that the power factor is equal to $\cos \phi$.

In specifying the maximum rating of many alternating-current devices, such as motors and transformers, it is common practice to state the maximum permissible apparent power rather than the true power. To avoid confusion it is usual then to quote apparent powers in V A or kilo-volt-amps kV A rather than in watts. The reason for this practice may be seen from a numerical example:

A transformer designed for connection to the 200 V mains is rated at 2 kV A. If the load connected to its secondary is such as to make the whole circuit behave resistively ($\phi = 0$), the maximum power that could be taken from the mains is 2 kW. The primary current is then

2000/200 = 10 A, and the heat dissipated in the windings of the transformer is the maximum that can safely be allowed. But suppose the secondary circuit is such as to make the power factor of the whole circuit 0.1. If we now allowed 2 kW to be taken from the mains, the apparent power would be 20 kV A, and the primary current would be 100 A, vastly more than the windings could tolerate without damage. Rather, we must ensure that the apparent power taken still does not exceed 2 kV A, which will keep the primary current below the allowed maximum of 10 A.

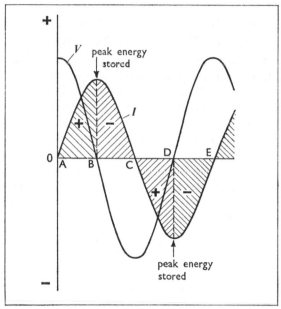

Fig. 12.7   The flow of energy when the current and p.d. are in quadrature

The special case of a phase difference of 90° between V and I is described by saying that the current and p.d. are *in quadrature*. In this case the power factor is zero, and the mean power taken from the supply is also zero; this applies even though a large current may be flowing at a large potential difference. The details of the flow of energy in this case merit closer study (Fig. 12.7). During the first quarter-cycle from A to B both V and I are positive, and the power supplied P ($= VI$) is also positive. During the next quarter cycle from B to C the current I is still positive but V is negative; thus the power supplied is negative, and the energy that was sup-

plied to the apparatus during the first quarter-cycle is returned again to the source of supply. This process recurs during the next half-cycle; from C to D energy is supplied to the apparatus (V and I both negative, and therefore P positive), and this returns to the source of supply from D to E. Although the mean power taken is zero, it is possible for considerable quantities of energy to be flowing into and out of the apparatus every quarter-cycle. In what form then is this energy stored in an entirely reversible manner? Now it can be seen from Fig. 12.7 that the *energy* stored in the apparatus reaches a maximum at the same moments as the *current*— namely, at the end of the interval AB, and again at the end of the interval CD. At these moments the magnetic field produced by the circuit is a maximum, and it is in the *magnetic field* that the energy is stored. This is in fact the situation that arises in an inductor.

If the curves for V and I are interchanged in Fig. 12.7, we have a similar situation; but now the *energy* stored reaches a maximum at the same moments as the p.d. This is the case in which the energy is stored in the *electric field* produced by the p.d., i.e. between the plates of a capacitor.

In the general case of a circuit containing resistance as well as capacitance or inductance there is a phase difference between the current and p.d. somewhere between 0 and 90°; the current leads on the p.d. in a capacitative circuit and lags in an inductive circuit. The phasor diagram in Fig. 12.8 shows the way in which the movements of energy may be analysed in the general case. The current I may be separated into two components:

$$I_R = I \cos \phi$$

in phase with the p.d. V; and

$$I_X = I \sin \phi$$

in quadrature with the p.d. The quadrature component I_X contributes nothing to the mean power supplied; it is responsible for the reversible flow of energy between the source of supply and the electric and magnetic fields of the apparatus; it is sometimes known as the *wattless* or *reactive component* of the current. The in-

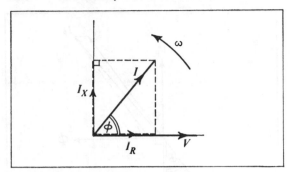

Fig. 12.8   The resistive and reactive components of the current in the general case

phase component I_R is responsible for the irreversible supply of power P to the apparatus; i.e.

$$P = VI_R$$

whose mean value $\bar{P}$ is given by

$$\bar{P} = \tfrac{1}{2}V_0 I_0 \cos\phi = V_{\text{rms}} I_{\text{rms}} \cos\phi$$

I_R is sometimes known as the *power* or *resistive component* of the current. The total current I is the *phasor sum* of the resistive and reactive components.

12.5   The theory of the transformer

This consists of two coils wound on a laminated closed iron core, as previously described on p. 82, Fig. 5.22. This design ensures that no free poles are produced when the core is magnetized; and there is very little demagnetizing effect. The maximum flux is therefore produced in the core for a given current. Also the magnetic coupling between the two coils is almost complete—i.e. nearly all the flux threads through all the turns of both coils. The a.c. supply is connected to one coil, called the *primary*. This is wound with a sufficient number of turns to reduce the no-load current I_0 in it to a low value.

The back e.m.f. E_p induced in the primary coil is nearly equal to the applied p.d. V_p.

$$E_p \approx V_p$$

The alternating flux in the core produced by the small primary current induces an e.m.f. E_s in the secondary coil. The same flux in the core threads equally through all the turns of both coils.

Therefore the e.m.f. induced per turn is the same throughout the transformer.

$$\therefore\ \frac{E_s}{E_p} = \frac{N_s}{N_p}$$

where N_s and N_p are the numbers of turns in the secondary and primary respectively. By suitable choice of the turns ratio (N_s/N_p), we can convert an alternating supply at one p.d. to a supply of the same frequency at any other p.d.—up or down. When current is taken from the secondary, there is a drop in the p.d. V_s across the coil because of its resistance; but, provided the current is not too great, this drop is small, and we can write

$$E_s \approx V_s$$

$$\therefore\ \frac{V_s}{V_p} \approx \frac{N_s}{N_p}$$

provided the transformer is not overloaded.

When a current I_s is taken from the secondary, by Lenz's law the effect of it must be to reduce the magnitude of the alternating flux in the core that caused it. If nothing further happened, this would lead to a drop in the primary back e.m.f. E_p. However, the current in any low-resistance coil always adjusts itself to such a value that the back e.m.f. induced in it is nearly equal to the applied p.d. Therefore a large increase I_p' occurs in the primary current so as to restore the alternating flux in the core to its original value. The increase in current I_p' is in exact antiphase with the secondary current I_s, and their effects on the flux in the core are equal and opposite. Now the flux produced by a coil wound on a given core is proportional both to the current and the number of turns.

$$\therefore\ N_s I_s = N_p I_p'$$

$$\therefore\ \frac{I_s}{I_p'} = \frac{N_p}{N_s}$$

In normal use the no-load current I_0 is small compared with the total primary current I_p, so that we can write

$$I_p \approx I_p'$$

$$\therefore\ \frac{I_s}{I_p} \approx \frac{N_p}{N_s}$$

This is the reciprocal of the turns ratio. If the p.d. is transformed up in a given ratio, then the current is transformed down in the same ratio (approximately), and vice versa.

The small losses in a transformer are classified under two headings according to the site at which the loss arises, and from which accordingly the heat must be carried away.

(*i*) *Copper losses.*   These are the ordinary heat losses that arise in the turns of any coil on account of its resistance. This loss ($I^2 R$) can always be sufficiently reduced by using thick enough wire.

(*ii*) *Iron losses.*   These are losses that occur in the iron core of the transformer; they arise from two different physical causes:

(*a*) *Eddy currents* (p. 91).   In the presence of an alternating magnetic field eddy currents will be generated in any solid lump of metal. By making the core out of sufficiently thin laminations this loss can always be reduced to any desired extent.

(*b*) *Hysteresis* (p. 121).   Some energy has to be expended in reversing the magnetization of the core. This appears in the iron as heat. The type of magnetic material for the core must be chosen accordingly.

★   The phase relations between the currents and p.d.'s for a loss-free transformer are shown in the phasor diagram in Fig. 12.9. The common factor between primary and secondary circuits is the flux Φ in the core, which links both coils equally; we therefore draw the phasor representing Φ first. When the secondary is on open circuit, the no-load current I_0 alone is responsible for the flux, and is therefore to be drawn in phase with Φ. Under these conditions the primary behaves simply as a high-inductance coil, and the applied p.d. V_p is therefore 90° ahead of I_0. The back e.m.f. E_p in the primary is nearly equal and opposite to V_p, and is therefore drawn in antiphase with it—i.e. lagging 90° behind the flux Φ. The same flux is also responsible for generating the secondary e.m.f. E_s, which is therefore in phase with E_p; the diagram in Fig. 12.9 is for a turns ratio of 1.5 : 1, so that $E_s = 1.5 \times E_p$.

When a current I_s is taken from the secondary, the phase difference between E_s and I_s depends on the nature of the load—resistive, capacitive or inductive. Fig. 12.9 is drawn for an inductive load, the phase angle being ϕ. The primary current now grows by an amount I_p', so as to maintain the flux in the core at its initial value against the action of I_s. Thus I_p' is drawn in antiphase with I_s (and of 1.5 times the magnitude). The total primary current I_p is the phasor sum of I_p' and the no-load current

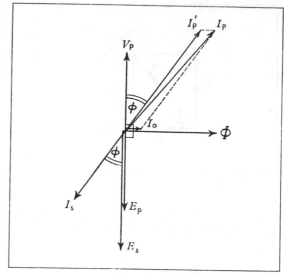

Fig. 12.9   The phasor diagram for a loss-free transformer. The diagram is drawn for the case of an inductive secondary load; with a resistive load ϕ would be zero

I_0. Under normal conditions I_0 is small, and I_p' and I_p are nearly equal both in magnitude and phase.

Since V_p is in antiphase with E_s, and I_p nearly in antiphase with I_s, it follows that the phase angle ϕ between current and p.d. is nearly the same for the primary and secondary circuits. For example, if a capacitor is joined across the secondary, the primary coil behaves also like a capacitor in the circuit in which it is connected. In general the power factors of primary and secondary circuits are always approximately equal.

The most important practical case is that of a resistive secondary load (power factor 1). The primary coil then also behaves like a resistance, and V_p and I_p are nearly in phase. Suppose the load joined to the secondary is R_s, then

$$\frac{V_s}{I_s} = R_s$$

Now

$$V_p = V_s \frac{N_p}{N_s}$$

and

$$I_p = I_s \frac{N_s}{N_p}$$

The primary coil then behaves as a resistance of magnitude R_p given by

$$R_p = \frac{V_p}{I_p} = \frac{V_s}{I_s}\left(\frac{N_p}{N_s}\right)^2$$

$$\therefore R_p = \frac{R_s}{(\text{turns ratio})^2}$$

12.6   Power transmission

As long as electric power is to be generated and used locally, losses in transmission lines are not a serious matter, and there is little point in using alternating current rather than direct current. In the early days of public electric supply under-takings, each town (and in the country the indi-vidual house!) had its own generating station, and d.c. systems were the rule.

The virtue of a d.c. system was that it enabled batteries of accumulators to be used to maintain the supply during off-peak periods, when it was best to switch off the generators. At other times of the day the generators produced surplus current, which was used to re-charge the accumulators.

However, there are many economic advan-tages in increasing the scale of an electric supply network. The fluctuations of power at different times of the day are smaller; the industrial centre may be switching off at about the same time as the nearby 'dormitory town' is switching on, and so on. Also, it is much cheaper to generate power in some places than in others; in a large system the power stations can be sited at the most economic points in the network—near ports or in places where hydroelectric schemes are possible. In addition, it is found that the larger the amount of power being generated the more efficiently it can be done.

Under modern conditions, therefore, enor-mous quantities of electric power have to be transmitted continually from one part of the country to another, and reduction of the losses in the transmission lines becomes of prime importance. The following analysis shows that it then pays to use *high p.d.'s* for the connecting links between different parts of the system.

Consider a supply system delivering average power $\bar{P}$ at a p.d. V_{rms}; suppose the total resistance of the pair of cables connecting the generator to the load is R (Fig. 12.10).

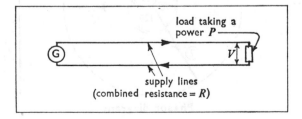

Fig. 12.10   The calculation of power loss in a pair of transmission lines

$$\therefore \ I_{rms} = \frac{\bar{P}}{V_{rms}}$$

$$\therefore \text{ the power loss in the cables} = \left(\frac{\bar{P}}{V_{rms}}\right)^2 . R$$

$$\therefore \text{ the percentage power loss} = \frac{\bar{P}R}{V_{rms}{}^2} \times 100$$

It is also important to keep the fluctuations of the supply p.d. small. When most of the load is switched off, the potential drop in the cables is suddenly reduced, and the supply p.d. may rise—with disastrous consequences in such equipment as is still switched on!

$$\text{the r.m.s. potential drop} = \left(\frac{\bar{P}}{V_{rms}}\right) . R$$

$$\therefore \text{ the percentage fluctuation of p.d. (r.m.s.)}$$

$$= \frac{\bar{P}R}{V_{rms}{}^2} \times 100$$

This is the same as the expression for the percent-age power loss. Therefore the same considera-tions apply in designing the system so as to keep both these quantities low. The maximum that can normally be tolerated is about 5% power loss and fluctuation of p.d. Both quantities are inversely proportional to the square of the p.d., and it is therefore essential to use high p.d.'s.

However, against this advantage must be set the increased capital cost of the installation that ensues. Insulation problems are immensely in-creased. Indeed at a p.d. of several hundred kV the loss from the surface of the cables by corona discharge (p. 201) becomes a significant factor. The power losses and fluctuations of p.d. can instead be reduced by increasing the diameter of the cable or by running several cables in parallel. The economic problem is one of achieving a nice balance between running costs and capital outlay for a given power to be transmitted.

The advantage to be gained from the trans-mission of power at high p.d.'s makes the use of alternating current inevitable. The practicable limit for d.c. generators is a few thousand volts. Alternators function readily at much higher p.d.'s; and then the use of a.c. transformers en-ables the p.d. to be changed to any desired value. Transformers are such efficient devices that the losses in them are relatively unimportant, even when a supply link includes several—some for transforming the p.d. up for transmission, and some for transforming it down again to the p.d.'s at which it is used. It is only in certain rather unusual circumstances that the transmission of power by direct current is preferable; such a case, for instance, is the cross-channel power link that connects England and France.

For many industrial and domestic applications there is little to choose between d.c. and a.c. supplies; lighting, heating, etc. are equally efficient on either; and a.c. motors are almost as satisfactory as d.c. ones. It is only for electrolytic processes, electronic apparatus, and for one or two other applications that direct current must be employed. However, even here efficient rectifiers and a.c./d.c. converters have been developed, so that there is no serious disadvantage in using an a.c. main power supply.

The three-phase system

This is the name given to the system in which a.c. is generated, transmitted and consumed in three separate circuits simultaneously. In the alternator, one rotor sweeps past three sets of coils in succession (Fig. 9.5, p. 151), so that the phases of the currents in the three circuits are equally spaced—i.e. the phase differences are 120°. We have already seen that this makes the action of the alternator smoother and more efficient (p. 150). It also leads to considerable economies in transmission, since a common return can be used for all three circuits; if the

three currents are then of equal amplitude, the current in the common wire is zero, and in principle it might be dispensed with altogether! Fig. 12.11 shows how this comes about. In the current/time graphs, the sum of the three currents at any instant is seen to be zero; thus, when one current is at its maximum value, the other two are each momentarily of half this magnitude in the opposite direction, etc. But the relationship may be proved easily from the phasor diagram representing the three currents; it is clear from this that the phasor sum of the currents is indeed zero. Even when the loads on the three circuits are not exactly equal, the phasor sum of the three currents is not large; and only a small balancing current needs to flow in the common return wire, which can therefore be of much narrower gauge than the main cables. Also this wire is normally earthed, so that no expensive insulators have to be used to support it. The advantage for power transmission of the three-phase system is therefore apparent; three cables (and the small common neutral wire) can be used where six cables of the same kind would be needed in a single-phase system. The capital cost of the cables is therefore approximately halved.

The phasor diagrams show that the same result holds also for any polyphase system; if the currents are equal, their phasor sum is zero, and no current flows in the common return wire. It

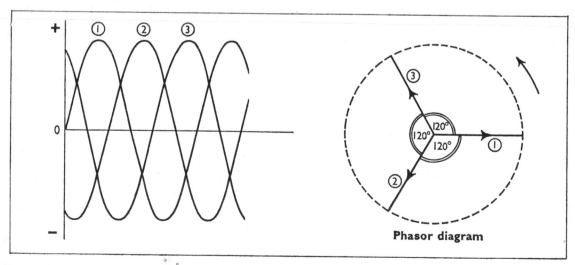

Fig. 12.11   Balanced currents in a 3-phase system; the current in the common return wire is zero at every instant

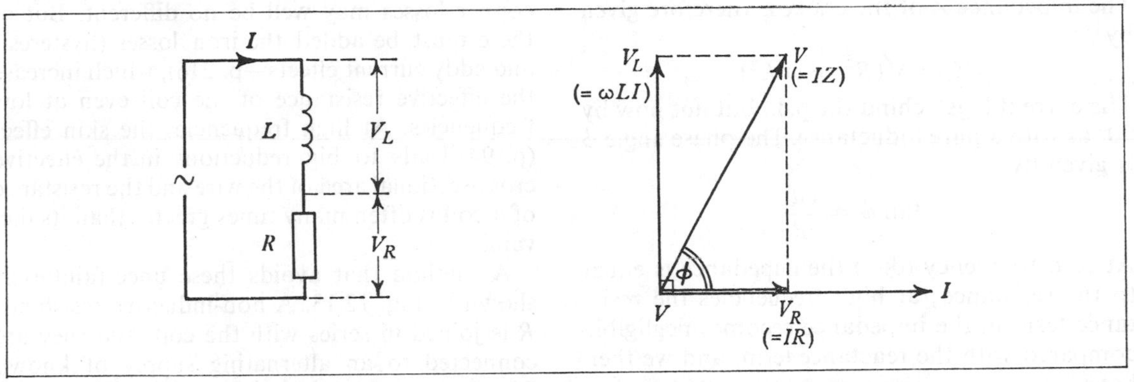

Fig. 12.12   The phasor diagram for a choke with appreciable resistance

seems therefore that there is nothing to be gained by using a three-phase rather than a two-phase system. However, the action of alternators (p. 150) and polyphase a.c. motors (p. 158) is only really smooth when some multiple of three phases is used; the same would not apply for a two- or four-phase system.

The loads on the three circuits need to remain fairly closely 'balanced'; otherwise the losses rise seriously. Large industrial consumers are normally supplied with all three phases; most large motors are best designed as three-phase devices, and it is then not difficult to keep the phases balanced for any given factory. A private house is supplied from one phase only; but by arranging the numbers of domestic consumers joined to each of the phases to be approximately equal in any given locality, the three loads remain about equal, subject to statistical fluctuations. The larger the number of people using the supply, the smaller the percentage variations of current that occur between the phases.

12.7   The general a.c. circuit

Earlier in this chapter we have analysed the behaviour of single circuit 'elements'—capacitors, coils and resistors. In the general case the analysis of phase relations and the calculation of currents and p.d.'s is best done with phasor diagrams. We shall in every case find that for a part of a circuit made up of capacitors, coils and resistors the r.m.s. values of current and p.d. are proportional to one another. Thus

$$\frac{V}{I} = Z$$

where Z is a constant known as the *impedance* of the arrangement; it has *resistive* and *reactive* 'components', and, like resistance and reactance, is measured in Ω.* In the special case of a coil or capacitor in which there are negligible resistive losses the impedance is equal to the reactance; and for a 'pure' resistor the impedance is equal to the resistance. In drawing phasor diagrams the first phasor to draw should always be that which represents the quantity which is the same for all the circuit elements concerned—i.e. the *current* for a number of elements *in series*; or the *p.d.* for a number of elements *in parallel*. Conventionally, this first phasor is drawn horizontal with the arrow pointing to the right. The other currents and p.d.'s are then drawn in relation to this.

(*i*)   *A choke with appreciable resistance.* In terms of 'pure' circuit elements this may be represented as an inductance L and resistance R in series (Fig. 12.12). In the phasor diagram we therefore draw the phasor representing the current I first. The p.d. V_R across the resistance is in phase with I, and is given by

$$V_R = IR$$

The p.d. V_L across the inductance leads 90° on I, and we have

$$V_L = \omega L I$$

where $\omega = 2\pi \times$ frequency. The total p.d. V across the choke is the phasor sum of V_R and V_L.

$$\therefore \ V = \sqrt{(V_R{}^2 + V_L{}^2)} = I\sqrt{(R^2 + \omega^2 L^2)}$$

* In the rest of this chapter we shall drop the awkward subscript in I_{rms}, V_{rms}, etc., and use the plain symbols I, V or I_L, V_C, etc., to refer to the r.m.s. values of the alternating quantities involved.

The impedance Z of the choke is therefore given by

$$Z = \sqrt{(R^2 + \omega^2 L^2)}$$

The current lags behind the p.d., but not now by 90° as with a pure inductance. The phase angle ϕ is given by

$$\tan \phi = \frac{\omega L}{R}$$

At zero frequency (d.c.) the impedance is equal to the resistance; at high frequencies the resistance term in the impedance becomes negligible compared with the reactance term, and we then have

$$Z \approx \omega L \quad \text{(i.e. } Z \propto \text{frequency)}$$

Thus a coil of suitable design has the useful property of offering high impedance to alternating currents and low resistance to direct current; it can therefore be used as a filter to separate an a.c. 'ripple' from the direct current on which it is superimposed. A practical application is to be found in the smoothing circuit used to filter the output of a rectifier (p. 262).

The impedance of a choke may be measured with the aid of an a.c. ammeter and a.c. voltmeter in much the same way as can be used for finding d.c. resistance (p. 49). If we can assume that the effective a.c. resistance is the same as that found for the coil by any d.c. method, then we have also the means of calculating the inductance L from the relationships given by the phasor diagram. But this assumption is often not justified. Thus in a coil with an iron core the copper losses may well be no different. But to these must be added the iron losses (hysteresis and eddy current effects—p. 216), which increase the effective resistance of the coil even at low frequencies. At high frequencies the skin effect (p. 94) leads to big reductions in the effective cross-sectional area of the wire and the resistance of a coil is often many times greater than its d.c. value.

A method that avoids these uncertainties is shown in Fig. 12.13. A non-inductive resistance R is joined in series with the coil, and they are connected to an alternating supply of known frequency. A high-impedance voltmeter (e.g. a cathode-ray oscilloscope) is joined in turn across the coil, the resistance, and then the whole combination. If the three readings obtained are V', V_R and V, their relationship is shown in the phasor diagram; from this the resistive and reactive components of V' can be worked out. In a practical measurement it is usually easiest to use graphical methods.

EXAMPLE   A non-inductive resistance of 200 Ω is joined in series with a coil and connected to the 50 Hz a.c. mains. The Y-plates of a cathode-ray oscilloscope are joined in turn across the coil, the resistance, and then the combination of the two; the heights of the traces obtained are 8.3 cm, 5.0 cm, and 12.2 cm, respectively. Calculate the inductance and a.c. resistance of the coil.

The phase relations of the p.d.'s are of the general form shown in Fig. 12.13 below. Since the p.d.'s are proportional to the impedances, and are also measured by the heights of the

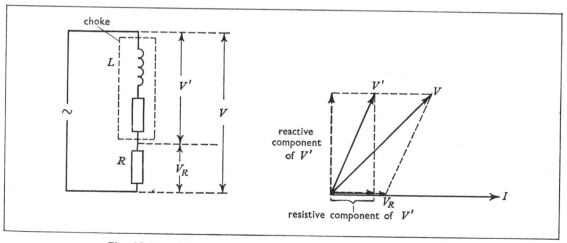

Fig. 12.13   Measuring the inductance and a.c. resistance of a choke

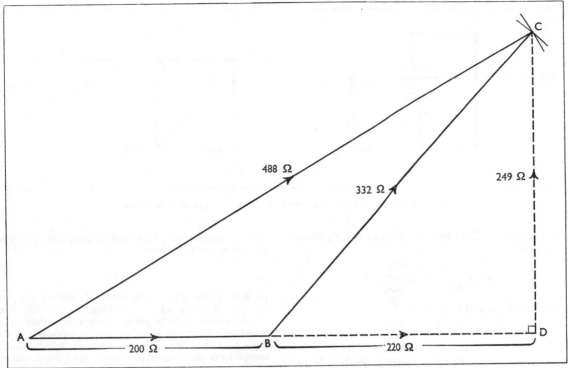

Fig. 12.14   The graphical solution of a phasor diagram problem

C.R.O. traces, we may draw the lines in the phasor diagram proportional to the C.R.O. measurements. Only half the parallelogram (i.e. a triangle) need be drawn to obtain all we want (Fig. 12.14). Choosing a scale of 1 inch to represent 2 cm (on the C.R.O. screen) we proceed as follows. Draw AB of length 2.5 inches; this can be taken to represent either the p.d. V_R or the 200 Ω resistance itself, since the two are proportional to one another. (The 'impedance scale' is then 1 inch to 80 Ω.) Then, with a compass, arcs of radii 6.1 inches and 4.15 inches are marked off centred on A and B, in order to find the point C. The line AC in now a measure of the impedance of the whole circuit (488 Ω); and BC is a measure of the impedance of the coil (332 Ω). Now produce AB, and drop the perpendicular CD onto it. The projection of BC on the resistance axis (i.e. BD), is a measure of the a.c. resistance of the coil; and its projection at right-angles to this (CD) is a measure of its reactance. By measurement on the diagram, we find

BD = 2.75 inches   CD = 3.11 inches

$\therefore$ the a.c. resistance of the coil = 2.75 $\times$ 80 Ω
$$= 220 \ \Omega$$
and its reactance = 3.11 $\times$ 80 Ω
$$= 249 \ \Omega$$

But,   inductive reactance = ωL
$$= 2\pi \times 50 \times L$$
$$\therefore \ L = \frac{249}{2\pi \times 50} \ \text{H} = \frac{2.49}{\pi} \ \text{H}$$
$$= 0.79 \ \text{H}$$

(*ii*) *Capacitor and resistor in series* (Fig. 12.15). In the phasor diagram we again draw the phasor representing the current I first. V_R ($= IR$) is in phase with this. The p.d. V_C across the capacitor lags 90° on I, and is given by

$$V_C = \frac{I}{\omega C}$$

The total p.d. V across the combination is the phasor sum of V_R and V_C.

$$\therefore \ V = \sqrt{(V_R{}^2 + V_C{}^2)} = I \sqrt{\left(R^2 + \frac{1}{\omega^2 C^2} \right)}$$

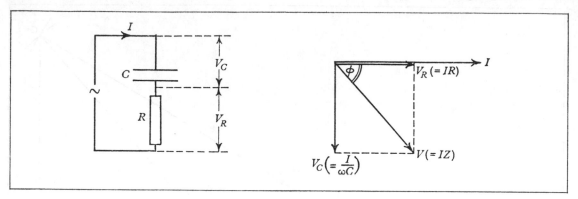

Fig. 12.15   The phasor diagram for capacitor and resistor in series

The impedance Z of the combination is therefore given by

$$Z = \sqrt{\left(R^2 + \frac{1}{\omega^2 C^2} \right)}$$

and the phase angle ϕ by

$$\tan \phi = \frac{1}{\omega CR}$$

At zero frequency (d.c.) the impedance is infinite; at high frequencies it is almost equal to the resistance R.

This arrangement can be used for separating an alternating p.d. from the steady p.d. on which it is superimposed. The capacitor is then called a *blocking capacitor*. A practical application is to be found in the coupling between two stages of an *R–C* coupled amplifier (p. 270).

★   More exactly, a combination of capacitor and resistor in series can be regarded as a frequency filter.

Thus the proportion of the applied alternating p.d. that appears across the resistor is

$$\frac{V_R}{V} = \cos \phi$$

At high frequencies this is nearly 1, and at low frequencies it drops off to zero (Fig. 12.16). The *R–C* coupling between the stages of an amplifier is then really a *high-pass* filter. The high-frequency components of the input pass the filter unaltered, while the low-frequency components are severely attenuated (and direct current is blocked altogether). There is no sharp cut-off frequency; but the 'knee' of the curve of Fig. 12.16 occurs roughly where

$$V_R = V_C$$

i.e. where $R = \dfrac{1}{\omega C}$

or $\omega = \dfrac{1}{CR} = \dfrac{1}{\text{time constant of the filter}}$

The time constant CR must therefore be adjusted so as to give little attenuation down to the lowest frequency that the amplifier is expected to handle.

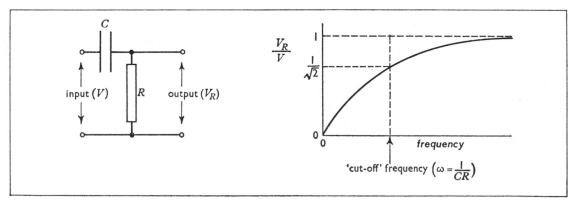

Fig. 12.16   The use of a capacitor and resistor as a high-pass filter

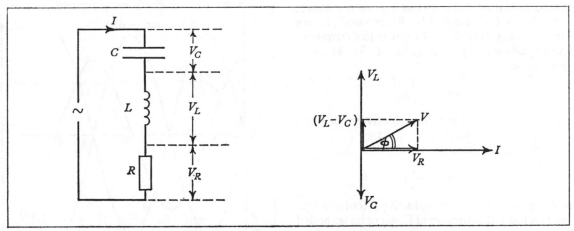

Fig. 12.17 The phasor diagram of an *L–C–R* series circuit

Just as the high-frequency components of the input appear mostly across the resistor, so the low-frequency components appear mostly across the capacitor. Thus a capacitor and resistor in series constitute a *low-pass* filter if the output is taken from across the capacitor. Combinations of low-pass and high-pass filters of this type are used to provide the tone controls of record players and radios.

12.8 Resonant circuits

The series resonant circuit

Fig. 12.17 shows a circuit containing capacitance, inductance and resistance, all in series; the phasor diagram is drawn alongside. In this case the p.d.'s V_L and V_C are in antiphase, and the phasor sum V of all three p.d.'s is given by

$$V = [V_R^2 + (V_L - V_C)^2]^{\frac{1}{2}}$$
$$= I\left[R^2 + \left(\omega L - \frac{1}{\omega C}\right)^2\right]^{\frac{1}{2}}$$

The impedance Z is therefore given by

$$Z = \left[R^2 + \left(\omega L - \frac{1}{\omega C}\right)^2\right]^{\frac{1}{2}}$$

and the phase angle ϕ by

$$\tan \phi = \frac{\omega L - (1/\omega C)}{R}$$

The impedance of the circuit drops to a minimum at the frequency for which the inductive and capacitative reactances are equal (and opposite); i.e. when

$$\omega L = \frac{1}{\omega C}$$

or $$\omega^2 LC = 1$$

At this frequency

$$Z = R \quad \text{and} \quad \phi = 0$$

The current I and p.d. V are then in phase, and

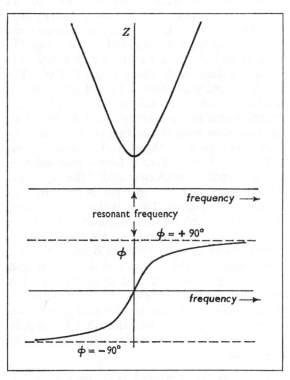

Fig. 12.18 The variation near the resonant frequency of the impedance Z and phase angle ϕ of a series resonant circuit

the combination behaves like a pure resistance; if the losses (represented by R) are small, a very large current may flow. At any other frequency the impedance is greater (Fig. 12.18). At lower frequencies

$$\omega L < \frac{1}{\omega C}$$

and the circuit behaves capacitatively (ϕ negative—i.e. current leading on p.d.).

At higher frequencies

$$\omega L > \frac{1}{\omega C}$$

and the circuit behaves inductively (current lagging on p.d.). If a mixture of frequencies is applied to the circuit, the current only builds up to a large value for frequencies near the one to which the circuit is 'tuned', given by

$$\omega^2 LC = 1$$

This is an example of *electrical resonance*. Like other cases of resonant systems in mechanics, sound, etc., this circuit is able to *act as a store of oscillatory energy*. The process is worth considering in detail. At the resonant frequency the p.d.'s across capacitor and inductor are equal in magnitude but in exact antiphase; the current is in quadrature with them (Fig. 12.19a). The energy stored in the electric field of the capacitor fluctuates with the p.d. across it; while the energy stored in the magnetic field of the inductor fluctuates with the current (Fig. 12.19b). At moments such as X the current is zero and the p.d. a maximum. There is then a maximum of energy stored in the electric field of the capacitor. At moments such as Y the p.d. is zero and the current a maximum. There is then a maximum of energy stored in the magnetic field of the inductor. At intermediate points the energy is partly in one and partly in the other. The total energy stored in the L–C system is constant, and is simply passed back and forth between the electric and magnetic fields. When the resonant current is first building up, this energy is drawn from the a.c. supply; but after that the supply need only make up the energy lost as heat in the resistance.

This may be compared with the similar behaviour of an oscillating mechanical system such as the pendulum of a clock. In this case the total

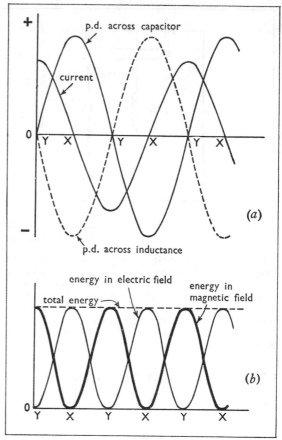

Fig. 12.19   The storage of energy in a series resonant circuit

energy is again constant, and is being continually changed from potential energy to kinetic energy and back again. The clock mechanism has only to supply the loss of energy (as heat) in the slight friction acting on the pendulum.

At the resonant frequency the p.d.'s across the two reactances build up to a value many times greater than the applied p.d. Thus at this frequency $Z = R$ and we have

$$I = \frac{V}{R}$$

$$\therefore \ V_L = \omega L I = \omega L \frac{V}{R}$$

$$\therefore \ \frac{V_L}{V} = \frac{\omega L}{R}$$

This ratio gives the *magnification* of p.d. produced by the resonance. Its value depends almost entirely on the design of the coil, since in practice the resistance of the circuit arises in this rather than in the capacitor. With a well-designed coil the magnification of p.d. at resonance can be 200 or more. It is known as the *Q-factor* of the coil (*Q* stands for 'quality'). Thus,

$$Q = \frac{\omega L}{R}$$

We can also regard *Q* as a measure of the extent to which the coil can be taken as a pure reactance. The phase angle ϕ between current and p.d. for a coil is given by

$$\tan \phi = \frac{\omega L}{R} = Q$$

For a perfect inductor, *R* would be zero, the *Q*-factor infinite, and ϕ exactly 90°.

The parallel resonant circuit

Consider now an arrangement of capacitor and inductor in parallel; the coil inevitably has both inductance *L* and resistance *R* (Fig. 12.20). Since this is a parallel arrangement, we draw first the phasor representing the common p.d. *V*. The current through the capacitor $I_C (= \omega CV)$ leads 90° on the applied p.d. The current through the inductor I_L lags on *V*, but by less than 90°, on account of the resistance *R*; the phase angle ϕ between I_L and *V* is given by

$$\tan \phi = \frac{\omega L}{R}$$

Also, $$I_L = \frac{V}{\sqrt{(R^2 + \omega^2 L^2)}}$$

The reactive component of $I_L (= I_L \sin \phi)$ is in exact antiphase with I_C. Resonance occurs when these two currents are equal in magnitude. Then

$$I_L \sin \phi = I_C$$

But $$\sin \phi = \frac{\omega L}{\sqrt{(R^2 + \omega^2 L^2)}}$$

∴ at resonance

$$\frac{\omega L V}{(R^2 + \omega^2 L^2)} = \omega CV$$

Now, $$Q = \frac{\omega L}{R}$$

$$\therefore \ R = \frac{\omega L}{Q}$$

Substituting for *R* and re-arranging,

$$\therefore \ \omega^2 LC \left(1 + \frac{1}{Q^2}\right) = 1$$

Q is usually a large quantity, so that for all practical purposes this reduces to the same condition as that for resonance with a series circuit with the same components:

$$\omega^2 LC = 1$$

We can regard the closed loop formed by the capacitor and inductor as a series resonant combination with the ends joined together. In a series circuit a large current flows for a given applied p.d.—i.e. the impedance of the circuit is very low; at the same time large p.d.'s are developed across the coil and capacitor. In the parallel circuit we still have a large current flowing in the coil and capacitor and a large p.d. developed

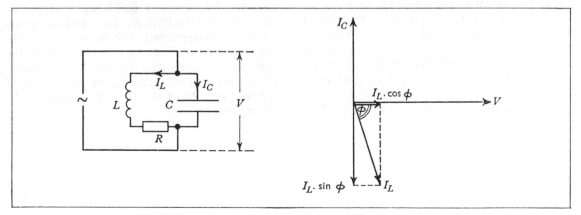

Fig. 12.20   The phasor diagram of a parallel resonant circuit

across them; but the current now *circulates* within the closed loop formed by the two components. The only current that enters the loop under the action of the applied p.d. V is the small resistive component $I_L \cos \phi$ needed to make up the losses occurring within the loop (in the resistance R). At resonance the current entering the loop is small and is in phase with V; the impedance Z' is therefore very large and entirely resistive. It is given by

$$Z' = \frac{V}{I_L \cos \phi} = \frac{(R^2 + \omega^2 L^2)}{R}$$

$$= R + \frac{\omega^2 L^2}{R}$$

$$= \omega L \left(Q + \frac{1}{Q} \right)$$

$$\therefore Z' \approx \omega L Q$$

since Q is normally large. At other frequencies the impedance is relatively small.

At lower frequencies

$$I_L \sin \phi > I_C$$

and the circuit behaves inductively (current lagging on p.d.).

At higher frequencies

$$I_L \sin \phi < I_C$$

and the circuit behaves capacitatively (current leading on p.d.).

The process of interchange of energy between the electric and magnetic fields is just as in the series resonant circuit. When the p.d. is a maximum, the current circulating in the loop is zero, and the stored energy is entirely in the electric field of the capacitor; and when the circulating current is a maximum, the p.d. is zero, and the energy is in the magnetic field of the inductor.

When a capacitor is charged or discharged through a low-resistance coil, the arrangement constitutes an L–C–R loop of just the kind we

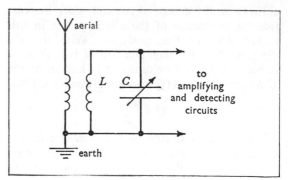

Fig. 12.21   The aerial circuit of a radio set

have been considering. The circuit therefore proceeds to 'ring' at its natural resonant frequency given by

$$\omega^2 LC = 1$$

There is now no external source of energy to maintain the oscillation, and the alternating current in the loop dies away as the energy is dissipated in the resistance of the coil (p. 176).

Resonant circuits may be used as filters to select one particular frequency from a mixture of many others. One example is the aerial circuit of a radio set (Fig. 12.21). Currents of many frequencies are induced in the aerial by the radio waves impinging on it. These flow between the aerial and earth through the small coil. By mutual induction currents of the same frequencies are induced in the tuning coil L of the resonant circuit. At most of these frequencies no resonance occurs and the p.d.'s developed across the combination remain small. But if a current is induced at a frequency close to the resonant frequency, it builds up a large circulating current within the loop and an appreciable p.d. is developed. The magnification produced is proportional to the Q-factor of the coil. The Q-factor also decides the 'selectivity' of the tuned circuit—that is, the narrowness of the frequency band effectively magnified by the arrangement.

13 Electrons and Ions

13.1 The properties of electrons

Electrons were first discovered in electric discharges through gases at low pressures. The discoveries were in fact made possible by the technical development of efficient vacuum pumps. We shall in due course consider these phenomena (p. 246). But the experiments concerned are not always easy to interpret, nor are they very safe on account of the X-rays that may be emitted incidentally. We shall deal first with safer and more straightforward demonstrations depending on effects which, historically, came later.

The thermionic effect

When a piece of metal is heated to a high temperature, it is found that something carrying negative charge emerges from its surface. This is known as the *thermionic effect*. It may be demonstrated with the apparatus in Fig. 13.1. A tungsten wire K is enclosed in a highly evacuated bulb; it can be heated by passing an electric current through it. Opposite the wire is a metal plate A. When the wire is cold, no current can pass through the vacuum from K to A. But when the wire is heated to incandescence a current can flow between the two electrodes—*in one direction only*. Presumably something carrying charge has emerged from the heated electrode into the empty space, and is now available to carry the current; the direction of this current is consistent with the idea that the charge emitted is negative. Thus, if A is made positive with respect to K, current flows as shown; in this case the electric field in the space would act on the negative charge to attract it towards A and so complete the circuit. But if the p.d. between A and K is reversed, no current flows; the electric field in this case would draw any negative charge back again into K, and there is no complete circuit.

It is customary to label the two electrodes in the same way as in an electrolytic tank. The electrode from which the negative charge emerges (K) is called the *cathode*; that towards which this charge travels (A) is called the *anode*. However, unlike the phenomenon of electrolysis, no chemical change occurs in either the cathode or anode. Whatever it is that carries the negative charge out of the cathode and into the anode is apparently a constituent of the two metals, and is taken by the electric current through the electrodes and round the rest of the circuit. The materials used in the apparatus can be varied to some extent; only high-melting-point metals can be used for the cathode. But with this limitation the effect is unaltered by such changes. Likewise, the nature of the residual gas in the bulb does not affect the phenomenon, provided the pressure is below about 10^{-3} N m^{-2} (so that the mean free path is much greater than the dimensions of the bulb). It is reasonable to conclude that the agent responsible for carrying negative charge from cathode to anode is also that which carries an electric current through a piece of metal. It is presumably a universal constituent of all metals (and perhaps of all matter).

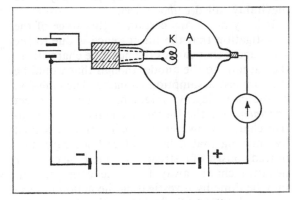

Fig. 13.1 The thermionic effect

When the p.d. between the anode and cathode is large enough, the glass of the bulb opposite the cathode *fluoresces* with a green colour. The anode seems to cast a shadow in the middle of this fluorescence, and the impression is at once

227

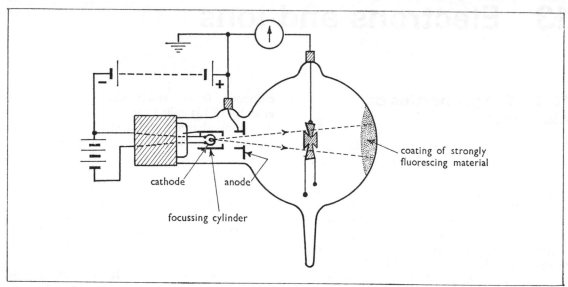

coating of strongly fluorescing material

cathode   anode

focussing cylinder

Fig. 13.2   The fluorescence produced by cathode rays; a shadow of the maltese cross is cast on the end wall

formed that the effect is due to some radiation emitted by the cathode. This is strikingly confirmed by using an electrode cut in a distinctive shape, like the maltese cross in Fig. 13.2. A shadow of the same shape appears in the fluorescence on the opposite wall of the tube; and its position and size clearly show that it is due to 'rays' of some kind travelling in approximately straight lines from the cathode. In the early development of the subject they were inevitably called *cathode rays*.

It is by no means certain at this stage of the investigation that the fluorescence is due to the same agent that carries the negative charge from the cathode to the anode, though that would be the simplest assumption to make. The emission of the cathode rays might be a separate effect *accompanying* the thermionic emission of negative charge. However, the point may be settled by showing that a beam of cathode rays does indeed consist of a stream of something carrying negative charge away from the cathode. This may be done by investigating the action of electric and magnetic fields on the beam of a 'cathode-ray tube' like that shown in Fig. 13.3. The anode A is a disc with a slit in it, through which a narrow beam of cathode rays passes. The beam passes obliquely over a sheet of fluorescent material, so that its path through the tube is made visible. When a p.d. is applied be-

tween the two metal plates P_1 and P_2, the electric field draws the cathode rays towards the positive plate and away from the negative one, and the beam is deflected accordingly. Likewise the cathode rays may be passed through a magnetic field—produced by a magnet or by a current flowing in suitably placed coils. The beam deflects as though the force acting on it is at right-angles to both the field and the line of the beam, just as we should expect for the magnetic force acting on an electric current. Thus, the beam could be deflected *downwards* in Fig. 13.3 by applying a magnetic field at right-angles to the plane of the diagram and directed away from the reader. If the beam carries a stream of *negative* charge, we must regard it as a current directed (in the conventional sense) *towards* the cathode. The direction of the force is then seen to be in agreement with Fleming's left-hand rule (p. 65).

Further evidence may be obtained by directing a beam of cathode rays onto an insulated electrode and collecting the charge carried. This may be done with the apparatus in Fig. 13.2. If a galvanometer or electrometer is joined between the maltese cross and the anode, a current is observed to flow in a direction corresponding with the collection of *negative* charge by the cross.

In experiments with cathode rays it is usual to work the tubes with the anode earthed. This

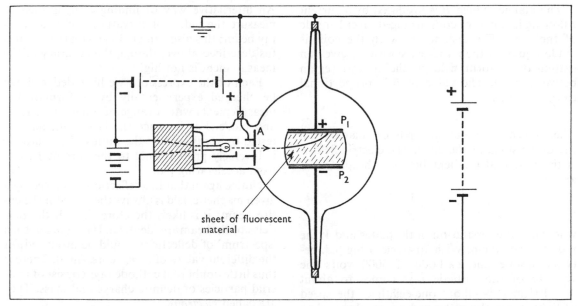

Fig. 13.3   The deflection of cathode rays by electric and magnetic fields

ensures that the region beyond the anode, in which the deflections are studied, is close to earth potential; the electric fields in the tube are not then much affected by movements of the observer's hand nearby. This procedure also makes for increased safety. But in any case such apparatus must always be handled with extreme caution, as the p.d.'s used can be lethal.

It is evident that a beam of cathode rays also carries energy. When it strikes the end wall of the tube, this energy is converted mostly into heat (the tube becomes warm); there is also light produced and a very small proportion of X-rays (p. 300). If the p.d. between anode and cathode is increased, the production of heat, light and X-rays goes up; at high p.d.'s the X-ray component also becomes more penetrating and is then dangerous. Some of the energy of the beam is always given up in ionizing the residual gas in the tube. If a cathode-ray tube is operated at a gas pressure of about 10^{-1} N m^{-2}, the ionization makes the path of the beam visible in a well-darkened room. This provides another means of studying the deflections produced.

If we assume that cathode rays are particles that obey the known laws of mechanics, the measurement of their deflections in electric and magnetic fields enables us to calculate the velocity v and the ratio of the charge e to the mass m_e of the particles; the latter quantity (e/m_e) is known as the *specific charge* of the particles.

One of the simplest ways of doing this is to arrange the electric and magnetic fields to act in opposition on the beam. In Fig. 13.3 this may be done by applying a magnetic field perpendicular to the diagram directed away from the reader, in addition to the electric field shown. By adjusting the strength of one or both fields they can be balanced so that the beam is undeflected (at any rate in the centre where the fields are uniform).

The force F_e with which an electric field of strength E acts on a particle of charge e is given by

$$F_e = eE \quad \text{(p. 184)}$$

If the flux density of the magnetic field is B, the magnetic force F_m acting on the particle is given by

$$F_m = Bev \quad \text{(p. 69)}$$

When the beam is undeflected

$$F_m = F_e$$
$$\therefore Bev = eE$$
$$\therefore v = \frac{E}{B}$$

The magnetic field B is produced by a current I flowing in a pair of coils arranged on either side of the tube. The distance between the coils is made equal to their radius, since this gives an extremely uniform field in the central region between them. The value of B is proportional to the current.

$$\therefore \ B = kI$$

where k is a constant that can be calculated from the dimensions of the coils. The electric field E is the potential gradient between the plates P_1 and P_2.

$$\therefore \ E = \frac{V}{d}$$

where d is the separation of the plates and V the p.d. between them. With an accelerating p.d. between anode and cathode of 5000 volts the velocity of the cathode rays comes to about 4×10^7 m s^{-1}—about one-eighth of the speed of light. Since the cathode rays are observed to remain in a single concentrated beam in this experiment, it is clear that all the particles must have the *same velocity*.

Once the velocity is known, e/m_e may be calculated from the value of the accelerating p.d. between anode and cathode. The simplest way of using the apparatus in Fig. 13.3 is to join the cathode to the deflector plate P_2 and use the same source of e.m.f. (with the same voltmeter) both for deflection and acceleration; the accelerating p.d. is then V also. The kinetic energy gained by a particle in accelerating through this p.d. is given by

$$\tfrac{1}{2}m_e v^2 = eV \quad \text{(p. 188)}$$

$$\therefore \ \frac{e}{m_e} = \frac{v^2}{2V}$$

If the electric field is switched off, and the magnetic field allowed to act alone, the force on the particle is entirely at right-angles to the motion at every point, and the beam is drawn into a circle (p. 68). The force F_m is now a centripetal force.

$$\therefore \ F_m = Bev = \frac{m_e v^2}{r}$$

where r is the radius of the circle.

$$\therefore \ r = \frac{m_e v}{Be}$$

An alternative way of finding e/m_e is then to measure the radius of curvature r of a cathode-ray beam; this also can be done with the apparatus described above, though the accuracy of the measurement is not high.

Each of the expressions we have deduced can be checked experimentally by performing the measurements with a range of different currents and p.d.'s. For instance the last result may be tested by verifying that the radius of curvature r is inversely proportional to the current I for a fixed accelerating p.d.

Notice again that the deflection produced by a given magnetic field is always the same for the entire beam; it is likely therefore that all the particles of the beam are identical. If it were not so, a 'spectrum' of deflections would be produced for the different values of e/m_e represented. There is thus little doubt that cathode rays consist of material particles of definite charge and mass. They are called *electrons*!

The currently accepted value of the specific charge of the electron is

$$\frac{e}{m_e} = 1.759 \times 10^{11} \text{ C kg}^{-1}$$

The specific charge of the electron can also be measured with the type of tube used in a cathode-ray oscilloscope (p. 231). In these only the point at which the beam strikes the end wall of the tube can be observed; and the theoretical analysis is rather involved. This is discussed further in section 13.7 below (p. 248). The best method for finding e/m_e with this kind of cathode-ray tube is to place it in an *axial* magnetic field; this causes the electrons to move in helical paths round the axis of the tube (p. 250), producing a rotation of the 'trace' on the screen.

The specific charge of the hydrogen ion (i.e. the proton) is given by electrolytic experiments; let its mass be m_p. We assume it carries a charge of the same magnitude e. Then

$$\frac{e}{m_p} = 9.58 \times 10^7 \text{ C kg}^{-1} \quad \text{(p. 36)}$$

Dividing,

$$\therefore \ \frac{m_p}{m_e} = 1836$$

The mass of the electron m_e is thus much less than the mass of even the smallest positive ion;

and the mass of the electrons in any piece of matter is a minute proportion of the total. The value of m_e is best calculated from the known values of e/m_e and e (p. 192):

$$\therefore \; m_e = \frac{e}{e/m_e} = \frac{1.602 \times 10^{-19}}{1.759 \times 10^{11}} \text{ kg}$$
$$= 9.11 \times 10^{-31} \text{ kg}$$

13.2   The cathode-ray oscilloscope

The deflection of an electron beam by electric or magnetic fields can be used to study variations in the p.d.'s or currents by which these fields are produced. An instrument employing this principle is called a cathode-ray oscilloscope; the construction of a typical cathode-ray tube for this purpose is shown in Fig. 13.4. Electrons are emitted by the heated filament F. The anodes A_1, A_2, A_3 are maintained at positive potentials with respect to the filament. The electron beam is therefore accelerated down the axis of the tube (which is highly evacuated). The shapes and potentials of the anodes are chosen so that the electric fields between them converge the beam into a fine spot on the fluorescent screen of the tube at S. The filament F is surrounded by a cylindrical electrode G, which is kept at a negative potential with respect to F. This potential controls the proportion of the emitted electrons that reach the hole in the first anode A_1, and so controls the brightness of the spot of light on the screen. (Since the electrode G fulfils the same function as the grid in a triode valve (p. 235), it is usually called the *grid* of the cathode-ray tube.)

In modern tubes the side walls beyond the last anode are usually coated with a conducting layer of graphite, which is connected to the anode; the electron beam is then moving in an enclosure at a constant potential and is unaffected by external electric fields. The tube is usually operated with the final anode and tube wall earthed, so that movements of earthed objects (such as the observer's hand) near the tube face do not affect the beam. The current in the electron beam is of the order of 0.1 mA. This current must be allowed to return from the screen to the final anode so as to complete the electric circuit. This can be done by coating the inside of the fluorescent screen with a conducting layer; but small tubes usually rely on another mechanism to provide the return current. When an electron strikes any surface sufficiently rapidly, it causes the emission of further electrons from it; the process is called *secondary emission*. The secondary electrons ejected from the fluorescent screen accumulate as a weak negative 'space charge' in the regions behind it. This cloud of electrons then drifts back slowly through the tube towards the anode.

The deflection of the electron beam is effected by the electric fields between the pairs of deflector plates—X_1 and X_2 for horizontal deflection, and Y_1 and Y_2 for vertical deflection. One of each pair of plates is connected to the final anode A_3 so as to avoid as far as possible stray electric fields between the deflector plates and the other parts of the tube. In a good modern tube the deflection of the spot of light is very closely proportional to the p.d. between a pair of deflector plates. The time of transit of the electrons through the tube is so small that the

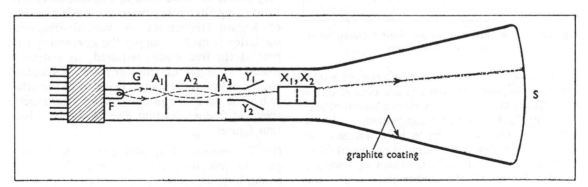

Fig. 13.4   A cathode-ray tube for an oscilloscope

spot faithfully follows the variations of X and Y p.d.'s up to very high frequencies of oscillation. The beam may instead be deflected magnetically using currents flowing in coils mounted round the neck of the tube. But this is not usually done in tubes designed for instrument purposes. In fact such a tube is normally equipped with a mumetal screen to shield it from stray magnetic fields produced by transformers, etc. in the associated apparatus.

Some uses of a cathode-ray oscilloscope

(*i*) *To measure peak p.d.'s.* For this purpose only one pair of deflector plates is used—normally the Y-plates, since these, being further from the screen, produce bigger movement of the spot of light for a given p.d.; the X-plates are both earthed. The alternating p.d. joined to the Y-plates deflects the beam up and down in a vertical plane; and the trace observed is a straight vertical line whose length is proportional to the *peak p.d.* The instrument may be calibrated by first connecting a known p.d. to the plates.

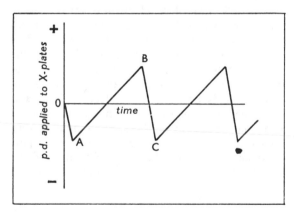

Fig. 13.5   The saw-toothed waveform required for an oscilloscope time base

(*ii*) *To study the waveforms of alternating p.d.'s.* For this purpose the X-plates are connected to an auxiliary circuit that generates a 'saw-toothed' p.d. (Fig. 13.5); such an arrangement is called a *time base.* The waveform to be investigated is joined to the Y-plates. During the interval from A to B the spot of light is drawn at a steady speed (ideally) in the X-direction across the screen, while in the Y-direction it follows the

p.d. under study. The trace is therefore a graph of this p.d. against time. During the interval from B to C the spot flies back quickly to its starting point. If the pattern observed is to be a steady one, it is necessary for the second and subsequent traces to be exactly superimposed on the first. This will only occur if the frequency of the time base is an exact submultiple of the frequency of the p.d. under test; this p.d. will then complete exactly a fixed number of oscillations in the time of one oscillation of the time base. The time base frequency therefore needs to be continuously variable. Even then, drifting of one or both frequencies is almost inevitable, and to obtain a truly steady pattern it is necessary to lock the two oscillations together. This is done by ensuring that the fly-back of the time base is always synchronized to occur at a given point of the wave pattern under study.

It is often necessary to amplify the waveform to be examined before it is applied to the Y-plates, since a peak p.d. of at least 50 volts is likely to be needed to give an adequate height of trace. Needless to say, the amplifiers used must be of the highest quality so that they do not introduce distortions of any kind into the waveforms being used.

(*iii*) *To compare frequencies.* For this purpose the time base is switched off, and the two p.d.'s whose frequencies are to be compared are joined one to each pair of deflector plates. If the frequencies are in a simple ratio, a steady trace is observed of the kinds shown in Fig. 13.6. Patterns of this sort are known as *Lissajous figures*; the ratios of frequencies may be deduced from them as shown in the diagram. This technique is very useful when we need to conduct an experiment using alternating currents covering a range of known frequencies. A variable-frequency oscillator is used to supply the alternating current at the frequencies required; its output is joined to one pair of deflector plates. A standard source of known frequency is joined to the other pair. The oscillator is then used at the exactly specified frequencies that give stationary Lissajous figures.

(*iv*) *To measure phase differences.* If the two pairs of deflector plates are supplied with sinusoidal alternating p.d.'s of the same frequency, but differing in phase, the trace obtained will in

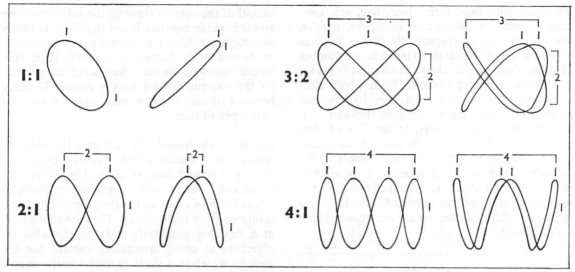

Fig. 13.6 Examples of Lissajous figures; the frequency may be obtained by comparing the numbers of loops along the top and down one side

general be an ellipse, as shown in Fig. 13.7. The shape of this figure may be used to give the phase difference. The variations of the x and y deflections with time t are given by

$$x = x_0 \sin \omega t$$

and
$$y = y_0 \sin (\omega t + \phi) \quad \text{(p. 210)}$$

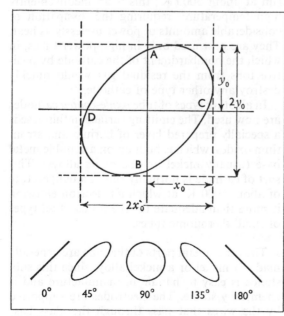

Fig. 13.7 The measurement of phase difference with a cathode-ray oscilloscope

where ϕ is the phase difference. The intercepts on the y-axis of the tube face are the values of y when $x = 0$; thus

$$\text{when } x = 0, \quad y = \pm y_0 \sin \phi$$

The distance between the two intercepts is therefore given by

$$AB = 2y_0 \sin \phi \quad \text{(Fig. 13.7)}$$

The peak-to-peak height of the trace is $2y_0$. We therefore have

$$\sin \phi = \frac{AB}{2y_0}$$

Using the intercepts on the x-axis we have likewise

$$\sin \phi = \frac{CD}{2x_0}$$

In the special case when ϕ is exactly 0° or 180°, the ellipse collapses into a straight line inclined to the axes (Fig. 13.7). When the two p.d.'s are in quadrature (phase difference 90°), the trace can be made into a circle by adjusting the amplitudes to be equal.

13.3 Thermionic tubes

There are a number of devices that make use of the thermionic effect to conduct an electric current through a high-vacuum tube. Such tubes are made with two, three, four, five, etc. electrodes; and are referred to as *diodes, triodes, tetrodes, pentodes,* respectively (using prefixes based on Greek numerals to indicate the number of electrodes). The diode is the simplest of these; it contains a heated cathode, from which electrons are emitted, and an anode. Its essential property is that current can flow through it in one direction only, namely, by the flow of electrons from cathode to anode; and this can occur only if the anode is positive with respect to the cathode. When the anode is negative, the electric field in the tube drives the electrons back to the cathode again, and no current flows. It is this behaviour that has led to its sometimes being called a *valve*.

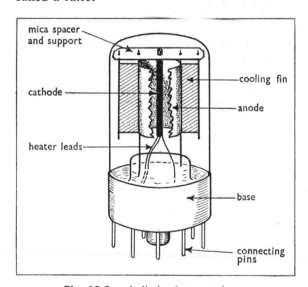

Fig. 13.8   A diode electron tube

The construction of a diode electron tube is shown in Fig. 13.8. It consists of a central cathode, surrounded by an approximately cylindrical anode. Two different types of cathode are used:

(*i*) *Directly heated.*   In this type the cathode is a single fine wire or *filament*, usually zigzagged up and down the tube. It is heated by a direct current passing through it from an independent supply. A disadvantage of this construction is that the potential is necessarily higher at one end of the filament than the other; this means that the p.d. between anode and cathode must vary from one part of the cathode to another. The

cut-off of the current through the tube when the anode is made negative is not therefore as sharp as it might be. Also, the cathode cannot normally be heated with alternating current, since this would impose a similar alternating component on the current flowing to the anode. Directly heated cathodes are now used only for specialized types of tube.

(*ii*) *Indirectly heated.*   In this case the cathode consists of a sheath which is heated by a wire zigzagged up and down inside it. The heater wire is packed around with a refractory material (some form of clay) and is thus electrically well insulated from the cathode. The cathode is now at a uniform potential; and there is also no objection to using alternating current for the heater—which is a great practical convenience. The indirectly heated cathode suffers from the slight disadvantage that it takes an appreciable time to reach its operating temperature. But it is the form of cathode now used in most small thermionic tubes.

At one time the only material used for making thermionic cathodes was tungsten because of its high melting point. However, to obtain adequate electron emission a tungsten cathode must be run at about 3000 K; this is an inconveniently high temperature, requiring the dissipation of considerable amounts of power uselessly as heat. They are now used only in high-power tubes, in which the bombardment of the cathode by positive ions from the residual gas would quickly destroy any other type of cathode.

In all other types of tube *oxide-coated* cathodes are now used. The emitting surface in this case is a specially prepared layer of barium and strontium oxides, which is built up on a suitable metal base (usually nickel or one of its alloys). This sort of cathode is normally run at a temperature of about 1000 K, at which its electron emission is more than adequate for the needs of all types of small thermionic tubes.

The other metal parts of the tube are generally made of nickel or a nickel alloy, since this substance is easy to handle in manufacture and is chemically stable. The electrodes are supported by the wires that pass through the glass base of the tube; accurately punched mica spacers are used to hold them rigidly in position.

For a thermionic tube to behave reliably it is necessary for a high vacuum to be maintained inside it—a pressure of less than 10^{-4} N m^{-2} is usually required. It is found that the glass and metal parts of the tube can retain an appreciable layer of gas molecules more or less permanently stuck to their surfaces; the phenomenon is known as *adsorption*. However, when the tube is heated the adsorbed gas evaporates off, destroying the vacuum. It is therefore necessary for all parts of the tube to be heated above their normal working temperatures during the process of evacuation. The glass envelope is heated in an oven to a temperature near its softening point; and the metal parts are raised to an even higher temperature by eddy currents induced in them by a high-frequency alternating magnetic field (p. 93). The cathode is effectively de-gassed by passing a large current through the heater. Even with these precautions small amounts of chemically reactive gases and vapours would be left in the tube, which would quite quickly contaminate the cathode and reduce the electron emission. To cope with this a small quantity of a chemically active metal (barium, magnesium, etc.) is placed in a gauze container mounted near the base of the tube. During the eddy current heating of the metal parts this is vaporized and deposited as a thin film on the inside of the glass bulb. Here it remains to combine with any chemically active substances present in the tube; the vacuum is thus maintained and even improved after the tube has been sealed off. This device is called a *getter*. After sealing off, the glass bulb is mounted on a plastic base; and the copper leads are soldered into the connecting pins.

If the p.d. between cathode and anode is V and the current flowing between them is I, the power P dissipated as heat is given by

$$P = VI$$

The effect of the p.d. across the tube is to accelerate the electrons in the space between the electrodes. In the first instance the power P exists as kinetic energy of the electrons; only on impact with the anode does this energy get changed into heat. In addition, the anode receives most of the energy radiated by the hot cathode. The anode must therefore have a large external surface area, and is often provided with cooling fins to enable it to dissipate the heat generated without excessive rise of temperature.

A small current is found to flow through a diode even when the anode is made slightly negative with respect to the cathode. This happens because some of the electrons are emitted with appreciable kinetic energy, and can reach the anode even against the action of a small opposing field. The anode–cathode p.d. must be reduced to about -0.5 V to prevent the flow of electrons completely.

The triode

The addition of a third electrode to a thermionic tube provides another means of controlling the current flowing to the anode. This electrode takes the form of an open wire spiral supported between the other two electrodes, but much closer to the cathode than the anode; it is called the *grid*. Otherwise the construction of a triode is very similar to that of a diode (Fig. 13.9). The electrons leaving the cathode are subject to the action of two electric fields—(*i*) that between the cathode and anode, and (*ii*) that between the cathode and grid. Normally these fields are arranged to act in opposition, with the anode positive with respect to the cathode and the grid negative. Provided the anode field is the greater of the two, electrons will still be drawn to the anode through the gaps in the grid; but no

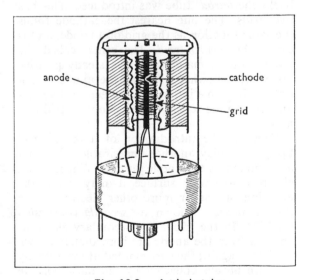

Fig. 13.9   A triode tube

electrons can be collected by the grid itself if it is more than 0.5 V negative with respect to the cathode. In this condition small changes in the grid potential cause considerable changes in the anode current. The triode tube can therefore be used as an amplifying device. It can be regarded as a generator of alternating current, in which the current at every instant follows exactly the variations of potential of the grid. But the power delivered comes from the H.T. supply joined between anode and cathode rather than from the grid circuit. The alternating p.d. applied to the grid is superimposed on the steady negative grid potential or *bias*, as it is called, so that the grid remains always negative; no current flows in the grid circuit, and negligible power is therefore dissipated in it. But the corresponding alternating current in the anode circuit can supply appreciable quantities of power to a load connected in series.

Multigrid tubes

One of the disadvantages of the triode tube is the appreciable capacitance that exists between its grid and anode. This amounts to about 10 pF, and is sufficient at frequencies above 10 kHz to feed back a significant fraction of the output p.d. into the grid circuit, affecting the amplification produced and possibly leading to uncontrolled oscillations. To overcome these drawbacks the *tetrode* tube was introduced. This has two grids. The one nearest the cathode fulfils the same function as the grid of a triode, and is called the *control grid*. The other, called the *screen grid*, is maintained at a steady positive potential somewhat less than the anode potential. As its name implies, the screen grid acts as an electrostatic screen between control grid and anode.

However the introduction of a screen grid brings into play another undesirable effect that also influences the anode current. Whenever an electron strikes a surface it may cause the emission of one or more other electrons from it—the process known as *secondary emission* (p. 231). In the triode the secondary electrons emitted from the anode are immediately drawn back to it again; they are emitted at low velocity and are always collected by the most positive electrode in their vicinity. The anode current is

not therefore affected by secondary emission. But in the tetrode the anode potential may in operation fall below that of the screen grid, and the secondary electrons will then be drawn to the latter. The behaviour of the tube is then very irregular. To avoid this trouble the pentode tube was developed. A third grid, called the *suppressor grid*, is mounted between the screen grid and anode. This is connected to the cathode. The electric field between this grid and the anode then acts so as to draw the secondary electrons back to the anode.

Tubes with even larger numbers of grids have also been made for special purposes. The *hexode* for instance has four grids and is used for mixing two signals together. It has two control grids (the first and third). The second and fourth grids are screens to shield the control grids from each other and from the anode.

13.4   The photoelectric effect

The emission of electrons from a metal may also be caused by illuminating its surface with light of sufficiently short wavelength—for most metals ultra-violet light is needed. This phenomenon is called the *photoelectric effect*. The photoelectric emission of electrons may be demonstrated with the apparatus in Fig. 13.10. A zinc plate is mounted on the cap of an electroscope so that it can be illuminated with ultra-violet light from

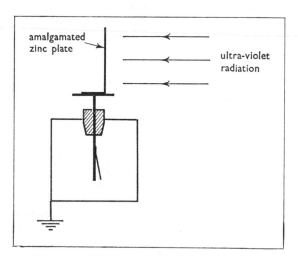

amalgamated zinc plate

ultra-violet radiation

Fig. 13.10   Demonstrating the photoelectric effect with an electroscope

a mercury-vapour lamp. The amount of emission varies considerably with small changes in the condition of the surface; it is found that the zinc surface for this experiment must either be freshly cleaned or else amalgamated with mercury. The zinc plate and electroscope are first given a *negative* charge; as soon as the light is directed at the plate the deflection starts to fall. The negative charge is evidently being lost from the illuminated surface. If the beam of light is intercepted with a sheet of ordinary glass (which absorbs ultra-violet light) the effect stops; while with an ultra-violet filter (opaque to visible light) the discharging of the electroscope continues unaffected. The rate of emission of electrons appears to be proportional to the intensity of the light. Thus, if the source of light is moved to half its original distance from the zinc plate, the illumination is increased four times; and the time for the deflection to fall a given amount is a quarter as much.

If now the electroscope is given a *positive* charge, no fall in the deflection occurs however bright the illumination of the zinc plate. We suppose that in this case the electrons emitted are at once drawn back to the plate, and there is no way by which the charge can escape to earth.

The electrons emitted by the zinc photocathode may be collected by another electrode, if this is at a high positive potential with respect to it. The current that flows between the electrodes may be detected with a sensitive electrometer.

More extended experiments show that for a given metal emission of electrons occurs only for light whose wavelength is less than a certain critical value. The cut-off wavelength varies from one type of surface to another; for most metals it is in the ultra-violet region of the spectrum. But for the alkali metals—sodium, potassium, caesium, etc.—the cut-off is in the visible part of the spectrum; for specially treated surfaces it is even in the infra-red. However, the alkali metals can only be used in a vacuum or in a tube containing an inert gas.

Photocells

The photoelectric effect is used in a number of devices for controlling an electric circuit with light. Fig. 13.11 shows a common design. The

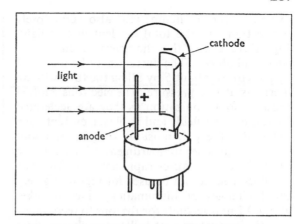

Fig. 13.11   A vacuum photocell

cathode is a semi-cylindrical plate coated with a photosensitive material. An alloy of antimony and caesium is satisfactory for the visible spectrum. For detecting the light from a tungsten filament lamp, a layer of caesium deposited on silver oxide is often used, since it responds well in the infra-red as well as in the visible spectrum. The anode consists of a single straight wire or loop, which is sufficient to collect the electrons emitted and does not much obstruct the incident light. The electrodes are enclosed in a highly evacuated glass bulb. To avoid damage to the photocathode the current is limited to very small values—about $10\mu A$ maximum is common. The sensitivity is about $30 \mu A$ per lumen of incident light; and the illumination of the cathode must therefore be controlled accordingly. The current is almost independent of the p.d. provided this

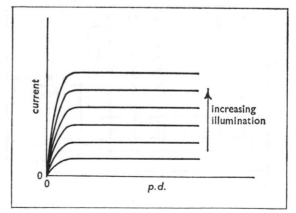

Fig. 13.12   The characteristics of a vacuum photocell

exceeds a certain figure; it is also very closely proportional to the total incident flux of light. Fig. 13.12 shows the characteristic curves for such a photocell. The current may be increased by a factor of about 5 by filling the tube with an inert gas at low pressure. The electrons emitted produce some ionization as they are acclerated by the electric field, and a bigger current then flows. But the p.d. must be kept below about 100 V, otherwise bombardment of the cathode by positive ions causes rapid deterioration. The gas-filled cell cannot be used for registering very rapid changes of illumination, since it takes about 0.1 milliseconds for the ions to recombine, and the current cannot follow the changes in a shorter time than this.

★   A very much higher sensitivity can be achieved by using the principle of *electron multiplication*. When an electron strikes a surface, it usually causes the emission of further electrons from it; this is known as *secondary emission*. By special treatment of the surface anything up to 4 electrons can be emitted for each incident particle. In the photomultiplier tube several intermediate electrodes (called *dynodes*) are inserted between the photocathode and anode so that the electrons must strike each of them in turn. By secondary emission the number of electrons in the beam is then multiplied by a factor of 4 at each impact. If there are 11 dynodes, the current in the beam is multiplied altogether by about 4^{11}, or nearly 5×10^6. The overall luminous sensitivity of the tube is thus about 10^8 μA per lumen, and it can detect very much weaker intensities than the human eye. Fig. 13.13 shows the construction of the Venetian blind type of photomultiplier tube, so named because of the slatted form of the dynodes. To accelerate the electrons between the electrodes the potential must increase by about 100 V from one dynode to the next through the tube. These potentials are supplied from a long resistance chain

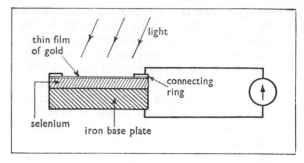

Fig. 13.14   A selenium photovoltaic cell

joined across a 1200 V supply. The practical difficulties in operating a photomultiplier tube are considerable, and it is only used where sufficient sensitivity cannot be gained by other means.

A photocell of a very different type is shown in Fig. 13.14. Unlike the cells described above, this actually produces a small e.m.f.; it is therefore sometimes called a *photovoltaic cell*. It consists of an iron base-plate on which is fused a layer of selenium, which is a semiconductor (p. 23). To make contact with the front surface of the selenium a thin transparent film of gold or aluminium is deposited on it; connection with this is made by a metal ring clamped round the edge. The density of electrons is much higher in the metal film than in the underlying layer of selenium, and initially electrons diffuse through into the surface layer of selenium. This therefore becomes negative with respect to the metal film, i.e. a contact p.d. arises between the two (p. 45); this grows until the electric field developed across the boundary prevents further diffusion. No net e.m.f. acts in the circuit, since opposing contact p.d.'s arise at the other junctions in it.

When light falls on the cell, some of the electrons which are usually firmly bound to selenium atoms in the boundary layer are given sufficient energy to become effectively 'free', and are then available to conduct a current. The electric field at once drives them across the boundary into the metal film. The cell thus behaves as a source of e.m.f. with the metal film on the front as its negative terminal. Some of the electrons always diffuse straight back across the boundary layer. But if the resistance of the galvanometer is low, the majority of them take the external path back to the base-plate, and the current is then proportional to the illumination. Joined across a low-resistance microammeter the cell forms an illumination meter useful for photometric purposes. The sensitivity is about 500 μA per lumen, much greater than that of a photo-emissive cell—and no additional source of e.m.f. need be used. If the external resistance is high, the current is approximately proportional to the *logarithm* of the illumination; i.e. doubling the illumination produces approximately the same increase of current at all parts of the range. Joined across a high-resistance microammeter the cell is then useful as an exposure meter for photographic purposes, where illuminations are handled in the same way—i.e. changing the stop from one '*f*-

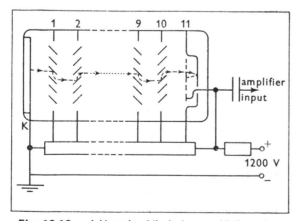

Fig. 13.13   A Venetian blind photomultiplier tube

number' to the next alters the amount of light collected by a factor of 2; and so the usual scale of *f*-numbers is also a logarithmic one. An exposure meter can thus be calibrated with *f*-numbers, and the points are about evenly spaced across the dial. Photovoltaic cells cannot be used to record rapidly fluctuating illuminations, since the capacitance between the electrodes is quite high and acts for alternating currents as a low-impedance shunt across the output. Other types of semiconductor light-sensitive devices are described on pp. 260 and 267.

13.5   The quantum theory

When the photoelectric effect was discovered (towards the end of the nineteenth century) it had long been established that light was a form of wave motion. Particle theories of light had been suggested, and were found incapable of explaining the facts; and Maxwell's electromagnetic field theory predicted waves of just the right kind in the electromagnetic fields in space (p. 292 *et seq.*). The behaviour predicted for such waves had been found at every point to be in exact agreement with the observed behaviour of light; all the laws of geometrical optics and the phenomena of polarization, interference and diffraction were completely explained in every detail by Maxwell's wave theory. But there seemed no way of bringing the wave theory into line with the facts of the photoelectric effect; and before long physicists found themselves forced into a radical revision of their views of the nature of light and matter.

It is true that an emission of electrons from an illuminated metal might be expected on the electromagnetic theory. The oscillating field should cause oscillations of the charged particles in the metal, and an electron might in this way be given sufficient energy to escape from the surface. But, beyond this, the predictions are completely at variance with the facts. On the wave theory we should expect to find that no emission of electrons occurs for very weak illuminations, since the electromagnetic field would not in this case be sufficiently strong. But in fact there is no threshold illumination for the photoelectric effect; some emission of electrons occurs even for vanishingly small intensities of light. Again, the wave theory would suggest that the *velocity* of the emitted electrons should depend on the intensity of the illumination. But in fact the velocities of the electrons are not affected by the illumination, but depend only on

the *frequency* of the light wave. The *illumination* decides the number of electrons emitted per second, i.e. the *current*.

The difficulties would be completely removed if we could regard a beam of light as a stream of particles. The intensity of the beam would now be measured by the number of particles arriving per second, and we should expect this to be proportional to the number of electrons emitted per second—which is what we observe. We should not now expect the velocity of the electrons to depend on the intensity of the beam, but rather on the amount of energy carried by each particle. If we suppose also that the position in the spectrum of a light 'particle' is decided by its energy then the dependence of the velocity of the emitted electron on the 'frequency' of the light is readily understood. Also it is then to be expected that particles of light which belong to the 'low-frequency' end of the spectrum would not have sufficient energy to cause the emission of electrons. The only objection to all this is that a particle theory will not explain the other laws of optics, with which Maxwell's wave theory is so admirably successful!

At the turn of the century there was also one other effect that defied explanation in terms of the classical wave theory. This was concerned with the nature of the light emitted by a hot furnace. In 1901 an explanation of this was put forward by Planck based on the idea that light was emitted and absorbed by the walls of the furnace in indivisible packets of energy or *quanta*, as they were called. The energy E of each quantum he supposed to be proportional to the frequency v of the light. Thus

$$E = hv$$

where h is a universal constant, which came to be called *the Planck constant*. Planck's theory was entirely successful; but there was still no need to suggest that the light, once emitted, consisted of anything other than the waves of Maxwell's theory.

It was left to Einstein (in 1905) to apply Planck's idea to explain the photoelectric effect. He suggested that the quantum of energy emitted by an atom continues as a concentrated, indivisible packet of energy until it is absorbed; it then gives up the entire amount to a single electron. Einstein supposed that in order to extract an electron from a given metal surface a minimum quantity of work would have to be done on it; equivalent to moving the electron through a potential difference Φ, which he called the *work function* of the metal. The work done might well need to be more than this for some electrons (for instance those relatively far below the metal surface). The quantum of energy supplied by the light must be at least equal to Φe, if there is to be any emission of electrons at all. The cut-off frequency ν_0 should therefore be given by

$$h\nu_0 = \Phi e$$

When light of higher frequency than this is used, the surplus energy is carried away as the kinetic energy of the electron. If v is the *maximum* velocity of the emitted electrons, then

$$\tfrac{1}{2}m_e v^2 = h\nu - \Phi e = h(\nu - \nu_0)$$

This prediction of Einstein's has been completely verified. It may be tested with a suitable photocell of the vacuum type. If the anode is made slightly negative with respect to the cathode, a retarding force acts on the emitted electrons; but some of them have sufficient kinetic energy to overcome this, and a small current continues to flow. If the reverse potential difference is V, the work done by the electron against the retarding forces of the electric field is Ve. The p.d. V is increased to the point at which the current is just stopped altogether. We then have

maximum K.E. of the electrons $= Ve = \tfrac{1}{2}m_e v^2$

$$\therefore\ Ve = h(\nu - \nu_0)$$

The stopping potential V is found for monochromatic light of a selected range of frequencies. A graph of V against ν should then be a straight line of the form shown in Fig. 13.15. The intercept on the frequency axis (where $V = 0$) gives the cut-off frequency ν_0; the gradient of the line is h/e. Since the electronic charge e is already known, we may use the experiment to measure the Planck constant h. The currently accepted value is

$$h = 6.625 \times 10^{-34}\ \text{J s}$$

The wave-particle duality

The quantum of energy in Einstein's theory has all the trappings of a definite particle; it is often

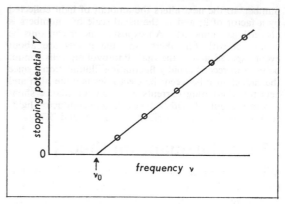

Fig. 13.15   Testing Einstein's photoelectric theory

referred to as a *photon*. At the same time we do not deny the wave picture of the light, since the energy of the photon is expressed in terms of the *frequency* of the wave, which is still assumed to be of the electromagnetic type described by Maxwell's theory. It appears that we are expected to regard the light sometimes as a particle and sometimes as a wave—or even as both things simultaneously!

At first sight this appears to be nonsense. But closer reflection shows that this is only because of our preconceptions of what particles and waves should be like. If someone talks of a particle, our thoughts turn to billiard balls, bullets, etc.; and, if waves are mentioned, we think of ripples on a pond and the like.

On the large scale of everyday experience particles and waves seem quite distinct sorts of things. But there is no reason why the behaviour of light should be limited by the mental images we may be able to form of the processes involved. Indeed it is not difficult to see that the words *particle* and *wave* are bound to have a restricted meaning applied to light. If we are asked to make a record of the path of a billiard ball on a table, we might take a motion picture of the events. But there is no way of doing the same sort of thing for a photon. Suppose for instance we have a point source of light on the axis of a lens, which forms a real image of the source on a screen the other side of it. There is no doubt that a photon emitted by the source will fall on the screen in the image spot. But what path has it followed in between? This is a question which, in the very nature of things, cannot be answered.

To decide whether the photon passed through a given point we should need to put something there (a photocell, say) to detect it. But this would absorb and destroy the photon, so that it never completes the journey at all. Likewise we might take a motion picture of the ripples on the surface of a pond, and we could thereby give an exact description of the progress of any given wavefront. But the same could not be done for a wavefront in the electromagnetic wave of a single quantum of light. The only way of deciding whether the wavefront is passing a given point is to detect it there; and this would absorb the quantum of energy concerned, so that the light no longer exists.

One interpretation of the wave-particle description of light is to regard the *intensity of the wave* at a given point as a measure of the *probability* of the photon appearing there. We cannot then predict the path of a photon with certainty; but we can state with certainty the probability of its following a given path. Whether or not it actually goes that way is only to be stated *after* the event.

As long as we are dealing with the vast numbers of photons emitted by ordinary sources of light the difficulty hardly arises, since statistical predictions become virtual certainties when the numbers are large enough. The intensity of the wave at a given point still gives the probability of photons turning up there; but this is now a 'certain' prediction of the number of photons arriving per second—i.e. of the intensity of the light. In such cases *wavefronts* and *rays* of light appear to be sharply defined concepts, in terms of which we may make exact predictions. But light interacts with atoms and electrons by the emission and absorption of single photons. In this context such phrases as 'the path of a single photon' or 'the shape of its wavefront' are essentially meaningless, since there is no conceivable way by which these lines could be traced. We must not then expect our mental images of waves and particles to mean very much.

Waves of probability and particles that can only be detected by making them vanish are certainly tenuous concepts! But this is not to say that we cannot calculate exactly what the behaviour of the light will be. We have found that Maxwell's wave theory is incomplete; to

give a full description of the known behaviour of light we need to use also Einstein's quantum hypothesis. But the equations describing the effects observed can then be written down with great precision. Whether the human mind is able to form a valid mental picture of the processes involved is really of no consequence at all!

Optical spectra

A chemical element in the form of a gas or vapour may be made to emit light either by heating it, as in a flame, or by passing an electric current through it, as in a discharge tube (p. 245). If this light is examined with a spectrometer, the spectrum is found to consist of a number of sharply defined *lines*, whose wavelengths are characteristic of the element concerned. Now the kinetic theory of gases leads us to believe that the molecules of a gas or vapour are very small compared with the average distance between neighbouring molecules (p. 2); except actually at the moments of collisions a molecule in a gas must remain almost entirely unaffected by the presence of other molecules. It therefore seems likely that the *line spectrum* of an element arises from the properties of its individual atoms. Because of this the spectrometer is one of the most powerful instruments at our disposal for analysing the structure of atoms. However, even a cursory examination of a few spectra shows something of the immense difficulties involved in unravelling the evidence. For instance the spectrum of iron vapour (in an electric arc between iron electrodes) shows over 250 strong lines in the visible part of the spectrum alone, and many more in the ultra-violet and infra-red. The spectrum of sodium, which is about the simplest there is, appears at first sight to be deceptively straightforward; if a sodium salt is heated in a Bunsen flame, the spectrum consists of just one pair of lines very close together in the yellow region. But when the sodium atoms are more vigorously excited in an electric discharge tube, many other lines are revealed in addition to the predominant yellow ones; and there is no obvious regularity in the pattern.

Bohr's theory of the atom

In 1913 Bohr propounded a revolutionary extension of the quantum theory to account for atomic spectra. It was clear that the classical laws of Physics could not be valid in their usual form for the particles inside an atom. He was therefore prepared to look for behaviour not allowed for in the established laws of mechanics and electromagnetism.

The existence of sharply defined line spectra suggested that a given atom was capable of emitting or absorbing only certain definite parcels of energy. He therefore suggested that an electron in an atom could exist only in sharply defined states of fixed energy (E', E'', etc.); the emission of light would then accompany the change from one such state to another, according to the established quantum relation

$$\text{energy emitted} = E' - E'' = h\nu$$

The frequency ν and wavelength λ are connected by

$$c = \nu\lambda \quad \text{(p. 288)}$$

where $c =$ the velocity of light. It remained to find the correct way of describing these fixed states so that the behaviour of an atom could be predicted.

In 1911 Rutherford had put forward the model of the atom with which we are now familiar—a massive, positively charged nucleus round which circulate the appropriate number of negatively charged electrons to make the atom as a whole electrically neutral. He suggested that electrostatic forces alone were responsible for holding the atom together. The evidence had steadily accumulated that Rutherford's picture of the atom was fundamentally correct. However, it raised considerable theoretical difficulties. It had long been established that an *accelerating* electric charge should generate electromagnetic radiation. This is precisely what occurs when a high-frequency alternating current flows in a radio transmitting aerial; the free electrons in the aerial are accelerating as they oscillate to and fro in the wire, and radio waves are emitted (p. 293). Now the electrons circulating round the nucleus of an atom have an acceleration towards the centre, and should therefore be emitting radiation. According to the established laws of electromagnetism, the steady loss of energy in this way by the electron should cause it to spiral in towards the nucleus; and it was calculated that a Rutherford atom must collapse completely by this process in about 10^{-10} s!

Either Rutherford's picture of the nuclear atom was wrong or else the laws of classical mechanics and electromagnetism broke down within the atom. Bohr boldly chose the latter alternative. He accepted Rutherford's model, and suggested that the fixed states of the electrons that he was looking for consisted of particular 'allowed' orbits. To fit the observed pattern of the hydrogen spectrum he proposed that the *angular momentum* of an electron in its orbit could only be an integral multiple of $h/2\pi$. Thus

$$\text{angular momentum} = n\frac{h}{2\pi}$$

where n is an integer. All other orbits he supposed to be impossible. He then assumed that the electron does not in fact emit radiation when in an allowed orbit; but that the emission or absorption of radiation occurs when its motion changes from one allowed orbit to another.

These assumptions appeared highly arbitrary. But the theory developed from them agreed with experiment in a remarkably detailed manner. In general the working out of the orbits for an atom containing many electrons is a problem altogether too complex to be contemplated. However, when there is only a single circulating electron, the working is fairly simple. Detailed predictions with Bohr's theory could therefore be made only for the spectra of hydrogen, deuterium, and for heavier atoms which are ionized to the point at which only one electron remains near the nucleus. In these cases the predicted wavelengths agreed with the measured values to within 1 part in 40 000! The result was so striking that there could be little doubt of the fundamental correctness of Bohr's ideas.

It is beyond the scope of this book to discuss the detailed analysis of the hydrogen spectrum according to Bohr's theory. But the mathematics is not in fact very difficult, and the student with a fair knowledge of mechanics would do well to look it up in a more advanced work.

Fig. 13.16 is a diagram showing the allowed energy levels of the single electron in a hydrogen atom. The groups of transitions represented by the arrows each correspond to well-known series of lines in the hydrogen spectrum. Although the orbits of electrons cannot be worked out for anything more complicated than a hydrogen

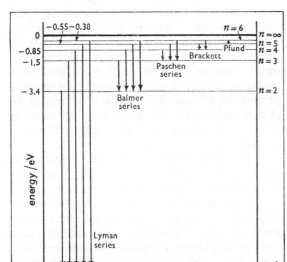

Fig. 13.16   The energy levels and transitions of the hydrogen atom

atom, the idea that the electrons in any atom have a limited number of allowed orbits and corresponding energy levels has turned out to be a universally consistent one. Even the most elaborate spectrum can be successfully analysed in terms of transitions between a relatively small number of atomic energy levels. Thus in the light of Bohr's theory the study of an apparently chaotic atomic spectrum becomes a means of measuring the energy levels of the atom; and energy level diagrams similar to Fig. 13.16 can now be constructed for all the elements.

Wave mechanics

We have seen that the description of light as a wave motion can be regarded as a way of expressing the inevitable uncertainty about the path followed by an individual photon. Now there is a similar uncertainty about the movements of an electron. Any experiment we might perform to ascertain the exact position of an electron would deflect it so violently as to destroy all knowledge of its previous or subsequent motion. Only when we are dealing with vast numbers of electrons (as in a cathode-ray beam) can we make accurate predictions on a statistical basis. For a single electron all we can state is the probability of its following a given path. In par-

ticular, inside an atom there is no possible way of determining the actual position of an electron in its orbit; any experiment to do this would certainly change the orbit or ionize the atom completely.

In 1923 De Broglie made the suggestion that the behaviour of electrons might be more satisfactorily described by treating them also as waves. The intensity of the wave at a given point can then be taken as a measure of the probability of an electron being there. Just as with light, the wave picture can be regarded as the proper expression of the inherent uncertainty in our knowledge of the behaviour of an individual electron. De Broglie suggested that the wavelength λ of the electron *wave* would be connected with the momentum ($m_e v$) of the *particle* by

$$\lambda = \frac{h}{m_e v}$$

where h = the Planck constant.

The physical reality of the De Broglie waves was soon confirmed when it was shown that beams of electrons showed diffraction effects when incident on the surface of a crystal or when passing through a thin foil; these effects were very similar to those observed with beams of electromagnetic waves in the same circumstances (p. 308), and could only be explained by supposing a beam of electrons to be a train of waves. The wavelengths measured in these experiments agreed with those predicted by De Broglie's equation. Since that time it has been shown that not only electrons but also many other 'particles' must be regarded as waves for the purpose of predicting their paths. Diffraction effects essentially similar to those of electrons have been demonstrated for hydrogen and helium atoms, and even for hydrogen molecules. In each case the measured wavelengths are correctly given by De Broglie's equation. There is now little doubt that there is an inherent uncertainty about the behaviour of all particles, so that we must predict their paths on a probability basis by considering the movements of the associated De Broglie waves. The branch of Physics that deals with the behaviour and interpretation of these waves is known as *wave mechanics*.

There is no contradiction in treating even massive particles in this way. With a large 'particle', such as a tennis ball, the wave trains would be so sharply defined as virtually to coincide with the ball itself at every moment. This is merely another way of saying that there is a negligible chance of the ball following a path detectably different from that predicted by Newtonian mechanics. Even with atoms the differences are rarely significant. Thus wave mechanics turns out to be a refinement of classical mechanics—a refinement we need only consider when dealing with the fundamental particles of matter.

The equation to be obeyed by the electron waves was worked out by Schrödinger; and we can now use this to predict the behaviour of the waves, just as we can use the equations of Maxwell's electromagnetic theory to predict the behaviour of light.

An electron wave is refracted in passing near a nucleus, and there is the possibility of a wave system becoming trapped so that it revolves permanently round the nucleus. In wave mechanics this is how we describe a stable electron orbit. Schrödinger showed that the only possible sort of trapped wave would be a *standing wave system*. This occurs when the path of the wave round the nucleus contains a whole number of wavelengths, so that the wave pattern 'joins up' continuously round the loop.

If the radius of a wave path round the nucleus is r, a standing wave could occur, if

$$2\pi r = n\lambda = \frac{nh}{m_e v}$$

$$\therefore \; m_e vr = \frac{nh}{2\pi}$$

Now, $m_e vr$ is the *angular momentum* of an electron moving in a circular orbit of this radius; and the above equation is therefore nothing more than Bohr's quantum condition for specifying the allowed electron orbits. Bohr's rather arbitrary rule is thus seen to be an inevitable consequence of the wave nature of the electron.

Scientists still tend to talk in terms of the Bohr picture of the atom, since this sort of language is move vivid than the 'hazy' wave mechanical picture. We refer to an electron 'rotating' in a certain 'orbit'; but this use of words can only be justified now if liberally sprinkled with inverted commas! It is really better to avoid talking about an electron in an atom as though it were a concentrated particle; the reality is rather the standing wave pattern, which in present terminology is referred to as the electron *orbital*. The position of the electron within the orbital can only be stated in terms of probabilities; and it is correct to regard its charge and mass as being 'smeared out' over the orbital in proportion to the intensity of the wave at each point.

The charge distribution is thus a stationary one. But we must not forget that a standing wave within an atom still represents an electron in motion round the nucleus; and the orbital may still have angular momentum and produce a magnetic field. In addition it is found that each electron has internal angular momentum, which is called its *spin*; and this causes it to produce a magnetic field quite apart from its motion round the nucleus.

The modern quantum theory is able to give a very detailed description of the atom. It predicts that only two electrons can occupy each orbital, and these must have opposite spin. (This is known as the *Pauli exclusion principle*.) Such a pair of electrons has no resultant angular momentum or magnetic field. Imagine an atom being built up with the electrons added to it one by one; each electron will fall into the orbital of lowest energy that is still unoccupied. In this way the electron structure forms a series of *shells*, in each of which the orbitals have approximately the same energy. These shells are usually labelled in ascending order of energy with the letters K, L, M, etc. The K-shell is complete with only 2 electrons. The L-shell can hold 8 electrons, and the M-shell 18. However, the electron orbitals interact with one another; and it can happen that an orbital in an outer shell has lower energy than one in a still incomplete inner shell. This happens to some extent with most atoms whose atomic number is greater than 18 (i.e. with more than 18 electrons). Thus argon with 18 electrons has the K- and L-shells complete, and 8 electrons in the M-shell; but in potassium (with atomic number 19) the extra electron occupies an orbital in the N-shell, although there is room for 10 more electrons in the M-shell. Only in atoms with 29 or more electrons is the M-shell completely filled up.

With the heaviest atoms there may be several incomplete shells; uranium with 92 electrons has 2 electrons in the Q-shell, while the O- and P-shells are very far from complete.

The spectrum and the chemical properties of an atom are found to depend to a large extent on the number of electrons in its outermost shell. Atoms with similar electron structures in this shell tend to have similar chemical properties and similar spectra. If the elements are arranged in the order of their atomic numbers, there is therefore a cyclic recurrence of atomic properties through the table; this feature of the chemical behaviour of the elements is displayed in the well-known periodic table. It is found that the number of electrons in the outermost shell can never exceed 8. Those atoms in which this shell is either complete or else contains exactly 8 electrons have particularly stable electron structures; these are the noble gases—*helium, neon, argon, krypton, xenon* and *radon*. Atoms which have one electron outside a complete shell (or a sub-shell of 8) likewise have similar properties; these are the alkali metals—*lithium, sodium, potassium, rubidium, caesium*. In the same sort of way the other groups of the periodic table arise from various recurring electron configurations.

The chemical bond

When two atoms are close together, the orbitals of their outermost shells become modified, and there exists the possibility of one or more orbitals being shared between the two atoms; these would represent electrons in motion round *both* nuclei. If the resulting electron configuration has a minimum total energy for a particular distance apart of the nuclei, then the atoms can be in equilibrium in this position, and a stable chemical bond can be formed between them.

In some cases the shared orbitals are so modified that one or more electrons are virtually transferred from one atom to another. One atom can thus acquire a negative charge, leaving a positive charge on the other. In this way ionic compounds are formed, in which the bond is largely due to the electrostatic force between the ions. But in general an orbital that provides a bond between two atoms is shared more or less equally between them, and there are no ionic

forces to affect it. Such is the case, for instance, with the hydrogen molecule (H_2). Here, the orbitals are shared between the two nuclei; the energy of this configuration is much less than for the separate hydrogen atoms, and a very stable bond is formed.

13.6   Electric currents in gases

For small potential differences a gas is an almost perfect insulator. An electric current can only pass through it if some independent means is used to ionize the gas (e.g. X-rays), or in some other way to produce free electrons in it (e.g. by the photoelectric effect).

In such cases the current is limited by the rate at which ions and electrons are being produced in the space between the electrodes (p. 176). But if the applied p.d. is sufficiently increased, a point is reached at which there is a considerable growth of current due to collision processes in the gas. The electrons initially present then gain enough energy between collisions to cause further ionization. The electrons so released are accelerated in their turn; and in this way a vast increase of ionization occurs. Each original ion pair leads to the creation of anything up to 10^5 fresh ion pairs, and the charge collected by the electrodes is increased by this factor. This form of magnified ionization current is known as a *Townsend discharge*, after its discoverer.

However, the current is still dependent on the original source of ionization—X-rays, ultra-violet radiation, etc.; and without this it stops at once. This sort of discharge is non-luminous.

In order to maintain a self-sustaining discharge it is necessary for the production of ions to keep pace with their collection by the electrodes. This can occur if the p.d. is increased still further. A luminous discharge then suddenly starts, known as a *glow discharge*. The p.d. needed to maintain this discharge falls considerably with the increase of current. This sort of discharge is therefore highly unstable, and the current will only reach a steady value if it is limited by a sufficiently high resistance in the rest of the circuit; in the absence of this it increases until either a fuse blows or the equipment fails in some other way.

The glow discharge is employed in a number

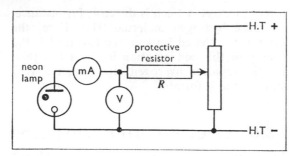

Fig. 13.17   Investigating the properties of a glow discharge tube

of important devices. Its essential characteristics may be investigated using a small neon lamp connected in the circuit of Fig. 13.17. These lamps are usually equipped with a built-in series resistance to limit the current to safe values; for this demonstration this must be disconnected. The potential divider across the d.c. supply is adjusted so that the p.d. applied to the lamp is slowly increased from zero. Up to a certain point no discharge is observed. (The Townsend discharge is non-luminous, and the current that might flow is far too small to be detected by the milliammeter.) But as soon as the p.d. reaches the striking value—usually about 180 V—the glow discharge starts and the p.d. across the lamp at once falls to a lower value. The p.d. supplied to the lamp and its series resistor R can now be varied by moving the potential divider. But this only alters the current through the lamp, and leaves the p.d. across it at almost the same value. To stop the discharge the applied p.d. must be reduced below this.

The appearance of a gas discharge takes many different forms as the pressure in the tube is varied. Colourful and fascinating effects may be observed in a discharge tube connected to a vacuum pump and pressure gauge. At one time it was usual to operate such a tube from an induction coil. However, this is not now considered safe, since the p.d.'s may be sufficiently high to generate dangerous quantities of X-radiation. A tube about a foot long can be worked satisfactorily using an E.H.T. supply of about 6 kV. A typical appearance of the discharge for a pressure of about 10 N m^{-2} is depicted in Fig. 13.18. Most of the tube is occupied by a bright luminous region called the

positive column; this starts close to the anode, and is usually broken up into a series of bands or striations. A region of weaker luminosity (usually of a different colour) is observed a short distance from the cathode; this is known as the *negative glow*. On either side of this there are dark regions, called the *Faraday dark space* and the *Crookes dark space* after the men who first observed them. There is usually also a further small glowing patch to be seen covering the surface of the cathode, but separated from it by a very thin dark layer (*Aston's dark space*). The colours of the discharge depend on the gas used; for air, the positive column is a brilliant pink, and the negative glow deep blue.

The electrons emitted from the cathode are accelerated by the electric field in the Crookes dark space; the length of this is approximately equal to the mean free path of the electrons in the gas. The negative glow arises from the ionization caused by these electrons. Most of the applied p.d. exists across the Crookes dark space. The field here is very intense and accelerates some of the positive ions to strike the cathode; this in turn stimulates the emission of electrons on which the existence of the discharge depends. In the Faraday dark space the electrons are further accelerated, and the positive column is caused by the excitation of the atoms in the rest of the length of the tube.

If a greater length of tube is used, the cathode end of the discharge is unaltered (for the same gas pressure); and the positive column extends to occupy the extra length. The applied p.d. must then be increased to maintain the field. A neon advertising sign consists of a lengthy discharge tube bent into the tortuous shape required by the lettering, etc. It is filled with neon at a suitable low pressure, and the positive column is arranged to occupy the entire visible portion. Owing to its length, p.d.'s of several thousand volts are normally needed to run the discharge.

As the pressure in a discharge tube is reduced, the mean free path of the electrons in the gas increases, and the Crookes and Faraday dark spaces and negative glow extend further down the tube; the positive column therefore contracts and becomes less intense. Eventually when the pressure is rather less than 1 N m^{-2}, the negative glow extends right to the anode and becomes much weaker. The walls of the tube are now

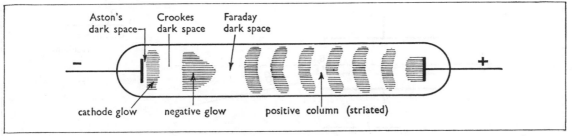

Aston's dark space   Crookes dark space   Faraday dark space

cathode glow   negative glow   positive column (striated)

Fig. 13.18   The features of a low-pressure gas discharge

seen to be emitting a green fluorescence arising from their bombardment with electrons emitted by the cathode. Indeed it was in discharge tubes at low pressures that cathode rays were first discovered. As the pressure is reduced still further, the p.d. needed to maintain the discharge rises rapidly; and at pressures below about 10^{-1} N m^{-2} the tube usually becomes a good insulator again.

A glow discharge may be modified by using the thermionic effect to provide the necessary emission of electrons from the cathode. If this electrode consists of a heated tungsten wire, the negative glow disappears, and the p.d. across the tube is reduced by the amount that would normally exist across the Crookes dark space. The positive column then fills nearly the whole space. This sort of discharge is used in the fluorescent tubes which are now common in electric lighting installations. In these tubes *both* electrodes are in the form of heated tungsten wires; they are normally designed to work on alternating current, so that each electrode in turn becomes the cathode. The current is controlled by a choke joined in series (p. 95). Once the discharge has

started, the current in the tubes is sufficiently strong to keep the two 'cathodes' at the right operating temperature; but initially they must be heated by passing a current through them. This is done with a special *starter* connected as shown in Fig. 13.19. It consists of a small discharge tube filled with argon. The electrodes are quite close together, and one of them is a small bimetallic strip. When the power is switched on, the discharge in the starter commences, and current therefore flows also through the two tungsten filaments joined in series. The heat generated in the starter causes the bimetallic strip to bend over; and within one or two seconds it makes contact with the other electrode. This short-circuits the starter, and the discharge in it stops. At once the bimetallic strip starts to cool, and the electrodes separate again. This switches off the current through the filaments and connects the full p.d. of the supply across the main tube. Provided the tungsten filaments have reached a high enough temperature the main discharge then starts, since the striking p.d. for the main tube is arranged to be less than that required for the starter discharge. If, however, the tungsten electrodes are still not hot enough to initiate a discharge, the starter strikes again; and the starting cycle is repeated until the lamp has lit up. The tube is filled with mercury vapour, which emits light of a blue-green colour. A bare mercury vapour lamp is sometimes used for street lighting; but even here its colour is far from satisfactory. It would be quite unacceptable in any other situation. However, it can be corrected by coating the inside of the tube with suitable fluorescent powders. The mercury vapour spectrum includes a considerable proportion of ultra-violet radiation. This is absorbed by the fluorescent coating and re-emitted as visible

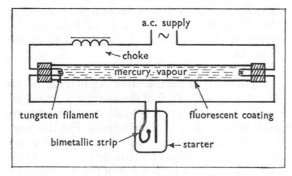

a.c. supply

choke

mercury vapour

tungsten filament   fluorescent coating

bimetallic strip   starter

Fig. 13.19   The circuit of a fluorescent lighting tube

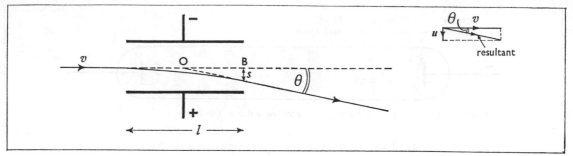

Fig. 13.20   The deflection of an electron beam in an electric field

light, chiefly in the red part of the spectrum where mercury light is deficient. By suitable choice of the fluorescent materials the light can be made to have almost any required tone value.

13.7   Electron and ion beams

Electric fields

Over most of the space between a pair of deflector plates the field is nearly uniform, with lines of force parallel to one another and normal to the plates. The path of an electron in such a field is therefore very similar to that of a projectile in a uniform gravitational field. In the motion of a projectile (in the absence of air resistance) the horizontal component of velocity is unaltered by the gravitational forces. But in the direction of the field (i.e. vertically downwards) the projectile experiences a uniform acceleration. The resulting path is then part of a parabola. In the same way an electron moves in a parabolic path between a pair of deflector plates (Fig. 13.20). If its initial path is parallel to the plates, its velocity v in this direction remains unaffected. But at right-angles to this its acceleration a is given by

$$a = \frac{F_e}{m_e} = \frac{eE}{m_e}$$

The length l of the path in the field and the time t spent in it are related by

$$t = \frac{l}{v}$$

The component of velocity u acquired in the direction of the field is therefore given by

$$u = \frac{eE}{m_e}t = \frac{eEl}{m_e v}$$

The angular deflection θ of the electron beam is found from the parallelogram of vectors:

$$\tan \theta \; (= \theta \text{ in radians, for small angles})$$

$$= \frac{u}{v} = \frac{eEl}{m_e v^2} \qquad \text{(I)}$$

$$\therefore \frac{e}{m_e v^2} = \frac{\tan \theta}{El}$$

The lateral displacement s of the electron beam on emerging from the field is given by

$$s = \tfrac{1}{2}at^2 = \frac{eEl^2}{2m_e v^2}$$

The final direction of the beam is as though the deviation θ had occurred *at the point O* where the final path produced back intersects the initial path (Fig. 13.20).

Now, $\tan \theta = \dfrac{s}{\text{OB}}$

Substituting for s and $\tan \theta$, we find

$$\text{OB} = \tfrac{1}{2}l$$

The point O is thus at the exact mid-point between the plates.

In a cathode-ray oscilloscope (p. 231) we can therefore measure the deflection θ by dividing the displacement of the spot of light on the screen by the distance of the screen from the point O.

Equation (I) gives the value of θ in terms of the quantity $e/m_e v^2$. To find the specific charge e/m_e we have to perform a subsidiary experiment to give the velocity v of the electrons. For this purpose we need to study the magnetic deflection of the beam. It might be thought that v could be calculated from a knowledge of the p.d. through which the electrons have been accelerated (i.e.

the p.d. between the cathode and final anode of the tube). But this calculation is only found to yield again the quantity $e/m_e v^2$. Thus, the kinetic energy of an electron that has fallen through the p.d. V between cathode and anode is given by

$$\tfrac{1}{2}m_e v^2 = eV$$

$$\therefore \frac{e}{m_e v^2} = \frac{1}{2V}$$

We can now substitute in equation (I) and eliminate e, m_e and v completely.

$$\therefore \tan \theta = \frac{El}{2V} \qquad \text{(II)}$$

The same deflection would therefore be obtained whatever the charge or mass of the particles in the beam. Indeed the study of electric deflections in a cathode-ray tube provides no certain grounds for supposing the charge and mass of all electrons to be the same. To establish this point and to find the value of e/m_e we need to investigate also the magnetic deflection of the beam.

Equation (II) shows that the sensitivity of an electrically deflected cathode-ray oscilloscope is inversely proportional to the accelerating p.d. V.

Magnetic fields

In general the path of an electron in a magnetic field is a *helix* (p. 69). But for the special case in which the motion is entirely at right-angles to the field the path is an arc of a circle; this is the arrangement normally used in deflection experiments. If the flux density of the magnetic field is B, then the radius of curvature r is given by

$$r = \frac{m_e v}{Be}$$

where v is the velocity of the particle (p. 230).

In a cathode-ray tube the deflection is usually produced by a pair of deflector coils mounted on either side of the neck of the tube. The magnetic field is not sharply bounded; but to a first approximation we may assume that it is uniform within the space enclosed by the coils and zero outside (Fig. 13.21). The beam follows a circular path in the field. It is clear from the geometry that the final direction of the beam is as though the deflection θ had occurred at the point C, the

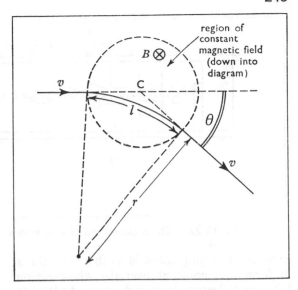

Fig. 13.21   The deflection of an electron beam in a magnetic field

centre of the coil system. If the length of the path in the field is l, we have

$$\theta = \frac{l}{r}$$

For small deflections we may take l as equal to the diameter of the coils. As with electrostatic deflections, we measure θ by dividing the displacement of the spot of light on the screen by the distance of the screen from the point C.

The cathode-ray oscilloscope may be used to make an estimate of e/m_e by comparing the electric and magnetic deflections of the beam. The methods available are the same in principle as those described earlier in this chapter, except that we cannot now observe the curvature of the path in the magnetic field directly. From either the electric deflection or a knowledge of the accelerating p.d. we may deduce the value of $e/m_e v^2$; and from the magnetic deflection we obtain the value of $e/m_e v$. The two results may be combined to give the values of v and e/m_e.

However the method is not very satisfactory because of the difficulty of knowing the exact regions over which the fields act. Fig. 13.22 shows a method employing a different principle that avoids this uncertainty. In this case a *longitudinal* magnetic field is used. A small cathode-ray tube (4 cm diam.) can be completely

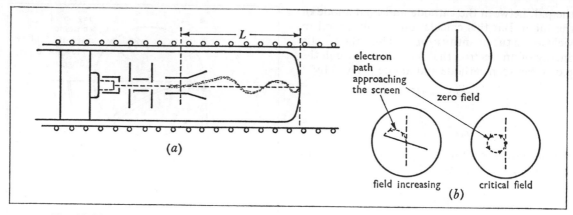

Fig. 13.22   The measurement of e/m_e with a cathode-ray tube in a longitudinal magnetic field

enclosed in a long solenoid so that the field is uniform and constant over the whole region between deflector plates and screen. As long as the beam is parallel to the magnetic field it is unaffected by it. But if a p.d. is applied to the Y-plates, the electrons acquire a component of velocity at right-angles to the magnetic field, and their subsequent path to the screen is a helix instead of a straight line. Viewed through the end of the tube, the electron beam would appear to be moving in a circular path as it approached the screen. If the magnetic field is adjusted so that the beam performs one complete revolution between deflector plates and screen, then the spot of light will be returned to the undeflected position. This adjustment is made by applying an alternating p.d. to the Y-plates sufficient to produce a straight-line trace covering the full height of the tube face; the frequency is of no importance. As the magnetic field is increased, the line rotates and gets shorter (Fig. 13.22b) until after a rotation of 180° it contracts to a point. The current in the solenoid required to do this is noted, and from this the magnetic field B is worked out. The velocity v of the beam depends on the p.d. V between final anode and cathode, and is given by

$$v^2 = \frac{2eV}{m_e} \quad \text{(p. 230)}$$

The time t taken to travel the distance L from the centre of the Y-plates to the screen is given by

$$t = \frac{L}{v}$$

$$\therefore t^2 = \frac{L^2}{v^2} = \frac{m_e L^2}{2eV}$$

Suppose the electrons acquire a component of velocity u at right-angles to v in passing between the deflector plates; then the magnetic force Beu is the centripetal force $(m_e u^2 / r)$ that deflects them into the helical path (of radius r).

$$\therefore Beu = \frac{m_e u^2}{r}$$

If the electrons perform one complete revolution in the time t, then

$$u = \frac{Ber}{m_e} = \frac{2\pi r}{t}$$

$$\therefore t = \frac{2\pi m_e}{Be}$$

(the same for all values of r)

$$\therefore t^2 = \frac{4\pi^2 m_e{}^2}{B^2 e^2} = \frac{m_e L^2}{2eV}$$

Re-arranging this we get

$$\therefore \frac{e}{m_e} = \frac{8\pi^2 V}{B^2 L^2}$$

If the magnetic field is increased beyond this point, the electrons perform more than one revolution, and the trace grows into a line again and rotates still further. It contracts into a point each time the field reaches an integral multiple of the value calculated above.

Refined modern techniques enable e/m_e to be measured with an uncertainty rather less than 1 part in 10^4. At these accuracies it is even possible to demonstrate the increase of mass of the electron with velocity. The theory of Relativity predicts that the mass m of a particle should vary with its speed v (relative to the observer) according to the relation

$$m = m_0\left(1 - \frac{v^2}{c^2}\right)^{-\frac{1}{2}}$$

where c = the velocity of light and m_0 = the mass at zero speed, called the *rest mass*. The mass should therefore *increase* and the specific charge *decrease* with increasing speed of the electron. The experimental results are in excellent agreement with this prediction at all speeds. The effect is by no means negligible even when quite moderate p.d.'s are used to accelerate the electrons. With the p.d.'s used in television tubes (16 to 20 kV) they reach about a quarter of the speed of light; and the corresponding increase of mass is then about 3%. With an accelerating p.d. of half a million volts (quite common in X-ray equipment) the mass of the electron is about doubled.

The mass spectrometer

In the course of the early experiments on gas discharges it was found that luminous 'rays' of some kind would emerge from the discharge through perforations in the cathode. They were known as *positive rays* (Fig. 13.23). It is not easy to demonstrate their deflection by magnetic

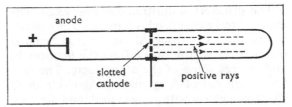

Fig. 13.23   Positive rays emerging through slots in the cathode

fields. Their specific charges are thousands of times smaller than that of the electron, and the magnetic fields to deflect them must be proportionately greater. However, with suitable fields it is possible to show that they consist of streams of positively charged particles; and we now understand that they are the positive ions formed in the passage of the electrons through the gas.

By the combined use of electric and magnetic fields it is possible to separate out the many different sorts of ions present in a positive ray beam. In one type of apparatus the fields are arranged to 'focus' all the ions of any one specific charge into a single line on a photographic film. A different line shows up on the developed film for each type of ion. The *mass spectrograph*, as the instrument is called, makes possible the precise measurement of atomic masses.

Another type of instrument is equipped to detect the presence of particular ions and measure their relative abundances. This is called a *mass spectrometer*.

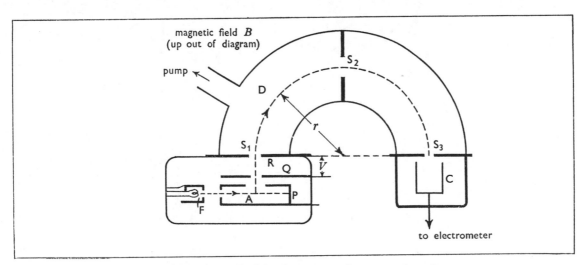

Fig. 13.24   The principle of one type of mass spectrometer

The principle is illustrated in Fig. 13.24. The material to be examined is introduced as a gas or vapour into the space A. An electron beam is directed across this space from the filament F, and ionizes some of the atoms there. The ions are drawn out of A by making the electrode Q slightly negative with respect to P, and are then accelerated by the much larger electric field between Q and R. Thus a narrow beam of ions emerges from the slit S_1 into the space D. If the p.d. between Q and R is V, the velocity v acquired by ions of charge e and mass m_i is given by

$$\tfrac{1}{2}m_i v^2 = eV$$

$$\therefore v^2 = \frac{2eV}{m_i}$$

In the space D the particles move in semi-circular paths under the influence of a magnetic field of flux density B directed at right-angles to the plane of the diagram. The slit system, S_1–S_2–S_3, is arranged to select particles whose orbits are of one particular radius r.

$$\therefore r = \frac{m_i v}{Be}$$

Eliminating v between these equations, the specific charge of the particles selected by the system is given by

$$\frac{e}{m_i} = \frac{2V}{B^2 r^2}$$

Any ions with this specific charge are collected by the electrode C, and the small current is registered by a sensitive electrometer. To make an analysis of the ions present in the apparatus, the ion current is measured as the p.d. V is varied, the magnetic field being kept constant. One type of ion after another is thus brought into the collector electrode C, and the relative sizes of the peaks of current indicate the proportions of different isotopes (and complex ions) present.

14 Electronics

14.1 The electronic theory of solids

The characteristic property of a conducting substance is that it contains 'free' electrons, i.e. electrons that are not fixed at any particular site in the crystal lattice, but are free to wander through it in any direction. Under equilibrium conditions these electrons diffuse at random rapidly in all directions and there is no net movement of electric charge. They form in fact a kind of 'electron gas' confined within the crystal structure. When the substance is placed in an electric field, the electrons gain momentum from the field and a slow drift of the electron gas through the material takes place; this is superimposed on the rapid random diffusion and constitutes an electric current. The average velocity of the charge carriers in the direction of the field is always quite small (p. 11), though the disturbance in the electric field that accompanies their motion is propagated through the material at almost the speed of light (p. 285), so that the charge is set in motion almost simultaneously at all points.

The conductivity of a material depends on two factors:

(*i*) the freedom of movement (or *mobility*, p. 24) of the charge carriers in it; i.e. the average velocity of drift acquired per unit electric field;

(*ii*) the number of charge carriers per unit volume (p. 24).

The latter quantity varies over an enormous range for different types of materials. In a metal about one valence electron per atom is effectively free to move through the crystal lattice and conduct electric currents; the number of conduction electrons does not change much with temperature. In an insulator there are virtually no free electrons—at least the number is less than in a metal by a factor of 10^{20} or more. Intermediate between these extremes lies the class of materials known as *semiconductors*. In a pure semiconductor at very low temperatures there are no free electrons and it behaves as an insulator. But as the temperature rises the material exhibits increasing conductivity as electrons are liberated by the thermal vibrations of the crystal. But even at room temperature only about one atom in 10^{10} has parted with an electron in this way.

We need now to consider how exactly an electron becomes 'free' in a crystal lattice, and why this spontaneous 'ionization' of the atoms occurs readily in some materials and not in others; while in semiconductors it is dependent on energy being available in the thermal vibrations of the lattice.

In an isolated atom the valence electrons can exist in a number of distinct orbitals each of sharply defined energy (p. 242 *et seq*.). Normally it exists in the *ground state*, which is the orbital of lowest possible energy; only when the atom absorbs energy from radiation or from a collision with another particle will the valence electron change to an orbital of higher energy—from which it quickly reverts to the ground state, with the emission of radiation. When two atoms come close together, the possible orbitals of their valence electrons are modified by the change in the electric field in which they move, and the energy levels are altered. If a stable chemical bond is formed (p. 245), the pattern of available energy levels is inevitably more complex than for the constituent atoms on their own; a large molecule has a very large number of closely spaced energy levels in any of which the valence electrons may exist, though they will always tend to occupy chiefly the states of lowest energy. An assembly of atoms in a crystal must be regarded as a single chemical unit, and in this case the energy levels become spread out into a series of bands. Each band consists of a very large number of levels (one pair of levels for each atom in the crystal); but these overlap to form in effect a continuous band of available energies. At sufficiently low temperatures (and in the absence of an external supply of energy such as radiation) all the electrons will exist in the states of lowest possible energy, and will therefore fill or partially fill the lowest available energy band. This is called the *valence band*. This situation represents a balanced electronic configuration in which there is no net movement of electrons in any particular direction, although the orbital of each electron is effectively spread out over a sizeable region of the crystal lattice. In order to conduct an electric current some of the valence electrons must be in an energy band that is only partly filled. This can happen either by some of the electrons being given sufficient energy to raise them into the next lowest energy band—called the *conduction band*—or by the valence band being only partly filled, as in the alkali metals that have only one valence electron per atom. If an electron is in a partly filled band, it is possible for its energy to be increased steadily in infinitesimal steps—in other words the electron can be accelerated by an applied electric field, thus gaining energy from it. In this way it can take part in the conduction of an electric current, and we may call it a 'free' electron.

The differences between the properties of the various classes of materials arise from the nature of the valence and conduction bands and the width of the energy gap between them.

(*i*) *Metals.* The majority of pure elements in crystalline form have metallic properties. In these the gap between the valence and conduction bands is non-existent, and in many cases the two energy bands actually overlap (Fig. 14.1); alternatively, the valence band is only partly filled. A metal therefore has relatively high conductivity at all temperatures. The proportion of valence electrons that are free to be accelerated increases slightly with temperature, and this tends to increase the conductivity. However, this effect is more than counteracted by the increasing interaction between the electrons and the crystal lattice as the temperature rises, which reduces the mobility of the electrons. The conductivity of a metal therefore falls with rising temperature (i.e. the resistivity rises, p. 25).

The freedom of movement of the valence electrons is also responsible for the high *thermal* conductivity of metals. Since the number of

effectively free electrons increases (slightly) with temperature, a temperature gradient through the material produces a gradient in the concentration of the free electrons in it. The rapid diffusion of the electrons then tends to equalize the electron concentration and conveys the surplus energy between the regions of high and low temperature. Electrical conduction consists of the diffusion of electric charge through the material under conditions of non-uniform electric potential; while thermal conduction consists of the diffusion of thermal energy under conditions of non-uniform temperature. In a metal both processes take place essentially by the same mechanism, and some connection between the two coefficients of conductivity is therefore to be expected. An examination of tables of physical properties shows that the ratio of thermal and electrical conductivities at a given temperature is approximately constant for a large range of metals. This is known as the *Wiedemann-Franz law*; it provides strong confirmation of the electronic theory of metals. The law does not hold, even approximately, for non-metals.

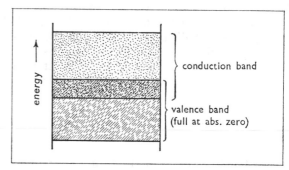

Fig. 14.1   The energy band diagram for a metal in which the conduction and valence bands overlap

In addition to the conduction of heat through a metal by the free electrons in it there is also a transference of thermal energy by the vibrations of the ions of the crystal lattice. The rate of conduction of heat on this account is comparable with that occurring in poor (non-metallic) conductors of heat, and it is normally swamped by the much more rapid process of electronic conduction. It therefore has little effect on the validity of the Wiedemann-Franz law.

(*ii*) *Insulators.* In this class of materials the valence band is completely filled and there exists

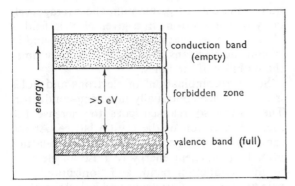

Fig. 14.2 The energy band diagram for an insulator in which there is a large forbidden zone between the valence and conduction bands

a large gap between the valence and conduction bands (Fig. 14.2). In good insulators the gap is 5 electron-volts or more. (This means that an electron must gain an amount of energy equivalent to being accelerated through a p.d. of 5 V in order to be transferred to the conduction band.) This is about 100 times as great as the average energy of vibration of an atom of the crystal lattice, and the chance of an electron being set 'free' for conduction is therefore very remote. However, if a sufficiently strong electric field is applied, the electrons may gain the requisite energy from it, and the insulation breaks down. The insulation strength therefore depends (amongst other things) on the width of the *forbidden zone* between the two energy bands.

The chance of electrons appearing in the conduction band increases when the temperature is raised. The conductivity therefore increases with temperature, and may even become relatively high. For instance, if a glass rod is heated to a temperature just below its melting point, a large enough current may be passed through it to maintain its temperature and even to melt it.

(iii) Semiconductors. In semiconductors, as in insulators, the valence band is completely filled when the electrons are all in their ground states. However, in this case the width of the forbidden zone between the valence and conduction bands is of the order of 1 electron-volt (Fig. 14.3). This means that at low temperatures there will be virtually no electrons in the conduction band and

the conductivity will be very low. But at normal room temperatures there is a small but significant chance of an electron gaining sufficient energy from the vibrations of the crystal lattice to raise it into the conduction band; and a very small proportion (typically about 1 in 10^{10}) of the atoms will be ionized in this way at any instant. The most commonly used semiconductors at present are *germanium* and *silicon*. In germanium the width of the forbidden zone is 0.72 eV; in silicon it is 1.12 eV. At a given temperature therefore silicon has fewer electrons in the conduction band, and its conductivity is considerably less than that of germanium.

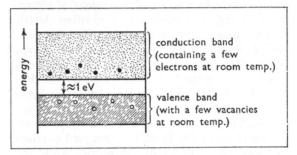

Fig. 14.3 The energy band diagram for a semiconductor in which there is a small energy gap between the valence and conduction bands

Because of the low level of 'ionization' in a semiconductor the conduction of a current through it can take place by two distinct mechanisms; Fig. 14.4 shows how this comes about. The black dots represent the few conduction electrons. Each of these has escaped from some atom, leaving behind it a positively charged *hole* in the electron structure of the crystal. If a p.d. is applied to the specimen, the free electrons move under the action of the electric field, as we should expect; and the current is partly

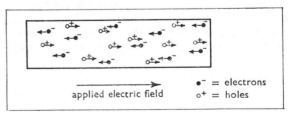

Fig. 14.4 The flow of electrons and positive holes in an intrinsic semiconductor

accounted for in this way. But it is also possible for a *bound* electron to move into a neighbouring hole; the electron still remains a bound one (in its new site), but the hole is transferred to the atom just vacated. Part of the conduction is thus by the movement of positive holes through the material; and it is found that these behave very much as though they were positively charged electrons.

In terms of the energy level structure of the material the effect may be explained as follows. The existence of a hole in the electron structure implies that there is an unoccupied energy level in the valence band. The valence band now possesses an *unbalanced* electronic configuration which can represent a net movement of charge in one direction. Under the action of an electric field the unoccupied energy level of the hole can be filled by a bound electron from a lower energy level in the valence band. In this way the energy represented by the hole is increased by the electric field; and this can be regarded as an acceleration of the positive charge of the hole in the direction of the field. Thus an applied p.d. causes the electrons in the largely unoccupied conduction band to drift in one direction, while the holes in the almost fully occupied valence band drift in the opposite direction.

The behaviour of electrons and holes in a semiconductor is analogous to the movement of water in a tilted tube. If a few drops of water are introduced, they will run down to the lower end of the tube gaining energy from the gravitational field. If the tube is sealed at both ends and is completely full of water, no net movement of water takes place when the tube is tilted. But if now a few 'holes' are introduced in the form of air bubbles, these will move *upwards* in the gravitational field. The movement of the bubbles upwards accompanies a net downward displacement of the water; but we can analyse the process by treating the bubbles as though they were droplets of negative weight (but, of course, of positive mass). Likewise the movement of holes in the valence band of a semiconductor is most conveniently analysed by treating them as positively charged electrons. In the tilted tube the resistance to the motion of the water droplets would doubtless be very different from that of the bubbles. So, likewise, the mobilities of electrons and holes in a semiconductor are usually

quite different. In germanium the average velocity of the electrons in a given electric field is about twice as great as that of the holes; and in silicon the electrons have more than four times the mobility of the holes.

Some 'recombination' of electrons and holes is taking place continually in a semi-conductor. The energy so released goes to increase the thermal vibration of the crystal lattice. At the same time there is a small chance of the lattice providing sufficient energy to raise an electron to the conduction band, and 'ionization' (i.e. pair production of electrons and holes) therefore occurs also at a small but steady rate. The number of current carriers is maintained at such a level that recombination and pair production proceed at equal rates. A rise in temperature increases the rate of production of electron-hole pairs and so increases the number of current carriers. On this account the conductivities of semiconductors rise very rapidly with temperature (p. 25).

A material in which there are equal numbers of electrons and holes is known as an *intrinsic semiconductor*. This only happens in an extremely pure material. Impurities tend to produce additional electrons or holes in the crystal structure. Since only 1 in 10^{10} of the atoms contributes a current carrier to the crystal in a pure semiconductor, an impurity level of 1 in 10^7 can be enough to increase the conductivity a thousandfold and to modify drastically the other electrical characteristics. Semiconductors acquire their useful properties from the controlled addition during manufacture of minute traces of impurities.

Germanium and silicon both crystallize in the same type of structure—the same in fact as that of diamond. Each atom has four valence electrons and is bonded to four neighbouring atoms in a tetrahedral arrangement. A double bond consisting of one pair of shared electrons thus exists between every pair of adjacent atoms. When an impurity atom of about the same size is incorporated in the crystal it simply occupies the site that would otherwise be taken by a germanium or silicon atom without affecting the structure of the lattice around it. Its effect on the electrical properties of the crystal depends on its valency.

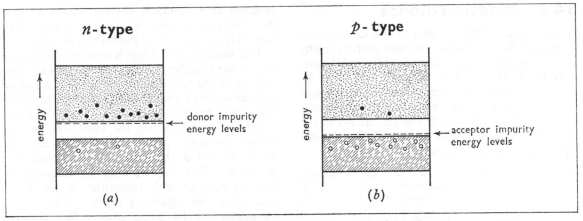

Fig. 14.5   Energy band diagrams for the two types of impurity semiconductor: (a) n-type, (b) p-type

Suppose an impurity is added having five valence electrons per atom, such as antimony or arsenic. In this case one electron per impurity atom will be left over after all the valence bonds with neighbouring atoms have been satisfied, and this will be free to move through the crystal with the other conduction electrons. These are known as *donor* impurities, and have the effect of greatly increasing the number of conduction electrons. Since the conduction is now mainly by the *negative* current carriers, the semiconductor is said to be *n-type*. In fact a small amount of energy is required to separate the spare electron from each donor atom; but unless the temperature is very low the available thermal energy of the lattice is sufficient to ensure that most of the donor atoms are ionized. From the point of view of the energy level diagram the donor impurity creates extra localized energy levels in the forbidden zone just *below* the conduction band (Fig. 14.5a). The gap between the impurity energy level and the conduction band is less than 0.1 eV, comparable with the thermal energy of an ion of the lattice at ordinary room temperatures.

It is also possible to add an impurity having three valence electrons per atom, such as *indium* or *gallium*. In order to complete the electron structure these atoms will tend to acquire an extra electron each from their neighbours. Each one thus becomes a centre of bound negative charge, while a positive *hole* is created nearby (p. 70). These are known as *acceptor* impurities, and have the effect of increasing the number

of holes in the crystal structure. Since the conduction is now mainly by the *positive* current carriers, the semiconductor is said to be *p-type*. Again a small amount of energy is needed to complete the electron structure round the impurity atom and so create the extra hole; but at room temperature this is readily available from the thermal energy of the lattice. On the energy level diagram an acceptor impurity creates an extra localized energy level in the forbidden zone just *above* the valence band (Fig. 14.5b).

In each type of semiconductor the addition of the impurity not only provides a large number of extra current carriers—the *majority carriers*, as they are called—but it also largely suppresses the carriers of opposite type—the *minority carriers*. Thus the addition of a trace of donor impurity might, say, increase the number of conduction electrons tenfold. But it also thereby increases the chance of recombination of electrons and holes by the same factor; and the net result is a tenfold reduction in the number of holes in the lattice. The minority carriers, holes in this case, then play a very small part in the conduction of current. Likewise, a trace of acceptor impurity produces an enormous preponderance of holes in the material. However, the proportion of minority carriers increases rapidly if the temperature rises because of the increased rate of production of electron-hole pairs; and this tends to set a limit to the current-carrying capacity of semiconductor devices whose functioning depends on maintaining the preponderance of majority carriers.

14.2 Junction diodes

There are a number of devices besides the thermionic diode (p. 234) that conduct current chiefly in one direction. These employ the special characteristics of junctions between two sorts of semiconductor or between a semiconductor and a metal. Because they have two terminals and fulfil a similar function in electric circuits to the thermionic diode, these semiconductor devices are commonly referred to as *diodes* also.

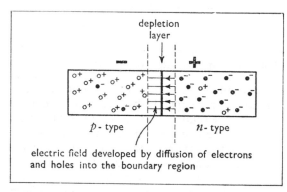

Fig. 14.6   The rectifying action of *p–n* junction

Such a diode may be made by forming regions of *n*-type and *p*-type semiconductor *within the same crystal* of germanium or silicon. At the junction electrons diffuse across from the *n*-type side and holes from the *p*-type side (Fig. 14.6). Recombination of these electrons and holes produces on either side of the boundary a narrow *depletion layer* (or *barrier layer*) relatively free of current carriers and therefore of high resistance. This leaves the *n*-type material positive with respect to the *p*-type material; the diffusion continues until the electric field developed across the depletion layer is sufficient to prevent further movement of charge carriers.

If now a p.d. is applied to the crystal making the *p*-type region positive (or less negative) with respect to the *n*-type region, the electric field in the depletion layer is reduced and a current can flow across the junction. But a p.d. applied the other way round increases the field in the depletion layer, and the flow of majority carriers across the junction is further inhibited. The only current is then that due to the very small proportion of minority carriers (electrons from the left, holes from the right, in Fig. 14.6). The

junction thus conducts current readily in one direction (the *forward* direction, it is called), but has a high resistance in the *reverse* direction.

To complete the diode it is of course necessary to provide metallic contacts to the *n*- and *p*-type regions of the crystal; and these junctions must themselves be of non-rectifying types. A junction between a metal and a semiconductor can form a barrier layer in much the same way as one between *p*- and *n*-type semiconductors, and this must be avoided where the connecting leads are joined on. This may be done by alloying the metallic contact with impurities of the same kind as in the adjoining region of the semiconductor—or by using solders that contain these. The impurities then diffuse into the immediately adjacent layer of semiconductor and produce there such a high concentration of current carriers that no depletion layer is able to form near the junction. This is called an *ohmic* contact. In general there will still be a contact p.d. (p. 45) between the metal and semiconductor arising from their different concentrations of charge carriers; but there is no barrier layer and the resistance is low for both directions of current flow through the junction.

One form of germanium diode is shown in Fig. 14.7. A wafer of *n*-type germanium is soldered onto a metal base, and a button of indium is then fused onto its top surface. The indium and germanium diffuse together in the boundary region, and the germanium there is thus converted into *p*-type. Connecting wires are soldered to the indium button and the base, and the whole unit is hermetically sealed in a light-proof capsule.

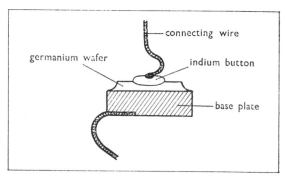

Fig. 14.7   A germanium diode

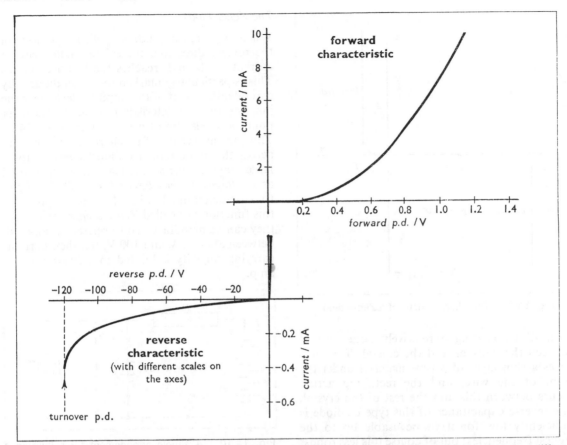

Fig. 14.8   The characteristics of a germanium diode; the forward and reverse characteristics are plotted on different scales, the latter covering a larger range of p.d.'s and a smaller range of currents

The electrical characteristics of a germanium diode are shown in Fig. 14.8. In the *forward* direction the current increases approximately exponentially with p.d. Very little current flows until the forward p.d. rises above about 0.2 V; it then increases very rapidly. The curve for the reverse direction is plotted separately in Fig. 14.8, using different scales on the axes, since the ranges of current and p.d. involved are quite different. For a large range of reverse p.d.'s the leakage current due to the minority carriers remains quite small. However, at a sufficiently high p.d. (the *turnover p.d.*) breakdown of the barrier layer takes place, and beyond this point the diode is likely to suffer damage unless the current is limited by some means. When breakdown occurs the electric field in the barrier layer is high enough for electrons to be lifted straight

into the conduction band. Thus, as the field reaches this critical value there is a sudden evolution of electron-hole pairs and the current rises abruptly; this is called the *Zener* or *avalanche effect*.

In the reverse direction a junction diode behaves like a leaky capacitor between its *p*- and *n*-type regions, the barrier layer being the dielectric. This means that the efficiency of the diode as a rectifier is greatly reduced at high frequencies, since it is effectively shunted by its own interelectrode capacitance. For rapidly varying currents it is usual therefore to employ an older design of diode known as a *point-contact diode*. This consists of a single *n*-type crystal of germanium on which a springy metal wire (called traditionally a *cat's whisker*) is made to bear. During manufacture the diode is

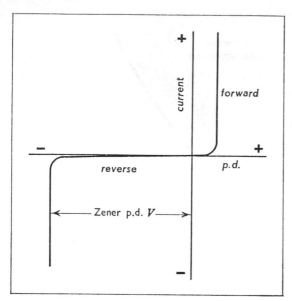

Fig. 14.9   The characteristic of a Zener diode

The Zener diode

When an increasing p.d. is applied to a diode in the reverse direction, there is a sudden rise in current as the p.d. reaches the turnover point. This is particularly marked in silicon diodes. By careful adjustment of the impurity level near the junction the characteristic beyond the turnover point becomes almost a vertical line (Fig. 14.9). Thus, in this region of its characteristic the p.d. across the diode remains almost constant for a large range of currents. It may therefore be used as a *voltage reference* device for stabilizing a p.d. to a predetermined value. Diodes designed for this function are called *Zener diodes*. At present they can be manufactured to operate at any p.d. between about 4 V and 100 V, but their current-carrying capacity is limited to a fraction of an amp.

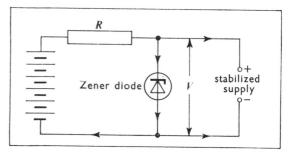

Fig. 14.10   A simple arrangement for stabilizing a battery supply using a Zener diode

'formed' by passing a relatively large current between the whisker and the crystal. This produces a thin layer of *p*-type material under the point of the wire, and the rectifying action occurs between this and the rest of the crystal. The reverse capacitance of this type of diode is sufficiently low for it to be usable up to the highest frequencies; but of course it is less robust than the usual junction type of diode.

The silicon diode is superior to a germanium one in some respects; it can be designed to stand peak reverse p.d.'s of about 1000 V, and the reverse currents are even smaller than in a germanium diode. But the forward drop in p.d. across it is rather greater.

The chief problem in designing germanium and silicon rectifiers to handle high powers is in arranging for the heat evolved in them to be conducted away rapidly enough. If the temperature of a germanium diode is allowed to rise much above 100 °C, irreversible changes take place in the crystal structure of the junction, and the diode is destroyed. It is therefore often necessary to bolt the diode to a *heat sink* consisting of a thick metal plate with cooling fins. The temperature of a silicon diode can be allowed to rise to about 200 °C, and the problem of arranging for heat dissipation is not as acute as with germanium.

Fig. 14.10 shows a simple arrangement for stabilizing the supply p.d. of, say, a transistor radio. The series resistance *R* is chosen so that, with a fresh battery and no current taken by the load, the maximum safe current (at the Zener p.d. *V*) flows in the diode. When the load draws a current, that through the diode falls by the same amount, so that the drop in p.d. across *R* remains almost constant. Thus variations in the load current affect the p.d. *V* supplied to it very little. The arrangement is also stabilized to some extent against variations in the battery p.d., provided this remains greater than the Zener p.d. of the diode and the current to the load is not excessive.

The photodiode

When a diode is biased in the reverse direction, the normal small leakage current is caused by the diffusion of minority carriers across the depletion layer; these carriers arise from the spontaneous formation of electron-hole pairs in and near the junction. If the junction is illuminated by light of sufficiently short wavelength, the energy of the photons (p. 240) can be sufficient to create extra electron-hole pairs. An increase in the reverse current thus occurs which is proportional to the flux of light falling on the sensitive area. A diode adapted to be used in this way is called a *photodiode*. If no bias is applied, a small e.m.f. is generated in it by the light in the same manner as in the selenium photocell (p. 238). But it is usually more satisfactory to operate it biased in the reverse direction by an external p.d. The incidence of light then causes an apparent fall in the resistance of the diode, and a current flows that may be used, for instance, to operate a relay or some other form of detector.

The solid-state particle detector

A suitable reverse-biased diode may also be used to detect other kinds of ionizing radiations, such as α- and β-particles (p. 314). The passage of one of these particles through the sensitive region round the junction of the diode produces many extra electron-hole pairs, and causes a pulse of current through the detector. The pulses may then be counted with a scaler or ratemeter (p. 322). The solid-state detector produces pulses whose amplitude depends on the amount of ion-

ization produced by the particle, and it may be used to discriminate between particles of different type. To detect α-particles the thickness of material on one side of the junction must be reduced to a minimum (p. 321). With β-particles and γ-rays it is best to increase the reverse bias as much as possible; this increases the thickness of the sensitive depletion layer round the junction, and gives a better chance of detecting weakly ionizing radiations.

14.3   Rectifier circuits

The ability of a diode to conduct current in one direction only enables it to be used as a rectifier for producing d.c. from an alternating supply. The simplest possible arrangement consists of a diode joined in series with the load (Fig. 14.11). (The word 'load' in this connection means that to which the rectifier is supplying power in the form of direct current; it is represented in Fig. 14.11 by the resistor.) The diode conducts only at those parts of the cycle when the point A is positive with respect to the point B. The current through the load is therefore uni-directional, but pulsating.

A better arrangement is that shown in Fig. 14.12. The connection of the *storage* or *reservoir capacitor* in parallel with the load radically alters the behaviour of the circuit. Suppose first that the load R is disconnected, so that there is no way for the capacitor to discharge when the

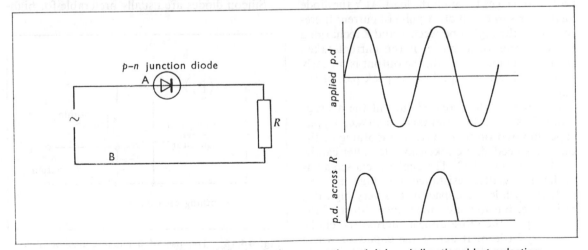

Fig. 14.11   A diode with a resistive load; the current through it is uni-directional but pulsating

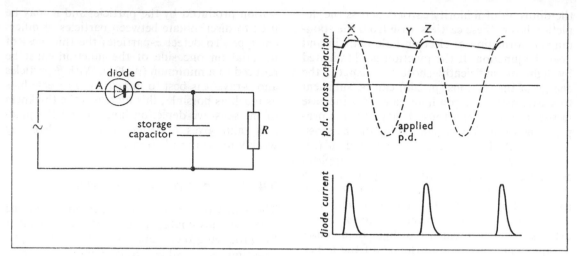

Fig. 14.12   The use of a *storage* or *reservoir* capacitor in a rectifier circuit

diode is non-conducting. Current flows through the diode during the positive half of the cycle charging up the capacitor with the upper plate positive. This continues in each cycle until the p.d. across the capacitor is nearly equal to the *peak* p.d. of the supply. When this has occurred, the point A can no longer become positive with respect to the point C, and no further current flows. With the load joined up, there is a continual steady discharge of the capacitor, and the output is of the form shown in Fig. 14.12. At X the capacitor is charged up nearly to the peak p.d.; the diode then ceases to conduct and the p.d. across the capacitor falls slowly as it partially discharges through the load. At Y the diode again starts to conduct. A pulse of current therefore flows through it between Y and Z recharging the capacitor once more. If the current taken through the load is small, the output is a nearly steady p.d. almost equal to the peak p.d. of the supply.

However, with increased current the average p.d. is less than this, and there is an a.c. 'ripple' superimposed on it. If further smoothing of the output is needed, a choke-capacitor filter can be used as in Fig. 14.13. The choke is chosen so as to have low d.c. resistance, but high a.c. impedance; while the capacitor has very high d.c. resistance, but low a.c. impedance. Most of the 'ripple' p.d. therefore appears across the choke and very little of it across the capacitor where the output is taken. A small part of the steady

p.d. across the storage capacitor is dropped in the resistance of the choke, but most of it is available across the load.

Because of the small size and simplicity of semiconductor diodes it is often preferable to use a full-wave type of rectifier circuit employing four diodes joined in a bridge (Fig. 14.14). On one half-cycle current passes through diodes P and S, charging the storage capacitor to near the peak p.d. of the supply; on the other half-cycle P and S remain non-conducting and current passes instead through Q and R. Further smoothing to reduce the a.c. ripple can be provided by the usual choke-capacitor filter.

Silicon diodes are usually preferable for high-

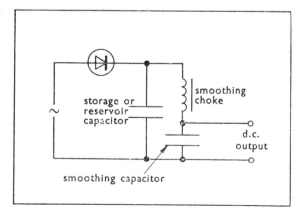

Fig. 14.13   The use of a choke-capacitor filter to obtain further smoothing

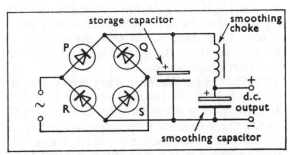

Fig. 14.14   A full-wave bridge type rectifier with storage capacitor and choke-capacitor smoothing

tension and high-power applications at low frequencies; germanium diodes are rather more efficient in low tension power units because their forward drop in p.d. for a given current is less than for silicon diodes. Germanium diodes are also particularly suitable for detector circuits in radio receivers (p. 298, Fig. 15.14); normally the point-contact type would be used for this purpose.

An a.c. peak voltmeter

An arrangement like that in Fig. 14.12 may be used to measure the peak value of an alternating p.d. In this case the load resistance *R* consists simply of a high-resistance d.c. voltmeter. The storage capacitor is chosen so that the time constant *CR* is very long compared with the period of oscillation of the p.d. There is then a negligible ripple on the rectified output measured by the voltmeter; and the reading gives the *peak value* of the p.d. However allowance must be made for the fact that a diode does not conduct significantly even in the forward direction until the p.d. reaches a certain minimum, about 0.3 V for a germanium diode and about twice as much for a silicon one. This zero error must therefore be added to the voltmeter readings to obtain the true peak p.d.

14.4   Transistors

A transistor consists of a thin layer of *n*- or *p*-type semiconductor sandwiched between two layers of semiconductor of opposite type—shown diagrammatically in Fig. 14.15*a* and *b*. The circuit symbols for the two types are shown alongside, *c* and *d*. (The symbols derive from the

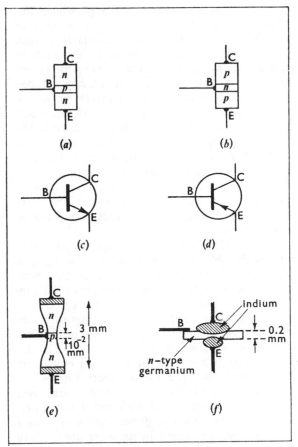

Fig. 14.15   Transistors: (*a*) *n–p–n* type; (*b*) *p–n–p* type; (*c*) and (*d*) are circuit symbols for these; (*e*) a grown junction transistor; (*f*) an alloy junction transistor

construction of an early form of transistor, consisting of two fine wires bearing on a wafer of semiconductor—a sort of double point-contact diode.) The three layers are not just in electrical contact but must be formed within a single crystal.

A number of different methods of construction are now employed. One of these is the *grown-junction* technique (Fig. 14.15*e*). In this type all three layers are formed within a single crystal as it is grown from molten germanium. A 'seed' crystal of germanium is lowered into molten germanium and then slowly withdrawn whilst being rotated, so that a rod of germanium is formed consisting of a single crystal. The type of semiconductor at each point of the rod can

be controlled by adding the appropriate impurities to the melt. To start with a trace of
arsenic is added so that *n*-type material is formed.
Then enough gallium is added to the melt to
convert it into *p*-type germanium. Almost immediately afterwards the melt is heavily doped
again with arsenic so that it reverts to *n*-type
material. The bar is then sawn into strips, which
are shaped to form the small unit shown in the
figure. Its length is about 3 mm, and at the
centre its cross-section measures about 1 mm ×
0.2 mm. Connecting wires are bonded onto the
three regions. The central *p*-type layer may be
only 10^{-2} mm thick, so that the fine gold wire
that makes connection with it must be positioned
with some care!

Fig. 14.15*f* shows another type of transistor
produced by the *alloy-junction* technique.
The central region in this type of transistor
is a wafer of *n*-type germanium about 0.2 mm
thick. Two buttons of indium are fused into
the sides of this, and sufficient indium dissolves
in the neighbouring layers of germanium to convert these into *p*-type semiconductor. The indium buttons are then the means of conveying
the current into the *p*-type layers formed on
their surfaces. Finally the transistor is enclosed
in a hermetically sealed, light-proof container.
All operations on hot germanium and silicon
have to be performed in an inert atmosphere
(e.g. argon), and contamination has to be rigorously excluded at all stages of manufacture.
Considering also the small scale and close toler-
ances of the work, it is clear that the production
of transistors demands manufacturing techniques of the highest sophistication.

These and many other processes (see below,
p. 266) have been used to produce both *n–p–n*
and *p–n–p* types of transistor. The two types
differ little in their properties, except for the
directions of flow of current in them. We shall
describe the action of the *n–p–n* type, which is
now rather more common.

The *n–p–n* transistor can be regarded as two
p–n junction diodes joined back to back. It is
operated with one junction (that between E and
B) biased in the forward direction and the other
in the reverse direction (Fig. 14.16). The layer E
(called the *emitter*) is a very heavily doped *n*-type
region, so that the current between E and B is
almost entirely by the movement of electrons

from E into B. The electrons entering the central
region (called the *base*) diffuse across it; and, if
it were thick, they would soon be neutralized
by the holes there. But by making the base layer
very thin a high proportion of the electrons will
diffuse to the far side of it. They then enter the
electric field at the junction between B and C
and are quickly swept into the third region C
(called the *collector*). In practice over 95% of
the electrons from the emitter reach the collector; and usually the proportion is more than
99%. The remainder recombine with positive
holes in the base layer and cause a small current
to flow in the base lead. [In the *p–n–p* type of
transistor the conduction process is similar, but
takes place instead by migration of *holes* from
emitter to collector, a few of which recombine
with *electrons* in the base. The battery connections must of course then be the other way
round to give the correct bias.]

In a typical small silicon transistor a current
of about 5 mA may flow through the emitter-
base junction when the p.d. across it is 0.7 V.
Only about 1% of this current (0.05 mA) passes
to the base lead, leaving 4.95 mA to flow out
through the collector lead. The collector current
is scarcely affected by changes in the collector

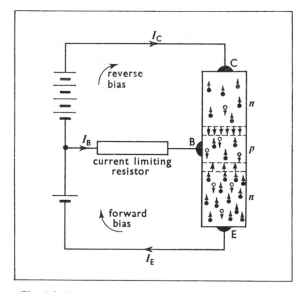

Fig. 14.16   The biassing of the junctions of an *n–p–n*
transistor; the current at the base-collector junction is
almost entirely by means of electrons that have diffused
from the emitter-base junction

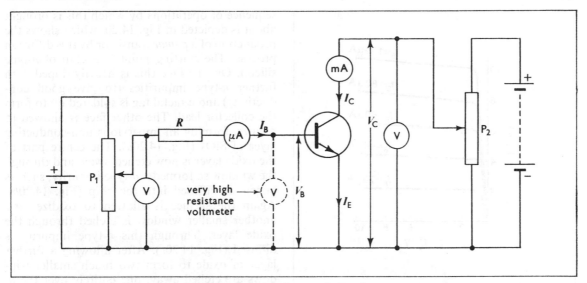

Fig. 14.17   Investigating the characteristics of an *n–p–n* transistor

potential, but is controlled almost entirely by the base current, and bears a nearly constant ratio to this (about 100 in the example above). Variations in the base current thus cause proportionately larger variations in the main current that flows through the transistor between emitter and collector.

If a small alternating current is fed into the base lead (superimposed on the steady bias current), it causes an amplified alternating current to flow from emitter to collector (superimposed on the steady collector current). The transistor may thus be used as a *current amplifier*. The input current flows between emitter and base, and the amplified output current between emitter and collector.

The characteristics of an *n–p–n* transistor may be investigated with the circuit shown in Fig. 14.17. The object is to find how the collector current I_C varies with the base current I_B and the p.d. V_C between collector and emitter. To represent the variation of all three quantities on one graph we can plot a set of curves showing the variation of I_C with V_C for a selected range of values of I_B. These curves are called the *collector characteristics*. The base current I_B may be adjusted to the required value by varying the p.d. V_B applied between base and emitter by means of the potentiometer P_1. The resistance R is chosen so as to limit the base current to the

maximum allowed value with the slider of P_1 set for maximum p.d. Thus, if the maximum current is to be 150 μA and a 1.5 V cell is used, a value of 10 kΩ would be suitable for R. If it is required to measure V_B as well, then an instrument must be used that takes a current small compared with I_B—e.g. an electrometer (p. 175). Alternatively an ordinary voltmeter can be joined to the left of the microammeter (in Fig. 14.17), but correction must then be made for the p.d. across the microammeter.

Having selected a suitable value for I_B, the p.d. V_C applied between collector and emitter is varied by means of the other potentiometer P_2, and the collector current I_C is read on the milliammeter. The base current I_B may be slightly affected by variations in V_C and must therefore be re-adjusted to the selected value before each observation. Fig. 14.18 shows the collector characteristic of a typical small silicon transistor. For values of V_C more than about 0.2 V the collector current I_C is decided almost entirely by I_B, and V_C has relatively little effect on it. In this region of the characteristics I_C bears a constant ratio to I_B; this ratio is known as the *current amplification factor* of the transistor, and is denoted by β. For the transistor whose characteristics are shown in Fig. 14.18 β is 100.

Fig. 14.19 shows the *input characteristic* of the same transistor, i.e. the curve showing the

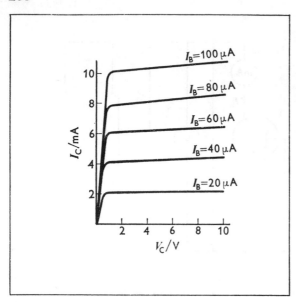

Fig. 14.18   The collector characteristics of a typical silicon transistor

variation of I_B with V_B for a constant value of V_C. No significant current flows until V_B rises above 0.5 V, and the current then increases rapidly and non-linearly with V_B. Over most of the operating range of the transistor the value of V_B is very close to 0.7 V; and in practical circuit calculations we can usually take this figure as the value of V_B for a *conducting* silicon transistor. (With a non-conducting transistor V_B may have any value less than 0.5 V.) With a germanium transistor the corresponding values of V_B are about a third as much.

In any transistor the charge carriers take a finite time to diffuse across the base layer. There is therefore a limit to the rapidity with which the current through a transistor can be changed, i.e. there is a limit to the frequency at which a given transistor can be used for amplification. The transistors of the types shown in Fig. 14.15 cannot readily be used outside the audio frequency range (i.e. above 20 kHz). New methods of manufacture have therefore had to be devised for making transistors that can be used at high frequencies. In one such method the base and emitter layers are produced by allowing the appropriate impurities to diffuse in turn into the face of a chip of silicon. The ingenious

sequence of operations by which this is brought about is depicted in Fig. 14.20, which shows the production of a *planar* transistor by this diffusion process. The starting point is a chip of n-type silicon. On one face this is heavily doped with further n-type impurities (to give good conductivity) and a metal tag is soldered on to form the collector lead. The other face is allowed to oxidize, forming an impervious non-conducting layer of SiO_2 (Fig. 14.20a). The centre part of the oxide layer is now etched away, and through the window so formed a p-type impurity, such as boron, is diffused into the chip (Fig. 14.20b). Again the surface is allowed to oxidize, and another smaller window is etched through the oxide layer. Through this n-type impurity is diffused (Fig. 14.20c). After allowing a further layer of oxide to form two much smaller windows are etched away, one entirely over the n-type region that is to form the emitter, and the other over the part where only p-type material reaches the surface. Metal is then evaporated onto these windows to form the emitter and base leads of the completed transistor (Fig. 14.20d). This form of construction provides a large surface area for the emitter/base junction, enabling large currents to be handled. It also ensures

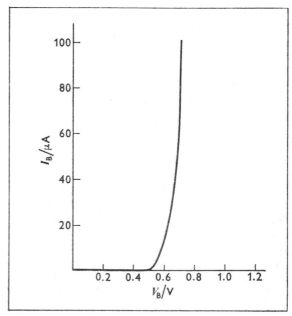

Fig. 14.19   The input characteristic of a typical silicon transistor

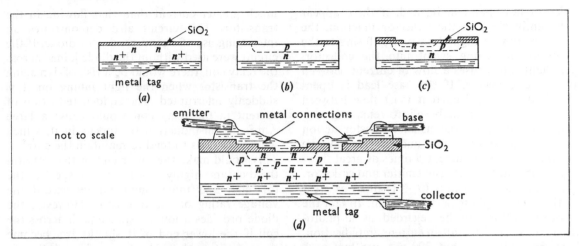

Fig. 14.20   The method of construction of a *planar* silicon transistor; the diagram is not to scale—in reality the emitter and base layers are very thin (less than 10^{-2} mm)

that the active junctions are entirely enclosed within the material. For it is found that when the junctions reach to the surface the energy levels there are modified, and this reduces the efficiency of the transistor action. But most important of all this method of construction enables the *p*-type base layer to be made exceedingly thin; thicknesses of a few nanometres (1 nm = 10^{-9} m) are common (Fig. 14.20 is *not* drawn to scale!). Also the diffusion process leads to the concentration of acceptor impurity varying through the base layer in such a way as to produce an electric field in the base that accelerates the electrons on their passage to the collector. The transit time of the charge carriers through the base layer is thus greatly reduced, and this enables this kind of transistor to be used at frequencies up to several GHz.

In another technique planar transistors are produced by *epitaxial* growth, i.e. the doped layers of silicon are condensed from the vapour phase directly onto a chip of substrate. This is done by heating the substrate in an atmosphere of silicon tetrachloride and hydrogen, when the silicon is deposited on the substrate surface together with whatever doping impurity was present in the vapour. By a succession of such depositions the complete transistor is grown by crystallization from the vapour phase.

By an extension of this technique it is possible to form complete circuits containing large numbers of transistors within a single chip of silicon;

these are known as *integrated circuits*. Resistors can be formed by laying down thin strips of silicon between the circuit elements they are to connect; capacitors are formed by depositing layers of insulating silicon dioxide between layers of conducting silicon; diodes appear at junctions of *p*- and *n*-type material; and so on. Each layer of material resembles a microscopic printed circuit. The intricate patterns required are produced by epitaxial growth, having first masked the surface by a photo-lithographic technique, so that the layer is deposited only on the unmasked parts. Many such layers go to make a complete integrated circuit. The manufacturing techniques needed are of great complexity. But once set up, they enable whole blocks of electronic circuitry to be fabricated in a single production process with enormous savings in costs. It is now possible to obtain single integrated circuits that contain all the parts of a complete radio (except the tuning capacitor and loudspeaker). Likewise a single integrated circuit can provide all the logical operations required for the functioning of a desk calculator. There seems to be no limit to the possibilities that are being opened up by integrated circuit technology.

The phototransistor

This is an ordinary *n–p–n* alloy-junction or grown-junction transistor with a transparent case. Light falling on the base region creates electron–hole pairs in it; and, if these appear sufficiently close to the collector junction, the electrons have a high chance of diffusing to the junction and being swept into the collector. The light thus causes a flow of current between base and collector. If the base lead is open-circuited, an equal current must flow between base and emitter if the base is to remain at constant potential. By the usual transistor action any current between base and emitter must be accompanied by a current β times as great flowing straight across between emitter and collector. The total current is thus $(\beta + 1)$ times as great as the original photoelectric current. In fact the phototransistor can be regarded as a photodiode (p. 260) with built-in current amplification. The sensitivity is about 300 mA per lumen as against about 8 mA per lumen for a photodiode and about 0.5 mA per lumen for a selenium cell (p. 238). In practice it pays to use the phototransistor wth a resistance joined between base and emitter, since this reduces the dark (leakage) current and gives greater stability of current if the temperature changes.

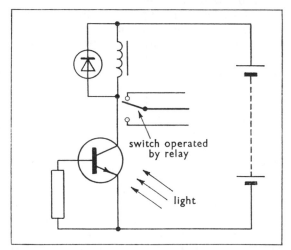

Fig. 14.21   A light-operated switch using a phototransistor

Fig. 14.21 shows a simple arrangement for a light-operated relay using a phototransistor. The increase of current produced by the light magnetizes the iron core of the relay; this acts on a small soft-iron armature and closes or opens a pair of contacts. In this way the light can be used to switch on or off some device that takes a very

much greater current than that flowing in the transistor. This circuit also demonstrates an interesting use of a semi-conductor diode. If the circuit were used without the diode joined across the relay coil, there would be a risk of damaging the transistor when the light falling on it is suddenly interrupted. The sudden reduction of current in the relay coil would cause a large e.m.f. of self-induction to appear in it, which would act so as to tend to maintain the current. This would make the lower end of the relay (in the diagram) highly positive with respect to the top end, and might momentarily exceed the voltage rating of the transistor. However, the diode provides a low resistance path across the coil if the lower end of it should ever become more positive than the top end, and thus it effectively prevents the p.d. across the transistor rising above that of the supply. At other times the lower end of the coil is less positive than the top end, and the diode remains non-conducting.

14.5 Amplifiers

There are many familiar applications in which a small alternating p.d. must be amplified to drive some piece of equipment. For instance a microphone or record player pick-up may produce an alternating p.d. of only a few millivolts at negligible power; this signal must then be sufficiently amplified to actuate a loudspeaker or the recording head of a tape recorder. Transistors provide the means of making such amplifiers.

A transistor is basically a current amplifier, i.e. the output current between emitter and collector is β times as great as the input current between emitter and base, where β is the current amplification factor. To make from it a distortionless amplifier of p.d.'s we need to arrange for the input current to the base to vary linearly with the input p.d. to be amplified; also for the output p.d. to vary linearly with the collector current. Fig. 14.22 shows how this may be achieved; this is in fact almost the same as the circuit of Fig. 14.17, but adapted to enable us to investigate the properties of a complete transistor amplifier. The base resistor R_B is needed in order to cope with the highly non-linear input characteristic of a transistor (see Fig. 14.19).

$$\text{p.d. across } R_B = V_{in} - V_B$$

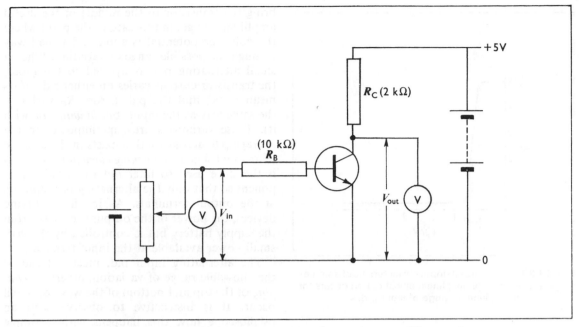

Fig. 14.22   Investigating the behaviour of a transistor amplifier circuit

where V_B = p.d. between base and emitter. Therefore the base current I_B is given by

$$I_B = \frac{V_{in} - V_B}{R_B}$$

Now over the useful operating range of the transistor V_B varies very little, and is approximately equal to 0.7 V.

$$\therefore I_B \approx \frac{V_{in}}{R_B} - \text{a constant}$$

And we therefore have the required linear relation between I_B and V_{in}. The approximation of taking V_B to be constant is valid in so far as the range of variation of V_{in} is much larger than the small variations in V_B; and R_B must be large enough to ensure that this is so.

The collector current I_C $(=\beta I_B)$ passes through the *collector load resistor* R_C, and it is in this that the variations of output p.d. arise.

$$\text{p.d. across } R_C = V_0 - V_{out} = R_C I_C$$

where V_0 = p.d. of the supply, which is constant. We therefore have a linear relation between V_{out} and I_C. The amplification of p.d. or *gain* achieved by the circuit is defined as the ratio of the change in output p.d. to the corresponding small change in input p.d.

$$\text{gain} = \frac{\delta V_{out}}{\delta V_{in}}$$

This can be calculated from the equations given above. Thus

$$\delta I_B = \frac{\delta V_{in}}{R_B}$$

$$\delta I_C = \beta \delta I_B = \frac{\beta \delta V_{in}}{R_B}$$

and

$$-\delta V_{out} = R_C \delta I_C = \frac{\beta R_C}{R_B} . \delta V_{in}$$

$$\therefore \text{gain} = -\frac{\beta R_C}{R_B}$$

Taking β as 100 and using the values of resistances indicated in Fig. 14.22, the amplification comes to -20.

The minus sign refers to the fact that the output is in antiphase with the input. Thus, when V_{in} *increases*, the base current *rises* and with it the collector current. The p.d. developed across R_C therefore *rises*, and the potential of the lower end of R_C *falls*. A single stage amplifier like this is thus an inverting amplifier.

It is clear that there is only a limited range of values of V_{in} for which the amplifier can function without distortion. If $V_{in} < 0.5$ V, the

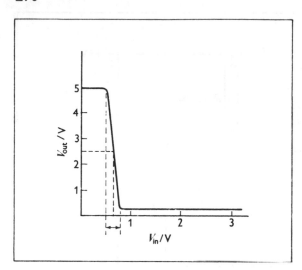

Fig. 14.23   The output-input characteristic of a single-stage transistor amplifier; linear amplification occurs for only a limited range of input p.d.'s

transistor is *cut off*, and no current flows through it. The p.d. across R_C is then zero, and V_{out} remains equal to V_0. If on the other hand V_{in} rises above about 0.95 V in the circuit given, the collector current rises to the point at which the full supply p.d. is developed across the collector load resistor, and the collector potential V_{out} falls to zero. The transistor is then said to be *bottomed*. In fact the collector characteristics (Fig. 14.18) show that V_{out} can never fall below about 0.2 V, since at this point the collector/base junction starts conducting in the forward direction and normal transistor action ceases.

   This behaviour of the transistor amplifier can be investigated with the circuit of Fig. 14.22. Over most of the range of possible values of V_{in} the transistor is either cut off or bottomed. Only over a small range of values is linear amplification possible (Fig. 14.23). In a practical amplifying circuit we must therefore arrange for a small steady current to flow into the base to hold the transistor in the middle of the useful amplifying range; this is called a *bias current*. The current arising from the small alternating input p.d. is then superimposed on this and causes the base current to oscillate above and below this mean value. Fig. 14.24 shows how this may be realized in a practical amplifying circuit. The *base bias resistor* R_s is adjusted to

bring the transistor to the middle of the useful amplifying range, in this case to the point where the collector potential is about 2.5 V, half-way through the possible range of variation. When a small alternating p.d. is applied to the input, the transistor current varies on either side of its mean value, and the p.d. across R_C varies in the same way as the input (but *in antiphase* with it). These variations are superimposed on the average p.d. across R_C that exists in the absence of an input signal. A *blocking capacitor C* (p. 222) is therefore used to filter out the steady component so that only the alternating part appears at the output terminals. As in all amplifying devices, the power of the output is derived from the supply battery but is controlled by the very small power available at the input terminals.

   The alternating input p.d. must not exceed the allowable range of variation, otherwise *clipping* of the top and bottom of the waveform will occur. It is instructive to observe with an oscilloscope how this happens with a simple amplifier fed with an alternating input of variable amplitude. As the amplitude is increased the onset of clipping is very obvious in the distortion of the waveform, and the amplifier is then said to be *overloaded*. If the base bias resistance is correctly chosen, both the top and bottom of the waveform will be affected simultaneously.

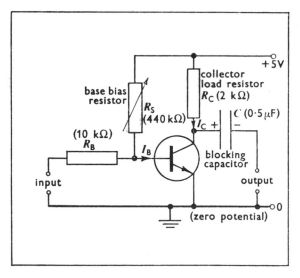

Fig. 14.24   A linear transistor amplifier for small a.c. inputs

Incidentally this provides a simple way of generating a square waveform from a sinusoidal one. We have only to make the amplitude of the sinusoidal input very much greater than that which gives clipping, and an approximately square waveform is obtained at the output.

A complete amplifier will usually consist of several amplifying stages joined in cascade, the output terminals of one stage being connected to the input terminals of the next. The output of the complete amplifier will be in phase with the input with an even number of such stages and in antiphase with an odd number. The coupling capacitors between the stages must be large enough to avoid loss of output at the lowest frequencies used. For an audio frequency amplifier designed as above a value of 0.5 μF might be suitable. For the sake of compactness an electrolytic capacitor would normally be used at this point; to provide the correct polarizing p.d. its positive terminal would need to be joined to the collector.

In the output stage of an amplifier, if more than a fraction of a watt of power is to be delivered, a special type of transistor is used that is able to pass currents of several amps. Such power transistors are made by the same techniques as other transistors; but the collector, in which most of the heat is dissipated, is fused onto a thick metal mounting plate forming one face of the unit. The transistor must be bolted down with the mounting plate in close thermal contact with a heavy metal sheet to act as a heat sink (p. 260).

Negative feedback

A simple amplifier like that described above suffers from a number of drawbacks, not least the way in which its behaviour depends so critically on the properties of the particular transistor used. The current amplification factor β can well vary over the range from 50 to 200 within the same batch of transistors. Also the properties of a transistor can change quite rapidly with temperature. It is therefore necessary to design amplifier circuits so that their behaviour is largely independent of the properties of the particular transistors they contain. The way this is done is by equipping the amplifier with feedback paths so that part of the amplified output signal is fed back to the input. The feedback must of course be in antiphase with the input, so that it partly cancels out the input; we call this *negative feedback*. (*Positive feedback* has the opposite effect, and is used deliberately in oscillator circuits for generating alternating p.d.'s—see p. 274.)

The effect of negative feedback is partly to reduce the gain of the amplifier. But at the same time it reduces the dependence of the gain on the properties of the transistors in the circuit; instead it now depends chiefly on the nature of the feedback network employed. Since this consists of stable and reproducible components like resistors and capacitors, the whole amplifier becomes stable and predictable in its performance.

How this comes about can be seen in the simple case shown in Fig. 14.25a. We represent the amplifier in this diagram by a single symbol. Its exact form is of no importance—it might well indeed be an integrated circuit of whose contents we are entirely ignorant. The only requirements are that it should be of high gain and of high input impedance, so that it draws negligible current at its input. Also of course it must be an inverting amplifier whose output is in antiphase with its input, so that we achieve *negative* feedback. The feedback in this case is provided by means of the pair of resistors R_F and R_B. (The simple circuit of Fig. 14.24 could be adapted in this way by joining a resistor R_F from the output to the base of the transistor; but the feedback path then includes also the capacitor, and the negative feedback does not exist at all for low frequencies or d.c.)

Since we can ignore the current at the amplifier input, it follows that the current I that flows into R_B from the input terminal flows also through R_F to the output terminal. Since also the amplifier is of high gain, the potential at its input can only vary over a very small range, and we can say that to a close approximation the point A remains at zero potential all the time. For instance, if the amplifier gain is -1000 and the output p.d. can vary between 0 and 5 V, then the potential at A can only vary one thousandth of this, namely over a total range of 5×10^{-3} V. Thus

$$\text{p.d. across } R_B = V_{in}$$
$$\text{p.d. across } R_F = -V_{out}$$

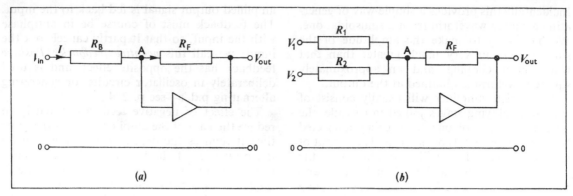

Fig. 14.25   A negative feedback amplifier: (*a*) to produce a precisely defined amplification R_F/R_B; (*b*) to produce a summing amplifier to mix two signals in any specified proportions

(The minus sign arises because V_{in} is negative when V_{in} is positive and vice versa.)

$$\therefore I = \frac{V_{in}}{R_B} = \frac{-V_{out}}{R_F}$$

$$\therefore \text{ overall gain} = \frac{V_{out}}{V_{in}} = -\frac{R_F}{R_B}$$

Thus if $R_F = 5R_B$, the overall gain is -5 regardless of the precise properties of the amplifier itself. The only effect of temperature that might matter is the variation of the resistances R_F and R_B. And if these resistors are of the same kind, even this will not matter, since changes of temperature will still not affect the ratio R_F/R_B. If $R_F = R_B$ we have a *see-saw amplifier*, in which the output and input potential remain always equal but move in opposite directions.

The same feedback system may be adapted also to provide a *summing amplifier*. This is shown in Fig. 14.25*b*. By the same reasoning the current through R_F is now equal to the sum of the currents through R_1 and R_2.

$$\therefore \frac{-V_{out}}{R_F} = \frac{V_1}{R_1} + \frac{V_2}{R_2}$$

If $R_F = R_1 = R_2$, then

$$V_{out} = -(V_1 + V_2)$$

and the output gives the sum at every instant of the two inputs. Clearly the principle can be extended to the addition of any number of inputs; and by choosing the resistances appropriately we may add the different inputs in any proportions we choose. This technique provides us with a *mixer* circuit for combining the signals from a number of different inputs—micro-

phones, pick-ups, etc.—in any way we choose. In this case the resistors would be made variable so that the balance of the different inputs can be adjusted at will.

The measuring amplifier

There are many purposes for which we need a measuring instrument that draws negligible power at its input; such instruments are called *electrometers* (p. 175). An electronic amplifier can be designed to make an excellent electrometer. The amplifier can be similar to those we have described above, but *direct* connection between the stages is employed, the usual blocking capacitors being eliminated; it thus becomes a *d.c. amplifier*. The p.d. developed at the output is then registered by a moving-coil meter. Heavy negative feedback is employed so that the output is strictly proportional to the input p.d. The feedback also helps to give stability to the amplifier, so that its readings do not wander with change of temperature. The input stage of the amplifier usually consists of a special thermionic triode (p. 235) designed so that the leakage of current from the grid is reduced to a minimum. The input current of such an amplifier is often less than 10^{-14} A, much less than could be obtained with a transistor input stage. Subsequent stages of amplification are provided by the usual transistor circuits.

Operational amplifiers

The feedback amplifiers we have been discussing above provide the means of making precise arithmetical combinations of the p.d.'s concerned. If these p.d.'s are chosen to represent the variables in some physical or engineering problem, we have the means of doing automatic calculations electronically. Feedback amplifiers used for this purpose are known as *operational amplifiers*. So far we have only considered ways in which we can perform multiplication by a constant, addition and subtraction (by adding the negative number). But by including capacitors in the feedback network we can also design circuits that perform the functions of differentiation and integration, and so enable us to 'solve' differential equations connecting the variables of the problem.

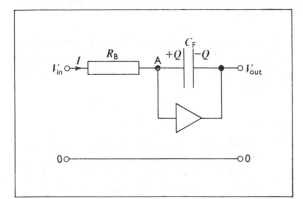

Fig. 14.26   An integrating circuit; the output is proportional to the time integral of the input p.d.

Fig. 14.26 shows how an integrating circuit may be made. In this case the current I that flows through R_B flows onto the capacitor plates; the current is equal to the rate at which the charge Q on the plates grows.

$$I = \frac{dQ}{dt}$$

By integrating we can express this equivalently as

$$Q = \int I . dt$$

Now $$Q = -C_F V_{out}$$

and $$I = \frac{V_{in}}{R_B}$$

$$\therefore V_{out} = -\frac{1}{R_B C_F} \int V_{in} . dt$$

Thus the output p.d. is proportional to the time integral of the input p.d. We may of course

EMP—K

differentiate this equation and express the same result as

$$V_{in} = -R_B C_F \frac{dV_{out}}{dt}$$

Alternatively, by interchanging the capacitor and resistor in Fig. 14.26 the opposite function is realized, namely the *output* p.d. is proportional to the rate of change of the *input* p.d. The proof of this is left as an exercise to the student.

Provided we can write down the differential equations that connect the variables of some problem, we can combine operational amplifiers to simulate the way in which the system will behave. We can for instance simulate the oscillations of a bridge or building under various stresses, making the p.d.'s at different points represent the displacements and velocities of the system. The behaviour of the electrical system thus provides an analogue of the behaviour of the mechanical system; an arrangement of operational amplifiers of this sort is therefore called an *analogue computer*. With it we may analyse the effect of changing the various parameters of the system simply by varying the resistors and capacitors of the computer; and so we can achieve the optimum design with the minimum of trouble and expense.

As an example, Fig. 14.27 shows how three operational amplifiers may be combined to simulate the behaviour of a ballistic galvanometer. The equation connecting the angular displacement θ, velocity $d\theta/dt$, and acceleration $d^2\theta/dt^2$ of this system was derived on p. 147.

$$J\frac{d^2\theta}{dt^2} + \lambda\frac{d\theta}{dt} + k\theta = 0$$

The student should have no difficulty in seeing how the analogue computer of Fig. 14.27 obeys the same differential equation. By suitable choice of the various resistors we can study the effect of changing the parameters J, λ, and k of the mechanical system represented. An oscilloscope may be joined to the output (θ)—or to any other point of the circuit—to reveal what happens when an electrical impulse is given to the circuit analogous to the angular impulse imparted to the ballistic galvanometer coil when a pulse of current passes through it. It is quite simple to arrange for this to be done by the electrical pulse associated with the fly-back of the time

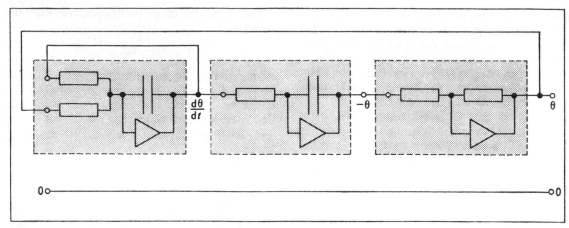

Fig. 14.27   Three operational amplifiers linked together to form an analogue computer to simulate the behaviour of a ballistic galvanometer

base of the oscilloscope. The pattern of oscillations resulting is then displayed as a steadily repeating trace on the oscilloscope screen, and we can watch how the pattern changes as the resistors of our analogue computer are modified in various ways. The student will appreciate that the set-up of Fig. 14.27 in fact represents *any* damped oscillatory system—a weight hanging on a spring, a capacitor discharging through an inductor, a tuning fork, or a multitude of other natural phenomena.

14.6 Oscillators

Any amplifier can become a self-maintained generator of oscillations if some of the output is fed back to the input in the right phase. Indeed quite unwanted oscillations can easily be generated in this way. In a public address system it sometimes happens that the sound fed back from the loudspeakers to the microphone is of greater intensity than the original sound to be amplified. The signal then builds up to a very large amplitude, and a piercing howl develops at some frequency to which the system is resonant. This is an example of acoustic feedback. But the same effect may occur if there is an electrical connection or any other sort of link between output and input. If there is sufficient feedback *in phase* with the input, the circuit is unstable and oscillations build up.

The simplest way to control the frequency is to include in the circuit a resonant combination

of inductance L and capacitance C; the natural frequency of the oscillator will then be close to the resonant frequency f of this combination. This is given by

$$\omega^2 LC = 1 \quad \text{(p. 223)} .$$
where
$$\omega = 2\pi f$$

Fig. 14.28 shows two ways of making an oscillator of this sort. In both the frequency is controlled by the resonant loop of inductor L and capacitor C in parallel. If there were no losses, an alternating current at the resonant frequency could circulate in this loop without ever decreasing. But in practice the inevitable resistance of the circuit would cause the oscillations to die away rapidly. The function of the transistor is to supply an alternating current to the resonant loop so that the oscillations can continue indefinitely.

In the *reaction oscillator* in Fig. 14.28a the resonant combination is in series with the collector. The small coil L' is inductively coupled to L so that the alternating current in the resonant loop provides the alternating input for the base. If L' is connected the right way round, the feedback will be positive, and the collector current will be in the right phase to increase the alternating current in the resonant loop. The oscillations then build up in amplitude until the power losses in the circuit are equal to the power that the transistor can supply. It is not necessary to do anything to start the oscillations; the random motions of electrons in the

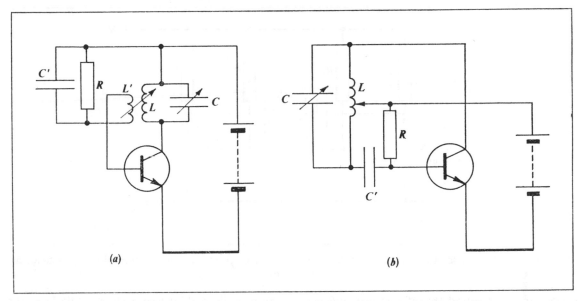

Fig. 14.28 Transistor oscillators: (*a*) a reaction oscillator, (*b*) a Hartley oscillator

coils are quite sufficient to provide the initial disturbance. Provided there is sufficient feedback between collector and base the oscillations start as soon as the circuit is switched on. To control the feedback the coupling between the two coils is made adjustable; this is represented by the arrow through the coils in the circuit diagram. Either the spacing of the coils is altered or a magnetic core linking them is moved in or out. The frequency may be altered by varying the capacitance C. The base bias current is provided as usual by the resistor R. A capacitor C' must be joined across R so as to provide a low impedance path for the alternating component of the base-emitter current. If this were not included, the e.m.f. induced in the reaction coil L' by mutual induction would not be able to drive a large enough alternating base current to maintain the oscillations.

The reaction oscillator works very well provided the losses in the circuit are not too great, otherwise it may be impossible to provide sufficiently close coupling to give enough positive feedback. A rather better arrangement is shown in Fig. 14.28*b*; this is known as the *Hartley oscillator*. In the reaction oscillator one end of the coil is kept at the fixed potential of the positive supply rail, and the potential of the other end oscillates about this. But in the Hartley

circuit the supply is joined to a point near the lower end of the coil. The potentials of the ends of the coil are now both oscillating about the supply potential, and are necessarily in antiphase. This ensures that the feedback will be positive. The capacitor C' serves to block the supply potential from the base, but allows the alternating potential at the end of the coil to be communicated to it. The resistor R provides the base bias current. The amount of feedback in the Hartley oscillator is altered by adjusting the position of the tapping point on the coil; it can readily be made much greater than in the reaction oscillator. The Hartley circuit can therefore work at much higher frequencies, at which the losses are liable to be greater.

These kinds of oscillator work well at radio frequencies, but in the lower audio frequency range the inductances required need to be rather large. It is then simpler to control the frequency by means of resistors and capacitors. If the feedback path between the output and input of an amplifier is provided by a network of resistors and capacitors, there is a phase shift through the network that depends on the frequency. The circuit can oscillate at the frequency for which the net phase shift through the amplifier and network is zero. For oscillations to occur the gain must of course be sufficient to make up the

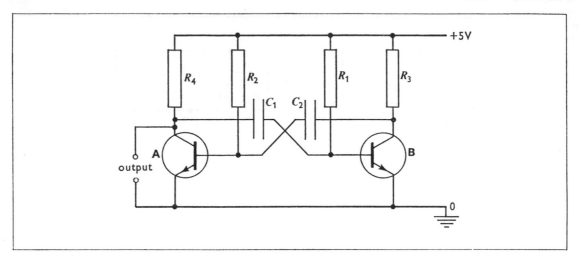

Fig. 14.29   A multivibrator or astable circuit—a type of oscillator that produces an approximately rectangular waveform

losses; and it is found that if it is *just sufficient* then the oscillations are truly sinusoidal. If there is excessive gain, then the transistors of the circuit are bound to be cut off or bottomed at some part of the oscillation, and the operation of such a circuit then resembles that of the multivibrator described below, and the waveform obtained is anything but sinusoidal. Most audio frequency oscillators nowadays consist of phase-shift circuits with some form of automatic control of the gain.

The multivibrator

Fig. 14.29 shows a form of oscillator in which the gain is deliberately designed to be vastly greater than is needed to make up for the losses of the circuit. It consists of a two-stage amplifier with the output fed directly back to the input. Each stage produces both amplification and phase reversal, so that overall there is considerable positive feedback. The circuit is therefore highly unstable. If the current is *increasing* in one transistor, the p.d. across its collector load will increase, and this will cause the current in the other transistor to *decrease*. This in turn acts to increase still further the current in the first transistor. When the circuit is switched on, it therefore goes rapidly into a condition in which one transistor is cut off and the other is bottomed. This condition is, however, only temporarily stable; and a short while later the condition reverses, the conducting transistor cuts off and its partner bottoms. The potential of either collector thus switches alternately between 5 V and zero (nearly), as the circuit rests for a short time in each of its two momentarily stable states. It is called an *astable* circuit.

Which way the circuit goes when it is first switched on is a matter of chance; suppose the conducting transistor is **A**. The instability of the circuit will have driven transistor **A** very suddenly into the bottomed condition and therefore will have caused a sudden fall of its collector potential from 5 V to 0.2 V. This fall of potential is communicated through the coupling capacitor C_1 to the base of **B** and holds this transistor cut off. The collector of **B** is therefore at the supply potential of 5 V. The circuit seems now to have reached a state of stability; but this is only short-lived, because the capacitor C_1 starts charging up through resistor R_1. If nothing further happened in transistor **B**, C_1 would eventually charge up with its right-hand plate at the potential of the positive supply rail. However, as soon as the right-hand plate reaches a potential of about 0.5 V it causes transistor **B** to start conducting again. The potential of the collector of **B** therefore falls. This is communicated to the base of **A** through the capacitor C_2 and starts to switch off this transistor. The regenerative action of the circuit quickly switches it into the

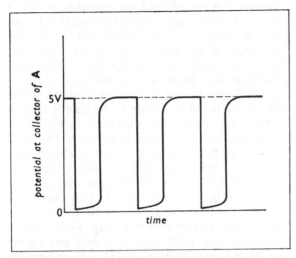

Fig. 14.30   The waveform of the oscillations produced by a multivibrator

opposite state in which **A** is cut off and **B** bottomed. Capacitor C_2 now charges through R_2, until the switching action is repeated as the base of **A** reaches a potential of 0.5 V. The period of oscillation is decided chiefly by the time constants of the two capacitor-resistor combinations and is approximately proportional to $(C_1 R_1 + C_2 R_2)$. The output of the oscillator may be taken from the collector of either transistor. The waveform is approximately a *rectangular* one (Fig. 14.30). This sort of waveform

can be analysed as a mixture of sinusoidal frequencies (harmonics) that are multiples of the fundamental frequency. In fact it generates a very wide range of harmonics; it was for this reason that the name *multivibrator* was coined for it. If the circuit is made entirely symmetrical, with $C_1 = C_2$, $R_1 = R_2$, etc., it becomes a square-wave generator, and the two parts of the waveform are of equal duration.

14.7   The transistor as a switch

The essence of the action of a transistor is that the collector current can be controlled by a small change in the base current. It may therefore be used in the same way as a relay switch (p. 268), in which the input current causes the switch contacts to close and so controls a much larger current in another circuit. Thus in Fig. 14.31, when the potential of A is zero, the transistor is cut off, and the output B will be at the potential of the positive supply rail, namely 5 V. Similarly when A is at 5 V the transistor is bottomed and the output B is at a potential close to zero (strictly 0.2 V). In this simple arrangement the output potential thus takes one of two values depending on the input potential applied at A. When A is at 5 V, B is at zero potential, and vice versa. The virtue of using a transistor to achieve this result is that the input potential can be

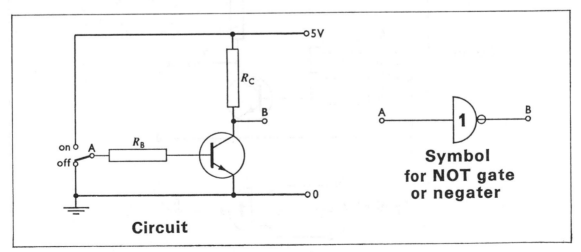

Fig. 14.31   A transistor used as a switch; the collector current is switched on or off by applying a potential of 5 V or 0 at A

provided by another electronic circuit, and so the 'position' of the switch can be made dependent on potentials derived from other points in the apparatus. Furthermore, electronic switches of this kind can be 'reversed' millions of times a second, and so enable interlinked switching operations to be carried out with great rapidity. It is this function of the transistor that has made possible the construction of high-speed digital computers.

Any process of computation consists in performing a sequence of operations on the data of the problem; at each stage in the computation the nature of the operation to be performed is decided partly by the pre-determined programme and partly by the outcome of earlier stages in the process. We therefore need 'switches' with multiple inputs, whose outputs are determined in specified ways by the condition of their inputs. These arrangements are known as *gates*, and mostly they are elaborations of the simple transistor switch of Fig. 14.31.

The gate in Fig. 14.32 is designed so that it will 'open' if any one (or more) of its inputs A_1, A_2, A_3 is switched 'on' (i.e. joined to a potential of 5 V). It therefore performs a function equivalent to the logical connective OR. When all the inputs are off (at zero potential), the

transistor is cut off, and the output potential at B is 5 V. But if any one of the inputs is joined to the positive rail, this is sufficient to raise the potential of the base to the point where the transistor conducts and bottoms; B is then nearly at zero potential. The output is of opposite polarity to the input that caused it. The circuit is therefore strictly to be described as a NOR gate. Underneath the circuit in Fig. 14.32 is shown the symbol used to represent this kind of gate in diagrams of the logical interconnections in a computer. The **1** indicates that the gate opens if only *one* of the inputs is 'on'. The circle on the output line shows that the pulses there are of opposite polarity to those at the input. The simple switch in Fig. 14.31 can be regarded as a NOR gate with only one input; or we can refer to it as a NOT gate or *negater*. The symbol for this is shown alongside. The triple input NOR gate in Fig. 14.32 could be converted into an OR gate by connecting a NOT gate to its output.

The gate in Fig. 14.33 is designed to open only if *all* the inputs are switched 'on' (i.e. at a potential of 5 V). It therefore performs a function equivalent to the logical connective AND. If any one of the inputs (say A_1) is at zero potential, the corresponding diode D_1 conducts, and

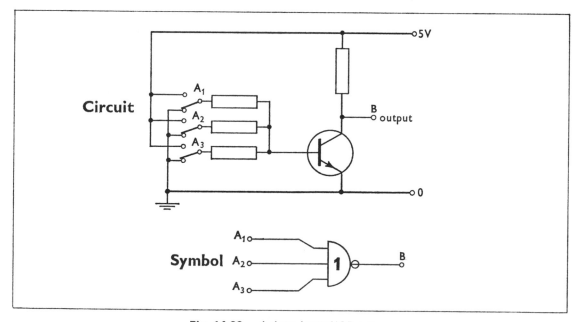

Fig. 14.32   A three-input NOR gate

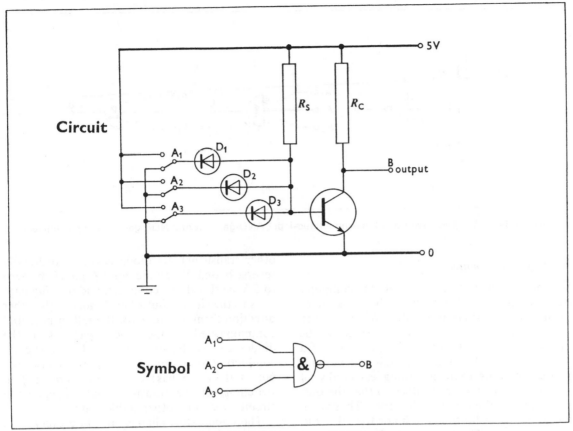

Fig. 14.33 A three-input NAND gate

this holds the base of the transistor at a low potential so that it is cut off. This remains so whatever happens to the other inputs. If A_1 is switched 'on' (potential 5 V), the diode D_1 is reverse biased and no current flows through it. But only if *all* the inputs are 'on' will it be possible for the base potential to rise so that the transistor conducts and bottoms. The output at B is again of opposite polarity to the inputs that caused it; and this arrangement is therefore described as a NAND gate. The symbol for it is shown underneath in Fig. 14.33. The **&** indicates that *all* of the inputs must be 'on' simultaneously for the gate to open. It can be turned into a three-input AND gate by joining a NOT gate to its output.

An alternative way of making an AND gate is to use a NOR gate with NOT gates (or negaters) joined to each of its inputs. This is shown in symbolic form in Fig. 14.34a; the neighbouring symbol, Fig. 14.34b, expresses the same function more briefly. Either by considering the logical meaning of this gate or by analysing its behaviour electronically we can see that this combination is equivalent to an AND gate (Fig. 14.34c). Thus, the NOR gate bottoms if any one of its inputs is at 5 V. With the negaters joined to the inputs, it bottoms if any of the inputs of the combination is at *zero* potential. Only if *all* the inputs are at 5 V will the NOR gate not bottom, i.e. its output will then be 5 V also. This is precisely the function required of an AND gate.

All the logical processes required for the operation of a digital computer can be provided by circuits of the few kinds we have described. But the example above shows that the logical interconnections must always be carefully analysed to discover the most economical way of expressing the logic electronically.

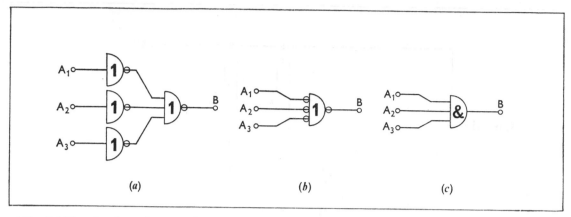

(a)

(b)

(c)

Fig. 14.34 An alternative way of making a three-input AND gate out of a NOR gate and three negaters

A binary scale counter

Arithmetical operations in a digital computer are normally carried out using a binary scale of arithmetic. This is because only two digits, 0 and 1, are required in this system, and these can be represented by the two possible 'positions' of a switch or gate. A counting circuit can be built up out of units each consisting essentially of two NOT gates joined together so that the output of each is the input of the other. Thus when one transistor is bottomed it holds the other off, and vice versa. The circuit is in fact closely related to the multivibrator of Fig. 14.29, except that it has direct coupling between the transistors and so is permanently stable in either of its two possible states. One of these states can represent the digit 0 and the other the digit 1. This is called a *bistable* circuit.

To make it perform its counting function we must arrange for it to change over from one state to the other when a suitable electrical pulse is fed into it. Fig. 14.35 shows a practical form of the circuit. The shaded parts of the circuit are concerned with steering the input pulses so that they switch off whichever transistor is conducting, so causing the circuit to change over from one stable state to the other. Considering first of all the bottomed transistor; the collector of this is at 0.2 V, while its base is at 0.7 V. The diode joined to its base is therefore slightly forward biased. The collector of the non-conducting transistor is at 5 V, while its base is at 0.2 V (being joined to the collector of the

other transistor). Its diode is therefore heavily reverse biased. When the input A goes from zero to 5 V neither diode conducts, and nothing happens to the circuit. But when A goes in the other direction from 5 V to zero, the fall in potential communicated through the capacitors to the diodes causes the one that is already slightly conducting to conduct strongly; this causes the potential of the base of its transistor to fall, switching off the transistor and changing the circuit over to its other stable state.

The circuit thus changes its state every time the input A is taken from 5 V to zero, and on each occasion the output terminal B changes its potential from zero to 5 V or the other way. If a multivibrator is joined to A, the bistable circuit changes over once in each cycle of oscillation. The output at B is of similar square waveform (Fig. 14.30), but is of *half* the frequency. This output may be applied to the input of another similar bistable circuit; the latter will then change over its state on every other zero-going pulse applied to A. A third bistable will change over once in every four such pulses arriving at A, and so on. A complete binary scale counter is made by joining up a sufficient number of such units, as in Fig. 14.36. Suppose we decide that the digit 0 is to be represented by the left-hand transistor conducting in each case, and 1 by the opposite state. The initial condition of the counter will then be as in the top diagram; the conducting transistor is shown shaded in each case. If we now feed in (say) 19 pulses at the input, the counter will reach the condition

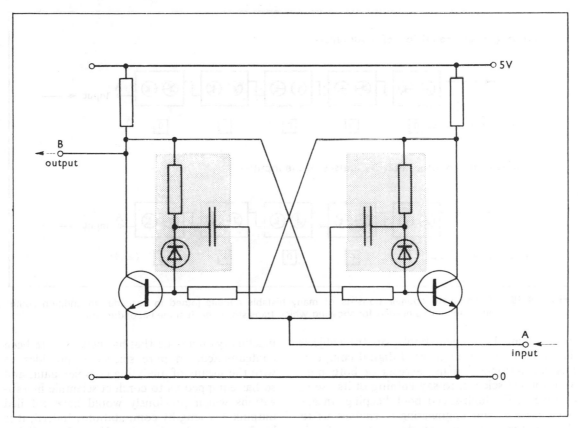

Fig. 14.35   One stage of a binary counter; this is called a bistable circuit

shown in the lower diagram, which represents 19 on the binary scale.

$$(1 \times 2^4 + 0 \times 2^3 + 0 \times 2^2 + 1 \times 2^1 + 1 \times 2^0 = 19.)$$

A digital computer contains many such counters, the condition of each one representing a number involved in the computation. Arithmetical operations of all kinds may be effected by performing the right sequence of logical processes on the data represented by the potentials in the counters. For this purpose gating circuits, such as those described above, are used.

A counter can also be regarded as a temporary 'memory' for the number that has been fed into it. As long as no further pulses arrive its condition provides a record of the original number of pulses concerned. However, a computer needs to store in its memory many thousands of numbers (and other coded data) besides those on which arithmetical operations are being per-

formed at any given instant; and it would be far too costly to use counters of this kind throughout. A variety of other techniques are therefore used to provide the different parts of the computer memory. One method is to use sets of small ferrite loops. These can be magnetized by pulses of current flowing in wires threaded through them. The two digits 0 and 1 are represented by the two possible directions of magnetization of a loop. Another method used when much longer access-time is permissible is to record the pulses representing the binary number on a magnetic tape; the technique is simply a refinement of the process used in an ordinary tape recorder.

It is beyond the scope of this book to consider any further the fascinating details of modern computer technology. Enough perhaps has been given to enable the student to perform experiments with transistors for himself and to follow

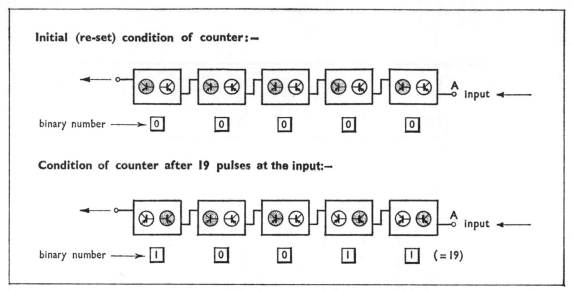

Fig. 14.36   A complete counter consists of many bistable circuits joined in cascade; in addition some
arrangement is needed for showing which transistor in each stage is conducting

more detailed technical books on the subject.
The advent of the high-speed digital computer
has revolutionized many aspects of both pure
and applied science, to say nothing of its use in
commerce for high-speed book-keeping. In applied science and engineering it enables us to
shorten the processes of design by performing calculations for which it would never be
economic to employ human beings. In industry
it makes possible the automatic control of

machinery on a scale that has never before been
contemplated. In pure science it provides us
with the means of analysing complex data, and
so has equipped us to conduct scientific investigations which previously would have entailed
unthinkably lengthy computations. Indeed, it is
hard to estimate what the effect of the digital
computer on human civilization may prove to
be by the time its potentialities have been
fully realized.

15 Radio

15.1 Sound reproduction

Sound consists of longitudinal wave motions transmitted through the air (or some other material medium). A microphone is essentially a device for producing an electrical signal whose time variation follows closely either the displacement or the pressure changes in the sound wave incident on it. In order to make use of these electrical signals we must arrange for them to be transmitted by means of wires or radio waves, or perhaps for them to be recorded to be available for future use. In the latter case the variations of the electrical signals are caused to produce corresponding physical changes in the selected recording medium—e.g. the magnetization of a magnetic tape or the displacement of a fine groove on a recording disc.

To transmit the signals to any distance we must make use of other kinds of technology. For a local telephone system we can conduct the electrical signals themselves along wires to the point of reception. But for trunk routes this is not an economical use of the expensive cables required, and we must arrange for a large number of transmissions to be carried by each pair of conductors. For instance a large number of high-frequency alternating currents can be transmitted simultaneously along the cable, and the instantaneous amplitude of oscillation of each current can be made to be in step with the original electrical signal from the corresponding microphone. At the far end of the cable the different high-frequency *carriers* must be separated from one another and the original electrical signals must be reformed from them and conveyed to the points of reception. The process is rather similar to that employed in radio transmission, where again the high-frequency radio wave is varied in step with the electrical signals from the microphone. The radio receiver is tuned to pick up only the frequency of the particular transmission required, and the original electrical signal is reformed in the receiver and reproduced as sound by the loudspeaker or earphone.

In this chapter we deal first with a few of the most used forms of microphone and loudspeaker. Then we shall consider the transmission of electrical signals along wires. Finally we analyse the nature of electromagnetic waves in free space and how these may be used for communication purposes. The scope of this book requires us to concentrate mainly on the physical principles involved in these processes. For the sophisticated techniques of the modern communications industry the student should consult more specialized literature. Likewise we shall not deal at all with the methods of sound recording and replay, for which many excellent technical books are available.

The movements of air in a sound wave are extremely small. A microphone will commonly be expected to produce a satisfactory signal with air displacements of only 10^{-10} m—less than the diameter of one atom. The corresponding pressure changes amount to about 10^{-3} N m^{-2}—not much more than the *total* pressure in the 'vacuum' of an electron tube. The power available to vibrate the diaphragm of the microphone may not be more than 10^{-12} W. It is not surprising therefore that the electrical signal produced is very small, requiring considerable amplification before it can be used to drive a loudspeaker or recording head. Furthermore a good microphone should respond uniformly to all frequencies to which the ear is sensitive—i.e. from about 20 Hz to 20 kHz, usually known as the *audio frequency* range. But this is an ideal not readily attained in practice.

The carbon microphone (Fig. 15.1a)

This consists of a small box B loosely packed with carbon granules G. An electric current from a small battery is driven through the granules between the carbon block C at the back and the

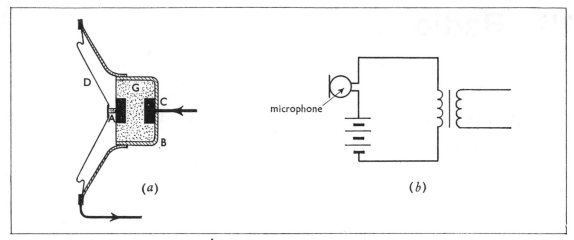

Fig. 15.1   (a) A carbon microphone; (b) the circuit of the microphone with battery and step-up transformer

carbon button A attached to the light aluminium diaphragm D at the front. When the granules are compressed, the resistance of the unit decreases, and the current in the circuit rises. Thus the oscillations of the diaphragm produced by the pressure changes in the incident sound wave cause corresponding oscillations of the current. The microphone and battery are joined in series with a high-ratio step-up transformer, so that the maximum alternating p.d. is produced at the output terminals (Fig. 15.1b). Although the signal generated in this way is controlled by the sound wave, its power is derived from the battery. The carbon microphone is thus an amplifying device, and is the only kind of microphone that can be used without additional electronic amplification. It is therefore very suitable for telephones. However, it is not satisfactory for high-quality sound reproduction because of the background 'hiss' produced by the loose contacts between the carbon granules.

The crystal microphone

In certain types of crystal, such as *quartz* and *Rochelle salt*, mechanical strains can produce electric polarization. Equal and opposite bound charges then appear on opposing faces of the crystal, and a p.d. is produced between two metal electrodes in contact with them. This is called the *piezo-electric effect*. A section from a single crystal can thus be used as a microphone if it is arranged so that the incident sound wave causes the appropriate type of strain in it.

The moving-coil or dynamic microphone

In this microphone a small coil is attached directly to the diaphragm. It is arranged to move in the strong radial field in an annular gap between the pole pieces of a permanent magnet (Fig. 15.2). When the diaphragm oscillates, the coil cuts the magnetic flux and an alternating e.m.f. is generated in it. The construction is very similar to that of a moving-coil loudspeaker (Fig. 15.3). In fact a moving-coil loudspeaker may be used as a microphone, though its bulk is a bit inconvenient for this purpose!

This microphone has a low impedance; and a relatively large current is generated in it at a small p.d. But most amplifiers are designed to be operated by variations of potential at a high input impedance. The signal is therefore applied

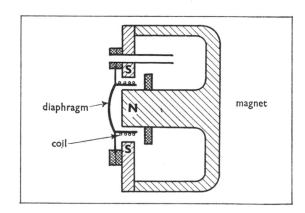

Fig. 15.2   A moving-coil (or dynamic) microphone

to the amplifier input through a high-ratio step-up transformer.

Modern technology has reduced the design of microphones and their associated amplifiers to something near perfection. The electrical signal obtained at the output can (at a price) be an almost perfect replica of the variations in the original sound wave. The art of recording sounds on disc or magnetic tape has also been developed to the point at which it often leaves little to be desired. The weak link in the chain of sound reproduction is the loudspeaker. Even the best loudspeakers are not free from resonances of one kind or another, and no one has succeeded in designing a single unit that responds uniformly to the complete range of audio frequencies. High-fidelity systems make use of two or more speakers covering different frequency ranges—'woofers' for low frequencies and 'tweeters' for high. Even if a perfect loudspeaker system could be evolved, the quality of the reproduction could be spoilt by using it in a room with inadequate acoustics. The 'perfect' loudspeaker would need a room with perfect acoustics to go with it.

The moving-coil loudspeaker (Fig. 15.3)

The moving part consists of a conical diaphragm D made of stiff paper. This is supported round its circumference by a flexible ring of cloth and at the centre by a weak spring strip. Also attached to the centre is a single-layer coil located in the radial magnetic field in the annular gap between the pole pieces of a permanent magnet. The signal current flows in the coil, and the electromagnetic forces move it to and fro in the gap, carrying the diaphragm with it like a piston, and generating sound waves. *Two* sound waves are in fact produced—at the front and back of the cone; and these are bound to be in antiphase. If the acoustic path from back to front is much less than half a wavelength, there is almost complete destructive interference of the waves in all directions. The baffle-board or cabinet is designed to reduce this effect, and must be regarded as an essential part of the loudspeaker system.

At high frequencies, when the wavelength is comparable with the diameter of the speaker, there is a tendency for the sound to be emitted chiefly along the axis. The frequency balance

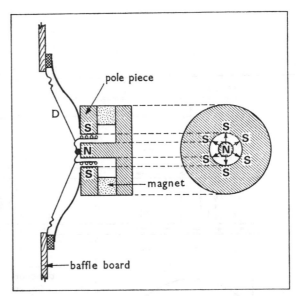

Fig. 15.3   A moving-coil loudspeaker

therefore varies with the position of the observer. This is partly mitigated by the behaviour of the cone; at high frequencies instead of vibrating as a single unit, complicated standing wave patterns are set up on it by the oscillations of the coil in the centre. This reduces the effective radiating diameter, but creates new resonance problems on its own account. All in all it is fortunate that the average human ear is so uncritical of the sounds that fall upon it!

15.2   Waves on wires

When an e.m.f. is first generated in an electric circuit, it takes time for the electrical disturbance to be propagated round it. The charged particles in the conductors of the circuit move at quite slow speeds even when very large currents are flowing (p. 11). But the electric and magnetic fields they produce are propagated round the circuit at very great speeds (approaching the speed of light), and this ensures that the charged particles are set in motion almost simultaneously throughout the circuit. With direct currents and with low-frequency alternating currents the transmission of the disturbance round a circuit is so rapid that we are not usually aware of the finite time taken. But at very high frequencies this time may become comparable with the period of oscillation of the current. We then

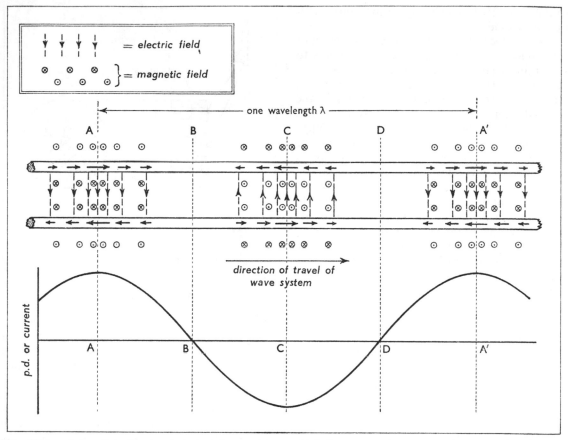

Fig. 15.4   A parallel-wire transmission line carrying a high-frequency electromagnetic wave, showing the distribution of currents and potentials at a given instant

have a situation in which the e.m.f. reverses before the original disturbance has travelled very far along the wire. In fact the high-frequency e.m.f. is then sending *waves* of current and potential round the circuit; these are accompanied by waves of magnetic and electric fields in the space surrounding the conductors. The transmission of energy along a cable by an alternating current is thus accompanied by the passage of *electromagnetic waves,* as they are called, through the insulating medium surrounding the cable. In this section we shall study the nature of such electromagnetic waves in the neighbourhood of high-frequency transmission lines. After that we shall proceed to consider the still more important process of the propagation of electromagnetic waves in empty space far removed from any conductors.

There are many different kinds of wave sys-

tem known to science, involving a great variety of physical processes. But they all tend to share certain common features, so that a clear understanding of one type of wave system is helpful in elucidating the functioning of another. Electromagnetic waves can be observed by their effects, as we shall see, but the detailed 'structure' of the waves has to be derived by theoretical reasoning and must be left largely to the imagination. The intangible nature of these waves means that they are by no means the most suitable type with which to *begin* a study of wave systems. It is therefore assumed below that the student already has a working knowledge of other kinds of waves. For instance, the properties of transverse waves on stretched cords and sound waves in tubes are in many ways closely analogous to those of electromagnetic waves on a transmission line; and we shall make use of these analogies in what follows.

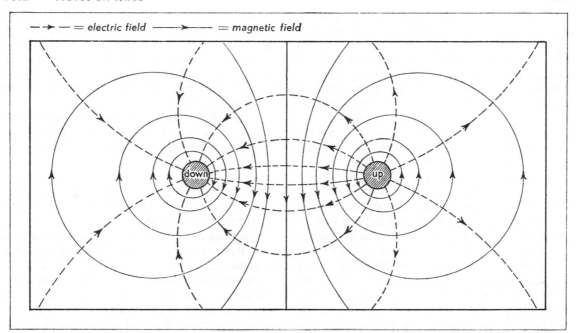

Fig. 15.5   The pattern of magnetic and electric fields at a given instant in a plane at right-angles to a parallel-wire transmission line; the current is approaching the reader in the right-hand conductor, and receding in the left-hand one

Fig. 15.4 shows the nature of the wave 'pattern' for an electromagnetic wave travelling along a transmission line consisting of a pair of parallel wires. In this case the wave pattern moves steadily forward along the line, and it is therefore called a *progressive wave*.

At a given instant there are certain points (such as A and A′) at which the current and p.d. have their maximum values in one direction (call this the *positive* direction). Intermediate between these are other points (such as C) at which, at the same instant, the current and p.d. have maximum negative values. At B and D the current and p.d. are momentarily zero.

An associated pattern of electric and magnetic fields exists in the medium surrounding the conductors. The lines of force of the electric field run across from one wire to the other, the electric field strength in any section of cable being proportional to the p.d. between the wires. The lines of force of the magnetic field form closed loops round the wires, and are therefore at right-angles to the plane of the diagram in Fig. 15.4; the magnetic field strength in any section of cable is proportional to the current in the wires. Fig. 15.5 shows the nature of the field pattern in a plane

at right-angles to the wires (at a position such as A or A′ in Fig. 15.4). Over a distance of one wavelength (e.g. from A to A′ in the figure) the p.d. and current vary over the complete range between their maximum positive and negative values; corresponding variations of the electric and magnetic fields occur over the same distance.

It may seem strange to find a situation in which the current can flow simultaneously in opposite directions in the same conductor, such as at A and C in Fig. 15.4. However, a little consideration will show that the pattern of currents in the figure is causing a flow of charge towards B in one conductor and away from it in the other, so charging up the capacitor formed by the part of the cable near B. A moment later this flow of charge will have caused the electric field to reach a maximum at B; likewise in the same time the flow of charge will have produced maximum electric field in the negative sense at D—and the electric fields at A and C will have died away to zero. Thus the system of currents shown causes a forward movement of the electric field pattern. Simultaneously the variation of potential along the wires is causing the pattern of *currents* to advance along the transmission line—and with it the associated pattern of magnetic field. For instance, at the moment depicted in Fig. 15.4, A is at a higher potential than C in the top conductor, and this is causing the current to grow in the section of cable between these points; the rate of growth of current near B depends on the inductance of this part of the transmission line. Likewise, C is at a lower potential than A' (in the top conductor), and this is causing the current to grow in the negative sense in the part of the cable near D. In this way the current wave also is made to move forward along the transmission line. As the positive peak of current at A moves towards B, the current between B and C first falls off to zero and then increases in the positive sense; when the current at B reaches a maximum, that at C is momentarily zero. At the same time the negative peak of current at C moves forward to D, and the current at A' falls momentarily to zero.

The energy of the wave is carried partly in the electric field produced by the p.d. between the conductors, and partly in the magnetic field associated with the current—in actual fact the energy is divided *equally* between these two forms. Thus, in Fig. 15.4, at A, C and A' the p.d. is a maximum in one direction or the other and energy is stored temporarily in the capacitance of the cable near these points—i.e. in the electric field. This is a form of potential energy, since it depends on the potential of the charges in the conductors. At the same set of points the current also has its maximum positive or negative values and energy is stored temporarily in the inductance of the cable—i.e. in the magnetic field. This is a form of kinetic energy, since it depends on the motion of the charged particles in the conductors. The individual charged particles oscillate to and fro as the current alternates, but the concentrations of electric and magnetic field move steadily forward together along the cable and carry the pulses of energy with them.

A similar sharing of the energy between potential and kinetic forms occurs in other types of wave system. Consider for instance the propagation of a sound wave. The wave can be described either in terms of the pressure variations in the air or in terms of the velocities of the air particles. At the positions of maximum or minimum pressure energy is carried in potential form; at the same positions the velocities also have their

maximum values in one direction or the other, and energy is carried in kinetic form. The individual particles of air vibrate to and fro along the line of propagation, but the regions of maximum pressure and velocity are transferred steadily forward, carrying the energy of the wave with them.

A detailed examination of the wave patterns in sound shows that the direction of propagation is that in which the air particles are moving at the points of *maximum* pressure. In a similar way it is clear from Fig. 15.4 that the electromagnetic wave is propagated in the direction in which the current flows at the positions of *maximum positive* potential.

The progressive electromagnetic wave we have described consists of two interlocking wave systems, the current wave and the potential wave; these remain always in phase at any given point, and each is associated with its own kind of field. Thus the patterns of the electric and magnetic fields also travel forward together and are in phase with one another at all points in any given plane at right-angles to the wires. The lines of force of the two kinds of field are in fact at right-angles to each other everywhere (Fig. 15.5) and are also at right-angles to the direction of travel of the waves along the cable. The associated electromagnetic wave is therefore of *transverse* type. (We are assuming in this section that the separation of the conductors is small compared with the wavelength; the situation is very much more complicated when this is not the case.)

The wavelength λ of the wave on the transmission line depends on its frequency f and velocity c. The relation between these quantities may be derived as follows. In the time taken for one complete oscillation of the current or p.d. the wave pattern advances a distance λ. In one second f oscillations take place and f complete waves pass any given point. The wave pattern therefore advances a distance $f\lambda$ per second.

$$\therefore c = f\lambda$$

One way of measuring the velocity c is to find the wavelength λ on the transmission line produced by an alternating e.m.f. of known frequency f; c is then given by the above expression.

At the ends of a transmission line in general some reflection of the wave occurs. The simple,

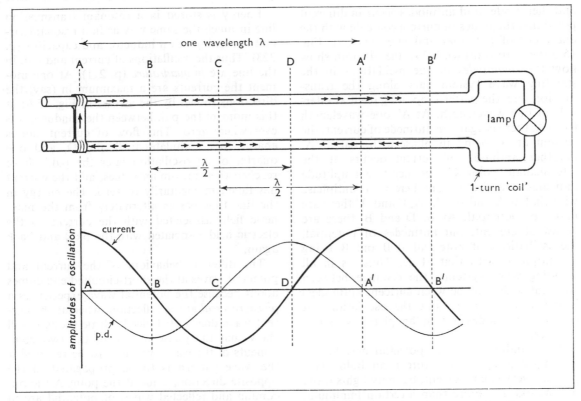

Fig. 15.6   The standing wave pattern on a resonant length of parallel-wire transmission line

progressive wave described above is obtained only if special steps are taken to absorb the energy arriving at the end of the line so that none is reflected. If this is not done, the oncoming and reflected waves interfere with one another producing a *standing wave* system on the line. In such a wave system at certain points the two waves are permanently in antiphase and the resultant amplitude is a minimum; these points are called *nodes*. At other points the two waves are permanently in phase, and the amplitude is a maximum; these points are called *antinodes*. If the oncoming and reflected waves are of equal amplitude, they cancel completely at the nodes, and there is also no net transfer of energy in either direction along the line.

Since reflection is liable to occur at both ends of the line, there are usually only certain sharply defined lengths of transmission line for which large amplitudes of oscillation can be built up for any given frequency. The line is then *resonant* to the frequency in question.

Fig. 15.6 shows an arrangement that may be used for studying a standing wave system of this kind. When the 'coil' of one turn at one end of the line is placed near the coil of an oscillator (working at 100 MHz or more), an e.m.f. is induced in it by mutual induction. A large current will flow in the coil only when the line is resonant to the frequency in use. Tuning may be effected by moving a short-circuiting strip along the line near its far end until the lamp indicates that a large current is flowing in the coil.

The current and potential waves behave in different ways when they are 'reflected' at the ends of the cable. The short-circuiting strip at A is of very low resistance. Although a large current may flow through it, the p.d. across it is always very small. The reflections therefore occur in such a way that the end A becomes an antinode of current, but a node of potential. Thus, although in a progressive wave system the waves of current and of potential are in phase at any given point, in a standing wave they are not, and their nodes and antinodes occur in different positions; the nodes of current coincide with the antinodes of potential and vice versa. In Fig. 15.6 the graphs drawn below the diagram show how the *amplitudes* of the oscillations in the standing wave system vary along the transmission line, the pattern repeating itself at intervals of one wavelength. At A′, one wavelength from A, there is again an antinode of current, the current at A′ being in phase with that at A. Another antinode of current occurs at the intermediate point C; the negative amplitude indicates that the current here is in antiphase with that at A and A′. At A, C and A′ there are nodes of potential. At B, D and B′ there are nodes of current, but antinodes of potential, the oscillations of potential at B and B′ being in antiphase with that at D. Thus, as in all standing wave systems, the distance between adjacent nodes (or between adjacent antinodes) is *half a wavelength*; and the oscillations at adjacent antinodes are in antiphase with one another.

The standing waves of potential may be detected by using a miniature neon bulb. This passes a very small current, but only lights if the p.d. across it is more than a certain minimum, usually about 80 V. As the neon bulb with its connecting clips is moved along the transmission line it reaches its maximum brightness at the positions of the potential antinodes, and does not light at all in the vicinity of the nodes. The wavelength is given by twice the separation of adjacent antinodes. If sufficiently high power is available from the oscillator, it is possible to observe the variations of the electric field strength between the wires using a large neon bulb without any leads attached to it. As the bulb is moved near an antinode of potential the electric field strength can be sufficient to ionize the low pressure gas inside the bulb, causing it to glow.

The standing wave of current may be detected with the aid of a small torch bulb attached to a pair of clips. The bulb is of low resistance and takes a relatively large current at a low p.d. It therefore acts in much the same way as the short-circuiting strip at A, which it may be used to replace. The bulb will be found to glow brightly at A, C and A′, where antinodes of current are located.

Energy is stored in a resonant transmission line in much the same way as in a resonant circuit consisting of an inductor and capacitor (p. 223). Thus the oscillations of current and p.d. in the line are *in quadrature* (p. 211). At one moment the currents are a maximum in (say) the directions shown by the arrows in Fig. 15.6; at this moment the p.d. between the conductors is everywhere zero. This flow of current causes charge to accumulate at B and D, and one quarter of an oscillation later the p.d.'s have reached their maximum values, and the current has fallen momentarily to zero. The energy in the line thus passes alternately from the magnetic field associated with the current to the electric field associated with the p.d., and back again.

The different behaviour of the current and potential waves at the short-circuited end comes about because the potential wave experiences a *phase reversal* at the reflection, whereas the current wave does not. Indeed a little thought will show that the phase of *one* of the two components of the wave system must be reversed if the wave pattern is to be propagated in the opposite direction. Thus at the point A the oncoming and reflected waves of potential are in antiphase and cancel; but the two waves of current remain in phase and produce maximum amplitude of oscillation.

A similar situation arises in many other kinds of wave systems. Consider, for instance, the waves that may be sent down a rope by waggling the end of it from side to side. If the rope is tied onto another cord of different mass per unit length, partial reflection of the wave occurs at the knot. The reflected wave suffers a phase reversal if it approaches the knot from the side of the less massive rope, but no phase reversal if approaching from the other side. Similarly it may be shown that a light wave travelling in air suffers a phase reversal when it is partially reflected at an air-glass boundary; but there is no phase reversal when it approaches the same boundary travelling in the glass.

Sound waves in a tube behave in a manner closely analogous to that of electromagnetic waves on a transmission line. In a progressive sound wave the velocity and pressure waves are in phase with one another, but in a standing wave system they are in quadrature—just as are the current and potential waves on a transmission line. Also, at a reflection, in both systems one component wave suffers a phase reversal where the other does not. Thus a sound wave reaching the open end of a tube is partially reflected back along it; in this case the pressure wave suffers a phase reversal, but the velocity wave does not. This means that the open end becomes a node of pressure but an antinode of velocity. The electromagnetic wave behaves in the same way at the short-circuited end of the transmission line—the potential is analogous to the pressure in the sound wave and the current to the velocity.

The behaviour of the two components of the sound wave is interchanged if the tube is terminated in a rigid boundary. The velocity wave then experiences a phase reversal but not the pressure wave; this forms a node of velocity at the boundary (where the air particles *cannot* move longitudinally) and an antinode of pressure. A similar form of reflection occurs with the electromagnetic waves if the end of the transmission line is left open-circuited. No current can then flow at this point, which therefore becomes a node of current and an antinode of potential. At an open-circuited termination the current must therefore experience a phase reversal, but not the potential. This also may be demonstrated with the apparatus of Fig. 15.6 with the shorting strip removed. To tune the transmission line we must now find a way of varying its length, or else the frequency of the oscillator must be adjusted, until resonance is achieved.

It is interesting to note the paradoxical fashion in which a quarter-wave length of line behaves. Consider, for instance, the section between B and A in Fig. 15.6. This is short-circuited at A, but behaves at B as though it were an open-circuit. Although there is maximum current flowing through the shorting strip at A, no current flows at B, in spite of the large alternating p.d. that exists at this point. Conversely an open-circuited quarter-wave length of line behaves like a short-circuit. At the open end the p.d. is a maximum and no current flows. But a quarter of a wavelength back along the line the current reaches its maximum value, although there is a potential node at this point.

It may be shown (see below) that, provided the resistance losses of a transmission line are small,

the velocity c of the electrical waves on it is given by

$$c = \frac{1}{\sqrt{\varepsilon_0 \mu_0}}$$

where ε_0 and μ_0 are the electric and magnetic constants (p. 192 and p. 104). Substituting

$$\varepsilon_0 = 8.84 \times 10^{-12} \text{ F m}^{-1}$$

and $\qquad \mu_0 = 4\pi \times 10^{-7} \text{ H m}^{-1}$

we obtain $\quad c = 3.00 \times 10^8 \text{ m s}^{-1}$

This is the same as the velocity of light, and provides a strong indication that light is an electromagnetic wave process. We shall show in the next section how it is possible for an electromagnetic wave to be propagated without the aid of the currents that accompany it in the case of a transmission line. The velocity of the waves turns out to be the same in both cases.

Once the theory of electromagnetic waves is satisfactorily confirmed by experiment, the measurement of their velocity (e.g. the velocity of light) provides the most accurate means of finding the electric constant ε_0. In the SI system of units the magnetic constant μ_0 is chosen arbitrarily in the course of defining the ampere (p. 107); ε_0 is then given by the theoretical relationship

$$\varepsilon_0 = \frac{1}{\mu_0 c^2}$$

★ *The velocity of waves on transmission lines*

A transmission line behaves as it does because of the inductance and capacitance that exists distributed along its length. It may thus be represented by the chain of inductors and capacitors shown in Fig. 15.7 (ignoring the resistance of the wires).

Over the length δx of the line from P to Q the total distributed inductance is δL, half of it in each wire. Let the capacitance distributed over the same length be δC. Quoting results derived in earlier chapters, the inductance and capacitance per unit length are given by

$$\frac{\delta L}{\delta x} = \frac{\mu_0}{\pi} \log_e (b/r) \quad \text{(p. 117)} \qquad (1)$$

and $\qquad \dfrac{\delta C}{\delta x} = \dfrac{\pi \varepsilon_0}{\log_e (b/r)} \quad \text{(p. 207)} \qquad (2)$

where b and r are the separation and radius of the conductors, respectively.

In the section of transmission line considered let the p.d. between the conductors be V_1 at P, and V_2 at Q, as shown. Taking into account the fall in potential along the distributed inductance between P and Q, we have

$$V_1 - V_2 = \delta L \frac{dI}{dt}$$

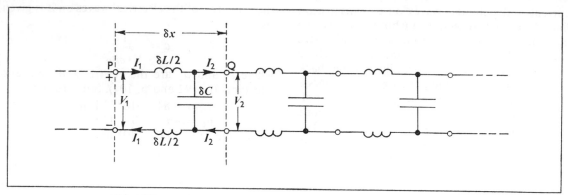

Fig. 15.7   Deducing the velocity of the waves on a transmission line

where I is the average current at any instant in the section between P and Q. The time taken for the wave to travel at speed c along the transmission line from P to Q is $\delta x/c$. In this time interval the p.d. at Q increases from V_2 to V_1, the value that previously existed at P. The rate of increase of the p.d. at Q is therefore given by

$$\frac{\mathrm{d}V}{\mathrm{d}t} = \frac{V_1 - V_2}{\delta x/c} = \frac{\delta L(\mathrm{d}I/\mathrm{d}t)}{\delta x/c} = c\left(\frac{\delta L}{\delta x}\right)\frac{\mathrm{d}I}{\mathrm{d}t} \qquad (3)$$

Likewise the current flowing into the distributed capacitance δC between P and Q is $\delta C(\mathrm{d}V/\mathrm{d}t)$. If the currents entering and leaving the section are I_1 at P and I_2 at Q, as shown, we have

$$I_1 - I_2 = \delta C\frac{\mathrm{d}V}{\mathrm{d}t}$$

The current at Q must increase by this amount n the time $\delta x/c$ taken by the wave to travel from P to Q along the line. The rate of increase of current at Q is therefore given by

$$\frac{\mathrm{d}I}{\mathrm{d}t} = \frac{I_1 - I_2}{\delta x/c} = \frac{\delta C(\mathrm{d}V/\mathrm{d}t)}{\delta x/c} = c\left(\frac{\delta C}{\delta x}\right)\frac{\mathrm{d}V}{\mathrm{d}t} \qquad (4)$$

We may eliminate $\mathrm{d}V/\mathrm{d}t$ and $\mathrm{d}I/\mathrm{d}t$ between equations 3 and 4 (by multiplying them together) giving

$$c^2\frac{\delta L}{\delta x}\frac{\delta C}{\delta x} = 1$$

Substituting from equations 1 and 2 we find the important result

$$c^2\varepsilon_0\mu_0 = 1$$
$$\therefore\ c = \frac{1}{\sqrt{\varepsilon_0\mu_0}}$$

A similar relation is found to exist between the inductance and capacitance per unit length of *any* air-insulated type of transmission line. A fuller analysis shows that the resistance losses in the line always have the effect of slightly reducing the speed of the wave. But the amount of this reduction is rarely significant; and we may normally assume that the velocity with which an electrical disturbance is propagated along an air-insulated cable is equal to the velocity of light, 3.00×10^8 m s^{-1}. With some other form of insulation (e.g. in a typical coaxial cable) the speed is considerably less because of the increased relative permittivity of the medium.

15.3   Electromagnetic waves without wires

In the previous section we have discussed how an electromagnetic wave accompanies the propagation of energy by an alternating current in a cable. In that case the electric field was produced by the alternating p.d. and the magnetic field by the alternating current. The possibility of electromagnetic waves being propagated through empty space far distant from any conductors was first mooted by Maxwell in 1865. He showed that if this is to happen there must exist mechanisms whereby

(*i*) an electric field can be produced by an alternating magnetic field; and

(*ii*) a magnetic field can be produced by an alternating electric field.

The first of these is nothing more than the familiar process of electromagnetic induction. When the magnetic field in a coil changes, an e.m.f. is induced in it. This e.m.f. is evidence that an electric field has been produced acting round the circumference of the coil and moving the free charges in it (p. 81); but the electric field exists even in the absence of any conductor for it to act in. Thus, a changing magnetic field produces an electric field at right-angles to it, with lines of force running in closed loops. The other mechanism is less familiar but occurs in much the same way—namely the production of a magnetic field by a changing electric field. When a capacitor is being charged (Fig. 15.8), the dielectric between its plates polarizes, i.e. the bound charges within it are displaced slightly from their normal positions (p. 193). While they are in motion these charges produce a magnetic field just like any other moving charge; but the displacement current only exists as long as the electric field in the dielectric is changing. Some magnetic field is of course produced by the pulse of current flowing in the connecting leads to the capacitor. But there is no doubt that the changing electric field in the dielectric also contributes to it; again the lines of force run in closed loops and are at right-angles to the electric field.

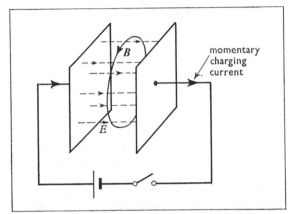

Fig. 15.8   A magnetic field is produced by a changing electric field between the plates of a capacitor

The two effects are exactly complementary. But there seems at first sight to be an important difference. Electromagnetic induction takes place with or without the presence of matter. A ferromagnetic core for the coil increases the effect, but it still happens in a vacuum. But in the second case the effect, as discussed so far, depends on having a material dielectric in the capacitor. In a vacuum there would be no bound charges and no reason to expect the 'dielectric' to contribute to the magnetic field.

Maxwell made the bold suggestion that there can be a displacement 'current' in a vacuum between two capacitor plates, producing a magnetic field just like any other form of current, though without any actual movement of charge. By this mechanism a changing electric field would produce a magnetic field even in a vacuum.

He could find no direct experimental test of his hypothesis, but it introduced a beautiful symmetry into the theory of electric and magnetic fields; and, if it were true, electromagnetic waves in a vacuum were a possibility. The theory showed that the waves should be of transverse type, and should therefore show polarization effects. It also predicted that the velocity c of the waves should be given by

$$c = \frac{1}{\sqrt{\varepsilon_0 \mu_0}}$$

$$= 3.00 \times 10^8 \text{ m s}^{-1}$$

This figure was remarkably close to the known velocity of light, which had already been shown to consist of transverse waves of some kind. Maxwell therefore suggested that light was an example of an electromagnetic wave motion.

It remained to show that electromagnetic waves could be produced by electrical means. The theory predicted that the waves should be emitted by any *accelerating electric charge*, since this would generate the necessary pattern of *changing* electric and magnetic fields. Some of the lines of force of the fields would then be formed into closed loops that would travel out into space carrying away some of the kinetic energy of the charge. To generate waves of given frequency the charge must perform oscillations at that frequency. In practical terms this means that a high-frequency alternating current must be made to flow in a suitably placed conductor—an *aerial* or *antenna*, as we now call it.

Possibly a high-frequency transmission line like that considered in the previous section might radiate away some of the energy it carries. But the currents in the two wires of a transmission line are necessarily in opposite directions at every instant, and the waves they radiate would be in antiphase; they would therefore almost entirely destroy one another by interference, and no radiation would be detected. Only if the conductors could be placed far apart compared with the wavelength was there any prospect of their producing electromagnetic waves. Thus, the length of the aerial and its distance from other conductors (such as the ground) needed to be comparable with the wavelength of the waves.

To detect the radiation it had to be shown that an alternating e.m.f. was produced in a 'receiving' aerial (or antenna) placed near the 'transmitter'. For, when the radiation flowed past the

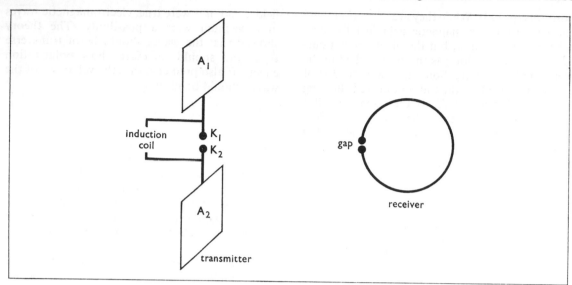

Fig. 15.9   Hertz's spark transmitter and receiver

second aerial, the alternating electric and magnetic fields should generate corresponding oscillations in it.

The electrical technology of 1865 could not provide the high-frequency generators that were needed for this demonstration. A frequency of perhaps 1 kHz might have been produced by mechanical means; but the wavelength would then be

$$\frac{3 \times 10^8}{10^3} = 3 \times 10^5 \text{ m} = 300 \text{ km}$$

and the aerials for this were unthinkable!

Eventually in 1888 Hertz produced very high-frequency currents in the oscillatory discharge of a capacitor (p. 179), and used these to generate electromagnetic waves. Maxwell's brilliant prediction was then confirmed in every detail. Hertz's apparatus is shown in Fig. 15.9. The 'aerial' consisted of two square metal sheets A_1 and A_2 arranged one above the other. These were joined to small metal spheres K_1 and K_2 connected to an induction coil. The two halves of the aerial thus formed a sort of capacitor. Each pulse of high p.d. from the induction coil charged up the capacitor until the insulation in the gap broke down; the ionized air then provided a low-resistance path through which the capacitor was rapidly discharged. Because of the

inevitable small inductance of the plates and connecting rods each discharge took the form of a damped oscillatory current of very high frequency. According to Maxwell's theory this arrangement should generate trains of electromagnetic waves with the electric vector parallel to the gap and magnetic vector at right-angles to it (i.e. with the lines of force in circles concentric with the connecting rods). This was duly detected by Hertz's 'receiver', which consisted of nothing more than a large loop of wire with a narrow gap at one point. When this was held in a vertical plane in line with the aerial, the alternating magnetic field in the radiation linked with the loop, generating an e.m.f. in it. A fine train of sparks was then observed crossing the receiver gap. The sparks ceased when the plane of the loop was turned through a right-angle, showing that the radiation was polarized in the way predicted. All the familiar properties of waves were demonstrated with apparatus of this kind. They were reflected from metal sheets and refracted by various materials, such as paraffin wax and pitch; and interference effects and standing wave patterns were readily produced. The wavelength used was found to be about 6 m, so that the frequency of oscillation in Hertz's transmitter must have been about 50 MHz.

Spark transmitters are now strictly illegal because of the extensive interference with radio reception they cause—a point that did not matter in Hertz's day. However, similar experiments may now be performed with more efficient transmitters and receivers that do not produce interference outside their own frequency limits; and in any case the experiments can be done in a laboratory at such low power that the radiation would be undetectable outside the building. But before embarking on any such tests the appropriate regulations should be consulted; for some purposes licences are required.

An electronic oscillator (p. 274) provides the ideal means of generating the high-frequency currents required in a radio transmitter. Fig. 15.10 shows a very simple arrangement. The alternating current in the tuned circuit of the Hartley oscillator induces an e.m.f. in the aerial coil coupled with it, and so feeds a high-frequency current into the aerial.

Waves of current and p.d. therefore travel along the aerial in much the same way as along a transmission line. The waves are reflected at the far end of the aerial, and a standing wave system is therefore set up on it with a node of current (and an antinode of potential) at the far end. Maximum current flows and maximum power is radiated when the aerial is resonant to the frequency in use. The shortest resonant length is a quarter of a wavelength. For maxi-

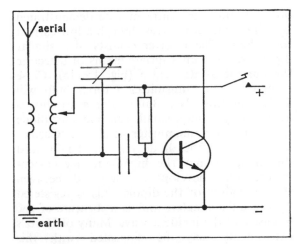

Fig. 15.10   A simple radio transmitter using a Hartley oscillator

mum efficiency either the length of the aerial or the inductance of the coupling coil must be adjusted so that this condition is effectively fulfilled.

At the highest frequencies the apparatus can take the form shown in Fig. 15.11a. This is coupled to the oscillator through a single-turn coil. The length of the aerial rod is arranged to be half a wavelength (i.e. it is two quarter-wave aerials back to back). Such an aerial is called a *half-wave dipole*. The current flowing up and

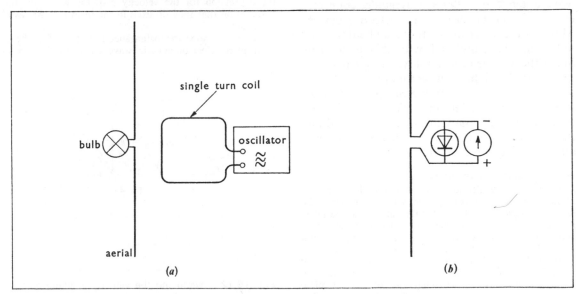

Fig. 15.11   A v.h.f. transmitter (*a*) and receiver (*b*), with which to simulate some of Hertz's experiments

down through its centre may be demonstrated by joining the two halves through a low-current torch bulb. The receiver consists of a similar dipole with the appropriate means of detecting the current at its centre (Fig. 15.11*b*). Close enough to the transmitter a torch bulb will suffice for this also. But greater sensitivity is achieved by using a small semiconductor diode (p. 258) and milliammeter as shown. Current flows through the diode only in the forward direction, so that the high-frequency current in the aerial develops a steady p.d. between the two halves of the dipole. This is registered by the milliammeter, and is a measure of the intensity of the incident wave. Many of Hertz's experiments may be simulated with this apparatus.

The higher the frequency employed (and therefore the shorter the wavelength) the more convincing becomes the demonstration of the wave nature of the electromagnetic radiation. It is worth experimenting with the special equipment now available using a wavelength of about 3 cm. In this the oscillations (of frequency about 10 GHz) are produced by a special electron tube called a *klystron*. This feeds a quarter-wave antenna (i.e. 7.5 mm long) located in a rectangular metal tube, called a *waveguide*. The radiation emerges from the flared end of this tube. The receiver consists of a similar quarter-wave antenna in a flared waveguide, connected to a diode and meter (or other detecting device). Many of the experiments that may be done with light may be simulated with this equipment. Since the wavelength employed is greater than that of visible light by a factor of about 10^5, the experiments can be conducted on a pleasingly massive scale! The demonstration of interference, diffraction and polarization leaves little doubt that we are dealing with a radiation of essentially the same nature as light.

★ *The 'plane of polarization'*

A notorious difficulty arises at this point in describing plane polarized electromagnetic waves, whether radio waves or light. The plane of polarization may be taken as the plane of either the electric or the magnetic vector. We know that the two vectors are at right-angles, and it is simply a matter of convention which we choose. Unfortunately there is no general agreement on which it should be. When polarization of light was first demonstrated, it was not known what kind of wave it might consist of; and the phenomenon of polarization by reflec-

tion was used to define the plane of polarization. When electromagnetic waves are reflected at the boundary between two 'transparent' media, the reflected waves are partially plane polarized; the plane containing the incident and reflected 'rays' was originally taken as the plane of polarization of the reflected waves. It turns out that this is in fact the plane of the magnetic vector. (The matter may be demonstrated by studying the reflection of very short wavelength radio waves from the surface of a block of paraffin wax.)

However, it appears that the electric vector is really the more significant of the two when we consider the interaction of electromagnetic waves with matter; and many writers have taken this vector as defining the plane of polarization. Indeed this is quite general in radio engineering—e.g. television aerials have to be aligned in the plane of the electric vector, and this is called the plane of polarization of the waves. Other writers (in textbooks on Light) have sought to avoid ambiguity by calling the plane of the electric vector the *plane of the vibrations*; and they do not then use the phrase 'plane of polarization' at all.

In the Author's opinion the student does well to avoid using both these ambiguous phrases, and to describe plane polarized waves by specifying the direction of either the electric or the magnetic vector. Thus Hertz's apparatus generates waves polarized 'with the electric vector in a vertical plane'; and light waves reflected from a glass plate are partially plane polarized 'with the magnetic vector in the plane of the incident and reflected rays'. This procedure leaves no loophole for ambiguity.

The velocity of electromagnetic waves in a vacuum

It is beyond the scope of this book to consider the detailed analysis by which Maxwell predicted the existence of electromagnetic waves. But we give here a deduction of the expression for the velocity c of the waves on the assumption that they do exist in the form that Maxwell predicted.

Fig. 15.12 shows the instantaneous pattern of the fields in a plane electromagnetic wave travelling in the *x*-

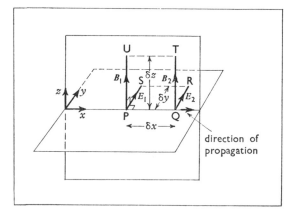

Fig. 15.12   Deducing the velocity of plane electromagnetic waves in a vacuum

direction. The wave is taken to be polarized so that the electric field lies in the y-direction and the magnetic field at right-angles to it in the z-direction. Neither field has any component in the direction of travel of the wave. Consider two points P and Q a short distance δx apart; let the two fields be E_1 and B_1 at P, and E_2 and B_2 at Q, as shown.

We shall apply first the law of electromagnetic induction (p. 87) to the rectangular path PQRS, which is of width δy. Now the electric field strength E is equal to the potential gradient. The p.d. between P and S is therefore $E_1 \delta y$; and that between R and Q is $-E_2 \delta y$. There is no potential gradient in the x-direction. The e.m.f. acting round the path in the sense $\overrightarrow{PSRQP}$ is therefore $(E_1 - E_2) \delta y$. [I.e. the e.m.f. is the *line integral* of the electric field strength round the path, p. 111.] This is equal to the rate of change of the magnetic flux linking the rectangle, which is the rate of change of $B.\delta x.\delta y$.

$$\therefore \text{ e.m.f.} = (E_1 - E_2)\,\delta y = \frac{dB}{dt}\,\delta x.\delta y$$

$$\therefore E_1 - E_2 = \frac{dB}{dt}\,\delta x$$

The time taken by the wave to travel from P to Q is $\delta x/c$. In this time interval the electric field strength at Q increases by the amount $(E_1 - E_2)$. The rate of increase dE/dt is therefore given by

$$\frac{dE}{dt} = \frac{E_1 - E_2}{\delta x/c} = c\,\frac{dB}{dt}$$

Similarly the *magnetomotive force* or *m.m.f.* (p. 128) acting round the rectangular path $\overrightarrow{PUTQP}$ is

$$\frac{1}{\mu_0}(B_1 - B_2)\,\delta z$$

According to Ampère's rule (p. 111) this should be equal to the electric current linked with the path. Since we suppose that the wave is travelling in a vacuum, there is no current in the ordinary sense anywhere in the neighbourhood. It was at this point that Maxwell introduced his hypothesis of the *displacement current*; without it no electromagnetic wave propagation is possible. To complete the symmetry of the equations we suppose that the m.m.f. acting round the closed path is equal to the rate of change of the total electric flux linking the rectangle. If the mean electric flux density is $\varepsilon_0 E$ (p. 195), the electric flux linking the rectangle PQTU is $\varepsilon_0 E.\delta x.\delta z$

$$\therefore \text{ m.m.f.} = \frac{1}{\mu_0}(B_1 - B_2)\,\delta z = \varepsilon_0\,\frac{dE}{dt}\,\delta x.\delta z$$

$$\therefore B_1 - B_2 = \varepsilon_0 \mu_0\,\frac{dE}{dt}\,\delta x$$

As before, the magnetic flux density at Q increases by this amount in the time $\delta x/c$ taken by the wave to advance from P to Q.

$$\therefore \frac{dB}{dt} = \frac{B_1 - B_2}{\delta x/c} = c\varepsilon_0 \mu_0\,\frac{dE}{dt}$$

Thus if the pattern of the electric and magnetic fields is to remain constant as it moves forward in the x-direction,

the rates of change of E and B must be connected by the two symmetrical relations derived above, which may be written

$$\frac{dE}{dt} = c\,\frac{dB}{dt}$$

and
$$c\varepsilon_0 \mu_0\,\frac{dE}{dt} = \frac{dB}{dt}$$

Dividing and rearranging we obtain the expression for c.

$$\therefore c^2 \varepsilon_0 \mu_0 = 1$$

$$\therefore c = \frac{1}{\sqrt{\varepsilon_0 \mu_0}} = 3.00 \times 10^8 \text{ m s}^{-1}$$

15.4   Broadcasting

In order to send messages with radio waves it is necessary to vary or *modulate* the wave in some way. An unmodulated *carrier wave*, as it is called, conveys no information except that the transmitter is working! The simplest form of modulation is to switch the transmitter on and off in time with the dots and dashes of the Morse code. A tapping key may be arranged so that the waves are only sent out when it is pressed down. For instance in the simple transmitter of Fig. 15.10 a tapping key may be inserted in the lead to the battery. To transmit speech or music the modulation must take the form of varying some feature of the wave in time with the electrical signal from the microphone. Either the amplitude or the frequency of the wave may be varied. Frequency modulation gives higher quality reception, but demands the use of a large band of frequencies on either side of the main carrier wave frequency. Such bands are only available at very high frequencies (v.h.f.).

Amplitude modulation is the simplest to transmit and receive and is used for most broadcasting and communication. Fig. 15.13 shows what is involved. The transmitter employs a master oscillator (such as that of Fig. 15.10) which generates oscillations of constant frequency and amplitude. One or more stages of power amplification are then used before the high-frequency oscillations are fed to the aerial. The modulation signal from the microphone is separately amplified and then applied to one of these stages in such a way as to vary its amplification. The current fed to the aerial is entirely a high frequency one, but its amplitude varies at the audio frequency of the signal from the microphone.

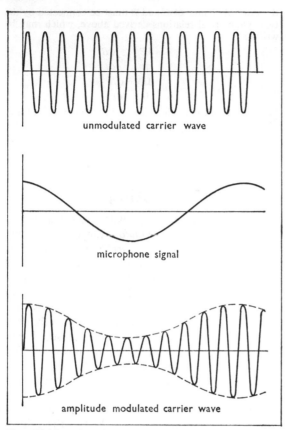

Fig. 15.13   Amplitude modulation of a carrier wave

At the receiving end the passage of the radio waves past the aerial generates small alternating currents in it. The audio frequency signal must now be recovered from the variations in amplitude of this current. Circuits for doing this are called *demodulators* or *detectors*. Fig. 15.14 shows a simple diode detector. The circuit should be compared with the power rectifier arrangement of Fig. 14.12 (p. 262). The parallel resonant circuit is tuned to the frequency to be received, and a magnified alternating p.d. is developed across it; this is applied to the diode and storage capacitor C in series. The time constant CR is made to be about 5×10^{-5} s. This is very long compared with the period of oscillation of the aerial current, and the storage capacitor therefore charges up to the peak p.d. across the tuned circuit. However, the time constant is short compared with the period of the audio frequency modulation. The p.d.

across C therefore follows faithfully the variations of amplitude of the aerial signal.

To give distortionless detection the diode must be able to pass some current for all amplitudes of the signal. Unfortunately, very little current flows in any kind of diode for p.d.'s less than about 0.4 V. This method of detection is not therefore suitable for small signals. However, with alternating p.d.'s of 10 V or more it is the best method available; it is regularly used in modern radio sets, in which there are several stages of high-frequency amplification before the detector stage. The load R in the detector could consist of a pair of high-resistance headphones, but this takes a rather large drain of power from the tuned circuit and so reduces its Q-factor (p. 225). It is better to develop the audio frequency signal across a high-resistance load and apply it to the high-impedance input of an amplifying circuit.

The ionosphere

One cause of poor radio reception is the rapid changes of signal strength that are liable to occur; this is known as *fading*. It arises because the waves can reach the receiving aerial by two different paths. One wave travels directly over the surface of the ground; diffraction effects enable it to follow the curvature of the earth and to be little affected by obstacles such as hills and buildings. Another wave travels up into the air and is reflected from layers of ionized particles in the upper atmosphere, called the *ionosphere*. Close to the transmitter the *ground wave* predominates, but at distances of 100 miles or so the *sky wave* may be of about the same amplitude;

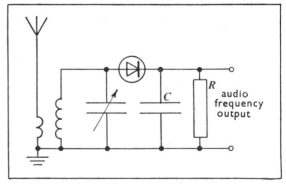

Fig. 15.14   A diode detector circuit

and interference then occurs between the two beams. Conditions in the ionosphere are changing continually. At one moment the receiving aerial may be at a maximum in the interference pattern and a strong signal is received. At another the resultant signal may drop off almost to nothing. To some extent fading may be countered by using Automatic Gain Control (A.G.C.) in the receiver; in this arrangement the average strength of the signal is made to control the amplification (or gain) of the receiver. When the signal increases, the gain is reduced; and vice versa.

The ionization in the upper atmosphere is produced by radiations of various kinds from the sun. It occurs in a number of separate layers at heights between 80 and 500 km; the layers fluctuate continually both in position and in density of ionization. In the medium waveband (500 to 1500 kHz) the conditions are such that the sky wave is of negligible intensity during the day-time; fading does not then occur, and reception of the ground wave is possible within a range of a few hundred miles. But when darkness falls the sky wave grows in strength, and fading and interference from adjacent channels are at their worst. Beyond the range of the ground wave reception is only possible by means of the sky wave. Even then fading can be serious, because the waves may reach the aerial by two or more different paths within the ionosphere. Sometimes the waves can reach the receiver after several reflections between the ionosphere and the ground, since the earth's surface, whether land or sea, is a fair conductor, and reflects electromagnetic waves strongly.

In the short waveband (from 3 to 30 MHz) conditions are rather different. The ground wave is of very short range (about 50 miles). But the sky wave is strongly reflected, at any rate for oblique incidence on the ionosphere; waves that strike it nearly normally pass straight through and are not reflected. Up to a certain range (known as the *skip distance*) no sky wave can therefore be received; but beyond this good reception is possible. By choosing the frequency of the transmitter to match the ionospheric conditions at a given moment reasonably reliable communication is possible with any part of the world. To increase the efficiency of short-wave communication and broadcasting it is common practice to *beam* the transmission towards the part of the world concerned. This is done by using several aerials in such a way that interference between the waves emitted concentrates the radiation at the required bearing and elevation. Many of these aerial systems are analogous to devices for producing interference in optics. Indeed some interference effects occur with almost all aerial systems. Even a simple half-wave aerial behaves like a Lloyd's mirror arrangement because of reflection by the ground; the radiation is therefore concentrated at particular elevations at which the waves from the aerial and its reflected 'image' in the ground happen to be in phase. A number of aerials side by side produce an interference pattern resembling that of a diffraction grating in optics; and by such means very narrow beams of radiation may be transmitted.

At still higher frequencies (>30 MHz) the ionosphere does not reflect the waves at all, even at oblique incidence; and the range is limited by the ground wave. However, this also limits the range from which interference can arise, and high-quality reception is possible.

16 X-rays

16.1 X-ray tubes

In 1895 Röntgen discovered that a very penetrating radiation was being emitted by a discharge tube in which cathode rays (p. 228) were passing at a high p.d. The radiation produced fluorescence on screens several metres away, and fogged photographic plates nearby, even when these were securely wrapped in dark paper. It was known that cathode rays themselves could not penetrate more than a few millimetres into air at s.t.p.; and it was therefore clear that a new kind of radiation was involved, to which Röntgen gave the name *X-rays*.

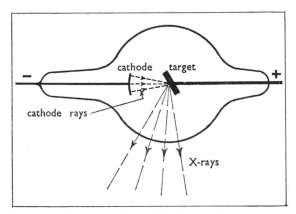

Fig. 16.1　A gas-filled X-ray tube

X-rays are produced whenever cathode rays (electrons) strike a piece of matter and so are brought to rest. In a discharge tube they are therefore emitted at the anode and from all parts of the glass wall reached by the cathode rays. In order to produce a localized source of X-rays it is necessary to focus the cathode rays into a small spot on the *target*. In the early days gas-filled tubes were used (at a pressure of about 10^{-1} N m^{-2}). The *cathode* was a concave plate whose surface was arranged to be concentric with the target spot on the *anode* or *anti-cathode* (Fig. 16.1). In this way some focusing of the cathode

rays is achieved since they tend to leave the cathode along lines normal to its surface. However, the p.d. across such a tube and the current through it depend critically on the gas pressure, and this tends to alter as the tube is used; it is therefore difficult to achieve reliable control.

A modern X-ray tube produces the cathode rays by thermionic emission from a heated tungsten cathode (Fig. 16.2). A high vacuum is used (less than 10^{-4} N m^{-2}), and the beam is focused by means of an electrode arrangement similar to that used in the electron gun of a cathode-ray tube (p. 231). The efficiency of the tube is increased by using a target material of high atomic number. Even so, only about 0.2% of the energy of the cathode-ray beam is converted into X-rays; the rest is changed into heat. The target is therefore made of a plug of tungsten (which has a high melting point); this is set in a substantial copper bar with external cooling fins. The maximum emission of X-rays occurs in the directions shown—approximately at right-angles to the cathode-ray beam. In all other directions the tube is heavily shielded with lead glass and sheets of lead. The *penetrating power* of the X-rays may be adjusted by varying the p.d. across the tube; the higher the p.d., the more penetrating the radiation. P.d.'s up to 100 kV are commonly employed; and for special purposes p.d.'s over a million volts are used. The *intensity* of the beam depends on the number of electrons striking the target per second—i.e. it depends on the current through the tube. This may be adjusted by varying the filament current, which controls the cathode emission.

One way of providing the high p.d. required to operate an X-ray tube is simply to connect it across the secondary of a high-tension a.c. transformer—T_1 in Fig. 16.3. The tube is itself a thermionic diode and conducts only on the half-cycles when the anode is positive with respect to the cathode. On the other half-cycles no current flows and no X-rays are emitted. The anode may need to be water-cooled, and it is usually con-

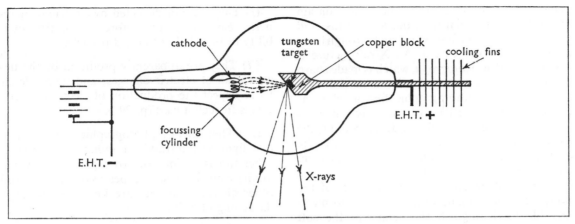

Fig. 16.2   A high vacuum X-ray tube

venient to earth this part of the apparatus. This means that the cathode and the whole secondary winding of the filament transformer T_2 are operating at very high potentials, and their insulation must be designed accordingly. This arrangement is satisfactory for some purposes; but when the tube must be operated at a steady known p.d., a separate rectifier employing a diode and storage capacitor must be used (p. 261).

X-rays are to some extent absorbed by all matter through which they pass; but their penetrating power is very great. They are therefore to be treated with the greatest possible respect; and

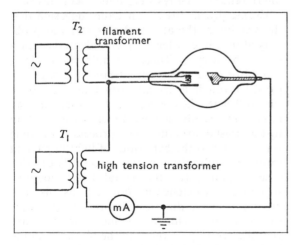

Fig. 16.3   A simple self-rectifying circuit for an X-ray tube

adequate, well-tested safety precautions must always be observed. Their effects on the human body are only partly understood; but they are known to produce deep-seated burns, destruction of living cells and unpredictable chemical changes. Most insidious of all is their tendency to produce serious diseases, sometimes many years after the time of irradiation; and they also produce genetic changes which may only become apparent in subsequent generations. All unnecessary exposure to X-radiation should be rigorously avoided; and the student is strongly advised never to use X-ray shoe-fitting equipment and similar apparatus. Discharge tubes operating at p.d.'s above 6 kV should never be used without adequate shielding. Television tubes (using p.d.'s up to 20 kV) should never be used without the screens designed by the manufacturers to go with them; however, often the front wall of the tube itself is made of glass that will give sufficient protection.

When X-rays pass through matter, their immediate effect is to produce ionization. Electrons may be ejected from all levels of the atoms, and the energy given to them is then dispersed through the rest of the matter by collisions. By this means the energy of an X-ray beam is mostly absorbed as heat. For a given thickness of screen the absorption is approximately proportional to the total number of electrons per unit volume. It is therefore greatest for dense substances containing elements of high atomic number.

The X-ray photography of the human body is

made possible by this effect. The bones contain much denser material than the soft organs, and therefore absorb the X-rays passing through them to a much greater extent. An X-ray picture is simply a *shadow photograph*, which is made sharp enough by using a very small source of X-rays and placing the photographic film as close as possible behind the part of the body under investigation. When the film is developed, all the tissues show up relatively pale against the darkened background; and the bones show up particularly white. Apart from the bones, the differences of density between the parts of the body are very slight, and it is not normally possible to distinguish the details of the soft organs—stomach, liver, brain, blood vessels, etc. However, in a number of cases doctors are able to introduce into an organ a harmless compound containing a dense element. For instance, in order to observe the stomach and intestinal tract, the patient is given a 'barium meal', consisting of a paste of barium sulphate and water. Each part of the alimentary canal in turn then shows up clearly as the barium meal reaches it. By techniques of this kind the radiographer can now take satisfactory X-ray pictures of many organs that could not otherwise be photographed.

When X-ray photography must be used for medical purposes, care is always taken to ensure that the dose is far below that which is definitely known to be harmful. Sometimes the destructive power of X-radiation is deliberately used for therapeutic purposes. It so happens that rapidly dividing cells are more readily damaged by X-rays than stable ones. A carefully controlled dose can thus destroy a growing tumour without doing irreparable harm to the surrounding tissue. Similar techniques are also used with the radiations from radioactive substances, considered in Chapter 17.

16.2   The nature of X-rays

When X-rays were first investigated, the conviction rapidly grew among physicists that they were a form of *electromagnetic radiation*—essentially the same as radio waves, light, etc., but of much shorter wavelength even than ultra-violet radiation. Several of the properties of X-rays suggest this conclusion:

(*i*) They are not deflected by electric or magnetic fields. We can therefore dismiss the possibility of their being charged particles.

(*ii*) They are apparently produced by the deceleration of electric charges; and theory leads us to expect the production of electromagnetic waves in such a case (p. 293).

(*iii*) They affect a photographic emulsion just like light and ultra-violet radiation. However, other things besides electromagnetic waves can do the same; for instance, beams of electrons and other charged particles are known to produce darkening of a film.

(*iv*) They produce ionization of the air through which they pass. It may be simply demonstrated by passing a beam of X-rays through the air inside or near a charged electroscope. Whatever the sign of the charge on the electroscope, there is a rapid fall of the deflection. The rate of discharge is proportional to the intensity of the X-ray beam, and is sometimes used as a means of measuring it.

However, none of this evidence was conclusive; and it was many years before a way was found of demonstrating a distinctive effect that would confirm the suggestion that X-rays were electromagnetic waves. They seemed to show no reflection or refraction like other sorts of waves, and the early attempts to demonstrate diffraction or interference of X-rays were quite inconclusive.

Eventually, in 1912, Von Laue suggested that the spacing of the atoms in a crystal was probably of the same order of magnitude as the wavelength of X-rays. Therefore diffraction effects should be observed when a beam of X-rays was passed through a crystal. Von Laue calculated that diffracted beams should be observed only in certain sharply defined directions in relation to the crystal structure. The process is in some ways similar to the behaviour of light in passing through a diffraction grating, except that in this case the grating (i.e. the crystal) has a repetitive structure that extends in *three* dimensions. The predictions were duly tested with the apparatus shown in Fig. 16.4*a*. A very fine beam of X-rays was allowed to pass through the pin-holes in the two lead screens S_1 and S_2. It then impinged on the crystal C, behind which was fixed a photo-

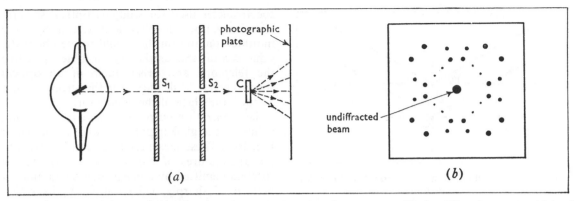

Fig. 16.4 The diffraction of X-rays by a crystal: (*a*) Von Laue's apparatus; (*b*) the diffraction spots obtained

graphic plate. The developed plate revealed a pattern of fine spots (Fig. 16.4*b*), which was in exact agreement with Von Laue's predictions. This clearly established the wave nature of X-rays, and showed that the wavelengths used were approximately in the range 10^{-11} to 10^{-10} m—about 1/10 000 of the wavelength of visible light. It was not surprising that previous experiments with ordinary diffraction gratings had not revealed anything.

However, subsequently the diffraction of X-rays has been successfully achieved, using ordinary optical diffraction gratings at very oblique incidence, so as to give maximum deviation of the diffracted beam. By this means the wavelengths of X-rays have been precisely measured; and this in turn has enabled the spacings of layers of atoms in crystals to be determined.

The wavelengths available from modern X-ray tubes now cover the full range from 10^{-8} m to 10^{-13} m—a range that overlaps appreciably with the wavelengths of ultra-violet radiation produced in discharge tubes. In the overlapping region the properties of the two types of radiation are found to be identical, a further confirmation of the electromagnetic wave nature of X-rays.

16.3 X-ray spectrometers

Soon after Von Laue's experiments Sir William Bragg suggested a much simpler way of analysing the diffraction of X-rays by a crystal lattice. When a beam of X-rays is incident on a single layer of atoms in one plane of a crystal, each atom scatters a minute proportion of the beam, i.e. it becomes the source of a small spherical wavelet of X-rays (Fig. 16.5). In most directions the wavelets destroy one another by interference. But in the direction for which the angle of incidence is equal to the angle of 'reflection' they reinforce one another. In other words the X-rays behave as though they were weakly reflected by the layer of atoms. The regularity of the arrangement of atoms in a crystal ensures that there are many such layers, parallel and equally spaced. In general therefore the beams reflected from successive layers will again destroy one another by interference, and only in certain sharply defined directions will constructive interference take place. These directions are those for which the path difference for the waves reflected from adjacent layers of atoms is a whole number of wavelengths. Referring to Fig. 16.6, this path difference is equal to PQ + QR. Strong reflection therefore occurs when

$$PQ + QR = n\lambda$$

where $\lambda =$ the wavelength, and n is an integer.

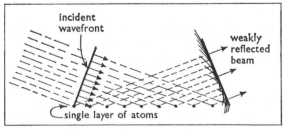

Fig. 16.5 The 'reflection' of X-rays by a single layer of atoms

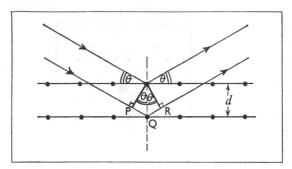

Fig. 16.6   The deduction of Bragg's law for X-ray reflection

If d is the separation of the layers of atoms and θ is the *glancing angle*, then there is strong reflection when

$$2d \sin \theta = n\lambda$$

a result known as *Bragg's law*. Notice that Bragg found it convenient to make his measurements in terms of the glancing angle θ, and not in terms of the *angle of incidence*—which would be measured from the normal and is equal to $(90° - \theta)$. For any given wavelength and crystal spacing Bragg expected to find several angles at which reflection occurred, corresponding to $n = 1, 2, 3$, etc. These he described as the *first-order reflection, second-order reflection*, etc.

Bragg's analysis was fully confirmed by experiment, and provided the practical means of measuring the spacing of atoms in crystals. It also enabled the pattern of X-ray wavelengths from any given source to be investigated. The instrument Bragg developed is therefore rightly called an *X-ray spectrometer*. The principle is shown in Fig. 16.7. A very narrow beam of X-rays was defined by the slits (less than 0.1 mm wide) in the two lead screens S_1 and S_2. The crystal C was mounted on a table that could be rotated about a vertical axis and carried a graduated scale, so that the glancing angle θ of the beam on its faces could be measured. The reflected beam (deviated through an angle 2θ, as shown) passed through a slit in another screen S_3 and into an ionization chamber with lead walls. This was connected to a suitable source of p.d. and an electrometer to measure the ionization current, which was taken as a measure of the strength of the reflected beam. The arrangement was rather similar to the familiar prism spectrometer used for studying optical spectra. The ionization chamber and screen S_3 were mounted on an arm that could rotate about the same central axis as the crystal table—just like the telescope and prism table of an optical spectrometer; while the slits in the screens S_1 and S_2 took the place of the collimator.

The ionization currents obtained for varying values of the glancing angle θ were as shown in Fig. 16.8. These measurements revealed two significant features of X-ray spectra. Firstly, the spectrum emitted by a given source contained a pattern of sharply defined wavelengths which were characteristic of the element used as the target of the X-ray tube. This can be compared with the characteristic line spectra emitted by gases and vapours in the visible region. For a given target element the wavelengths of the X-ray 'lines' were unaffected by varying any other factors, such as the p.d. in the X-ray tube—except that when the p.d. was too low a particular line of the spectrum might not be excited at all. Secondly, a continuous background spectrum was apparent containing all possible wavelengths within a certain range. This can be compared with the continuous spectrum emitted in the visible region by a hot solid (such as a lamp filament). The character of the continuous spec-

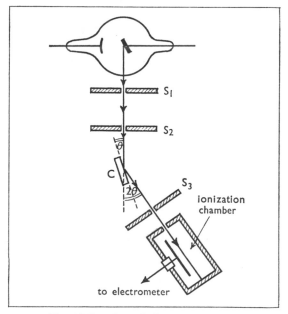

Fig. 16.7   Bragg's X-ray spectrometer

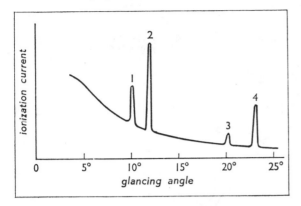

Fig. 16.8   The variation of ionization current with glancing angle

trum was unaffected by changing the target material, but was found to vary in a characteristic way with the p.d. used in the tube.

The same wavelength can appear in a number of different peaks in the diffraction pattern. Thus, peaks Nos. 3 and 4 (in Fig. 16.8) are due to the same pair of wavelengths as peaks Nos. 1 and 2; but 1 and 2 are the first-order reflection, while 3 and 4 are the second-order. This may be proved by showing that the values of $\sin \theta$ for the second pair are twice those for the first—in agreement with Bragg's law.

When the experiment is repeated with different crystals, the same pattern is found in every case, but the values of θ are shifted in accordance with the lattice spacing d of the particular crystal. This enables us to compare the lattice spacings of different crystals. Thus, for a particular wavelength in the first-order spectrum we have

$$d \propto \frac{1}{\sin \theta}$$

Even within a single crystal there are several sets of planes which are rich in atoms. Fig. 16.9 shows the positions of the centres of the atoms in a simple cubic crystal lattice. First of all there are planes such as ABCD or AEFB parallel to the faces of the cube. Then there are planes such as ADGF or AEGC inclined at 45° to the cube faces. Other planes rich in atoms are those parallel to ACF or EBG. The separation of adjacent planes can be calculated in each case in terms of the distance between neighbouring

atoms. For the three types of planes in a simple cubic lattice, the separations are in the ratio

$$1 : \frac{1}{\sqrt{2}} : \frac{1}{\sqrt{3}}$$

Other forms of crystal lattice give different ratios for their plane separations. By studying the diffraction patterns for different orientations of a crystal we can measure these ratios and establish exactly which type of crystal structure we are dealing with. If we also know the wavelength of the X-rays used, we can go further and actually measure the spacing of the atoms (or ions) in the crystal. Conversely, if we know the atomic spacing in any particular crystal, we can use this to measure any required X-ray wavelength.

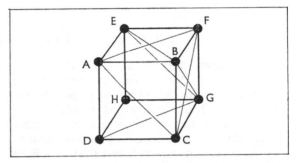

Fig. 16.9   The positions of the centres of the atoms in part of a simple cubic crystal

A 3 cm wave model

It is not generally possible to perform X-ray diffraction experiments in a school laboratory, since the sources of X-rays that can safely be allowed for this purpose are too weak to give measurable diffracted beams. However an analogous experiment can be carried out using electromagnetic waves of about 3 cm wavelength (p. 295), and this can be very instructive. Fig. 16.10 shows how a model of Bragg's X-ray spectrometer may be built with this apparatus. The transmitter is mounted at the focus of a plano-convex wax 'lens' so that a parallel beam of radiation is thereby produced. The 'crystal' is placed on a large turn-table centred in this beam. The receiver is similarly mounted at the focus of a wax lens, and is carried (with its lens) on an arm that rotates about the same vertical axis as the turn-table. A metal screen is usually needed to shield the receiver from the edges of

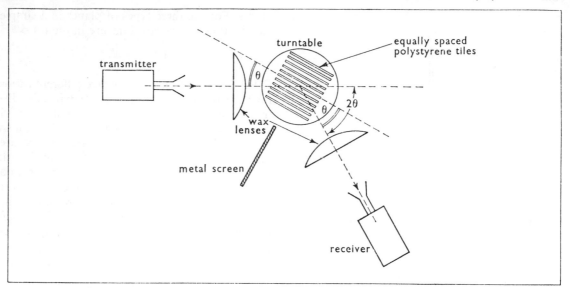

Fig. 16.10   A 3 cm wave analogue of Bragg's X-ray spectrometer

the direct beam from the transmitter. Diffraction by a stack of partially reflecting layers may be demonstrated with the aid of a set of polystyrene ceiling tiles. These are arranged parallel to one another on the turn-table, using spacing rather greater than the wavelength of the radiation (3 cm). In general it will be found that very little radiation can be detected at the receiver, even when this is placed so that the angles of incidence and reflection are equal, since for most angles of incidence the beams reflected from the different tiles destroy one another by interference. Only at certain clearly defined positions is there a significant reflected beam.

The most effective way of demonstrating this is to rotate the turn-table at a steady speed (one or two revolutions per second). The output of the transmitter may be modulated at a suitable audio frequency and the receiver connected to an amplifier and loudspeaker. An audible signal is then heard whenever the radiation reaches the receiver. The receiver arm is moved round very slowly while the turn-table revolves. In most positions little or nothing is received, but at a few clearly marked values of 2θ (Fig. 16.10) strong reflections are received twice in every revolution of the turn-table. The various values of $\sin \theta$ so obtained are found to be related according to Bragg's law

$$2d \sin \theta = n\lambda$$

showing that these are indeed Bragg's first, second, (and perhaps) third-order reflections.

The same apparatus may be used with 'crystals' of polystyrene balls mounted on the turn-table. Crystal models may be constructed with these balls according to the different arrangements of atoms that are possible. Each atomic structure gives its own characteristic pattern of reflections. For instance a simple cubic crystal mounted with the cube faces vertical (such as ABCD in Fig. 16.9) will give strong reflections from these main cube faces for one particular value of 2θ; there will be *four* such reflections in each revolution of the turn-table. At a larger value of 2θ there will again be a set of reflections, this time from the diagonal planes (such as ADGF)—four in each turn-table revolution. The two values of $\sin \theta$ will be found to be in the ratio $\sqrt{2} : 1$, in agreement with the geometrical relationship described above. To receive reflections from the third type of planes (such as ACF) the crystal must of course be remounted so as to bring these planes vertical.

The Avogadro constant

An experiment to measure the spacing of the atoms in a crystal is equivalent to counting the number of atoms in a given amount of the substance. It is therefore in effect an experiment to

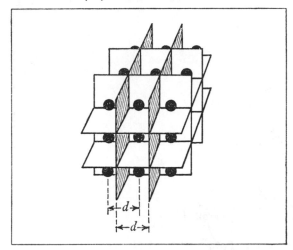

Fig. 16.11   Calculating the number of ions per unit volume of a simple cubic crystal

measure L, the Avogadro constant (p. 37). This may best be seen by repeating the kind of calculation that Bragg first performed with his data. Bragg turned his attention first to the crystals of halogen salts—potassium chloride and sodium chloride (rock salt). He soon established by the method outlined above that these substances crystallize in a cubic system. Imagine the material divided up by sets of planes parallel to the cube faces and passing midway between the atoms—as in Fig. 16.11. If the distance between adjacent atoms (or, more strictly, ions) is d, then each cubical box in the figure is of volume d^3 and contains just *one* ion. We shall consider the case of the rock salt crystal. This contains equal numbers of the two species of ion Na^+ and Cl^-. The volume occupied by one *pair* of such ions is therefore $2d^3$.

Referring to sets of physical tables we find the following data: the relative atomic masses of sodium and chlorine are 23.0 and 35.5 respectively.

∴ The molar mass of NaCl

$$= 58.5 \times 10^{-3} \text{ kg mol}^{-1} \quad \text{(p. 36)}$$

The density of rock salt

$$= 2.18 \times 10^3 \text{ kg m}^{-3}$$

∴ The molar volume of NaCl

$$= \frac{58.5 \times 10^{-3}}{2.18 \times 10^3} \text{ m}^3 \text{ mol}^{-1}$$

$$= 2.68 \times 10^{-5} \text{ m}^3 \text{ mol}^{-1}$$

X-ray diffraction experiments give the value of d the ionic spacing for rock salt as 2.81×10^{-10} m. The Avogadro constant L is the number of pairs of ions per mole of NaCl.

$$\therefore L = \frac{2.68 \times 10^{-5}}{2(2.81 \times 10^{-10})^3} \text{ mol}^{-1}$$

$$= 6.04 \times 10^{23} \text{ mol}^{-1}$$

There are a number of simple crystals for which similar calculations may be performed after we have measured their atomic or ionic spacing by X-ray spectroscopy. And this technique has provided the most accurate means we have of finding the Avogadro constant. The currently accepted value is

$$L = 6.022 \times 10^{23} \text{ mol}^{-1}$$

As a result of Bragg's work X-ray spectroscopy was established on a firm footing, and has proved to be one of the most powerful techniques of modern science. It has made possible the analysis of the arrangement of atoms in even the most complicated molecules. It has also provided an additional weapon for investigating further the inner structure of the atom.

Powder photography

There are many materials for which it is hard or impossible to obtain single crystals suitable for use in a Bragg X-ray spectrometer. Most solids have a polycrystalline structure in which there are myriads of microscopic crystals orientated at random within the material. However, X-ray diffraction can be observed with such materials also. Because of the large numbers of small crystals involved, some of them are bound to lie with their planes at the correct glancing angle θ to the incident beam to give strong reflections. The reflected beams leave the specimen in directions making an angle 2θ with the incident beam. If the specimen is small, the reflected beams will therefore be concentrated on the surface of a cone of semi-angle 2θ, as shown in Fig. 16.12. The figure shows the principle of an X-ray *powder camera*. A strip of photographic film (wrapped in opaque paper) is placed round the inside of a cylindrical drum; there are gaps in the cylinder through which a strong narrow beam of X-rays enters and leaves the chamber. The specimen is mounted at the centre of the

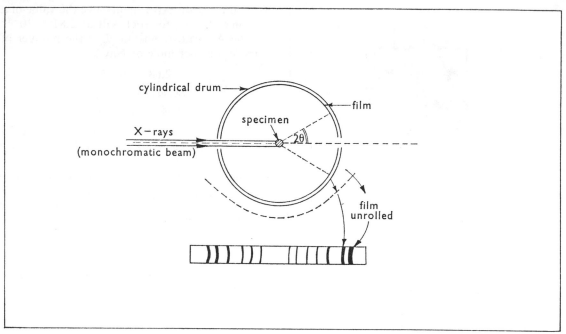

Fig. 16.12   An X-ray powder camera

drum. It usually takes the form of a small tube packed with the finely ground powder of the material under investigation. But in some cases it can simply be a rod or wire of the material. After exposure the developed film shows a pattern of slightly curved lines that mark the intersections of the film with the cones into which the X-rays were diffracted. The various possible values of 2θ are calculated from the positions of the lines and the radius of the camera cylinder. A single exposure thus provides a complete record of the possible Bragg angles of the material for the wavelength of X-rays used. The method is quick and simple, though the subsequent analysis of the results can be very time consuming.

With a non-crystalline material such as glass the powder photographs obtained are of quite a different character. Only a few diffracted lines are observed, and these are fuzzy and ill-defined. The atoms in glass are arranged in intersecting networks of a random nature, which have no regularly repeating structure. Round any given atom there is some measure of order in the way the neighbouring atoms lie; hence the existence of the few lines that appear in the powder photograph. But this order does not extend any

appreciable distance through the material. A similar pattern is observed in the 'powder' photographs of many liquids; water is an example. The few lines on the photographs reveal some measure of local ordering, but beyond this the molecular structure is a random one—and in a liquid is continually changing. Glass can be regarded in fact as a sort of 'congealed liquid'.

Electron diffraction

Powder cameras are items of equipment that are not likely to be found in a school laboratory because of the health hazards involved. But the diffraction of *electrons* can readily be observed using special electron tubes made for this purpose. In a previous chapter we discussed De Broglie's theory of the wave nature of the electron, according to which the path of an electron is to be described by the behaviour of the associated electron wave (p. 243). In De Broglie's theory the wavelength λ of this wave is related to the momentum $m_e v$ of the particle by the relation

$$\lambda = \frac{h}{m_e v}$$

where h is the Planck constant. For a typical cathode-ray tube operated at a p.d. of a few thousand volts the wavelength involved proves to be of the same order of magnitude as for X-rays. (The two sorts of waves are of course quite different; only their wavelengths are similar.) It is therefore to be expected that diffraction effects should be observed for cathode rays in much the same way as for X-rays. Indeed the prediction of just such effects was the special achievement of De Broglie's theory.

Fig. 16.13 shows the design of an electron diffraction tube; it is similar to the cathode-ray tubes described in Chapter 13 (e.g. Fig. 13.3, p. 229). In this case the hole in the anode is covered with a thin specimen (e.g. a film of graphite); if the electrons are to pass through this, it must in fact be so thin as to be optically transparent. The specimen is a polycrystalline one; and for a given Bragg angle θ that gives strong reflection of the electron waves the diffracted beams leave the specimen along the surface of an imaginary cone of semi-angle 2θ—just as for X-rays. On the fluorescent end-wall of the tube we therefore observe a central undiffracted spot, together with one or more circles formed by the impact of the diffracted beams. From the geometry of the tube the angle 2θ is given by

$$2\theta = \frac{x}{l}$$

The kinetic energy ($\frac{1}{2}m_e v^2$) of an electron is gained from the electric field between anode and cathode.

$$\therefore \tfrac{1}{2}m_e v^2 = eV$$

where $V =$ p.d. between anode and cathode

$$\therefore \text{momentum} = m_e v = \sqrt{2em_e V}$$

The De Broglie wavelength λ is therefore given by

$$\lambda = \frac{h}{m_e v} = \frac{h}{\sqrt{2em_e V}}$$

Applying Bragg's law we have

$$2d \sin \theta = n\lambda = \frac{nh}{\sqrt{2em_e V}}$$

where $n =$ the order of reflection. In practice the angle θ is probably small enough for us to replace

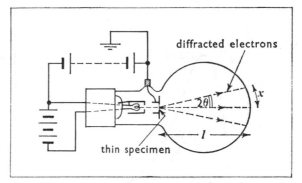

Fig. 16.13   An electron diffraction tube

$\sin \theta$ in this equation by θ (in radians). Substituting for 2θ we then have

$$d . \frac{x}{l} = \frac{nh}{\sqrt{2em_e V}}$$

i.e.
$$x \propto V^{-\frac{1}{2}}$$

This result may be tested for a suitable range of p.d.'s, and provides direct confirmation of the details of De Broglie's theory. If two or more diffraction circles are visible, we can also confirm that for a given accelerating p.d.

$$\sin \theta \propto n$$

Taking accepted values of the atomic constants h, e and m_e, we can then use the measurements to give a value of d, the layer spacing in the material.

16.4 X-ray spectra

(i) *The continuous spectrum.*   The distribution of energy in the continuous background spectrum of an X-ray tube is found to depend only on the p.d. across the tube. The results obtained for different values of the p.d. are as shown in Fig. 16.14. As the p.d. is raised, the total intensity increases (for a given tube current), and X-rays of shorter wavelength are emitted—which are therefore more penetrating.

An important feature of these curves is that for any given p.d. there is a sharply defined *minimum wavelength* emitted by the tube. According to classical electromagnetic theory any accelerating (or decelerating) charged particle should generate electromagnetic waves (p. 293). It is therefore to be expected that radiation would be emitted by electrons brought suddenly to rest on striking

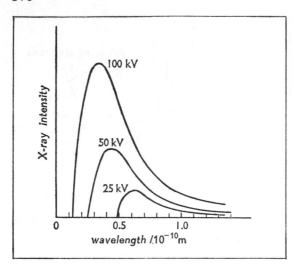

Fig. 16.14   The continuous X-ray spectrum for various values of the p.d. across the tube

the target of an X-ray tube. However, the classical theory is quite unable to explain why there should be any *sharp* cut-off wavelength of the kind observed. Once again the quantum theory (p. 239) comes into its own in explaining this effect. According to this theory the emission or absorption of electromagnetic radiation can take place only in packets of energy (or *photons*) of amount given by

$$\text{energy} = h\nu \quad \text{(p. 239)}$$

where $h =$ the Planck constant

and $\nu =$ the frequency of the waves.

Now, the electrons in an X-ray tube are all accelerated through the same p.d. and therefore all reach the target with practically the same energy, given by

$$\text{electron energy} = eV$$

where $V =$ the p.d. across the tube, and $e =$ the charge on an electron.

This is also the maximum energy that can be imparted to a photon, if we assume that each photon arises in the deceleration of a single electron. We should therefore expect to find a *maximum* X-ray frequency ν_{max} given by

$$h\nu_{max} = eV$$

and a minimum wavelength λ_{min} given by

$$c = \nu_{max}\lambda_{min}$$

where $c =$ velocity of light.

$$\therefore \lambda_{min} = \frac{hc}{Ve}$$

Thus the quantum theory predicts that the cut-off wavelength λ_{min} should be inversely proportional to the p.d. V across the X-ray tube; and this is in good agreement with experiment. Most of the electrons that generate photons give up only part of their energy in this way, and most of the X-radiation is therefore of longer wavelength than λ_{min}.

The cut-off wavelength λ_{min} can be measured with great precision with an X-ray spectrometer. Since c and e are already known to high accuracy, this measurement provides one of the best ways of finding the value of the Planck constant h. The result is in excellent agreement with the value found from the photoelectric effect.

(*ii*) *The line spectrum.*   Like the optical spectra of gases and vapours the line spectrum of an X-ray tube is characteristic of the element used for the target. However, unlike optical spectra, which are of great complexity, the X-ray spectra of the elements are all strikingly similar. A typical pattern for an element of medium atomic mass is shown in Fig. 16.15a. The group of lines of shortest wavelength is known as the K-series. Usually two lines only of this series can be detected (though there are more); these are known as the K_α and K_β lines, in order of decreasing wavelength. The K_α line is always the strongest in the spectrum. At longer wavelengths there is another group of lines known as the L-series (L_α, L_β, L_γ, etc.). The pattern is essentially the same for all elements. But as we proceed through the periodic table the whole pattern shifts progressively towards shorter wavelengths with increasing atomic mass. For the heaviest elements a third series, the M-series, can be detected; while for the lightest elements the L-series is beyond the range of our instruments, and only the K-series can be observed.

In 1913 Moseley measured the wavelengths of the K and L lines for a large range of elements. Fig. 16.15b shows the results obtained for the K-series of a selection of neighbouring elements in the periodic table. It is apparent that the elements can be arranged in a definite sequence on the basis of the wavelengths of their K lines.

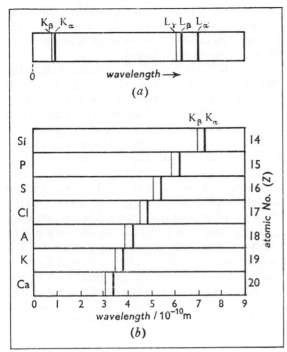

(a)

(b)

Fig. 16.15 (a) The characteristic X-ray spectrum of an element. (b) The K-series of a set of neighbouring elements of the periodic table

(The same sequence is obtained using the wavelengths of other lines in the X-ray spectra.) Now this order coincides exactly with that obtained on the basis of the chemical properties of the elements—expressed in the periodic table. To some extent the same order is given by arranging the atoms in order of ascending atomic masses. But this sequence differs in detail from that given by the X-ray wavelengths or the chemical evidence. Fig. 16.15b shows one example of this. On the basis of their X-ray spectra the order should clearly be:

chlorine–argon–potassium

and this agrees with the sequence in the periodic table. But in order of atomic masses the sequence is:

chlorine (35.5)–potassium (39.1)–argon (39.9)

When Moseley numbered the elements according to their positions in the periodic table, he found there was a linear relation between the *atomic number* (as he called it) and the *square root of the frequency* of the K_α line. It was thus established that both the chemical properties and the X-ray

spectrum of an element were decided by its atomic number rather than its atomic mass. It remained to be settled what feature of the structure of an atom the atomic number described.

Bohr's theory of the atom (p. 241) was propounded in the same year in which Moseley did his work on X-ray spectra. In the first instance the theory was put forward to account for the features of optical spectra; but it was also immediately successful in explaining Moseley's results. The theory showed that the *optical* spectrum of an element arises from changes in the outermost electron orbitals of the atom. Thus, if one of the outermost electrons of the atom is raised into an orbital of higher energy than normal, it subsequently returns to its ground state, emitting the surplus energy as a quantum of radiation (light). The frequency ν of the radiation is given by Planck's relation

$$h\nu = E' - E'' \quad \text{(p. 242)}$$

where E' and E'' are the energies of the initial and final electron orbitals. The optical line spectrum of the element contains the frequencies corresponding to the possible transitions of the outermost electrons of the atom. The energy jumps are in all cases relatively small; and so the frequencies are low and the wavelengths long— in fact, in or near the visible region of the spectrum.

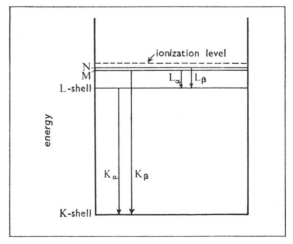

Fig. 16.16 The energy levels and transitions involved in X-ray spectra

In an X-ray tube we suppose that the energy of the incident electrons is so high that they are able to ionize even the innermost shells of the atoms of the target. When this happens a vacancy is created in one of these shells, which can then be filled by a transition from a shell of higher energy. Because the innermost shells are very tightly bound to the nucleus, the energy differences between them are considerable. The photon emitted in such a transition is therefore of high frequency and short wavelength—in fact, in the X-ray region of the spectrum.

Fig. 16.16 shows the energy level diagram for the inner shells of an atom. When there is a vacancy in the K-shell, transitions are possible from all the other shells. A transition from the L-shell produces the K_α line; that from the M-shell gives the K_β line. The K_β transition involves a larger change of energy, and this line is therefore of shorter wavelength than the K_α line. However, it is a less probable transition; and the K_β line is therefore weaker than the K_α. Likewise the L-series arises from transitions following ionization of the L-shell. These are of lower energy than the K-series transitions, and are thus of much longer wavelength.

It is now clear why all X-ray spectra show the same pattern. The innermost shell structure of all atoms is the same, since in all cases these shells are completely filled in the ground state. When electrons are ejected from the inner shells, the same pattern of transitions is possible whatever the type of atom involved. However, if one atom carries a larger charge on its nucleus than another, its shells are more tightly bound and are therefore of lower energy. The corresponding changes of energy are then of larger amount, giving a spectrum in which all the lines are of shorter wavelength. It therefore seemed that the atomic number, which Moseley showed to be of such significance in fixing the properties of the atom, was to be identified with the *total charge* on the nucleus. It is equal to the number of orbiting electrons in the neutral atom.

17 Nuclear Physics

17.1 Radioactivity

In 1896 the French scientist Becquerel discovered a hitherto unknown type of radiation which was emitted spontaneously by any sample of the element uranium or its compounds; the phenomenon was called *radioactivity*. Becquerel had been investigating an effect in which certain substances emit light weakly for a time after being exposed to a strong beam of light; this is known as *phosphorescence*. It occurred to him that there might be a connection between this effect and the *fluorescence* observed on the walls of a cathode-ray tube, which was known to be accompanied by the emission of X-rays. Accordingly he set out to find whether X-rays were emitted by phosphorescent materials. This he did by wrapping photographic plates in black paper and standing on them a mineral that was phosphorescing through being exposed to sunlight. The results obtained were negative until phosphorescent compounds of uranium were used. The developed photographic plates then showed intense blackening. However, it soon became clear that the effect had nothing to do with the phosphorescence, the investigation of which had prompted the experiment. The blackening of the plates was found to continue even when the uranium compound had been kept strictly in the dark; also it occurred with compounds of uranium that did not exhibit phosphorescence and for metallic uranium as well.

At first it appeared that the new radiation was similar to the X-radiation discovered by Röntgen. Thus, it was very penetrating; also it was strongly absorbed by a sheet of metal. A metal screen placed between the uranium compound and the photographic plate cast a 'shadow' on the plate, just as would occur with X-rays. However, fuller investigation showed that the effect was a complex one, in which several different kinds of radiation were involved; some of these were found to have properties very different from those of X-rays.

Becquerel's discovery led to an intense search by many scientists for other radioactive materials. Madame Curie and others found that *thorium* and its compounds were also radioactive. She then discovered that the mineral known as *pitchblende*, which contains uranium and thorium, showed about four times as much activity as could be explained by its content of these elements. None of the other elements known to be present in pitchblende had been found to be radioactive. Madame Curie therefore deduced that the mineral must contain one or more hitherto undetected substances; since they were presumably present in very small quantities, it was clear that they must be vastly more radioactive than uranium or thorium. After years of tedious chemical separations she and her husband succeeded in identifying two new radioactive elements, which they called *polonium* and *radium*. The latter is more than a million times as radioactive per unit mass as uranium. Within a few years a long list of radioactive elements had been discovered by various scientists; many of these turned out to be isotopes of already known elements, and nearly all of them were of large atomic mass. In recent years the list has been enormously extended following the discovery of techniques for making artificial isotopes not found in nature; most of these are radioactive. Indeed it is now possible to find radioactive isotopes of almost every element in the periodic table, light elements as well as heavy ones. The same techniques have also given rise to new radioactive *elements* with atomic masses greater than that of uranium, the heaviest element found in nature.

Radium is still one of the most useful radioactive sources for laboratory purposes, since a small quantity of it strongly emits radiations of several different kinds simultaneously. But it is sometimes convenient to have a source that provides a single kind of radiation only. For this purpose certain artificial radioactive isotopes are particularly suitable; those usually encountered in school laboratories are *plutonium* (a 'trans-uranic' element), radioactive *strontium*, radioactive *cobalt* and radioactive *caesium*.

The chief effect produced by the radiations from radioactive substances is the ionization of the matter through which they pass. This may be demonstrated by placing a radioactive source inside or near a charged electroscope, when the deflection at once starts to fall. The rate of decrease of the deflection is a measure of the ionization current (p. 176), and for a given type of source may be taken as an indication of the strength of the radiation emitted. A more detailed investigation of the ionizing effect of the radiations may be conducted using an ionization chamber connected to an electrometer (Fig. 17.1). For this purpose the outer metal cylinder of the chamber can form one electrode and the insulated central rod the other. A high p.d. is applied between the outer cylinder and one terminal of the electrometer input. The 'circuit' is then completed by a movement of ions inside the chamber. The ionization current is a measure of the rate at which ions are formed by any radiation passing through the chamber.

α-rays and β-rays

The most strongly ionizing component in the radiations emitted by radium has a very short range—only a few centimetres in air, and proportionately less in a denser substance. One way of demonstrating this is to equip the ionization chamber with a gauze cover (Fig. 17.1). When a radium source is held a short distance above the gauze, a steady ionization current is collected; this can only be due to that part of the radiation which has passed through the gauze and caused ionization in the electric field inside. When the source is raised, the current decreases, and finally falls to a very small value when the radium is about 60 mm from the gauze. If the radium is placed once more close to the gauze, it will be found that even a single thick sheet of paper is sufficient to absorb the radiation almost completely.

The absorption of the radiation by matter may be studied by placing thin foils of aluminium under the source. A foil 10^{-2} mm thick reduces the current to about a quarter of its initial value.

With three or four such foils the current falls to about 1% of its original value. However, it then becomes apparent that the radium is also emitting a second kind of radiation, which is both more penetrating and also produces less ionization. The thickness of absorbing foil over the source may be doubled or trebled without much affecting the residual 1% of ionization current; and a sheet of aluminium 1 mm thick is barely sufficient to halve it. These two kinds of radiation are called *α-rays* and *β-rays*. The α-rays are stopped by a few centimetres of air or a thin film of solid material; but within this short range they cause heavy ionization. To stop the β-rays requires a thickness of about half a millimetre of lead or about half a centimetre of wood; their range is about 10 times that of the α-rays and the ionization they produce in a given distance is about a tenth as much.

It is found that the stopping power of a slab of material is approximately proportional to its mass per unit area of cross-section. Thus

$$\text{the density of air} = 1.3 \text{ kg m}^{-3}$$

and

$$\text{the density of aluminium} = 2.7 \times 10^3 \text{ kg m}^{-3}$$

$$\therefore \text{ the ratio of densities} = \frac{2.7 \times 10^3}{1.3}$$

$$\approx 2 \times 10^3$$

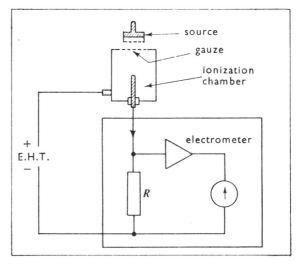

Fig. 17.1   Measuring the ionization current produced by a radium source

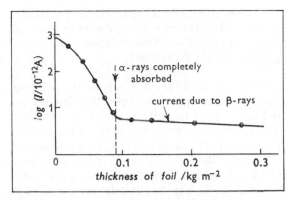

Fig. 17.2 Absorption curve showing the existence of at least two kinds of radiation from a radium source

and this is also the ratio of the stopping powers of the two materials for a given thickness. For the most energetic α-rays emitted by radium,

$$\text{range in air} = 60 \text{ mm}$$

$$\therefore \text{ range in aluminium} = \frac{60}{2 \times 10^3}$$

$$= 3 \times 10^{-2} \text{ mm}$$

The mass per unit area required to stop these rays

$$= 60 \times 10^{-3} \times 1.3 \text{ kg m}^{-2}$$

$$\approx 0.08 \text{ kg m}^{-2}$$

A graph of ionization current against effective thickness of stopping foil is of the form shown in Fig. 17.2. (Since the current covers a large range of values, it is convenient to plot this sort of graph using a logarithmic scale of current.) At the point where there is a total thickness equivalent to 0.08 kg m^{-2} there is a pronounced discontinuity in the graph. At this point the α-rays have been completely absorbed and the small remaining ionization current is caused by the β-rays only. The absorption of β-rays may be studied in the same way, though we must now use a very much thicker range of foils. The ionization current due to the β-rays is approximately halved for an additional absorbing thickness of about 3 kg m^{-2}; and the absorption is complete when the thickness approaches 8 kg m^{-2} (Fig. 17.3).

Using even thicker absorbing screens than these it is possible to show that a radium source emits yet a third kind of radiation, called γ-rays.

These are even more penetrating than β-rays, and produce much less ionization—about 1/1000 of that produced by β-rays in the same distance. With the radium sources available in schools the ionization current produced by the γ-rays will not be much more than 10^{-13} A, and a very sensitive electrometer is required. However, there are other means of detecting the radiations, such as the Geiger-Müller tube (p. 320) and the solid-state particle detector (p. 261); these respond satisfactorily to all three types of radiation without the need for very sensitive measuring instruments. Thus, with several centimetres thickness of wood between a radium source and the detector—quite sufficient to stop the α- and β-rays completely—it will be found that some form of ionizing radiation is still reaching the detector; and this is what we call γ-radiation. To cut down the γ-radiation to even half its intensity requires an absorption thickness of about 100 kg m^{-2} (about 1 cm thickness of lead); this is some 30 times as much as is required to produce the same reduction for β-radiation.

Safety precautions

The effects on living organisms of the radiations from radioactive substances are in many ways similar to those of X-rays (p. 301). They can cause burns and destruction of living cells, and are capable of initiating serious diseases and even genetic changes. Also the effects are to some extent cumulative, so that a very small daily dose of radiation may add up in the course of a year

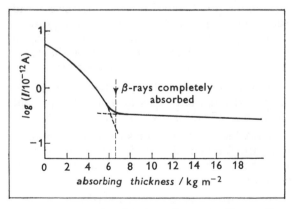

Fig. 17.3 The β and γ absorption curve for a radium source

or so to a total dose sufficient to cause irreparable harm. Radioactive sources must therefore be treated with the greatest possible care—particularly because, unlike an X-ray tube, they are never 'switched off', and it is all too easy to forget their presence.

The damage caused by the radiations arises in the first instance from the ionization they produce. This in turn leads to many different kinds of chemical changes that can affect the functioning of living cells. α-*rays* produce intense ionization, but their range is so short that they would not penetrate even the surface layers of skin. There is therefore little danger from α-active materials provided they are not absorbed into the system through the stomach or lungs. In particular, radium sources must be carefully sealed to prevent any escape of the α-active gas *radon* that is generated continually by radium. β-*rays* are less strongly ionizing, but their penetration is rather greater. However, a layer of wood or perspex 5 mm thick gives complete protection against external β-radiation. The main hazard from β-active materials is again the possibility of absorbing them into the system through the stomach or lungs. γ-*rays* constitute the chief danger in handling radioactive materials, owing to their great penetrating power. Even a layer of lead 1 cm thick is only sufficient to absorb about 50% of the γ-radiation incident on it, and so it is difficult and expensive to provide adequate screening. All γ-sources must therefore be as weak as possible, and must be kept at a sufficient distance from people in the laboratory to reduce the radiation dose well below the acceptable limit—allowing for whatever lead screening is in use. It is usual to insist on the following rules being observed in a school laboratory in which experiments of the types described in this chapter are performed:

(*i*) All sources must be weak (less than 10 μCi, see p. 341), and must be sealed in foil to prevent the escape of radioactive material.

(*ii*) All sources must be stored in a cupboard in a little-frequented part of the laboratory; they must be kept and transported in a suitable lead container.

(*iii*) Sources must only be handled with forceps, and should always be held well away from the body.

(*iv*) No eating, drinking or smoking must take place in a laboratory where radioactive materials are in use; also the licking of labels or sucking of pipettes should be avoided.

Having said all this, it is only fair to point out that by observing these rules the dangers are indeed reduced to quite negligible proportions. To give further confidence it is always possible to carry a pocket dosimeter (p. 178), which gives a record of the total dose received by the experimenter. These instruments are usually calibrated in *röntgens* (R), a special unit devised for measuring the total dose of X- or γ-radiation by means of the ionization produced.

The **röntgen** is the quantity of X- or γ-radiation that produces in one kg of air at s.t.p. ions of either sign carrying a total charge of 2.58×10^{-4} C.

The obscure origins of this medical unit of radiation need not concern us. Other units used for measuring an absorbed dose are the *rad* and the *rem*. With the types of radiation likely to be encountered in schools these can both be taken as equal to the röntgen. The total dose received by anyone in one year should not normally exceed 50 mR (50 milliröntgens); and the dose rate at any given time should not exceed 0.25 mR per hour. Actual destruction of living tissue requires a total dose of several thousand röntgens.

17.2   Detecting individual particles

The spark counter

The ionization produced in the air by α-rays is sufficiently intense to trigger off a spark discharge in an electric field that would not otherwise be strong enough for a spark to pass. This provides a useful means of detecting α-rays. One form of spark counter consists of a fine wire stretched about 1 mm above the surface of a metal plate, and carefully insulated from it (Fig. 17.4). The p.d. between wire and plate is adjusted to a value slightly less than that required to produce sparking. At this p.d. there is usually a weak corona discharge from the wire, which may be slightly luminous in the dark. When a radium source is held close to the wire, sparks are seen to pass between the two electrodes. The sparks occur at

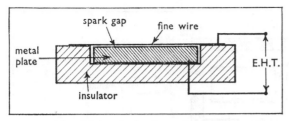

Fig. 17.4   A simple spark counter

irregular intervals and appear to be distributed at random along the wire. They are clearly due to the α-rays only, since no sparks occur when the source is moved to more than 60 mm from the wire or if a sheet of paper is placed to intercept the α-rays. The β-rays evidently do not produce sufficient ionization to initiate a spark. By watching the spark counter we get the impression that the α-rays do not form a steady stream of radiation but rather a random 'rain' of particles. This impression, as we shall see, is confirmed by other means available for detecting α-rays. We thus have the means of *counting* the individual α-particles (as we shall call them) that arrive in the vicinity of the wire.

By testing a radioactive source first with a spark counter and then with an ionization chamber and electrometer we can analyse the nature of the emissions it gives. A plutonium source produces discharges in the spark counter, and must therefore emit α-particles. However, if it is placed near the ionization chamber and covered with sufficient foil to stop the α-particles altogether, no ionization current can be detected; we therefore assume that in this case there is no emission of β-rays. Using the same technique we can show that a radioactive strontium source emits β-rays but not α-rays. A cobalt-60 source appears to emit neither α-rays nor β-rays; but by using a Geiger-Müller tube or a solid-state particle detector we can show that γ-rays are emerging from it. (Actually the cobalt also emits β-rays, but these are of such low energy that they are absorbed completely by the piece of foil enclosing the source.)

Scintillation counters

The radiations from radioactive substances can also be detected by the fluorescence they produce in certain types of materials known as *phosphors*.

This effect may be observed with an instrument called a *spinthariscope* (Fig. 17.5). It consists of a screen covered with a suitable phosphor, such as zinc sulphide, with a speck of radium mounted on a needle a short distance from it. The screen is viewed through a magnifying glass. When the eye has got adapted to the darkness inside the instrument, minute flashes of light, called *scintillations*, are seen coming from points on the screen. It is difficult again to avoid the conclusion that we are observing the impact of individual particles ejected from the source in an irregular and random way. The scintillations cease if the source is covered by a thin absorbing sheet, and it is therefore clear that they are due to the impact of α-particles.

It is now known that any ionizing radiation can produce scintillations in a suitable phosphor, though with β-rays and γ-rays the flashes of light are too weak to be detected by the human eye. They are, however, within the range of a photomultiplier tube (p. 238). The combination of phosphor and photomultiplier tube is called a *scintillation counter*; it has become one of the most useful instruments for detecting ionizing radiations. It has the special merit that the strength of the flash of light depends on the type and energy of the particle that caused it. Employed with suitable electronic circuitry a scintillation counter can be used to discriminate between different sorts of ionizing particle. To detect α-particles a zinc sulphide phosphor is used. This is deposited on a perspex plate mounted on the end of the photomultiplier tube; the phosphor is covered with a thin aluminium foil, which allows α-rays to pass through, but

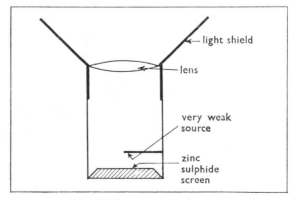

Fig. 17.5   A spinthariscope

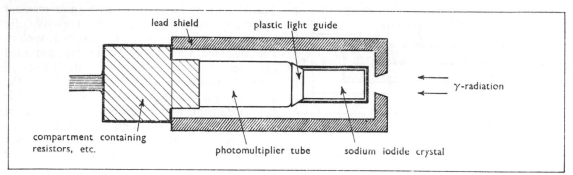

Fig. 17.6   A scintillation counter for γ-rays

serves to exclude light from the system. For β-particles a crystal of anthracene is commonly employed as a phosphor. Fig. 17.6 shows a typical design of scintillation counter for detecting γ-rays. The phosphor in this case is a crystal of sodium iodide containing a trace of thallium. The tube and phosphor are enclosed in a light-tight case (which must also be air-tight, since the crystal would deliquesce in contact with moist air). The crystal is a dense material, and a high proportion of the incident γ-rays are absorbed and detected by this instrument.

The scintillations caused by the α-particles in zinc sulphide are used to provide luminous dials for clocks and watches. The numbers and other markings on the dial are made of a mixture of the phosphor with a suitable radioactive substance. The luminosity is the combined effect of thousands of scintillations. This is clearly seen if part of the dial is viewed in the dark through a low-power microscope.

The cloud chamber

In this device the actual tracks of the α-rays are made visible, thus providing an even more vivid demonstration of their particulate nature. A common form of cloud chamber is shown in Fig. 17.7. The walls and top of the chamber are made of a transparent plastic. Round the top of the walls is fixed a strip of felt which is kept soaked with a suitable volatile liquid, such as ethanol. The air in the chamber is therefore saturated with the ethanol vapour. The floor of the chamber consists of a metal plate, which is cooled by a layer of 'dry ice' (solid carbon dioxide) packed beneath it. The ethanol condenses on the plate, and there is thus a steady diffusion of vapour from the top to the bottom of the chamber. A short distance above the metal plate the vapour passes through a narrow region in which it becomes supersaturated. In this condition it will condense on any ions that happen

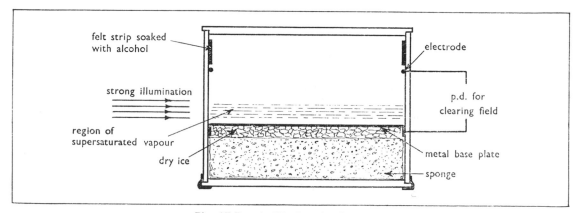

Fig. 17.7   A diffusion cloud chamber

to be present. A string of droplets therefore forms along any track in which ionization occurs, and in this way the paths of any ionizing particles are made visible. To see the tracks clearly it is necessary to illuminate brightly the active region in the chamber, and the base-plate needs to be painted black to cut out background reflections. Also the ions must be swept out of the chamber by an electric field so that the space is continually being cleared for the detection of fresh ionizing particles. The clearing field may be provided by joining a suitable p.d. between the base-plate and another electrode mounted near the top of the chamber. Alternatively, the plastic top may be rubbed with a duster, when the static charge often produces a sufficient clearing field.

In another design of cloud chamber the super-saturation is produced by a sudden reduction of pressure. To effect this the chamber is connected to a cylinder and piston (a bicycle pump with the valve reversed is often used). When the piston is pulled back, the sudden expansion causes a fall in temperature, and the vapour becomes super-saturated. In this design of cloud chamber the tracks are only visible for an instant just after the expansion.

When a suitable radioactive source is mounted in the chamber, the tracks of α-particles are strikingly shown up. Their range in air and the manner in which they are stopped by thin foils shows that the tracks are indeed those of α-particles such as we have already detected by other means. The ionization produced by an α-particle is always very heavy; detailed measurements show that each particle produces about

3000 ion pairs per mm of its path in air at s.t.p. It is therefore not surprising that the energy of the particle is rapidly dissipated, bringing it to rest in the short distance that we observe. The paths of the particles are seen to be almost exactly straight, though small deflections of about 1° are fairly frequent (Fig. 17.8a and Plate A.1). The patient observer, who watches a chamber for several minutes, may be rewarded by seeing an α-particle deflected through a much larger angle (90° or more). The study of the large-angle deflections of α-particles provided Rutherford with the evidence he needed to propound the nuclear theory of the atom (p. 332).

It is also possible to use a cloud chamber to observe β-rays. Again we find a series of clearly marked tracks, and it seems that β-rays also must be regarded as particles. The ionization along the track of a β-particle is much less heavy than for an α-particle, and the path is only thinly marked out by a line of droplets; it vanishes within a fraction of a second of being formed. But with a magnifying glass focused on the right region of the chamber an alert observer will manage to see the occasional track. Satisfactory observation is really only possible by photographic means. The appearance of β-particle tracks is shown in Fig. 17.8b and Plate A.2. The paths of the particles are very far from straight; they seem to suffer frequent small deflections and occasional large deflections of 90° or more. The deflections become more and more frequent as the speed of the particle falls; and the end of the track is usually very tortuous.

A beam of γ-rays shows up in a cloud chamber in the same manner as a beam of X-rays

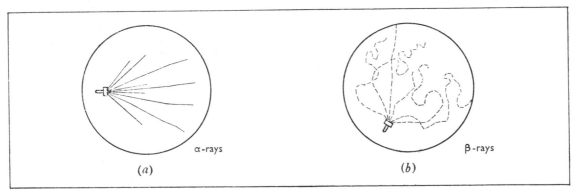

Fig. 17.8   Cloud chamber tracks: (a) α-rays; (b) β-rays

(Plates A.2 and A.3). In this case there is no clear line of droplets marking the path of the beam; but a number of short tracks are observed, resembling those of β-particles, each track originating in the line of the beam. This is quite different from the tracks of α- or β-particles, which only rarely show 'whiskers' of the kind that characterize the paths of γ-rays. It is difficult to observe these tracks in the conditions of a school laboratory.

Detection by photographic films

The tracks of α, β and γ-rays may also be studied in special photographic films. The emulsion reduces the range of α-particles to about 0.02 mm, and even β-particle tracks are only about 1 mm in length. A microscope must therefore be used to observe them. With ordinary photographic plates this method is not satisfactory, since the grains of silver bromide are not sufficiently close together to show up the tracks clearly. But special 'nuclear plates' are obtainable, in which the emulsion is thicker and contains a higher density of silver bromide grains. These have been used extensively in the investigation of radiations of various kinds. The method has the advantage of automatically making a permanent record of the 'events' studied. Also the tracks are so small that an enormous number of events can be recorded on one plate—sufficient sometimes to keep an experimenter busy at the eyepiece of a microscope for months on end!

17.3   The Geiger-Müller tube

This instrument is probably the most versatile

and useful of the devices available for detecting radiations from radioactive substances. Like the ionization chamber and the spark counter it is actuated by the ionization of the gas it contains; but it is far more sensitive for this purpose than either of the other devices. It is essentially a form of discharge tube, containing gas at a pressure of about 10^4 N m^{-2}; it is operated at a p.d. somewhat less than that which would produce a continuous discharge in it. A typical design is shown in Fig. 17.9a. The anode consists of a fine wire which runs along the axis of the cylindrical cathode. A large electric field is therefore produced in the immediate vicinity of the anode; and in this region any free electrons are sufficiently accelerated to cause further ionization. The process is cumulative, and a small amount of initial ionization can give rise to a considerable 'avalanche' of electrons. The electrons, being very light, are collected almost at once by the anode, leaving behind a space-charge formed by the more massive and slow-moving positive ions. In a short time ($\sim 10^{-6}$ s) the space-charge becomes sufficiently dense to cancel the electric field round the anode; the ionization process then ceases, and the positive ions are drawn away by the field to the cathode. Thus any ionization of the gas in the tube triggers off an appreciable pulse of current. When the right p.d. is used, the charge collected by the electrodes is independent of the amount of the original ionization—a single ion pair may be sufficient to initiate a full-scale pulse.

The G-M tube (as it is usually called) is connected in the circuit shown in Fig. 17.9b. When an ionizing particle enters the tube, the resulting pulse of current causes a corresponding pulse of

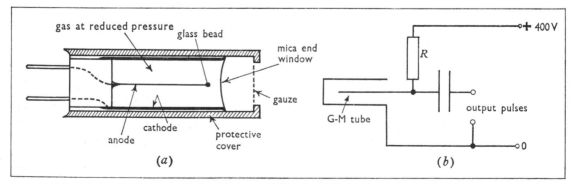

Fig. 17.9   (a) An end-window Geiger-Müller tube; (b) the circuit in which the tube is used

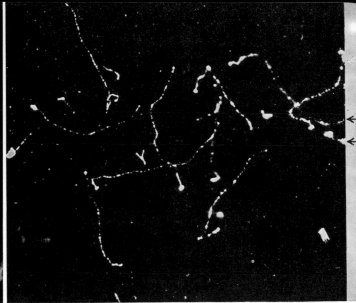

A.1 A typical fan of α-ray tracks in a cloud chamber. The tracks are almost all straight, and heavy ionization is produced along them (p. 319). (C. T. R. Wilson)

A.2 Cloud chamber tracks produced by a beam of X-rays entering the chamber from the right. The tracks start in the path of the beam and are identical with those of weak β-rays (p. 319). Similar tracks are caused by γ-rays. Their lengths depend on the wavelength of the radiation involved. In this case the K-series X-rays from silver were used, and the tracks are about 1·5 cm long (energy 24 keV). (C. T. R. Wilson; print prepared by W. H. Andrews)

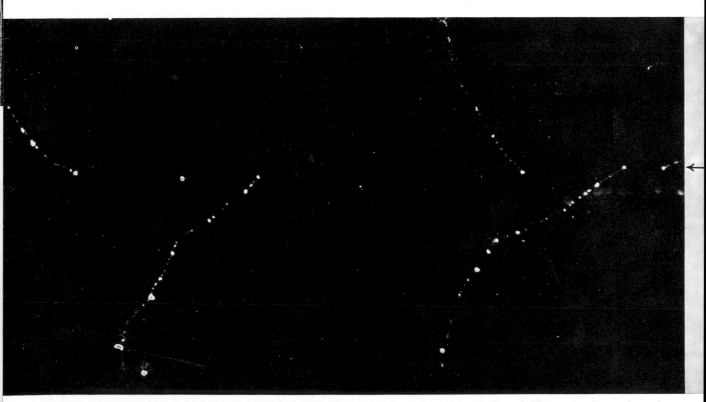

A.3 In this photograph a very narrow beam of X-rays enters the chamber from the right, as indicated. The tracks observed are those of electrons ejected from the molecules of the air in the path of the beam. There is nothing to mark the path of an X-ray photon between the source and the point at which it gives up its energy to a single electron (p. 331). (C. T. R. Wilson (1923), *Proc. Roy. Soc. A*, **104**, plate 5)

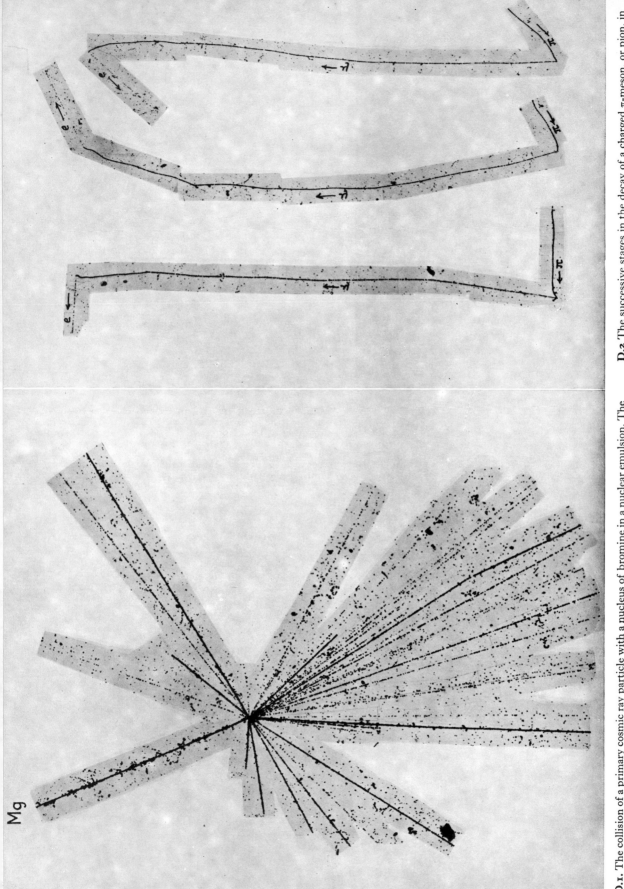

D.1 The collision of a primary cosmic ray particle with a nucleus of bromine in a nuclear emulsion. The incoming particle is probably a magnesium nucleus of energy about 15 GeV (p. 346). This is completely dissolved into its component nucleons, and the bromine nucleus is shattered into several fragments. At still higher energies collisions of this sort produce π-mesons, or pions (p. 352). (Dainton & Kent, Bristol University, 1950)

D.2 The successive stages in the decay of a charged π-meson, or pion, in a nuclear emulsion. First the pion decays into a muon and a neutrino (which, being uncharged, leaves no track). Then the muon decays into an electron and two further neutrinos (p. 352). (Bristol University, 1949)

p.d. across the resistance R in series with it. This is amplified and registered by a suitable detecting device.

It is obviously important that only one pulse should be registered for each ionizing particle entering the tube. Unfortunately, the positive ions formed in the gas may well reach the cathode with sufficient energy to eject secondary electrons from it; these then cause further avalanches of electrons, and a whole train of pulses may be triggered off by only one original particle. One way of preventing this is to make the resistance R in series with the tube very large ($10^9 \Omega$, say); the time taken for the p.d. across the tube to return to its original value after a pulse of current is then relatively long (10^{-2} s)—long enough for all the ions to be collected by the cathode while the p.d. is still at a low value. However, this technique is not very satisfactory, as the tube remains insensitive for such a long time after each pulse; it is therefore only suitable for low rates of counting. A better method is to include in the filling of the tube a small quantity of a suitable vapour that will absorb the energy of the positive ions before they can cause secondary electron emission; such substances are known as *quenching agents*. Various organic compounds have been used—e.g. ethanol. These are slowly destroyed by the action of the tube, and the quenching action becomes ineffective after some 10^8 counts. Such counters require operating p.d.'s of about 1500 V. Another method is to include a small quantity of a halogen vapour as quenching agent—e.g. chlorine or bromine. A counter of this kind works on a p.d. of about 400 V, and there seems to be no definite limit to the number of counts it can make. The interval during which these tubes are insensitive to the arrival of further particles is about 10^{-4} s, a quantity known as the *dead time* of the counter. However, the time taken for the tube to recover completely from a discharge is usually about 3×10^{-4} s. This is known as the *recovery time*; during this interval any pulses delivered by the tube are likely to be of reduced amplitude and may not be registered by the recording equipment. This means that there is always a proportion of the particles passing through the tube that are not counted. At low rates of counting this effect is insignificant. but when the counting rate is more than a few thousand per minute a correction for the recovery time should be made to the readings.

It is possible to design a Geiger-Müller tube to detect α-particles, but this is difficult because of the extreme delicacy of the window that must be provided for the particles to enter. An effective thickness of about 0.02 kg m^{-2} is all that can be allowed; and even supported by a metal grill it is very easily damaged. To detect β-rays, a rather thicker window can be used— 0.3 kg m^{-2} is a common figure. It is usually made of mica or glass, and is still thin enough to be treated with great care.

A modified design of G-M tube is often used when it is required to detect γ-rays only. The thin window can be dispensed with, since γ-rays penetrate readily through any part of the tube. Indeed the cathode and the walls of the tube are often deliberately made thicker than necessary in order to provide the maximum chance of a γ-ray ejecting an electron to initiate the discharge. Even so the γ-rays are so weakly ionizing that they have only a small chance of producing any ionization at all inside the tube. In most cases not much more than 1% of the γ-rays passing through the tube are actually counted. A scintillation counter has a very much better performance in this respect.

Recording electrical impulses

(*i*) *With headphones.* The G-M tube may be connected to an amplifier and a pair of headphones or loudspeaker. Each pulse is then detected as an audible click. At very low counting rates it is possible to note the individual clicks; but at higher speeds they merge into a continuous roar, and the method is then only of use for qualitative observations—we can at least detect the presence or absence of radiations by this means.

(*ii*) *With a scaler.* The exact number of pulses occurring in a given time interval may be recorded by using an electronic counting device, known as a *scaler*. Many different forms of circuit have been used for this purpose; but one of the most commonly encountered is that employing a special gas-filled discharge tube, known as a *decatron* (Fig. 17.10). The anode of the tube

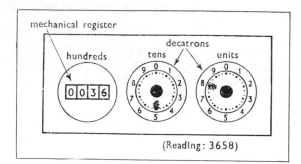

Fig. 17.10   The panel of an electronic scaler

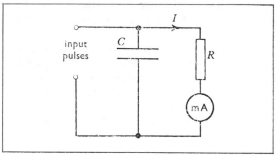

Fig. 17.11   The principle of a ratemeter

is the circular metal plate in the centre of its end face. This is surrounded by 10 separate cathodes, each consisting of a single metal wire. The current through the tube is limited to such a value that the discharge can cover only one of the cathodes at any one time. The glow in the tube shows clearly which cathode is in use. The counting process consists in causing the discharge to transfer from one cathode to the next for each pulse delivered to the apparatus. This is effected by applying the pulse to sets of intermediate cathodes, called *guide electrodes*. The mimimum transfer time is about 10^{-4} s, and this limits the speed of operation; but the circuit works reliably at rather more than a thousand counts per second. The 'zero' cathode is brought out to a separate terminal. When the discharge reaches this, a pulse of p.d. is produced, which can be used to operate a second decatron unit. One pulse is thus passed on to the second decatron for every 10 pulses delivered to the first. At the output of the second decatron the pulses are usually coming sufficiently slowly to operate an electro-mechanical register (with a scale resembling the milometer of a car); and this is more economical than using further sequences of decatrons. The first two digits of the total count are then read off from the positions of the glows in the decatrons, and the remaining digits (hundreds, thousands, etc.) from the mechanical register (Fig. 17.10).

(*iii*) *With a ratemeter.* Sometimes it is only necessary to know the *average rate* at which pulses are delivered by the G-M tube; and a ratemeter is designed to give a direct reading of this figure. The pulses are first amplified, and an electronic device is employed to ensure that the

pulses at the output of the amplifier are all of the same amplitude; these are then fed to a capacitor–resistor combination known as an *integrating circuit* (Fig. 17.11). Each pulse delivers a fixed quantity of charge to the capacitor C, and this then leaks away slowly through the resistor R. The time constant of C and R is arranged to be large compared with the probable interval between pulses. The p.d. across the capacitor then builds up to a steady value which is proportional to the number of pulses arriving per unit time. Thus, if there are n pulses per second, each delivering a quantity of charge Q to the capacitor, then the average current I registered by the meter in series with R is given by

$$I = nQ$$

It is usual to calibrate the meter to read the average number of counts per minute directly. Because of the random nature of the counting process there is always a certain amount of 'wandering' of the meter, due to statistical fluctuations of the count rate. This is more pronounced when the time constant (CR) is low (1 s, say). Steadier readings are obtained by increasing the time constant; but it is then more tedious taking the readings, since we have to wait correspondingly longer for the meter to settle down to its equilibrium reading.

The characteristic curve of a G-M tube

It is important to operate a Geiger-Müller tube at the correct p.d. This is found experimentally as follows. The tube is mounted at a fixed distance from a suitable radioactive source; we can then assume that a constant number of particles enters the tube per minute (subject to statistical fluctuations). The count rate is recorded for

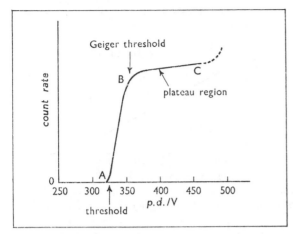

Fig. 17.12   The characteristic curve of a G-M tube

different values of the applied p.d. Up to a certain value of the p.d. (known as the *threshold p.d.*) no counts are recorded at all, since the amount of gas amplification in the tube is not enough to give pulses of sufficient magnitude to be detected. Just above the threshold the count rate steadily increases with applied p.d. In this region of p.d.'s (between A and B in Fig. 17.12) the magnitude of the pulse developed in the tube depends on the initial ionization, and this varies from one incident particle to another; only some of the particles give pulses of sufficient magnitude to be recorded by the scaler or ratemeter. However, beyond a certain point (the *Geiger threshold*) there is a range of p.d.'s in which the count rate is almost constant; this is known as the *plateau region* (between B and C in Fig. 17.12). The pulses are now all of the same magnitude, and every incident particle that produces any ionization at all is being registered. This is the right range of p.d.'s in which to operate the G-M tube. If the p.d. is raised above the plateau region, the quenching of the pulses is no longer fully effective, and one incident particle may start a whole train of pulses. The count rate therefore increases abruptly. The p.d. should not be raised much above this point, or there is a danger of a continuous glow discharge starting, which could damage the tube permanently. In a halogen quenched tube the plateau region usually extends over about 100 V, so that any operating p.d. in the middle of this range may be used. Actually there is a slight rise of count rate

through the plateau region—usually between 5 and 10%; this is apparently due to effects arising in the non-uniform fields at the ends of the anode wire. It is therefore important to maintain the tube p.d. constant during any particular experiment, although the exact value is not critical.

The background count

Even without any special radioactive source in the neighbourhood a stream of background radiation is recorded by a Geiger-Müller counter. This is due partly to radioactive contamination of the apparatus and its surroundings and partly to cosmic radiation entering the earth's atmosphere from outer space. The background count usually amounts to between 20 and 50 particles per minute, though with adequate lead shielding it can be reduced to about 10 per minute. When we are investigating the emissions from a radioactive source, it is necessary first to record the background count, and then to subtract this from all the subsequent readings. To obtain satisfactory results we must ensure that the total count in a given interval of time is substantially greater than the background count in that time. At high rates of counting (over 5000 per minute, say) the contribution from the background radiation is obviously negligible, and this correction can be ignored.

Statistical fluctuations

Because of the random nature of the emission of particles by a radioactive source, the count recorded by a Geiger counter for a given time interval fluctuates from one reading to another. The amount of these fluctuations is described in statistical theory by the *standard deviation* of the readings. This is a figure that indicates the deviation from the average count to be expected in any one reading. It is shown in works on statistics that for a random process such as we are considering the standard deviation σ in the count obtained in a given time interval is given by

$$\sigma = \sqrt{N}$$

where N is the average count obtained in that time. This result is readily confirmed by experiment, and serves to establish that the emission of the particles is indeed a random process.

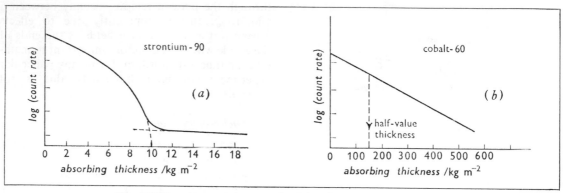

Fig. 17.13   Absorption curves for (a) a β-source, (b) a γ-source

Thus, suppose the background count on some occasion averages 50 per minute. The standard deviation for a count of 50 is expected to be $\sqrt{50}$, i.e. about 7. Thus the actual values obtained in a series of 1-minute counts will mostly lie between 43 and 57, though about one-third of the readings will lie outside this range. However, a deviation more than twice the standard value will be quite rare—it should occur for only about 1 in 20 of the readings. If now a radioactive source is held near the tube so that the count rate averages 10 000 per minute, the standard deviation for 1-minute counts should now be $\sqrt{10\,000}$ or 100; and most of the readings will now be found to lie between 9900 and 10 100. If enough 1-minute counts are made in each case, the correct statistical procedure may be applied to calculate the standard deviation; and the results will be found to agree with the predicted values.

These inevitable statistical fluctuations limit the accuracy that can be obtained in any observation of the count rate. Thus, if the total count recorded is only 100, we must expect a standard deviation of 10; and the accuracy of the measurement cannot exceed 10%. To obtain an accuracy of 1% the count must be continued long enough for the total count to reach at least 10 000. An accuracy of 0.1% can only be achieved by extending the total count to 1 000 000. In general

$$\text{fractional error} \approx \frac{\text{standard deviation}}{\text{total count}}$$

$$= \frac{\sqrt{N}}{N} = \frac{1}{\sqrt{N}}$$

The behaviour of a Geiger-Müller counter provides clear evidence that both β-rays and γ-rays are emitted by a radioactive source in an irregular and random manner—rather than as steady streams of radiation. Just as with α-rays it is difficult to avoid the conclusion that we are again dealing in both cases with *particles* ejected at random by the source.

Absorption curves

A G-M tube and scaler or ratemeter provides a convenient means of investigating the absorption of β- and γ-rays by layers of matter. Fig. 17.13 shows the absorption curves obtained with two sources that emit one kind of radiation only. A strontium-90 source evidently emits β-rays only, while only γ-rays emerge from a cobalt-60 source. Sources that emit both β-rays and γ-rays—such as radium—show a discontinuity in their absorption curves, where the radiation detected changes from the mixture of β- and γ-radiation to γ-radiation only.

There is also an important difference in the *character* of the absorption curves for β-rays and γ-rays shown in Fig. 17.13. The β-rays are *completely* absorbed in all cases for a total thickness of about 10 kg m^{-2}. But γ-rays are never completely absorbed no matter how thick the layer of absorbing material may be. There is no maximum 'range' for γ-rays as there is for α- and β-radiation. We find instead that a given thickness of absorbing material always reduces the intensity of the γ-radiation by a certain fixed proportion. Thus the intensity of the cobalt γ-radiation is reduced by one-half by a layer of lead 1.25 cm thick. Twice this thickness reduces it to a quarter of the original intensity; three times to one-eighth, etc. This is another way of saying that the intensity I decreases *exponentially* through the layer of absorbing material, and may be expressed by the equation

$$I = I_0 \, e^{-\mu x}$$

where I_0 = the initial intensity, x = the thickness, and μ = a constant known as the *linear absorption coefficient* for the given γ-radiation. This coefficient is approximately proportional to the density of the material; it is therefore usually convenient to express the 'thickness' x in kg m^{-2} of material, and the absorption coefficient so obtained is then called the *mass absorption coefficient*, and is denoted by μ_m.

Taking logarithms on both sides of the above equation, we have

$$\log_{10} I = \log_{10} I_0 - \mu_m x \log_{10} e$$

A graph of log I against x should therefore be a straight line, which indeed is confirmed by experiment (Fig. 17.12b). The value of μ_m can be calculated from the gradient of the line. It is found that μ_m varies with the type of γ-radiation; for a high-energy γ-ray μ_m is smaller (i.e. the ray is more penetrating) than for a low-energy γ-ray.

The absorption of β-rays does not strictly follow an exponential law; and a graph of log I against absorbing thickness is not a straight line. However, the central part of such a curve is usually approximately straight, and we may use this to find an absorption coefficient for β-rays also. The value of μ is not in this case an exactly defined quantity; but this measurement sometimes provides a useful means of comparing the absorption of β-rays in different cases.

Once the absorption characteristics of the β-particles from a given source are known, we may use the results to measure the thickness of an absorbing sheet placed between source and detector. This is the principle of the *β-particle thickness gauge*, which is widely used in factories for monitoring the thickness of thin metal or plastic sheets. The source is placed on one side of the sheet of material emerging from the mill and a Geiger counter and ratemeter on the other. The ratemeter may be calibrated to read thickness directly. A continuous record of the output thickness is thus obtained without any part of the apparatus actually coming in

contact with the material. Similar instruments have also been designed using γ-radiation; these can be used with metal sheets several inches thick.

The inverse square law

It is found that the intensity of γ-radiation near a small source varies inversely with the square of the distance from the source. This may be tested by measuring the γ count rate I with a G-M tube at varying distances d from the source. As always the background count must be subtracted from each reading to give the true γ count rate. Thus, we find

$$I \propto \frac{1}{d^2}$$

and a graph of I against $1/d^2$ is a straight line through the origin.

It is of some interest to consider why the inverse square law holds for γ-radiation. A similar law holds also for some other emission processes, e.g. light, radio waves and sound, provided the distribution of the 'radiation' is not modified by reflection, refraction, etc. A little thought shows why this must be so. Consider a point source O emitting P units of radiation per unit time uniformly in all directions (Fig. 17.14). At a distance r from the source this radiation is spread over the surface of a sphere of this radius

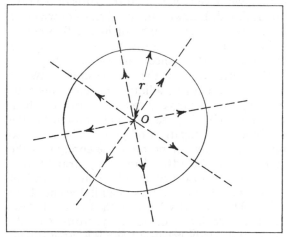

Fig. 17.14   The basis of the inverse square law for the intensity of some sorts of radiation

(provided there has been no absorption). The intensity I is the rate of incidence of the radiation per unit area. Therefore at the distance r

$$I = \frac{P}{4\pi r^2}$$

i.e.
$$I \propto \frac{1}{r^2}$$

which is the inverse square law.

The result depends upon three conditions being satisfied:

(*i*) The source must be small compared with the distances involved.

(*ii*) There must be negligible absorption of the energy in the intervening medium.

(*iii*) There must be rectilinear propagation of the 'radiation'—i.e. the 'rays' must proceed in straight lines. (The law could still hold for radiation that diffused outwards following random paths, but only if we were dealing with a source radiating uniformly in all directions. If the radiation is confined to a narrow beam, it must travel radially, otherwise the energy will not remain in the beam; and the inverse square law breaks down.)

For example, the law holds fairly precisely for a small source of light in air (away from reflecting objects). The result may be tested with a small bulb and photocell. But it breaks down if the light has been refracted in a lens or scattered from a white ceiling, so that the paths of the rays are deviated. Likewise the law breaks down close to a large source of light such as a fluorescent lighting tube; and certainly it does not apply in a strongly absorbing medium, such as a fog!

The verification of the inverse square law for narrow beams of γ-radiation therefore depends on the γ-rays being very weakly absorbed, which we know to be the case; but it also establishes the rectilinear propagation of γ-rays. The law also holds for α-rays at points close enough to the source, since they travel in nearly straight lines and are not absorbed at all before the end of their range. But for β-rays in air it does not hold at all. This is partly because the tracks of these particles deviate considerably from straight lines; and partly because there is a large spread of energies represented in a β-source, and the slower particles are completely absorbed close to the source.

In a vacuum the inverse square law holds for β-radiation also. Both the absorption of β-particles and the tortuous tracks they normally follow arise from their interactions with matter. In a vacuum the tracks are straight and there is no absorption; the β-rays do not just come to a certain point and peter out. The deflections they normally suffer, and their eventual absorption are due to collisions with the molecules of the air.

17.4   Identifying the particles

We have seen how the three kinds of radiation from radioactive substances may be distinguished from one another by the different ways in which they are absorbed in passing through matter. All three kinds give the impression of being particles. In order to establish what these particles consist of we need to investigate more of their properties; and in particular we must find out how they are deflected in magnetic and electric fields.

β-rays

A preliminary experiment shows that β-rays are readily deflected by a magnetic field. One way of doing this is with the arrangement in Fig. 17.15. A β-emitting source (strontium-90) is mounted at the end of a tube whose walls are thick enough to stop the β-radiation completely.

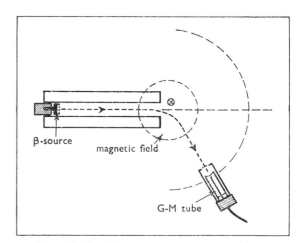

Fig. 17.15   The magnetic deflection of β-rays

A Geiger-Müller tube is supported in a clamp so that it can be moved to various positions along an arc, as shown. The β-radiation is thus virtually confined to a narrow beam near the axis of the tube—which may be confirmed by observing the count rate for various positions of the G-M tube. Outside the beam the count rate falls to a low value about equal to the background count. However, when a magnetic field is applied at the mouth of the tube (at right-angles to the plane of the diagram), the beam is deflected to one side; the count rate falls to a low value in the central position and the β-particles may now be detected by moving the tube to the right position on one side or the other. The direction of the deflection suggests that the β-rays are *negatively* charged particles. Thus in Fig. 17.15, if the magnetic field is *down* into the paper, the beam is deflected towards the *lower* edge of the diagram. Fleming's left-hand rule (p. 65) shows that the beam constitutes an electric current directed (in the conventional sense) *towards* the source; the particles emerging from the source must therefore be negatively charged. Quite moderate magnetic fields are found to produce large deflections. A flux density of 10^{-2} T is quite sufficient for this demonstration (see also Plate B.1).

The β-particles may also be deflected by an electric field, though this is harder to demonstrate, since rather large fields must be used. Again the deflection is consistent with the particles being negatively charged. Thus, when the beam is passed between two metal plates maintained at a large p.d., the deflection is towards the *positive* plate.

It is also possible simply to collect the β-particles on a suitable conductor and show that this has acquired a negative charge. But to do this the conductor must be completely embedded in an insulating material so that the air ionized by the β-radiation does not come in contact with it. With *strong* radioactive sources the charge collected can be identified with a suitable electrometer.

In these experiments there is not much point in making precise measurements of the deflections, since the β-particles are being slowed down and scattered by collisions with air molecules along their paths; and little useful information is gained. But using vacuum apparatus such

measurements enable the velocities and specific charges of the particles to be found. We shall not describe the methods here, since they are not readily repeated in a school laboratory; and in any case the principles involved are the same as those used in studying cathode rays (p. 229) and mass spectra (p. 248). The results show that:

(*i*) The β-particles from a given source have all possible velocities from zero to a definite maximum.

(*ii*) The specific charge of the β-particles is found to *decrease* slightly with increasing velocity. This variation is entirely consistent with the assumption that the charge of the particles remains constant, while their masses *increase* with velocity in the manner predicted by the theory of *Relativity* (p. 251). At low velocities the specific charge is identical with that of the electrons produced in the thermionic or photoelectric effects. There is therefore little doubt that β-particles are *electrons* emitted with speeds up to two-thirds the velocity of light. Their energies may be anything up to several million electron-volts.

α-rays

When radioactivity was first discovered, it was concluded that the least penetrating component of the radiation—the α-rays—could not be deviated by electric or magnetic fields. Fields sufficient to produce large deflections of cathode rays and β-rays could give no detectable deflection of the α-rays. However, Rutherford eventually succeeded in devising very sensitive techniques for observing the minute deflections that do actually occur, and thus managed to show that α-particles are relatively massive and carry a *positive* electric charge. One piece of apparatus he used is shown in Fig. 17.16. A layer of radium was placed at the bottom of a cavity in the base of a gold-leaf electroscope. Above the radium was fixed a stack of parallel metal plates, so that the α-rays could enter the electroscope only through the narrow gaps (about 1 mm wide) between them. The rate of fall of the leaf was observed, and could be taken as a measure of the ionization current produced by the α-rays. A strong magnetic field was

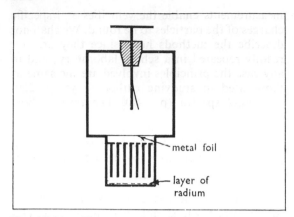

Fig. 17.16   The magnetic and electric deflection of
α-rays

then applied at right-angles to the plane of the
diagram, and it was found that the ionization
current decreased; the greater the magnetic field,
the more slowly the gold leaf collapsed. Presumably, many of the particles that had previously passed through the gaps were now being
deflected to strike the plates. To determine the
direction of the deflections the tops of the gaps
were each masked by a small projection on one
side, as shown. It was then found that a much
greater reduction in ionization current occurred
with the magnetic field directed *down* into the
diagram (in Fig. 17.16) than in the reverse direction. By Fleming's left-hand rule (p. 65) the
stream of particles emerging from the source
must be *positively* charged (see also Plate
B.2).

The same apparatus was modified to detect
the deflections of the particles by electric fields.
Alternate plates were connected together like an
air capacitor, and a p.d. was applied between the
two sets of plates. The electric field so produced
would certainly have interfered with the functioning of the electroscope. To prevent this the
floor of the case above the plates was covered
with a fine metal foil, which acted as an electrostatic screen; but this was thin enough to allow
the α-particles to pass through. Again a reduction of the ionization current was observed
when the electric field was applied.

Knowing the magnitudes of the fields used in
these experiments, Rutherford was able to make
preliminary calculations of the specific charge
and velocity of the α-particles. These enabled
him to design apparatus to work in a vacuum
for measuring more precisely the deflections
of the particles. In principle the methods were
similar to those employed for finding the specific
charge of cathode rays (p. 230); but because of
the small deflections of the α-particles the technical difficulties were considerable.

A qualitative experiment of this kind can be
performed in a school laboratory using the
apparatus shown in Fig. 17.17. It serves at least
to show how little α-particles are deflected by
even a large magnetic field; also it gives the sign
of the charge they carry. Some reduction of
pressure is needed in the system so as to increase the range of the α-particles sufficiently for
the path length required. The particles can be
detected with a solid-state particle detector,
which can be used without damage in a vacuum
system provided the changes of pressure to
which it is subjected are not too sudden.

In the absence of the magnetic field the flexible
tube is bent into a shallow arc just sufficient to
cut off the α-particles from the detector. The
magnetic field is now applied at right-angles to
the curved part of the tubing; if the field is

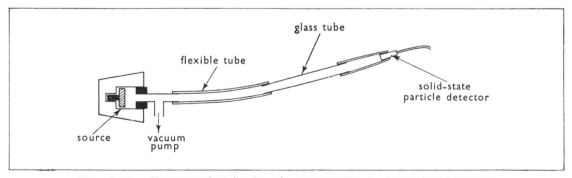

Fig. 17.17   The magnetic deflection of α-rays travelling in air at reduced pressure

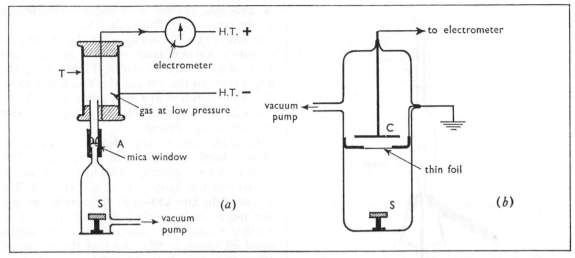

Fig. 17.18   Measuring the charge on the α-particle: (a) counting the particles emitted; (b) measuring the total charge carried

strong enough and is in the right sense, the detector will once more register the arrival of α-particles. To achieve the direction of deflection indicated in Fig. 17.17 it will be found that the field needs to be *down* into the diagram. By Fleming's left-hand rule (p. 65) the beam of α-particles from the source must therefore carry *positive* charge.

Rutherford's detailed experiments yielded two important results:

(*i*) With some radioactive sources *all* the α-particles were found to be emitted with the same velocity; and even when this was not the case, the α-particles always fell into one of a few groups, each having a sharply defined velocity. This is sometimes described by saying that the α-particle energies form a *line spectrum*.

(*ii*) The specific charge of the α-particles from all sources was found to be about 5×10^7 C kg^{-1}. This is just half the value of the specific charge of the hydrogen ion obtained from experiments in electrolysis (p. 36). Rutherford therefore concluded that the α-particle was either a singly charged atom of mass number 2 or else a doubly charged atom of helium (of mass number 4).

To decide between these possibilities it was necessary to find some way of measuring the actual charge carried by an α-particle. This was

done by Rutherford and Geiger using two pieces of apparatus shown in Fig. 17.18. The tube T in (a) was an early form of Geiger counter, the prototype of the modern G-M tube. The α-particles from the source S entered the counter through a small aperture A covered with a thin film of mica. The count rate with their equipment had to be limited to a few pulses per minute, but knowing the area of the aperture A and its distance from the source the measurements enabled the total number of α-particles emitted into the space above the source to be calculated. This figure is half the total number of particles emitted, since the same number presumably pass downwards into the supporting plate. The same source was then mounted opposite the collecting plate C in an evacuated enclosure (Fig. 17.18b). A magnetic field was applied across the apparatus sufficient to deflect any β-particles back into the source. The only charge collected by the plate C was therefore then carried by the α-particles, and this was measured by a sensitive electrometer. The charge q carried by one α-particle was then given by

$$q = \frac{\text{charge collected per second}}{\text{no. of particles emitted per second}}$$

The value obtained was 3.1×10^{-19} C, which is almost double the charge on a hydrogen ion. Rutherford therefore decided that the α-particle must be a doubly charged helium atom. The

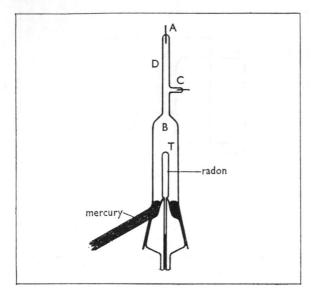

Fig. 17.19   Rutherford and Royd's experiment to show that α-particles are helium nuclei

matter was settled by the celebrated experiment of Rutherford and Royd depicted in Fig. 17.19. A small quantity of the radioactive gas *radon* was placed in the central tube T; this had sufficiently thin walls to allow the α-particles emitted by the radon to pass through into the evacuated outer compartment B. Any gas which collected in this space could be compressed up into the discharge tube D by raising the level of the mercury. After six days enough gas had collected in B to enable a discharge to be passed between the electrodes A and C. The light from this was examined with a spectrometer, and the usual spectrum of helium was revealed. There remained the possibility that the helium was present already mixed with the radon in the tube T and had diffused through into the outer compartment. However, when the tube T was filled with helium instead of radon, no helium could be detected in the discharge tube even when several days had been allowed for possible diffusion to occur.

A further confirmation of this result is obtained by studying the tracks of α-particles in a cloud chamber filled with helium instead of air. An occasional collision is observed of an α-particle with a helium nucleus. The helium nucleus is projected forwards while the incident α-particle

is deflected, giving a Y-shaped appearance to the tracks (see Plate C.2). The tracks in the two arms of the Y both appear to be of identical density compared both with one another and with other α-particle tracks. Furthermore the angle between the arms of the Y is found to be 90°. Consideration of the dynamics of colliding particles shows that this only happens with particles of *equal* mass.

Collision tracks may also be observed in cloud chambers filled with other gases. For instance with hydrogen, collisions occur between α-particles and hydrogen nuclei (protons). The tracks of the knocked-on protons are longer and are more thinly marked by droplets than α-particle tracks; and measurement of the angles involved shows in this case that the masses of the colliding particles are in the ratio 4 : 1, in agreement with Rutherford's supposition (see Plate C.1). With a school cloud chamber it is not possible to use gases other than air. But the tracks of protons knocked on by α-particle collisions can be observed by covering an α-particle source with a sheet of polythene (Fig. 17.20). This material is rich in hydrogen atoms, presenting plenty of scope for the type of collision we are seeking. The source is placed to one side of the cloud chamber space completely enclosed in a container, one side of which is formed by the polythene sheet. The sheet must be thick enough just to stop the α-particles completely. Proton tracks will then be observed coming from the polythene. If a pin-hole is made in the polythene, the occasional α-particle track will be seen emerging from this; the α-particle and

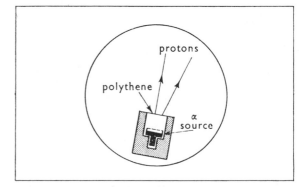

Fig. 17.20   Observing the tracks of protons knocked on by α-particles

proton tracks can then be compared both in appearance and length.

γ-rays

Early in the investigation of radioactivity it was suggested that γ-rays were identical with short wavelength X-rays; and this supposition has been fully confirmed by later work. The difference between γ-rays and X-rays seems to be one of origin rather than of nature; γ-rays arise in the nucleus, whereas X-rays are generated in the inner shells of the electron structure outside the nucleus, But the properties of the two forms of radiation appear in every respect to be identical:

(*i*) Neither form of radiation shows any deflection by electric or magnetic fields; nor is there any other evidence suggesting that they carry electric charge.

(*ii*) Both forms of radiation cause ionization of the air through which they pass; and in both cases the ionization is relatively slight compared with that produced by α-rays or β-rays.

(*iii*) The diffuse tracks they cause in a cloud chamber or a nuclear photographic plate are similar in character (p. 319, see also Plate A.2).

(*iv*) Both forms of radiation are very penetrating and show the same kinds of absorption characteristics in passing through matter.

(*v*) Both X-rays and γ-rays may be diffracted in the same way by crystals (p. 302); and their wavelengths may be found by this means.

There is thus very little room for doubt that γ-rays, like X-rays, are very short wavelength electromagnetic waves. While γ-rays are generally of shorter wavelength than X-rays, there is in fact a considerable overlap between the X-ray and γ-ray parts of the spectrum. From a given γ-source only certain sharply defined wavelengths are observed, which are characteristic of the type of nucleus concerned. In other words, like the visible and X-ray spectra of the *atom*, the γ-radiation from the *nucleus* forms a 'line' spectrum. This implies that the particles in the nucleus, like the electrons in the atom, can exist only in certain particular states, each of well-defined energy.

As with optical and X-ray spectra, the energy and frequency v of the radiation are connected by

$$\text{energy} = hv$$

where h = the Planck constant (p. 239). The quantities of energy liberated in nuclear processes are generally far larger than those arising from re-arrangements of electrons in the outer parts of an atom; a γ-ray photon is therefore of very high frequency (and so of very short wavelength). The energy carried by the photon is quite sufficient for it to be detected individually by a Geiger-Müller tube (although it may pass right through the tube without causing any ionization). The behaviour of a Geiger counter in a beam of γ-radiation is therefore a convincing demonstration of the particulate nature of electromagnetic radiation (discussed on p. 240 *et seq.*).

Like other forms of electromagnetic radiation, a γ-ray photon is able to give up its energy to a charged particle in its path; and it is this process that causes the ionization of the matter through which the radiation passes. By analogy with the equivalent optical effect this process is called *photoelectric emission*. The quanta of energy in γ-radiation are large enough to eject electrons from any level of an atom with a considerable amount to spare. An ejected electron then behaves just like a β-particle. This is shown up in a cloud chamber through which the γ-radiation is passing. The tracks revealed are those of β-particles each originating in the path of the beam. Most of these tracks represent the *complete* absorption of a γ-ray photon (see Plate A.3).

This may be contrasted with the process by which α-particles and β-particles are brought to rest. An α-particle has a relatively high probability of ejecting an electron from an atom, but loses only a minute proportion of its energy in the process; it may produce 10^5 ion pairs before its speed becomes too small to cause further ionization; and there is little variation in the lengths of the tracks of α-particles of given energy. A β-particle has a lower probability of ejecting an electron, but the events are sufficiently frequent for the line of droplets that marks its path in a cloud chamber to appear almost continuous. However, the path of a γ-ray is not marked at all—until the point at which it is absorbed by an electron. We can then infer its path in a straight line from the source to the start of the observed β-particle track.

It is in fact possible for the energy of a γ-ray photon to be only partially absorbed in a collision with an electron. The photon that emerges from the collision is then of lower energy (and therefore lower frequency) than the incident photon. This is known as the *Compton effect* after its discoverer. Thus when a γ-ray (or X-ray) beam is passed through matter the beam is to some extent scattered to the side; but in the process the frequency (and wavelength) of the scattered radiation is changed. This is a purely quantum effect that has no parallel in the classical theory of electromagnetic radiation.

A γ-ray photon is very penetrating, i.e. the probability of its being absorbed or scattered is very low. We can describe this by specifying the thickness of matter in which a γ-ray photon has an even chance of being absorbed; on average, half the photons emerge through this thickness. Thus, the intensity of the γ-rays from cobalt-60 is reduced to a half in passing through about 1 cm of lead; 2 cm of lead reduce it to a quarter, and so on. An α-particle has a certain maximum range because it is extremely unlikely not to make the 10^5 collisions that will bring it to rest within this distance. But there is no maximum range for γ-rays; all we can say is that a given thickness of matter will reduce the intensity in a certain proportion.

X-rays are produced when fast electrons are brought to rest in a target. It is therefore to be expected that β-particles will generate X-rays in any matter through which they pass. These X-rays are of such high energy that we may well class them as γ-rays. The effect is apparent when we are investigating the absorption characteristics of β-rays from a given source. There is a pronounced 'tail' at the end of the curve (Fig. 17.13, p. 324). This is due to the 'γ-rays' generated by the β-particles within the absorbing material. To obtain the true β-absorption curve it is necessary to produce back the tail of the curve and subtract the part of the total count rate caused by these γ-rays.

17.5 The nucleus

It is known that atoms are of such a size that in a piece of solid material they are more or less tightly packed together. When therefore an α-particle traverses a thin piece of metal foil, it must actually pass right through the interior of a large number of atoms. Any changes of direction that occur in the process give some indication of the nature of the forces that act on a charged particle *inside* the atom. In general the tracks of α-particles in cloud chambers and nuclear emulsions are very nearly straight; but a close inspection shows that they suffer frequent small deflections of 1° or less—particularly towards the ends of their paths, where their speed is much reduced. These deflections, small though they are, cannot be accounted for by collisions with the electrons in the atoms. The α-particle is more than 7000 times as massive as the electron, and the maximum possible deflection in such a collision is less than half a minute of angle. The deflections observed are therefore evidence of the interaction of the α-particle with the positively charged part of the atom, which is presumably associated with most of its mass. Acting under Rutherford's guidance, Geiger and Marsden therefore embarked on a detailed investigation of the deflection of α-particles in passing through thin metal foils. Most of the particles were scattered through angles of about 1°; but a small proportion (about 1 in 8000) were deflected through angles of more than 90°—in fact they appeared to be *scattered back* from the metal foil. In due course the same effect was found also in cloud chamber tracks. The same small proportion of α-particle tracks showed sharp deviations of 90° or more in collisions with atoms of gas in the chamber.

If the force acting on an α-particle was electrostatic in nature, it was evident that the field in at least some part of the atom was an exceedingly intense one. At that time (1910) it was generally supposed that the atom consisted of a sphere of uniformly distributed positive charge, which carried most of the mass of the atom; the electrons were supposed to be embedded within this. It was thought that the electrons must be stationary, since otherwise they would radiate energy until they came to rest (p. 242). This was affectionately called the 'plum pudding model' of the atom; and there was a certain amount of experimental evidence to support it. However, it was quite incapable of explaining the large deflections observed by Geiger and Marsden. The maximum possible electric field in the plum pudding atom would occur at the surface of the positively charged sphere; and this would only be sufficient to account for a deflection of about 1°. The chance of 100 deflections of this magnitude all occurring in the same direction in succession is inconceivably remote. Besides, cloud chamber photographs show that the large-angle deflections occur as single events, not as the result of multiple small-angle scattering. Rutherford therefore suggested that the positive charge of the atom and most of its mass must be concentrated in a small central nucleus. Assuming that the inverse square law for electrostatic forces holds for such short distances, the electric field close to the nucleus could then be of sufficient magnitude to account for the deflections observed. Rutherford worked out the consequences of his hypothesis, and deduced a mathematical formula predicting the proportion of α-particles that should be scattered through any given angle. Geiger and Marsden then set out to test the formula with the apparatus in Fig. 17.21. The radioactive source A was enclosed in a container with a slit S in the side; the emerging α-particles were thus confined to a narrow beam that fell normally on the metal foil F. The particles deflected through any given angle θ were detected by looking for the scintillations on the zinc sulphide screen Z attached to the microscope M. The box B was evacuated through the tube T so that the α-particles would not be affected by collisions with molecules of air. The source and foil were fixed to the central tube T, while the box B and the microscope could be rotated in an airtight joint so as to vary the angle θ. The results of the measurements agreed in every detail with Rutherford's theory; and there was therefore very little doubt that the nuclear model of the atom was essentially correct.

One of the quantities on which the proportion of scattered particles depends is the charge carried by the nucleus. Once the validity of the theory had been established the experiment became a means of finding this quantity directly. Geiger and Marsden's experiment yielded only low accuracy in this measurement; but later experiments on the same lines showed that the charge on the nucleus, expressed in electronic charge units, was exactly equal to the *atomic number*. This number had been introduced in the

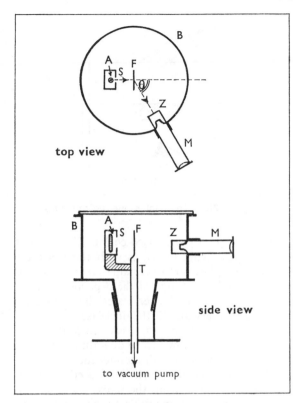

Fig. 17.21   Geiger and Marsden's investigation of the large-angle scattering of α-particles

first place to specify the position of an element in the periodic table, and it is thus a means of indicating the chemical properties of the element. Moseley's work had shown that the X-ray spectrum of an element also depended on its atomic number (p. 311). Rutherford's discoveries had finally demonstrated that the atomic number was to be identified with the *charge on the nucleus*. This charge presumably controlled the number and distribution of electrons in the atom, and in this way fixed its spectrum and chemical properties.

The scattering of β-particles

It is not easy to investigate the scattering of α-particles in a school laboratory. But the equivalent effect for β-particles is readily demonstrated with the arrangement in Fig. 17.22. When the metal plate is in the position shown, a substantial proportion of the β-particles emitted by the source are scattered back, and the G-M tube

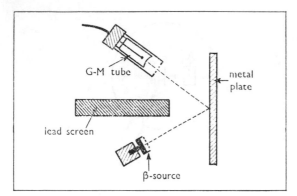

Fig. 17.22   The back scattering of β-rays

17.6 Nuclear transformations

With the picture of the nuclear atom clearly established, we are in a position to see what the process of radioactive emission of particles consists of. It seems likely that the source of the particles is actually the nucleus of the radioactive atom. The β-particles could conceivably arise from the electron structure of the atom, external to the nucleus; but the α-particles, which are helium nuclei, could not possibly do so. However, the point can be settled by considering the transmutations that should accompany the emission of these particles from a nucleus. If a nucleus ejects an α-particle, which carries a charge of 2 electronic units, its atomic number should *decrease* by 2; at the same time its mass number should *decrease* by 4, since that is the mass number of helium. Likewise, the emission of a negatively charged β-particle should *increase* the positive charge on the nucleus by one unit, and so raise its atomic number by one. But the mass of the β-particle is very small compared with the mass of the proton (or neutron)—less in fact than the small difference between the *relative atomic mass* of an isotope and the nearest whole number (which is its *mass number*). The emission of a β-particle therefore leaves the mass number of the atom *unchanged*.

Thus a radioactive element should be transmuting its atoms into those of another element according to the rules outlined above. There is abundant evidence that this is exactly what happens.

We shall consider how this works out by analysing the radioactive processes that occur in a radium source. The atomic number of radium is 88, and the mass number of its principal isotope is 226. Using the usual chemical symbol for radium, this information can be expressed symbolically:

$$^{226}_{88}\text{Ra}$$

The subscript indicates the atomic number, and the superscript the mass number. Since an α-particle is a helium nucleus, it can be described in the same notation by

$$^{4}_{2}\text{He}$$

Strictly speaking the information carried by the subscript is redundant; if the chemical element is known, a copy of the periodic table suffices to

registers a greatly increased count. The amount of back scattering increases with the thickness of the metal plate up to a certain maximum (equal to about half the range of the β-particles in the metal). The use of thicker plates than this does not then produce any further increase.

As with α-particles, the back scattering of β-particles can be shown to be due almost entirely to their interactions with atomic nuclei.

The scattering of β-particles is used in an instrument for measuring the thickness of a thin sheet, when it is impossible to have access to both sides of it—e.g. the wall of a tube or a layer of paint. One design is shown in Fig. 17.23. The β-emitting material is arranged in an annular groove in the brass plate P. The G-M tube G is mounted centrally to collect the particles scattered along the normal from the layer under test. A ratemeter is used to register the count rate, and its scale can be calibrated to read thicknesses direct.

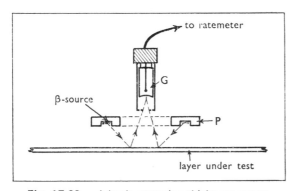

Fig. 17.23   A back-scattering thickness gauge

tell us also the atomic number. To specify a nucleus it is only necessary to state the element and its mass number. Sometimes therefore the above two isotopes are simply described as radium-226 and helium-4.

When a nucleus of radium-226 emits an α-particle, its atomic number falls to 86 and its mass number to 222. It therefore forms a nucleus of the noble gas, *radon*.

$$^{226}_{88}Ra \xrightarrow{\alpha} \,^{222}_{86}Rn$$

The evolution of radon from a quantity of radium is readily demonstrated, though the experiment is not safe enough to attempt in a school laboratory. If a solution of a radium salt is stored in a sealed vessel, within a few days a small bubble of radon appears at the top. This may be collected and analysed (by observing its spectrum in a discharge tube). Subsequently the radium salt will generate a fresh supply of radon. In fact the radium could be used as a source of radon for as long as it lasts—i.e. for some thousands of years.

Radon itself is a radioactive element which emits an α-particle, and so changes into an isotope of *polonium*—one of the radioactive substances discovered by the Curies,

$$^{222}_{86}Rn \xrightarrow{\alpha} \,^{218}_{84}Po$$

Polonium in its turn emits an α-particle and decays to form a radioactive isotope of *lead*.

$$^{218}_{84}Po \xrightarrow{\alpha} \,^{214}_{82}Pb$$

This lead isotope is a β-emitting substance, which therefore increases its atomic number by one when it decays, forming bismuth-214.

$$^{214}_{82}Pb \xrightarrow{\beta} \,^{214}_{83}Bi$$

The original radium-226 is in fact the start of a sequence of radioactive changes, involving 8 intermediate isotopes and terminating eventually on the stable isotope lead-206. A sample of originally pure radium, left for a long enough time, is found to contain all of these elements. Several of them are present in sufficient quantity to be detected by chemical means, though the existence of the shorter-lived substances can only be inferred from the radiations they emit.

The nature of the radiations observed from a radium source depends on whether the first decay product, radon, remains trapped in it or not. In practice care is taken to seal a laboratory source, as radon-222 and its decay products would be most harmful if they escaped into the air of the laboratory. Usually a radium source consists of a piece of foil of a radium-silver alloy. Almost all the radon then remains trapped in the foil. What we call a 'radium source' is really a mixture of isotopes of nine elements, including the parent radium. Between them they emit α-, β-, and γ-radiations of many different energies. A freshly purified sample of radium emits α-particles of one range only, together with some γ-radiation. But if it is kept with its decay products it soon produces β-particles also, and its α- and γ-emission increases by a factor of about 4 as well.

Since a γ-ray consists of electromagnetic radiation, it does not alter the charge of the nucleus from which it comes and affects its mass very little. It therefore leaves the composition of the nucleus unaltered. A γ-ray is found always to *follow* the emission of either an α-particle or a β-particle. When one of these particles emerges, it may leave the resulting nucleus in an excited state. Shortly afterwards the nucleus emits the surplus energy as a γ-ray. In a few cases it happens that the emission of an α- or β-particle takes away *all* the available energy, so that the resulting nucleus is left already in its ground state. There is then no emission of γ-radiation. But this is exceptional.

Artificial transmutations

The close agreement between the results of Rutherford's scattering experiments and his predictions on the basis of the nuclear model of the atom made the acceptance of the nuclear theory inevitable. However, when such experiments were performed with materials of low atomic number, discrepancies were revealed. It appeared that in a sufficiently close collision between an α-particle and a light nucleus the inverse square law for the electrostatic force between them broke down. Rutherford supposed that in such a collision the α-particle actually penetrated the nucleus; and so there was the possibility that an artificial nuclear transmutation might be brought about by this means.

The first effect of this kind was discovered in

the bombardment of nitrogen atoms by α-particles. It was found that the passage of α-particles through nitrogen generated a new kind of 'radiation' which was about four times as penetrating as the α-particles themselves. The 'radiation' caused scintillations on a fluorescent screen, though the flashes of light were noticeably weaker than those produced by α-particles. Radiation with similar properties had previously been observed when α-particles were passed through hydrogen. In this case there was little doubt that it consisted of *protons* (hydrogen nuclei) knocked on by the impact of the α-particles (p. 330). Rutherford therefore concluded that some of the incident α-particles had reacted with nitrogen nuclei, forming a new nucleus and causing the ejection of a proton.

$$\underset{\text{(α-particle)}}{^{14}_{7}\text{N} + {^{4}_{2}\text{He}}} \longrightarrow ? + \underset{\text{(proton)}}{^{1}_{1}\text{H}}$$

The requirement of balancing the atomic numbers and mass numbers leads us to expect a nucleus of oxygen-17 ($^{17}_{8}\text{O}$) to be formed in this process—an isotope that is known to form a small part of naturally occurring oxygen. Cloud chamber photographs confirmed Rutherford's supposition. With nitrogen in the chamber about one α-particle track in 50 000 was found to be branched as in Fig. 17.24. The α-particle track comes to an end at the fork. Of the two tracks emerging from this point, one is more thinly marked than an α-particle track, and is that of a proton; the other is short and thick, and is taken to be that of the residual oxygen nucleus (see Plate B.3).

The above nuclear reaction is one of a class

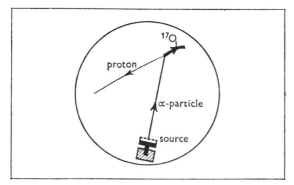

Fig. 17.24   A nuclear transmutation of a nitrogen-14 nucleus into an oxygen-17 nucleus

referred to in the shorthand notation of nuclear physics as an (α, p) reaction. This means that it is caused by an incident α-particle, and results in the emission of a proton (p). In this notation the complete nuclear reaction is specified by stating first the target nucleus, then the class of reaction, and finally the resulting nucleus. Thus the transmutation Rutherford discovered is described as the reaction

$$^{14}_{7}\text{N}(\alpha, \text{p})^{17}_{8}\text{O}$$

Many other examples were subsequently discovered of the transmutation of nuclei by bombardment with α-particles. In some cases the residual nuclei turned out to be radioactive isotopes not found in nature.

The neutron

Up to this point in the development of the subject the composition of the nucleus had remained something of a mystery. It seemed almost certain that the *proton* was one ingredient; the presence of this particle in the nucleus presumably accounted for its positive charge. Since, however, the mass number of a nucleus is usually more than double its atomic number, it must contain other particles besides protons. Rutherford suggested the electron as another ingredient. This seemed reasonable enough, since electrons were known to be emitted in the process of β-decay. On this basis an α-particle would consist of 4 protons and 2 electrons; its net positive charge would thus be 2 units, while its mass would be 4 units. However, this theory ran into difficulties.

The uncertainty was eventually cleared up with the discovery of yet another kind of 'radiation'; this was generated in the element *beryllium* by bombardment with α-particles. The nature of this effect was elucidated by Chadwick in 1932 with the apparatus shown in Fig. 17.25. The α-emitting source S and the beryllium target B were mounted in an evacuated enclosure. The beryllium disc was thick enough to stop the α-particles completely. In the absence of any other sheets of matter between the beryllium and the ionization chamber C a very small ionization current was registered. The amount of ionization was comparable with that produced by γ-radiation; but the insertion of absorbing sheets showed that the new radiation was even more

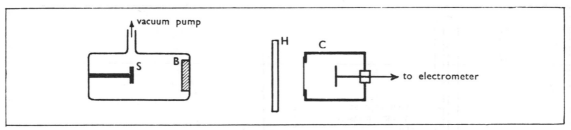

Fig. 17.25   The discovery of neutrons by Chadwick

penetrating than normal γ-radiation. It was not likely to consist of charged particles of any kind, since these would cause much greater ionization than was observed. The real surprise came, however, when a sheet of paraffin wax H (or any other material containing a large amount of hydrogen) was inserted in the path of the beam. The ionization current was then found to increase by a factor of two or more. It was soon established that the increased current was due to the ejection of protons from the paraffin by the new radiation. It seemed that the radiation had no interaction with the electrons in the matter through which it passed but only with the atomic nuclei. This was confirmed by cloud chamber photographs. No electron tracks were observed in a chamber through which the radiation passed. Instead, occasional heavy straight tracks were found that were interpreted as those of nuclei projected forward by the impact of some particle.

Chadwick assumed that the action of α-particles on beryllium caused the emission of a hitherto undetected particle, which he supposed to be uncharged. It was named the *neutron* (n). We describe this as an (α, n) reaction

$$^{9}_{4}Be(\alpha, n)^{12}_{6}C$$

By careful measurement of the energies of nuclei struck by the neutrons Chadwick estimated that the mass of the neutron was about equal to that of the proton. It is now known that the neutron is in fact slightly more massive.

Mass of the neutron $= 1.00867$ u
$\qquad\qquad\qquad\quad = 1.6749 \times 10^{-27}$ kg
Mass of the proton $\;= 1.00728$ u
$\qquad\qquad\qquad\quad = 1.6726 \times 10^{-27}$ kg

The discovery of the new particle settled the problem of the composition of the nucleus. The previous difficulties were removed if it could be assumed that the number of protons in the nucleus was equal to the atomic number, and that the additional mass was made up by neutrons. On this basis the α-particle must consist of just 2 protons together with 2 neutrons. However, there were new difficulties in explaining the process of β-emission. On the neutron theory this must consist of the transformation of a neutron into a proton and the creation of an electron to carry away the surplus negative charge. At the time this seemed rather unlikely; but subsequently this very process has been observed in neutrons outside the nucleus. In the free state the neutron itself is a *radioactive particle*, which decays into a proton by β-emission; its half-life is 12.8 minutes (p. 338). In fact according to modern understanding protons and neutrons in a nucleus are continually changing one into the other, so that it is not really appropriate to talk about them as though they could be distinguished from one another (p. 3). In this connection they are both referred to as *nucleons*; only when one of them emerges from the nucleus can we say whether it is a proton or a neutron.

We can now see more precisely the reason for the balancing of the mass numbers before and after a nuclear reaction. The masses of the nucleons usually change very slightly in a nuclear reaction, but the *total number of nucleons* does not alter, although protons may change into neutrons and vice versa. The mass number therefore represents the *number of nucleons* in a nucleus, rather than their exact mass.

By using beams of neutrons many new kinds of nuclear reaction can be produced. A *charged* particle entering a nucleus must have sufficient energy to overcome the intense electrostatic repulsion. Protons and α-particles must therefore have large energies to do this. But the neutron, being uncharged, can pass readily *through* a nucleus, whatever its energy; and it may in the

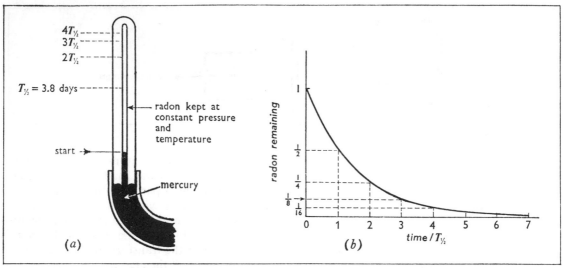

Fig. 17.26   (*a*) The radioactive decay of a sample of radon-222; (*b*) exponential decay curve

process be absorbed by it. Many of the artificial radioactive isotopes in current use are produced by irradiation of suitable substances with the intense beams of neutrons obtained in nuclear reactors (p. 353).

One such reaction, which we class as an (n, α) reaction, provides one of the most convenient modern methods of detecting neutrons:

$$^{10}_{5}\text{B}(n,\ \alpha)^{7}_{3}\text{Li}$$

The boron-10 nucleus absorbs neutrons very readily, and then disintegrates into an α-particle and a lithium nucleus—both of which are heavily ionizing particles. The ionization produced by neutrons in most gases is so slight that an ordinary Geiger counter scarcely responds to them at all. But by filling the tube with boron trifluoride gas a high proportion of the incident neutrons may be detected.

Because of their high penetrating power neutrons can constitute a serious health hazard. The most effective substances for stopping them are those which contain as many light nuclei as possible per unit volume. This means that protective screens round neutron sources are best made of compounds of hydrogen and other light nuclei, e.g. water or paraffin wax, rather than lead.

17.7   Radioactive decay

For many radioactive substances the decay rate is so slow that there is no perceptible reduction in the quantity of the material within the lifetime of one observer! But in all cases that can be investigated in a reasonable time the decay is found to follow the same pattern. For instance a sample of radon-222 may be collected over mercury in a capillary tube (Fig. 17.26*a*). The radon may be kept at constant pressure and temperature while its volume is measured at intervals during the next few days. Its decay products are solid substances; so the volume of the gas decreases as the radon decays. This is found to happen according to an *exponential law*, as shown in Fig. 17.26*b*. In 3.8 days half the radon has decayed; this is called the *half-life* of the isotope. In another 3.8 days half of what remains decays; after a third period of 3.8 days only $\frac{1}{8}$ of the original sample remains; and in about seven times the half-life the radon has virtually disappeared. The walls of the tube are then lined with the decay products—which may be detected by the radiations they emit. Similar exponential decay curves are found in all cases that have been investigated. All such processes are completely described by stating the half-life of the radioactive isotope concerned.

The **half-life** of a radioactive isotope is the time taken for any quantity of the isotope to decay by one-half of the quantity of it initially present.

An exponential curve of the type shown in Fig. 17.26 may be expressed by the equation

$$m = m_0\,\text{e}^{-\lambda t}$$

where m_0 is the mass of the isotope initially present, and m is the mass remaining after the passage of time t. The constant λ is a measure of the decay rate, and is known as the *decay constant*. The half-life $T_{\frac{1}{2}}$ may be expressed in terms of λ as follows:

From the definition of the half-life $T_{\frac{1}{2}}$,

$$\text{when } t = T_{\frac{1}{2}}$$
$$m = \tfrac{1}{2}m_0$$
$$\therefore\ e^{-\lambda T_{\frac{1}{2}}} = \tfrac{1}{2}$$
$$\text{or}\ \ e^{\lambda T_{\frac{1}{2}}} = 2$$
$$\therefore\ \lambda T_{\frac{1}{2}} = \log_e 2 = 0.693$$
$$\therefore\ T_{\frac{1}{2}} = \frac{0.693}{\lambda}$$

The radon-222 from the decay of natural radium has harmful and relatively long-lived decay products. However, another isotope radon-220, formed as one of the decay products of thorium-232 can safely be handled in a school laboratory. About 25 g of thorium hydroxide form a convenient source of this isotope. The thorium reaches approximate equilibrium with its decay products in about 12 years, and may then be used to produce radon-220 (sometimes called *thoron*). The 'vintage' thorium hydroxide is kept in a polythene bottle that can be con-

nected to an ionization chamber by flexible tubing in such a way that the radon is confined within the closed circuit of tubing. Inside the polythene bottle the thorium hydroxide is held in a dust-proof bag that acts as a filter to prevent the escape of the fine powder. Radon-220 is formed continually, and may be transferred to the ionization chamber by simply squeezing the bottle. The mass of radon expelled from the bottle is probably less than 10^{-19} kg—quite undetectable by chemical means. However, some 10^5 atoms may enter the ionization chamber, and these will decay initially at a rate of about 600 per second, which provides an ionization current that may be measured with an electrometer (Fig. 17.27a).

The ionization current at any instant is proportional to the quantity of radon still remaining in the chamber, and can therefore be used as a measure of this quantity. A graph of ionization current against time is thus of exponential form (Fig. 17.27b), and the half-life of radon-220 may be calculated from it.

Another decay process that may conveniently be observed in a school laboratory is that of the isotope of protactinium (^{234}Pa) formed as one of the decay products of uranium-238. Any compound of uranium therefore contains trace amounts of protactinium-234, which may be

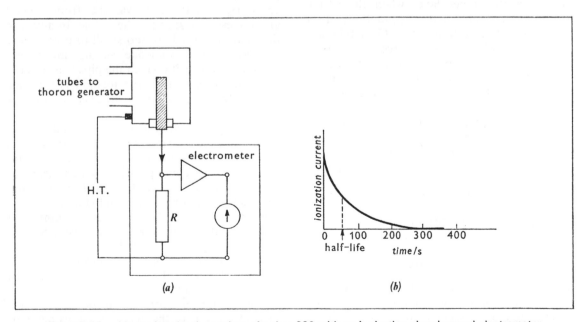

Fig. 17.27   Measuring the decay time of radon-220 with an ionization chamber and electrometer

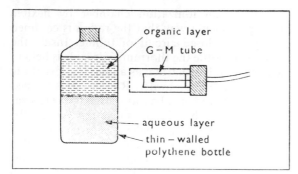

Fig. 17.28   Measuring the half-life of protactinium-234

extracted from it by chemical means. The protactinium decays by β-emission into another long-lived isotope of uranium (^{234}U) which is itself α-emitting. The β-activity at any instant of the extracted solution can therefore be used as a measure of the quantity of protactinium still present in it.

A simple practical arrangement is that shown in Fig. 17.28. A thin-walled polythene bottle is filled with equal volumes of an acid solution of uranyl nitrate and a suitable organic reagent (e.g. pentyl ethanoate). When the liquids are shaken up together the organic reagent removes most of the protactinium present, which then remains in the upper layer when the liquids have once more separated. The β-activity of the protactinium is observed with a G–M tube and ratemeter, and the count-rate is recorded at (say) 20 s intervals. A graph of count rate against time is of the usual exponential form, when due allowance has been made for the background count of the G-M tube.

An exponential decay law is best tested by plotting the variables in such a way as to obtain a straight-line graph. Thus the exponential decay of the count rate A above is represented by

$$A = A_0\,e^{-\lambda t}$$

where λ = the decay constant.

Taking logs (to the base e), we have

$$\log_e A = \log_e A_0 - \lambda t$$

and a graph of $\log_e A$ against t should thus be a straight line whose gradient is $-\lambda$. From this the half-life may be calculated.

Radioactive tracers

In both the above experiments we use quantities of radioactive isotopes which, by other standards, would be regarded as vanishingly small. However, the numbers of atoms in each sample are large enough to ensure a measurable decay rate, and this is what matters. This shows up a feature of work with radioactive isotopes. It is possible to follow an isotope through a physical or chemical process, and its presence at any point can be detected by the radiations it emits; but, it is only necessary to use sufficient material to ensure that the count rate obtained is significantly greater than the background count.

Numerous practical applications of such techniques have been developed. For instance water mains may be tested for leaks by filling them with water containing minute quantities of a radioactive isotope of sodium (^{24}Na), usually in the form of sodium bicarbonate. If there are any leaks in the pipe, the 'labelled' solution flows out into the surrounding soil. The pipe is then emptied and washed through with plain water. Sodium-24 emits high-energy γ-radiation, which is readily detected near a leak by a G–M tube drawn along the inside of the pipe or lowered into small holes made in the ground beside suspected joints. The half-life of the isotope is only 15 hours, so that it soon vanishes from the soil and leaves no hazard for the public. In any case the quantities used are too small to present any danger, even if someone should mistakenly drink the water with which the pipe was initially filled. In spite of this, the method is sensitive enough to detect leaks as little as 10 cm³ per hour!

There are many branches of pure and applied science in which the use of *radioactive tracers* has become an indispensable tool of research. For instance it has become possible by this means to unravel some of the complex sequences of chemical reactions that take place in living organisms. The organism may be 'fed' with a compound labelled with a suitable radioactive isotope. The subsequent growth of radioactivity in other compounds present in the organism reveals the pattern of chemical reactions into which the original 'food' enters.

It is outside the scope of this book to go further into details of the many ingenious methods that have been developed for using radioactive tracers. The student who is interested should consult books of a more specialized nature.

The decay of long-lived isotopes

The half-lives of the known radioactive substances cover an enormous range—from less than 1 microsecond to more than 10^{15} years. The direct measurement of the decay rate by (say) chemical analysis is unthinkable in most cases. The half-life of radium (1620 years) is trifling compared with that of uranium (4.5×10^9 years). Even so the amount of radium in a given sample would have decreased by only 4% during the time since this element was discovered by the Curies; and this cannot be made the basis of a measurement of its half-life. However, each particle that emerges from a source represents the decay of one nucleus; and it is possible by counting the emitted particles to estimate the number of nuclei decaying per second. If we can also estimate the total number of atoms in the source, we can then calculate the half-life of the specimen. Thus, if N_0 is the number of atoms initially present in a source, and N is the number present after time t, then

$$N = N_0 \, e^{-\lambda t}$$

where λ = the decay constant.

Now the *activity* A of a radioactive source is defined as the rate of disintegration within it (i.e. the number of disintegrations per second). Thus

$$A = -\frac{dN}{dt}$$

Differentiating, we therefore have

$$A = \lambda N_0 \, e^{-\lambda t} = \lambda N$$

$$\therefore \lambda = \frac{A}{N}$$

$$= \frac{\text{the activity of the source}}{\text{no. of atoms in the source}}$$

The Avogadro constant (6.02×10^{23} mol^{-1}) is the number of atoms in 1 mole of any element (p. 37). The denominator of the above expression may therefore be calculated if the mass of active isotope in the source is known. Then λ and the half-life may be found.

EXAMPLE   A radioactive source contains 1 μg of plutonium (mass number 239). The source is estimated to emit a total of 2300 α-particles per second in all directions. Calculate the half-life of plutonium.

0.239 kg of plutonium contain 6.02×10^{23} atoms.

$$\therefore \ 10^{-9} \text{ kg contains } \frac{10^{-9} \times 6.02 \times 10^{23}}{0.239} \text{ atoms}$$

$$= \frac{6.02}{2.39} \times 10^{15} \text{ atoms}$$

Therefore the decay constant λ is given by

$$\lambda = \frac{2300 \times 2.39}{6.02 \times 10^{15}} \text{s}^{-1} = 9.15 \times 10^{-13} \text{s}^{-1}$$

$$\therefore \ \text{half-life} = \frac{0.693}{\lambda}$$

$$= \frac{0.693}{9.15} \times 10^{13} \text{ s}$$

$$= 7.57 \times 10^{11} \text{ s}$$

$$= \frac{7.57 \times 10^{11}}{3600 \times 24 \times 365} \text{ years}$$

$$= 24\,000 \text{ years}$$

The curie

In using a radioactive source we are not generally concerned so much with the actual mass of active material present as with the rate at which it emits particles, i.e. its activity. Thus what matters about the plutonium source in the above example is that its activity is 2300 s^{-1}.

There is also a traditional unit of activity that remains in use for this purpose; it is called the *curie*. This was fixed originally as the activity of 1 g of radium. But it is now redefined on a numerical basis.

The **curie** is the activity of a radioactive source in which the disintegration rate is 3.7×10^{10} s^{-1}.

In school laboratories 10 microcuries (10 μCi) is usually considered the maximum safe activity for any one source; and for most purposes the sources can well be much weaker than this. The strength of the plutonium source appearing in the example above is therefore

$$\frac{2300}{3.7 \times 10^{10}} \text{ Ci}$$

$$= 6.2 \times 10^{-2} \ \mu\text{Ci}$$

For more advanced work sources in the millicurie range are used; safety precautions must then be of a much more stringent kind than those we have described. For medical purposes radium sources of a few curies are sometimes used; and γ-emitting sources of cobalt-60 of more than 2000 curies have been used for deep therapy. Such sources could well give lethal doses of radiation in quite a short space of time.

It is found that no physical or chemical process is able in any way to affect the rate of disintegration of a given isotope. It may be exposed to extremes of temperature or placed in large electric and magnetic fields. It may be fixed in any number of chemical compounds—gaseous or solid—or it may be ionized in a high-temperature discharge. But in no case has any change in the disintegration rate been detected. In other words no action *outside* the nucleus is able to affect the moment at which any particular nucleus disintegrates. Since we cannot know what is going on inside the nucleus, we can only describe the behaviour of the nucleus by stating the *probability* of its decay in a given interval of time.

When we are dealing with very large numbers of atoms, the probability P of decay of a particular atom within an interval of one second is equal on average to the fraction of the atoms that do in fact decay in this time.

$$\therefore P = \frac{\text{no. of disintegrations per second}}{\text{total no. of atoms in the source}}$$
$$= \frac{\lambda N}{N} = \lambda$$

The true significance of the decay constant λ is thus apparent; it is equal to the probability of a given atom decaying in one second, and is a constant for the atom that is unaffected by any external conditions. The universal validity of the exponential type of decay laws shows also that λ does not depend on the *age* of the atom. Any variation of the probabilities with time would yield a different decay curve.

17.8 Radioactive equilibrium

The radioactive decay of radium is the start of a series of disintegrations, ending eventually with a stable isotope of lead. Several such radioactive

17.8 Radioactive equilibrium

The uranium series

Isotope	Symbol	Emissions	Half-life
Uranium-238	$^{238}_{92}U$	α, γ	4.5×10^9 y
Thorium-234	$^{234}_{90}Th$	β, γ	24 d
Protactinium-234	$^{234}_{91}Pa$	β, γ	1.2 min
Uranium-234	$^{234}_{92}U$	α, γ	2.5×10^5 y
Thorium-230	$^{230}_{90}Th$	α, γ	8.0×10^4 y
Radium-226	$^{226}_{88}Ra$	α, γ	1620 y
Radon-222	$^{222}_{86}Rn$	α	3.8 d
Polonium-218*	$^{218}_{84}Po$	α	3.1 min
Lead-214	$^{214}_{82}Pb$	β, γ	27 min
Bismuth-214*	$^{214}_{83}Bi$	β, γ	20 min
Polonium-214	$^{214}_{84}Po$	α	1.6×10^{-4} s
Lead-210	$^{210}_{82}Pb$	β, γ	19 y
Bismuth-210*	$^{210}_{83}Bi$	β	5.0 d
Polonium-210	$^{210}_{84}Po$	α	138 d
Lead-206	$^{206}_{82}Pb$	Stable	—

* In these isotopes alternative modes of decay occur in a small proportion of cases. For example
(i) Polonium-218 may decay by β-emission to astatine-218; this then decays by α-emission to bismuth-214.
(ii) Bismuth-214 may decay by α-emission to thallium-210 which decays by β-emission to lead-210.
(iii) Bismuth-210 may decay by α-emission to thallium-206; this decays by β-emission to lead-206.

series are known. Indeed the series produced from radium is itself only part of a much longer series starting with the most abundant isotope of uranium (^{238}U). The complete uranium series is displayed in the table.

The half-life of radium is very short compared with the age of the earth, so that any radium originally present in the earth must long since have decayed away. The radium we now have is that which has arisen as a daughter product of uranium and has not yet decayed into radon, etc. Radium and the other members of the uranium series are therefore found naturally in uranium-bearing ores (such as pitchblende). Given a sufficient period of time the quantity of each isotope in the ore builds up to the point at which its rate of decay is equal to the rate at which it is formed from the decay of the previous member of the series. When this condition is reached, the proportions of the radioactive isotopes in the rock do not change with time, and we say that the uranium is in *radioactive equilibrium* with its daughter products. The half-life of uranium is so long (4.5×10^9 years) that it takes about 50 million years for the quantity of it in the rock to change by even 1%. Its disintegration rate is therefore almost constant, and this is equal to the disintegration rate of each member of the series in equilibrium with it. It is also the rate at which the stable end-product (^{206}Pb) is being formed. The net result is therefore the conversion of ^{238}U to ^{206}Pb at a steady rate, while the quantities of the intermediate members of the series remain constant.

The amounts of daughter products in equilibrium with a given amount of the parent substance may be calculated from a knowledge of their half-lives. Suppose the decay constants of the members of the series are $\lambda_1, \lambda_2, \lambda_3$, etc. Let the numbers of atoms of each isotope present be N_1, N_2, N_3, etc. The corresponding rates of disintegration are $\lambda_1 N_1, \lambda_2 N_2, \lambda_3 N_3, \ldots$; and in equilibrium these rates are all equal.

$$\therefore \quad \lambda_1 N_1 = \lambda_2 N_2 = \lambda_3 N_3 = \ldots$$

i.e.
$$N \propto \frac{1}{\lambda}$$

for a given series.

Now the half-life $T_{\frac{1}{2}}$ of an isotope is given by

$$T_{\frac{1}{2}} = \frac{\log_e 2}{\lambda}$$

$$\therefore \quad N \propto T_{\frac{1}{2}} \text{ for the series}$$

The mass m of any isotope present is proportional to its relative atomic mass A_r and the number of atoms N.

$$\therefore \quad m \propto A_r T_{\frac{1}{2}}$$

for a given radioactive series.

EXAMPLE (*i*)   A quantity of ore is found to contain 1.0 kg of uranium. Estimate the mass of radium in the ore.

The half-life of uranium-238 is 4.5×10^9 years and that of radium-226 is 1620 years.

Let the mass of radium in the ore be m. Using the relation deduced above (assuming the ore to be in radioactive equilibrium),

$$\frac{m}{1 \text{ kg}} = \frac{226 \times 1620}{238 \times 4.5 \times 10^9}$$

$$\therefore \quad m = \frac{2.26 \times 1.62}{2.38 \times 4.5} \times 10^{-6} \text{ kg}$$

$$= 0.34 \text{ mg}$$

EXAMPLE (*ii*)   One of the decay products of uranium-238 is lead-210 which is a β-emitting substance. The lead-210 extracted from a quantity of uranium ore is found to emit 1.5×10^5 β-particles per second. Calculate the mass of uranium-238 in the ore. (The half-life of uranium-238 is 4.5×10^9 years; the Avogadro constant $= 6.0 \times 10^{23}$ mol^{-1}.)

The decay constant λ of uranium is given by

$$\lambda = \frac{0.693}{4.5 \times 10^9 \times 365 \times 24 \times 3600} \text{ s}^{-1}$$

$$= 2.78 \times 10^{-18} \text{ s}^{-1}$$

Assuming that the lead-210 is in equilibrium with the uranium in the ore, the disintegration rate of the uranium is the same as that of the lead.

$\therefore$ the no. of atoms of uranium present

$$= \frac{\text{the disintegration rate}}{\lambda}$$

$$= \frac{1.5 \times 10^5}{2.78 \times 10^{-18}}$$

$\therefore$ the amount of ^{238}U present

$$= \frac{1.5 \times 10^5}{2.78 \times 10^{-18} \times 6.0 \times 10^{23}} \text{ mol}$$

Now 1 mol of uranium consists of 0.238 kg.

$\therefore$ the mass of uranium in the ore

$$= \frac{0.238 \times 1.5 \times 10^5}{2.78 \times 10^{-18} \times 6.0 \times 10^{23}} \text{ kg}$$

$$= 21.3 \times 10^{-3} \text{ kg}$$

$$= 21 \text{ g} \quad \text{(to 2 significant figures)}$$

Radioactive dating

The existence of radioactive substances with long half-lives provides us with a means of measuring time intervals on a geological scale. If

a radioactive isotope and its decay products remain trapped in a rock, the measurement of the
proportion of the parent isotope that has decayed enables the age of rock to be determined.
This can be done for instance with uranium-
bearing rocks. Uranium-238 reaches equilibrium
with its decay products in about 2 million years.
After this time the net result of the disintegration
processes is the production of ^{206}Pb from
^{238}U, while the intermediate products remain in
constant concentrations. The measurement of
the ratio of ^{206}Pb to ^{238}U enables the age of the
rock (since it first solidified) to be established.
By this means ages of rocks between 50 million
years and 4000 million years have been measured with fair reliability. Some assumption must
be made about the proportion of the ^{206}Pb that
was in the rock when it was first formed, and
this introduces an uncertainty. It is usually
desirable to check the measurement by several
similar methods. For instance the less common
isotope of uranium (^{235}U) is also the start of a
radioactive series that terminates with ^{207}Pb;
and the same sort of measurement can be performed for these two isotopes. Sometimes a
check can also be provided by analysing the
proportions of lighter elements that have radioactive isotopes. For instance about 30% of the
element rubidium exists as a radioactive isotope,
rubidium-87. This decays by β-emission into
strontium-87, which is a stable isotope. The
half-life of this decay process is 4.3×10^{10}
years. The measurement of the relative concentrations of these isotopes in a rock thus provides
another means of estimating its age. Measurements on the oldest rocks have established that
the age of the earth must be about 4500 million
years. The ages of meteorites are sometimes
found to be much greater than this, showing
that many of these pieces of matter must have
solidified before the earth was formed.

Carbon-14 dating

The half-lives of the naturally occurring radioactive substances discussed above are far too
great to be of much value for dating archaeological finds, whose ages range up to a few
thousand years only. Fortunately it has been
established that a radioactive isotope of carbon
(^{14}C) is formed continually by the action of

cosmic radiation in the earth's atmosphere. The
cosmic rays cause some nuclear disintegrations
from which neutrons emerge. Some of the
neutrons react with nitrogen nuclei forming
carbon-14 by the (n, p) reaction:

$$^{14}_{7}N(n, p)^{14}_{6}C$$

Because of this a small proportion of the carbon
in the atmosphere is in this radioactive form.
Provided the intensity of cosmic radiation has
not changed during the last few thousand years
the proportion can be assumed to have been
constant during this period. There is a continual
interchange of carbon (in the form of carbon
dioxide) between plants and the atmosphere;
and animals (including man) feed on the plants
and on each other. All living creatures therefore
contain the same small proportion of ^{14}C.
However, when a creature dies, the interchange
of carbon with the atmosphere ceases, and the
^{14}C content of the remains starts to decay.
The half-life is 5600 years, which is ideal for
archaeological purposes. Also carbon is an
ingredient of many materials of archaeological
interest—wood, paper, peat, household refuse,
etc.

The technical difficulties of dating objects by
this means are enormous. Very often only small
quantities of material can be spared for the
measurement. In any case even in contemporary
specimens only about 1 part of 10^{12} of the carbon is in the form of the active isotope. This is
equivalent to about 15 disintegrations per
minute in 1 g of carbon. Also the β-particles
emitted by ^{14}C are of low energy; so the thickness of the specimen must be kept to a minimum.
In fact it is usual to convert the carbon into a
suitable gas (carbon dioxide or acetylene) and
introduce it into the counter itself. Even so, the
count may not greatly exceed the normal background count of a Geiger-Müller tube. Elaborate
precautions must therefore be taken to reduce
the background radiation as much as possible.
Thick lead shielding is employed; and as far as
possible all radioactive substances are eliminated from the materials employed in constructing the apparatus. For instance potassium
must be excluded from the glass of the tube,
since a small proportion of natural potassium is
in the form of the radioactive isotope^{40}K.

This technique has enabled ages of suitable finds to be measured from 600 years up to more than 10 000 years old; often this can be done with an uncertainty of less than 100 years. It has been possible on a number of occasions to check the method by measurements on objects whose age can be independently established from historical records.

17.9   Mass and energy

It is convenient to express the energies of particles emitted by radioactive sources in *electron-volt* (p. 188). For example, the α-particle emitted by radium has an energy of 4.8 MeV. This means that it has the same energy as that acquired by an electron accelerated through a p.d. of 4.8×10^6 V. It is also useful to be able to express the energies of γ-ray photons in the same unit.

EXAMPLE   Calculate the frequency and wavelength of a 2-MeV γ-ray.

Taking the charge on an electron as 1.6×10^{-19} C, the energy gained by an electron accelerated through a p.d. of 2×10^6 V is

$$1.6 \times 10^{-19} \times 2 \times 10^6 \text{ J} = 3.2 \times 10^{-13} \text{ J}$$

and this is the energy of the photon. Taking the Planck constant h as 6.6×10^{-34} J s, we have

$$\text{frequency of γ-ray} = \frac{3.2 \times 10^{-13}}{6.6 \times 10^{-34}} \text{ Hz}$$

$$= 4.85 \times 10^{20} \text{ Hz}$$

Now the velocity of light $= 3.0 \times 10^8$ m s^{-1}

$$\therefore \text{ the wavelength} = \frac{3.0 \times 10^8}{4.85 \times 10^{20}} \text{ m}$$

$$= 6.2 \times 10^{-13} \text{ m}$$

One of the most important conclusions of the Special theory of Relativity is that energy and mass are equivalent. The theory predicts that a body gaining an amount of energy E thereby increases in mass by an amount m given by

$$E = mc^2$$

where $c =$ the velocity of light. The quantities of energy handled in everyday life are far too small to involve detectable changes of mass. Thus 1 kg of water absorbs 4.2×10^5 J of energy to raise its temperature from 0° C to 100 °C. The theory of Relativity therefore predicts that its mass will

increase accordingly by an amount m, given by

$$m = \frac{4.2 \times 10^5}{9 \times 10^{16}} \approx 5 \times 10^{-12} \text{ kg}!$$

In chemical reactions also the changes of energy are of this order of magnitude, involving at the most a few eV of energy per atom; and the associated changes of mass are again quite undetectable.

However, in nuclear processes the quantities of energy released are of the order of a few MeV, and are sufficient to produce significant changes in the masses of the particles concerned. The mass equivalent of 1 MeV of energy may be calculated as follows:

$$1 \text{ MeV} = 1.6 \times 10^{-19} \times 10^6 \text{ J}$$
$$= 1.6 \times 10^{-13} \text{ J}$$
$$\therefore \text{ mass equivalent} = \frac{1.6 \times 10^{-13}}{9 \times 10^{16}} \text{ kg}$$
$$= 1.78 \times 10^{-30} \text{ kg}$$

This is almost exactly twice the mass of an electron; i.e. a 1-MeV γ-ray carries this quantity of mass away from the nucleus that emits it. In the same way a 1-MeV β-particle has a total mass *three* times the rest mass of a stationary electron (its rest mass + the mass of its kinetic energy).

Thus the processes of radioactive decay involve appreciable changes in the total masses of the nuclei and particles concerned, though the changes are still small compared with the mass of a nucleon, so that the relative atomic masses of all atoms are always close to whole numbers. The changes in mass can readily be measured by modern techniques of mass spectroscopy (p. 251).

Consider for instance the α-decay of radium-226. The relative atomic mass of $^{226}_{88}\text{Ra}$ is 226.0254; i.e. the actual mass m_{Ra} of one atom of this isotope is given by

$$m_{\text{Ra}} = 226.0254 \text{ u}$$

where $1 \text{ u} = 1.66 \times 10^{-27}$ kg   (p. 5).

Similarly the masses of one atom of the decay product $^{222}_{86}\text{Rn}$ and of helium are given by

$$m_{\text{Rn}} = 222.0175 \text{ u}$$
$$m_{\text{He}} = \underline{4.0026 \text{ u}}$$
$$\therefore m_{\text{Rn}} + m_{\text{He}} = 226.0201 \text{ u}$$

This is less than the mass of the parent radium atom by 0.0053 u, a quantity known as the *mass defect* of the reaction. The energy equivalent of this quantity of mass appears as the kinetic energy of the emitted particle. Thus

$$\text{surplus energy} = \frac{0.0053 \times 1.66 \times 10^{-27}}{1.78 \times 10^{-30}} \text{ MeV}$$

$$= 4.95 \text{ MeV}$$

The α-particle carries away 4.8 MeV of this energy, the remainder being emitted as a γ-ray shortly afterwards. In every case of radioactive decay the large quantities of energy released are found to be provided at the expense of the masses of the nuclei and particles involved, as predicted by the Special theory of Relativity. Indeed we may say that a given nuclear change could not take place unless the combined masses of the daughter products were less than the mass of the parent nucleus, so that a surplus of energy can be made available to carry away the decay products. In general, the greater the mass defect of any possible nuclear change, the more probable the change becomes. This is another way of saying that a radioactive substance emitting a very high-energy particle will be very short lived.

★ *Kinetic energy in the Special theory of Relativity*
In Newtonian mechanics the kinetic energy E_k of a particle of mass m moving with velocity v is given by

$$E_k = \tfrac{1}{2}mv^2$$

The Special theory of Relativity approaches the calculation of kinetic energy from a different point of view. The total energy E of the particle is given in all circumstances by

$$E = mc^2$$

This includes the rest mass energy that the particle has even when at rest. When the particle is moving with velocity v, its mass increases according to the relation

$$m = \frac{m_0}{\sqrt{1 - v^2/c^2}}$$

where m_0 = the rest mass.

The kinetic energy of the particle is accounted for by the increase in its mass $(m - m_0)$, and is therefore given by

$$E_k = (m - m_0)c^2$$

Substituting for m, and expanding the bracket in powers of v^2/c^2, we have

$$E_k = m_0c^2\left[\left(1 - \frac{v^2}{c^2}\right)^{-\frac{1}{2}} - 1\right]$$

$$= m_0c^2\left[1 + \frac{1}{2}\frac{v^2}{c^2} + \frac{3}{8}\frac{v^4}{c^4} + \ldots - 1\right]$$

$$= \tfrac{1}{2}m_0v^2\left[1 + \frac{3}{4}\frac{v^2}{c^2} + \ldots\right]$$

For sufficiently small velocities (i.e. $v^2 \ll c^2$), this expression agrees closely with the Newtonian formula, ($E_k = \tfrac{1}{2}m_0v^2$). But at speeds approaching that of light the Newtonian theory breaks down completely.

For example, according to the Newtonian formula the kinetic energy of a particle moving at half the speed of light is

$$\tfrac{1}{2}m_0\frac{c^2}{4} = \tfrac{1}{8}m_0c^2 = 0.125m_0c^2$$

According to the Special theory of Relativity the mass of the particle at this speed becomes

$$m = \frac{m_0}{\sqrt{1 - 1/4}} = \frac{2}{\sqrt{3}}m_0$$

$$\therefore E_k = (m - m_0)c^2 = \left(\frac{2}{\sqrt{3}} - 1\right)m_0c^2$$

$$= 0.155m_0c^2$$

In this case the Newtonian calculation gives an error of about 25%.

17.10 High-energy accelerators

Most of our knowledge of the nucleus has come from collision experiments, in which we study the effects of firing high-speed particles at a piece of matter. Much of it was gained using as projectiles the α-particles from radioactive substances. But the maximum energy obtainable in this way is less than 10 MeV, and there is no control over the direction of emission of the particles. To obtain a reasonably strong collimated beam a dangerously large source must be used with all its attendant difficulties.

Cosmic rays

Nature provides also another source of high-speed particles in the cosmic radiation that enters the earth's atmosphere from outer space. Here the energies available are of quite a different order of magnitude. The majority of primary particles have energies of about 10^4 MeV; but energies have been observed up to a maximum of about 10^{13} MeV. The primary radiation consists mostly of protons with a small proportion of other light nuclei and of electrons. However, all these particles are lost in collisions with nuclei in the upper atmosphere, and only the secondary products (mostly electrons) can be observed at ground level. To experiment with the primary radiation the equipment must be flown in a balloon high in the atmosphere. In spite of the difficulties several vital discoveries have been made with this technique. But the intensity of the radiation is very low—fortunately for us—and the gathering of detailed statistical data about the behaviour of the particles is an impossibly lengthy business. It has become clear since the 1930's that progress depends on the development of artificial means of producing very high-energy particles (see Plate D.1).

Electric field accelerators

The most obvious way of accelerating a charged particle is to use a very high p.d. between two electrodes in an evacuated tube. Several techniques for doing this were developed in the 1930's. Cockcroft and Walton adapted conventional rectifier circuits to produce high p.d.'s

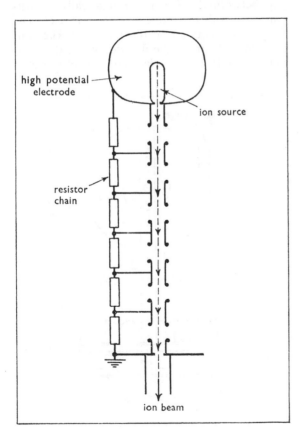

Fig. 17.29   An electric field accelerator tube for use with a high-voltage generator

from an a.c. input. Eventually they reached p.d.'s up to 2×10^6 V by this means. At the same time Van de Graaff was developing his electrostatic generator (p. 202). In both machines the maximum p.d. depends on the quality of the insulation surrounding the charged electrode. By encasing the whole apparatus in a pressure tank containing *freon* gas Van de Graaff generators have been made to work at p.d.'s up to 7×10^6 V.

The ions to be accelerated are produced in an ion source contained inside the charged electrode; a hydrogen tube gives protons, heavy hydrogen deuterons, helium α-particles, etc. Some of the ions then pass through a small hole into the main accelerating tube. This is lined with a number of cylindrical electrodes maintained at intermediate p.d.'s (Fig. 17.29). The electric field is thus concentrated in the gaps between the cylinders, and has the effect of focusing the ions into a narrow beam.

The energies achieved were less than those of natural α-particles; but the strong, collimated beams of exactly known energy enabled a wide range of new experiments to be performed. Also for the first time it was possible to observe artificial nuclear reactions produced by protons and deuterons.

The linear accelerator

In this machine high energies are produced without employing particularly high p.d.'s by using what is known as the principle of *synchronous acceleration*. The electrode system consists of a series of coaxial cylinders which increase in length towards the target at the far end (Fig. 17.30). A high-frequency alternating p.d. is applied to the electrodes, as shown, the odd-numbered cylinders being joined to one terminal of the supply and the even-numbered ones to the other. Any positive ions that reach the gap between cylinders A and B at a moment when B is at a high negative potential with respect to A will be strongly accelerated. But once inside the hollow electrode B they are travelling in a field-free region and are unaffected by the changing potential of the electrode. The length of each cylinder is so chosen that the time taken by the ions to move through it is equal to half the period of oscillation. The

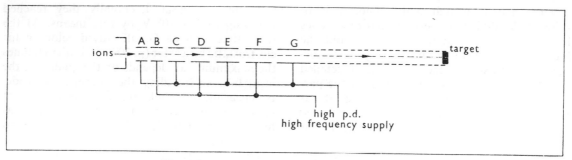

Fig. 17.30   The principle of a linear accelerator

electric field in the next gap (between B and C) will then be in the right phase to accelerate the ions again as they pass through it; and the same applies at each gap right down the tube. Suppose the peak p.d. of the supply is 200 kV. Then at each gap the ions acquire about 200 keV of energy; and the final energy is limited only by the number of electrodes that can be accommodated inside a tube of reasonable length. Very high frequencies are employed so that the electrodes can be as short as possible.

This machine was developed first for accelerating heavy ions. But it has also been adapted for electrons. In this case the electrode system is in some ways rather simpler. No particle can travel faster than the speed of light; and at 2 MeV an electron has already reached 98% of this speed. Scarcely any acceleration is then possible, and any gain in energy comes rather from the increase in mass of the particle. Thus after passing through the first few electrodes the electron is virtually moving at constant speed, and all subsequent electrodes can be of equal length. However, the technical problems of getting the p.d. to each gap at the right moment in the right phase are very considerable.

The cyclotron

This machine also employs the principle of synchronous acceleration; but excessive length is avoided by using a magnetic field to wrap the paths of the particles into tight spirals. The high-frequency p.d. is applied to two hollow **D**-shaped electrodes (Fig. 17.31). The whole system is placed in an evacuated box between the poles of a very large magnet. The positive ions are injected near the centre and are accelerated each

time they cross the gap between the **D**'s. Inside the **D**'s there is no electric field, and they move in semicircular orbits. As a particle gains speed, the radius of its orbit in the magnetic field increases proportionately, but its period of revolution remains constant. If this is also the period of oscillation of the alternating p.d., the field in the gap between the **D**'s is automatically in synchronism with the motion of the particles. The frequency *f* required for this may be worked out as follows. Suppose the flux density of the magnetic field is B, and that the particles are of

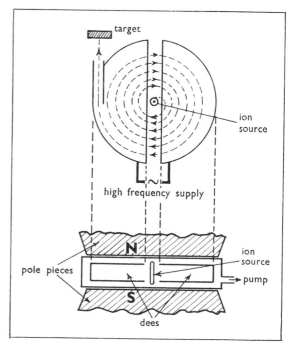

Fig. 17.31   The principle of the cyclotron

charge e and mass m. Then the radius r of the orbit of a particle of velocity v is given by

$$Bev = \frac{mv^2}{r} \quad \text{(p. 230)}$$

$$\therefore \frac{r}{v} = \frac{m}{Be}$$

The period of revolution T is given by

$$T = \frac{2\pi r}{v} = \frac{2\pi m}{Be}$$

$$\therefore f = \frac{1}{T} = \frac{Be}{2\pi m}$$

After performing about a hundred revolutions the particles reach the edge of the system and enter a subsidiary electric field that deflects them out of the circle to strike the target.

In the earliest forms of cyclotron the energy attainable was limited by the relativistic increase of mass of the particles. At 20 MeV the mass of a proton is about 2% greater than its rest mass, and beyond this point it becomes very difficult to preserve the synchronism between the motion of the particles and the alternating p.d. However, much higher energies than this have been produced by varying the frequency during the acceleration process. A high frequency is used at the moment of injection; but as the particles gain energy and move out towards the edge of the system the frequency is steadily reduced to maintain the synchronism. Machines employing this principle have produced energies up to 600

MeV. They are known as *synchrocyclotrons*. The maximum energy is decided by the strength of the magnet and the radius of its pole pieces. Magnets weighing hundreds of tons have been used in the largest machines, and the economic limit has undoubtedly been reached.

The synchrotron

In this machine the particles are accelerated inside a large annular ring (Fig. 17.32). The magnetic field is then only required over a limited region, and a ring-shaped magnet can be used to hold the particles in the channel. Since the acceleration is to take place at fixed radius, the magnetic field must be steadily increased as the particles gain energy. At one point of the ring the particles pass through a hollow cylindrical electrode connected to a high-frequency generator, and are accelerated both on entering and leaving it. For this to happen they must enter the electrode at the moment when it is at a large negative potential (with respect to the earthed frame of the apparatus), and they must leave it after precisely half an oscillation of the potential. The frequency must therefore be steadily increased as the particles speed up, and this increase must be kept in step with the variations of magnetic field.

There seems to be no limit in principle to the energy that might be achieved with machines of this kind, except that the cost soon becomes too much for the resources of any one nation. The

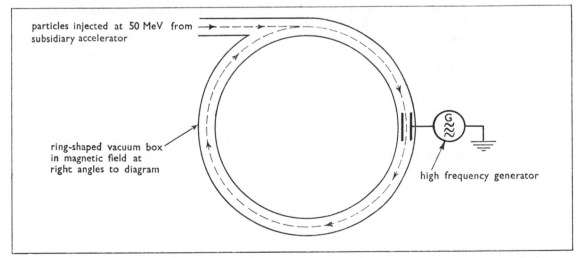

particles injected at 50 MeV from subsidiary accelerator

ring-shaped vacuum box in magnetic field at right angles to diagram

high frequency generator

Fig. 17.32   The principle of the synchrotron

largest machine in operation in Europe is the internationally owned one at Geneva, which can accelerate protons to energies of 30 000 MeV (or 30 GeV).

To keep a true perspective in these matters it is well to remember that the highest energies so far produced by man-made accelerating machines are of the same order of magnitude as the *average* energy of cosmic-ray particles. The highest energies recorded in cosmic rays are greater than this by a factor of about 10^9, but experiments with these rare particles can only be conducted above the protective blanket of the earth's atmosphere.

17.11   More particles

The years since 1930 have seen the discovery of an ever-increasing number of 'fundamental' particles. At the time of writing over 100 are known, and the pattern revealed is one of great complexity. The scope of this book does not allow anything beyond a sketch of the more important discoveries.

The positron

In 1930 Dirac produced a theory of the electron which seemed to show that there should be a positive electron as well as the familiar negative one; and a few years later the new particle, called the *positron*, was duly discovered. It was found that a very high energy γ-ray passing near an atomic nucleus may give rise to an electron–positron pair (see Plate B.1). The discovery of *pair production* provided one of the most convincing demonstrations of the equivalence of mass and energy implied in the theory of Relativity. On this theory the minimum energy of a γ-ray that can cause pair production should be $2m_e c^2$, where m_e is the mass of an electron (or positron). This is equal to 1.02 MeV, which is in good agreement with the experimental results. At very high energies pair production accounts for most of the absorption of γ-rays by matter. A positron has a short life, since it soon annihilates itself in combination with an electron. The energy liberated is carried away by two γ-ray photons; each of these is of energy 0.51 MeV.

In 1934 the Joliots discovered a positron form of β-disintegration in the isotope phosphorus-30.

This isotope they produced in an (α, n) reaction with aluminium:

$$^{27}_{13}\text{Al}(\alpha, \text{n})^{30}_{15}\text{P}$$

The phosphorus-30 then decays to silicon-30 by β^+-emission with a half-life of 3 minutes.

$$^{30}_{15}\text{P} \xrightarrow{\beta^+} {}^{30}_{14}\text{Si}$$

This was the first example to be discovered of artificially induced radioactivity. Since then numerous other β^1-emitting substances have been found. Generally they are produced by bombarding stable nuclei with *protons*, since this tends to give nuclei with too many protons for stability. Bombardment by *neutrons* tends to produce nuclei with excess neutrons; and these usually revert to the more stable isotopes by β^--emission, which in effect changes one of the neutrons into a proton.

Electron-capture

In some radioactive nuclei, as an alternative to β^+-emission, the nucleus may capture one of the electrons from the K-shell of the atom. This is called *K-capture*. (L-capture and M-capture also occur, but are less probable than K-capture, of course.) No particles are emitted; and the only evidence for the process is the appearance of the daughter product and the emission of the characteristic K-lines of its X-ray spectrum as another electron falls into the K-shell. Sometimes also the nucleus is left in an excited state, and a γ-ray is emitted. An example of this process is the decay of beryllium-7 into lithium-7:

$$^{7}_{4}\text{Be} \xrightarrow{\text{e}^-\text{capture}} {}^{7}_{3}\text{Li}$$

This is equivalent to a β^+-emission, for which in this case there would be insufficient energy available.

It now appears that Dirac's theory applies more widely than just to electrons and positrons. There seems to be complete symmetry in the table of fundamental particles; to every particle there is an *anti-particle* of opposite charge (if any), but otherwise with similar properties. A particle and its anti-particle are mutually annihilated when they combine together, and the energy liberated appears in some other form. To produce a nucleon–antinucleon pair requires about 2000 times as much energy as for an electron–positron pair; and the full confirmation of the theory therefore had to await the development of giant accelerators in the 1950's. In a few cases a particle and its anti-particle are identical; this applies for instance to the photon. But generally, even with uncharged particles, the two particles are distinct and capable of mutual annihilation.

The neutrino

In the 1930's an unexplained feature of the process of β-decay came to light. The energy spectrum of β-particles from the nucleus was known to be a *continuous* one (p. 327). With α-particles and γ-rays only a limited number of sharply defined energies are found; and this is what the modern atomic theory leads us to expect. A decay process involves a definite reduction in the mass of the nucleus, and by the mass/energy relationship a fixed quantity of energy should be released. This applies to β-emission as much as to any other process; and yet the β-particles may actually be emitted with *any* energy up to a certain maximum. To account for this Pauli made the suggestion in 1933 that a second particle was emitted at the same time as the β-particle; the total energy carried away by the two particles could then be constant, but might be divided between them in any way whatever. The conservation of charge required the new particle to be uncharged, and it was labelled the *neutrino*.

It is now believed that its anti-particle, the *antineutrino*, also exists; in fact this particle is thought to be the one that emerges in β⁻-decay, while the *neutrino* is emitted in conjunction with a positron in β⁺-decay. In modern nuclear theory electrons and neutrinos are grouped together in a class of light particles, called *leptons*. It is believed that the total number of leptons remains unaltered in any nuclear reaction. (For this purpose the number of anti-particles is counted as negative.) In nuclear equations the neutrino is represented by the Greek letter ν, and the antineutrino by the same letter with $\sim$ over it, $\tilde{\nu}$. Thus the process of β⁻-emission is really the simultaneous production of two particles; e.g. with lead-210 (p. 342):

$$^{210}_{82}\text{Pb} \xrightarrow{\text{e}^-, \tilde{\nu}} {}^{210}_{83}\text{Bi}$$

Counting the antineutrino as 'minus one' particle, the total number of leptons is zero before and after the reaction. Likewise when a neutron decays by β⁻-emission (p. 337) there are really three particles produced:

$$n \rightarrow \text{p}^+ + \text{e}^- + \tilde{\nu}$$

The β⁺-decay of phosphorus-30 consists in the emission of a neutrino as well as a positron:

$$^{30}_{15}\text{P} \xrightarrow{\text{e}^+, \nu} {}^{30}_{14}\text{Si}$$

and the process of electron-capture can be described in the notation of nuclear reactions as an (e, $\tilde{\nu}$) reaction; e.g.

$$^{7}_{4}\text{Be}(\text{e}, \tilde{\nu})^{7}_{3}\text{Li}$$

The neutrino is thought to have *zero rest mass*. The theory of Relativity requires that such a particle should only exist at all if it is travelling with the speed of light! In this respect it resembles the photon, which also has zero rest mass. But unlike the photon, the neutrino has very little interaction with matter—so little that it can pass right through the earth with only a slight chance of interacting with any of the particles in its path. But in spite of this its interaction with matter has been directly observed. This was achieved in 1956 by Cowan and Reines with the intense stream of radiations emerging from a nuclear reactor. Using a large quantity of an organic liquid near the reactor, they were able to demonstrate a number of cases of the simultaneous production of a neutron and positron in the following ($\tilde{\nu}$, e⁺) reaction

$$\tilde{\nu} + \text{p} \rightarrow n + \text{e}^+$$

This is essentially the reverse of the neutron decay process described above.

Mesons

In 1935 the Japanese scientist Yukawa sought to explain the very large forces that operate between nucleons by postulating the existence of yet another group of particles, which came to be called *mesons*. He supposed that the circulation of mesons round the nucleons provided the means of binding the latter particles together. This way of accounting for forces is a familiar one from other parts of atomic theory. For instance the chemical bond between atoms arises from the sharing of electron orbitals between them (p. 245). The mesons were supposed to fulfil the same function for the nuclear forces.

To explain the observed nature of the forces it had to be assumed that the mass of the meson was about 200 times that of the electron. The theory predicted that the meson would be unstable and have a half-life of the order of 10^{-6} s.

Not long afterwards a particle was discovered in the cosmic radiation that appeared at first to fit this description. Its mass was 207 electronic masses, and it was found both in positive and negative forms. It was unstable, and decayed (with a half-life of 2×10^{-6} s) into an electron (or positron) and *two* neutrinos. It later came to be known as the μ-meson or *muon*. However, the muon was found to be very penetrating, and could pass through great thicknesses of matter without any interactions with atomic nuclei. Yukawa's particle by its very nature would need to be a strongly interacting particle.

Eventually in 1947 another particle was discovered by Powell in nuclear emulsions exposed to cosmic radiation at high altitudes; this appeared to have the required strong interaction with matter. It was produced in collisions of the primary cosmic ray particles with atomic nuclei in the upper atmosphere (or in Powell's emulsions); and it could cause disintegration in another nucleus that it then struck. It has come to be called the π-meson or *pion*. Its mass is about 270 electronic masses, greater than that of the muon. Both positive and negative forms were discovered, and these were shown by Powell to decay into muons (and neutrinos) with half-lives of about $2.5 + 10^{-8}$ s (see Plate D.2).

$$\pi^+ \longrightarrow \mu^+ + \nu$$
$$\pi^- \longrightarrow \mu^- + \tilde{\nu}$$

A neutral pion was also found. This has a shorter existence than the other two, having a half-life of about 10^{-15} s. It decays into two γ-rays.

Before very long evidence began to accumulate for yet another kind of meson, less frequently produced than the pion. This became known as the K-meson or *kaon*. It has a mass of 966 electronic masses. Like the other mesons it decays with a very short half-life; its modes of decay are varied and complex. Thus Yukawa's theory earned the distinction of having provoked the search for new types of particle, where none was previously suspected. But the pattern then

revealed by experiment has turned out to be vastly more complex than could have been predicted.

Hyperons

The giant accelerating machines have made experiments possible with particles of ever-increasing energies. One by one whole new *families* of particles have been discovered in the range of masses heavier than the proton or neutron. Their modes of decay and their interactions with other matter are even more elaborate than those of the mesons.

17.12   Nuclear energy

Fission

In 1934 Fermi discovered that the bombardment of uranium with neutrons produced radioactive substances with half-lives that could not be matched with those of any known isotopes of neighbouring atomic number. He concluded that he had produced new elements of atomic number greater than that of uranium. This interpretation was partly correct. It is now known that such a reaction does occur in uranium-238.

$$^{238}_{92}\text{U} + \text{n} \longrightarrow ^{239}_{92}\text{U}$$

^{239}U is a β-emitting substance that decays with a half-life of 24 minutes into the first transuranic element, *neptunium*. This in turn decays by β-emission with a half-life of 2 days into *plutonium* (p. 314).

$$^{239}_{}\text{U} \xrightarrow{e, \tilde{\nu}} ^{239}_{93}\text{Np}$$

$$^{239}_{93}\text{Np} \xrightarrow{e, \tilde{\nu}} ^{239}_{94}\text{Pu}$$

Then in 1939 Hahn and Strassmann found by chemical analysis that a number of elements of medium atomic mass were also formed when uranium was bombarded with neutrons. They suggested that it might be possible for a uranium nucleus to break up into two or more fragments when it absorbs a neutron, a process now known as fission. The hypothesis was soon confirmed in cloud chamber photographs. It was found that the energy released in the fission of one uranium nucleus was about 200 MeV! The nucleus may break up in many different ways; usually the two main fragments are unequal. Most of the immediate fission products are radioactive.

Just before the second world war it was discovered that, in addition to the two main fragments, several high-speed neutrons were emitted at the moment of fission. There was thus the possibility of starting a chain reaction in a piece of uranium. The fission of one nucleus might produce enough neutrons to cause the disintegration of several more; these in their turn could trigger off the fission of an even larger number; and so on. In this way it might be possible to tap some of the vast amount of energy stored in the material (200 MeV per nucleus).

Fission has been found to occur in a number of heavy elements. Both of the chief uranium isotopes show the effect. In ^{238}U it can only happen with high-energy neutrons ($>$1 MeV); at lower energies the most probable result is the absorption of the neutron to form uranium-239. As most of the neutrons evolved at the moment of fission are of lower energy than 1 MeV, there is no possibility of a chain reaction developing in this isotope (over 99% of natural uranium). But in ^{235}U fission is caused chiefly by slow neutrons; and the same applies to plutonium-239. Both these materials can support chain reactions. All that is necessary to start the reaction is to bring together enough of the material in one confined space. With a small quantity of fissile material too many of the neutrons escape from the surface before colliding with another nucleus. But there is a certain critical size above which enough neutrons remain in the material to support the reaction. The *atomic bomb* consists essentially of an arrangement for shooting together suddenly two subcritical masses of fissile material. There are always a few neutrons present to initiate the reaction, which then spreads with great rapidity through a substantial fraction of the material—with effects that need no description.

In the *nuclear reactor* the chain reaction is made to proceed at a steady controlled rate. This is achieved by absorbing a proportion of the neutrons in some non-fissile material, so that on average *one* neutron from each fission event remains in the reactor to produce further fission. Any increase in the number of neutrons beyond this point causes the reaction rate to rise. In one kind of reactor natural uranium is used as the 'fuel', though only the small proportion of ^{235}U

is actually consumed. To prevent most of the neutrons being absorbed by the heavier isotope the fuel is arranged in the form of slender rods which are embedded in a material of low atomic mass known as a *moderator*. Most of the neutrons escape at once from the rods into the moderator, where collisions quickly reduce their energy to that of the atoms around them (about 0.03 eV). When they re-enter the fuel rods they are travelling at speeds at which they are very strongly absorbed by ^{235}U and scarcely at all by the heavier isotope. By this means the more abundant isotope is prevented from interfering with the reaction. However, a certain proportion of the neutrons are inevitably absorbed by the ^{238}U, so that there is a steady production of plutonium in the reactor. At intervals the fuel rods are removed, and the plutonium is extracted from them by chemical means. The reactor can therefore be used as a source of pure fissile material (plutonium). This is an alternative to the lengthy process of separating the two isotopes of uranium to obtain pure ^{235}U; this can be done, but the chemical separation of plutonium from uranium is a much simpler technique. The reaction rate is adjusted by means of control rods made of materials that absorb neutrons very strongly; cadmium and boron are found to be suitable for this. The control rods lie in channels running through the reactor core. By withdrawing them slightly the absorption of neutrons is decreased, and the reaction rate goes up. Automatic devices are of course provided to move the rods deeper into the core if an excessive temperature rise occurs. The moderator must be of a material that does not itself absorb neutrons to any extent. Carbon (in the form of graphite) and heavy water have both been used successfully.

One of the most important uses of nuclear reactors has been the production of radioisotopes. Many of these are formed as the fission products of the uranium, and are extracted when the fuel rods are changed. Caesium-137 and strontium-90 are familiar examples. Others are produced by irradiation with neutrons. To do this a suitable compound is drawn into the centre of the core, where the flux of thermal neutrons may well be more than 10^{16} m^{-2} s^{-1}. Sodium-24 (p. 340) is manufactured in this way.

$$^{23}_{11}\text{Na} + \text{n} \longrightarrow \, ^{24}_{11}\text{Na}$$

Neutron irradiation has also been used as a means of non-destructive chemical analysis—a technique of great value to archaeologists. The constituents of the object to be analysed react in different ways with neutrons and can be identified by the radiations they subsequently emit. In this way the compositions of ancient coins have been measured with sufficient precision to establish the mine from which the metal was taken.

Most of the energy liberated in the reactor appears in the form of heat, which must be carried away by a cooling fluid. In a nuclear power station this heat is used to generate steam which drives turbines and alternators of conventional design. The great merit of this type of power station is the small fuel consumption. A few kilograms of uranium are sufficient to produce hundreds of megawatts of power for a whole day. Fuel transport costs are therefore trivial, and nuclear power stations can be sited in remote places where a conventional power station would be uneconomic. A serious practical problem encountered is the safe disposal of the large quantities of radioactive 'waste'.

Once a sufficient stockpile of plutonium-239 is available, it is possible to operate what is known as a *fast breeder reactor*. No moderator is used, and the core consists of a set of plutonium rods. They are enclosed in a container of natural uranium. The fission proceeds in the core for fast and slow neutrons alike. The neutrons that escape from the core are absorbed by the ^{238}U of the containing wall, thereby generating further supplies of plutonium in it. The reactor therefore 'breeds' more fuel for itself in the containing wall while it is using up the fuel in the core.

Fusion

In the fission process energy is released because the fragments of the explosion are of much lower mass than the original nucleus. But there is also a considerable reduction of mass when light nuclei are fused together. Such reactions are readily observed in experiments with high-energy beams of protons and deuterons (d). For instance

$$p + p \rightarrow d + e^+ + \nu$$
(protons)

and $\qquad$ $d + d \rightarrow {}^3_2He + n$
$\qquad$ (deuterons)

In each case the products are of lower mass than the original particles, and considerable quantities of energy are released (as kinetic energy of the particles). Numerous such reactions have been discovered. In each case one condition for the reaction to proceed is that the two reacting particles should approach one another at sufficient speed to overcome the coulomb repulsion between them. To achieve this on a large scale it is only necessary to raise the temperature of the ingredients to the point at which some of the particles will have the required speed. For this reason these are known as *thermonuclear reactions*; unfortunately temperatures of about 10^8 K are needed to bring them about.

The energy of the sun and stars is evolved by cycles of thermonuclear reactions, whose net effect is the conversion of hydrogen into helium. The formation of each helium nucleus liberates 27 MeV of energy, more than 0.8% of the mass of the original particles being lost in the process. The sun loses some 4×10^6 tonnes of matter per second in this way; and there seems no reason why it should not continue in this profligate style for the next 10^{10} years! Paradoxically a larger star than the sun uses up its greater resources of hydrogen fuel in a shorter time; and the largest stars may effectively consume all their hydrogen in about 1 million years.

The world contains almost unlimited resources of hydrogen in the oceans. But so far, mankind's only successful attempt to tap this store of energy has been the *hydrogen bomb*. In this weapon an ordinary fission bomb is used as a detonator to bring the other ingredients to the temperature at which thermonuclear reactions start. There seems to be no theoretical limit to the size of such a bomb or the devastation it might produce. If on the other hand a way could be found of maintaining a controlled thermonuclear reaction, a source of cheap and abundant energy would be available—to the very great benefit of the human race. The resources of several nations have for many years been directed towards this goal, but in 1974 the prize still remains tantalizingly out of reach.

Questions

Additional data required:
Charge of an electron $= -1.60 \times 10^{-19}$ C
Mass of an electron $m_e = 9.1 \times 10^{-31}$ kg
Unified atomic mass constant m_u
$= 1.66 \times 10^{-27}$ kg $= 1$ u
Specific heat capacity of water
$= 4.2 \times 10^3$ J kg^{-1} K^{-1}

A

1 What current flows when a total electric charge of 150 C passes in 30 s?

2 What is the p.d. across a piece of equipment to which 15 J of electrical energy are supplied by the passage of 5 C of electric charge?

3 What is the power of a piece of equipment that supplies 300 J of energy in 20 s?

4 The current taken by a small torch bulb is 0.25 A. What total charge passes a point in the circuit in 15 minutes? How many electrons pass the point in this time?

5 A total charge of 5.0 C passes through a piece of equipment during a pulse of current lasting 10.0 ms. What is the average current during the pulse?
If the energy supplied during the pulse is 600 J: (a) estimate the p.d. applied to the piece of equipment during the pulse, (b) calculate the average power supplied.

6 The e.m.f. of a small cell is 1.5 V. How much energy does it produce when it drives a total charge of 160 C round a circuit? What energy would be produced if the same charge was driven by a battery of three such cells joined (a) in series, (b) in parallel?

7 How much energy is produced by a 60 W electric light bulb in 5 minutes?

8 A power station generates 7.2×10^{12} J of electrical energy in a certain 10-hour period. What is the average power generated? If the overall efficiency of the generating process is 40%, how much heat must be dissipated from the power station every hour?

9 A p.d. of 12 V is maintained between the ends of a wire. What quantity of electric charge must flow through it to supply 900 J of energy? If it takes 1 minute for this charge to pass, what is the current?

10 Two copper wires of diameter 2.00 mm and 0.50 mm respectively are joined end to end. What is the ratio of the average electron velocities in the wires when a steady current flows through them?

11 The average velocity of the electrons in a piece of wire of cross-sectional area 2.0×10^{-6} m^2 is 2.5×10^{-4} m s^{-1} when a current of 4.0 A is flowing. Calculate the number of 'free' electrons per unit volume of the material.

Other questions: Ch. 3: 1, 9 Ch. 10: 1, 6 Ch. 13: 1, 3 Ch. 16: 2 Ch. 17: 5

B

12 The electron beam in a television picture tube travels a total distance of 0.50 m in the evacuated space of the tube. If the speed of the electrons is 8.0×10^7 m s^{-1} and the beam current is 2.0 mA, calculate the number of electrons in the beam at any one instant.

13 It is usually reckoned that the maximum safe current density for a piece of copper wire is 1.0×10^7 A m^{-2}. If there are 1.0×10^{29} 'free' electrons per m^3 of copper, calculate the mean drift velocity of the electrons when the current reaches this value.

14 A charged thunder cloud is producing raindrops of average diameter 3 mm, each carrying a charge of 3×10^{-15} C. How heavy must be the rainfall (in cm per hour) for the electric current density to the earth to be 5×10^{-11} A m^{-2}?   (*L*)

15 It is estimated that the average quantity of electric charge transported in a lightning flash is 30 C. If the energy liberated is 2×10^{10} J, what is the p.d. involved? In a typical thunderstorm lightning flashes strike the ground at intervals of about 3 minutes. Over the whole surface of the earth the current carried in this way between the atmosphere and the ground amounts to 1800 A. Calculate the average number of thunderstorms taking place at any one instant over the whole earth.

16 An electric kettle (rated accurately at 2.5 kW) is used to heat 3 kg of water from 15 °C to boiling point. It takes 9.5 minutes. How much heat has been lost? Outline briefly the explanation of this loss of heat.   (S)

17 An immersion heater is placed in 0.16 kg of water, and a current of 4.0 A is passed through it for 2 minutes. The temperature is found to rise by 5.0 K. What is the p.d. across the heater?

18 An aluminium electric kettle weighing 0.5 kg contains 1 kg of water at 10 °C. It is fitted with a 1 kW heating element. How long will it take the kettle to come to the boil, and how much longer will it take to boil dry? Assume that there is no heat loss. (Specific heat capacity of aluminium
$$= 0.92 \times 10^3 \text{ J kg}^{-1} \text{ K}^{-1}$$
Specific latent heat of steam = 2.27×10^3 J kg^{-1}.)
                                                          (*Ox. Schol.*)

19 Given that 1 kg of copper contains approximately 10^{25} atoms, and that the density of copper is 9×10^3 kg m^{-3}, make a very rough estimate of the diameter of the copper atom, stating any assumptions you make.   (*O & C*)

20 Copper is an element of *relative atomic mass* 63.5 and *atomic number* 29. It has two naturally occurring *isotopes* of *mass numbers* 63 and 65. Give the meanings of the terms printed in italics and estimate the relative abundance of each isotope.   (*N*)

21 Given that the unified atomic mass constant m_u is 1.66×10^{-27} kg, calculate the mass of one atom of carbon-12. Hence calculate how many atoms there are in 1 kg of carbon-12.

If the density of graphite is 2.3×10^3 kg m^{-3}, what is the volume per atom of carbon making up a crystal of graphite?

Other questions: Ch. 2: 33, 34, 41 Ch. 4: 17, 18, 19 Ch. 13: 15 Ch. 16: 7

C

22 Find the momentum acquired by the electrons in 10 cm of wire when a current of 1 A starts to flow.   (*Cam. Schol.*)

CHAPTER 2
Resistance

A

1 A current of 0.20 A flows in a wire when the p.d. applied to it is 5.0 V. What is its resistance?

2 The p.d. across a 1.5 V cell falls to 1.3 V when a current of 0.4 A is taken from it, what is its internal resistance?

3 The current in a 10 Ω resistance must not exceed 0.4 A. What is the maximum p.d. that may be applied to it? What then is the power dissipated?

4 A p.d. of 5 kV is applied to a resistance of 20 MΩ. What current flows? What power is dissipated?

5 What is the resistance of (i) a 240 V, 60 W bulb, (ii) a 220 V, 100 W bulb?

6 Two cells each of e.m.f. 1.5 V and internal resistance 0.4 Ω are connected (i) in series, (ii) in parallel. What is the e.m.f. and internal resistance of the battery formed in each of these ways? What current would be driven by each of these batteries through a resistance of 1.6 Ω?

7 Resistances of 2 Ω and 3 Ω are joined (*a*) in series, (*b*) in parallel. What is the combined resistance in each case?

8 How many 240 Ω resistances must be joined in parallel to give a combined resistance of 40 Ω?

9 A battery of e.m.f. 1.5 V and internal resistance 0.5 Ω is connected to a resistance of 7.0 Ω. What current flows? What is the p.d. across the battery?

10 A small torch bulb is marked '2.5 V, 0.3 A'. (*a*) What is its resistance at its normal working temperature? (*b*) What resistance would you join in series with it to run it from a 6 V battery? (*c*) What resistance would you join in parallel with it so that the combination could be run in series with a 200 V, 100 W bulb from a suitable mains supply (with both bulbs at normal brightness)? (*d*) When the torch bulb is run directly from a battery of e.m.f. 3.0 V, the correct p.d. of 2.5 V is produced across it. What is the internal resistance of the cell?

11 A tape recorder is operated by a battery of five 1.5 V cells in series, each of internal resistance 0.2 Ω. What is the combined e.m.f. and internal resistance? What is the p.d. across the battery when a current of 150 mA is being taken?

12 A nominal 2 Ω resistor is found on test to have an actual resistance of 2.040 Ω. What length of constantan wire of resistance per unit length 8.5 Ω m^{-1} should be joined in parallel with the resistor so that the combined resistance is 2.000 Ω?

13 Equal lengths of iron and copper wire of the same diameter are connected first in parallel and then in series. If a potential difference applied between the ends of each arrangement in turn is gradually increased from a small value, it is found that in the first instance the copper begins to glow before the iron; in the second instance the reverse occurs. Explain this.   (*N*)

14 The r.m.s. value of a sinusoidal alternating current flowing in a circuit is 2.00 A. What is its peak value?

15 A sinusoidal alternating p.d. is applied to the terminals of a piece of equipment between which the insulation breaks down at 1000 V. What is the maximum r.m.s. value of the p.d. that may be applied?

16 An electric power line has a total resistance of 5.0 Ω. If the power input is 2×10^4 kW at 1.0×10^5 V, find (a) the percentage loss of power in transmission, (b) the drop in p.d. on the line.

If the figure of 1.0×10^5 V refers to the r.m.s. value of a sinusoidal alternating p.d., what will be the maximum p.d. for which the line must be insulated? 						(O & C)

17 A power cable consists of one strand of steel and six of aluminium, each of diameter 3.0 mm. Calculate its resistance per kilometre length.
(Resistivity of steel = 9.0×10^{-8} Ω m
Resistivity of aluminium = 2.7×10^{-8} Ω m.)
						(Cam. Forces)

18 An electric hot-plate has two coils of manganin wire, each 20 m in length, and of cross-sectional area 2.3×10^{-7} m². Show that it will be possible to arrange for three different rates of heating, and calculate the power dissipated in each case, when the heater is supplied from 200 V mains. The resistivity of manganin is 4.6×10^{-7} Ω m. (O & C)

19 An electric heating element to dissipate 450 W on 250 V mains is to be made from nichrome ribbon of width 1.0 mm and thickness 0.05 mm. Calculate the length of ribbon required.
(Resistivity of nichrome is 1.1×10^{-6} Ω m.) (L)

20 When a p.d. of 100 V is applied between opposite faces of a slab of polythene 5.0 mm thick and 0.50 m square, a current of 2.5×10^{-12} A is found to flow through it. What is the resistivity of polythene? What is its conductivity?

21 Calculate the resistance between opposite edges of a square film of carbon deposited on an insulating plate; the film is 5.0×10^{-7} m thick, and the resistivity of the carbon is 4.0×10^{-5} Ω m. If a similar film is deposited on the outside of an insulating rod 3.0 mm in diameter, what length of film on the rod will have a resistance of 100 Ω?

22 The heating 'element' of an electric radiator is rated as 1 kW on a 240 V supply, and the temperature of the element may be taken as 1000 °C. When the element is at 0 °C its resistance is 50.0 Ω. Calculate the mean temperature coefficient of resistance of the metal over the range 0 °C–1000 °C. 			(L)

23 An alternating current passes through a wire of resistance 8.4 Ω immersed in oil and it is found that the temperature rises at the rate of 3.0 K min⁻¹ when the temperature of the calorimeter and its contents is equal to that of the surroundings. If the heat capacity of the calorimeter and its contents is 840 J K⁻¹, calculate a value for the r.m.s. current. 							(L)

24 An electric fan heater is marked '2 kW, 50 Hz, 250 V, a.c. only'. The current I taken by the device varies with time t according to the equation $I = I_o \sin \omega t$. Find the value of I_o, ω, and the maximum instantaneous power consumed by the device. (Assume that the fan behaves as a pure resistor.)

Comment on the following statements made concerning the heater:

(a) 'It has a fan as well, so it must give much less heat than a 2 kW radiator.'
(b) 'It says 250 V, so a cable which will stand 300 V is good enough.'
(c) 'It says just a heater, so it must work on d.c. as well as a.c.'
(d) 'I never bother to connect the third wire . . .'
						(C)

Other questions: Ch. 1: 7, 9 Ch. 3: 10 Ch. 4: 1, 2 Ch. 8: 4 Ch. 9: 1 Ch. 11: 10 Ch. 13: 6, 7, 8

B

25 A circuit consists of a battery and two 1 Ω resistances. When the resistances are connected in parallel the battery delivers 1.5 A, and when they are in series the current through them is 0.6 A. Calculate the e.m.f. of the battery and its internal resistance.
						(Ox. Schol.)

26 A cell has an e.m.f. of 1.5 V and a resistance of 1.8 Ω. It is connected in series with a rheostat R and a milliammeter of resistance 3 Ω. A 2 Ω coil is then joined in parallel with the meter. What value of R will cause the meter to read 30 mA? What is the p.d. between the terminals of the cell in these circumstances? 						(L)

27 Current from 200 V mains is sent through resistances of 400 Ω and 600 Ω connected in series. What will be the p.d. between the ends of the 400 Ω coil? If an accurate voltmeter of resistance 1000 Ω is connected in parallel with the 400 Ω coil, what will it read? 						(O & C)

28 Current is passed from a battery having negligible internal resistance through resistors of 5000 Ω and 3000 Ω in series. An accurate voltmeter of resistance 2000 Ω reads 6 V when joined across the 3000 Ω coil. What is the e.m.f. of the battery, and what would the voltmeter read if joined across the 5000 Ω coil? 						(Cam. Forces)

29 Six resistors, AB, BC, CD, DE, EF, FA each having resistance 1000 Ω, are joined to form a closed circuit, and the points A and E are joined by a battery having e.m.f. 20 V and negligible internal resistance. What will be the reading of a voltmeter having resistance 2000 Ω joined between (a) A and C, (b) A and F, (c) F and C? 						(Cam. Forces)

30 A certain wire has resistance 27.30 Ω when free from tension. What will its resistance become if tension is applied so that the length increases by 1% and the diameter decreases by 0.4%, if the resistivity is unaltered? *(C)*

31 The resistance of a piece of steel wire of diameter 0.6 mm is reduced to a third of its value by coating it with copper. What is the thickness of the copper?

(Resistivity of copper $= 1.8 \times 10^{-8} \Omega$ m
Resistivity of steel $= 1.98 \times 10^{-7} \Omega$ m.) *(S)*

32 A uniform wire, of length 2 m and cross-sectional area 1.0×10^{-5} m², carries a steady current of 1.04 A. The resistivity of the material is $4.8 \times 10^{-6} \Omega$ m. What is the potential drop down each centimetre of the wire?

What current would be needed to give a potential drop of 5 mV down the whole length of the wire, and what resistance would be needed in series with the wire and a 2 V driver cell (of negligible resistance) to get this current? *(O)*

33 Copper contains 10^{29} 'free' electrons per m³ and its resistivity is $1.72 \times 10^{-8} \Omega$ m. A potential difference of 10 mV is set up between two points 10 cm apart on a uniform wire. Calculate the average drift velocity with which electrons will move through the wire, given that the charge on an electron is 1.60×10^{-19} C. *(O & C)*

34 Give a short account of the conduction of electricity through a metallic conductor.

Calculate the mean speed of the carriers of current in a silver wire under a potential gradient of 1 V m⁻¹ given the following data:

108 kg of silver contains 6.02×10^{26} atoms,
density of silver $= 10.5 \times 10^3$ kg m⁻³,
resistivity of silver $= 1.62 \times 10^{-8} \Omega$ m,
electronic charge $e = 1.60 \times 10^{-19}$ C.

State clearly any assumptions which you need to make.

Under a potential gradient of 1 V m⁻¹ the silver ion in aqueous solution moves much more slowly, the mean speed being about 10^{-8} m s⁻¹. Comment on the difference in the speeds of the carriers in the two cases. *(O & C)*

35 A current of 0.75 A passes through a lamp filament of diameter 0.12 mm. Find the rate of heat loss per unit area of surface of the filament.

(Resistivity of filament material $= 1.8 \times 10^{-8} \Omega$ m.) *(Cam. Oversea)*

36 A surge suppressor is made of a material whose conducting properties are such that the current passing through is directly proportional to the fourth power of the applied p.d. If the suppressor dissipates energy at a rate of 6.0 W when the potential difference across it is 240 V, estimate the power dissipated when the potential difference rises to 1200 V. *(L)*

37 Given that the resistivity of aluminium is twice that of copper, and that the density of aluminium is one-third that of copper, find the ratio of the masses of aluminium and copper conductors of equal length and equal resistance. *(N)*

Other questions: Ch. 1: 16, 17, 18 Ch. 3: 11, 12, 13, 14 Ch. 4: 8, 14, 17, 18, 19 Ch. 8: 9, 10, 11 Ch. 9: 5 Ch. 10: 14(*b*) Ch. 12: 6, 12 Ch. 13: 17, 18

C

38 Discuss the factors which govern the final steady temperature attained by a conductor carrying a current. How is the heating effect applied in the case of a fuse wire? A fuse wire breaks a circuit when a current of 5 A is passed through it. If the diameter of the fuse wire is doubled, all other factors remaining the same, what will now be the maximum current that can be passed? Show that your result does not depend upon the assumption of any particular law for the way in which the heat lost depends upon the excess temperature. *(W)*

39 Compare (*a*) the radii, (*b*) the lengths of the filaments of a 100 V 60 W tungsten lamp and a 230 V 100 W tungsten lamp, assuming that the operating temperatures of the two filaments are the same.

(O & C)

40 State Ohm's law and define *electrical resistance*. Discuss the application of your definition to (*a*) a metal wire, (*b*) a battery.

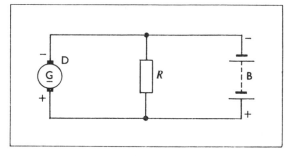

Fig. Q1

A circuit is joined as shown. D is a dynamo, B a battery, and R a load resistor, the following being the relevant data:

Dynamo: e.m.f. $= 17.0$ V, and internal resistance $= 0.50 \Omega$
Battery: e.m.f. $= 12.6$ V, and internal resistance $= 0.10 \Omega$
Load resistance $= 2.00 \Omega$.

Calculate the current through the load resistor and that through the battery, and state whether the latter is being charged or discharged. (*Cam. Forces*)

41 Assuming that a metal contains n 'free' conduction electrons per unit volume, explain what occurs when an electric field is set up in the wire by connecting a battery across its ends. Indicate the physical meaning of the concept of *resistance*.

If the value of n for copper is 10^{29} m^{-3}, and its resistivity is 1.6×10^{-8} Ω m, calculate the mean drift velocity of the electrons under an electric field of 100 V m^{-1}.   (*O & C*)

Other question: Ch. 13: 20

CHAPTER 3
Electrolysis and Thermoelectricity

Additional data required:
Charge of an electron $= -1.60 \times 10^{-19}$ C
Avogadro constant $L = 6.02 \times 10^{23}$ mol^{-1}
Faraday constant $F = 9.65 \times 10^4$ C mol^{-1}

A

1 What is the electric charge carried by a zinc ion? How many ions of zinc are required to transport 8.0 C of electric charge in electrolysis? If this process of transport takes 20 s, what is the current?
(Charge number of zinc = 2.)

2 An electric current of 10.0 A flowing for 1.00×10^3 s is found to deposit 11.2 g of silver in electrolysis. What is the average specific charge of the silver ions involved?

3 What is the mass of 6.0 mol of helium atoms (relative atomic mass 4.0)? How many atoms are there in this sample?

4 What is the *amount of substance* in 4.0 mg of caesium (relative atomic mass 133)? How many atoms of caesium are there in this sample?

5 A speck of plutonium dust contains 2.5×10^9 atoms. What is the *amount of substance* involved?

If the relative atomic mass of plutonium is 239, what is the mass of the speck of dust?

6 A steady current is passed for 50 minutes through a solution of copper sulphate and 5.5 g of copper are deposited on the cathode. Find the value of the current.
(Relative atomic mass of copper = 63.6
Charge number of the copper ion = 2.)

7 An article of total surface area 0.020 m^2 is to be plated to a depth of 0.020 mm with silver. How long will this take using a plating current of 3.5 A?
(Relative atomic mass of silver = 108
Charge number of the silver ion = 1
Density of silver = 10.5×10^3 kg m^{-3}.)

8 One kilogram of copper is to be purified by electrolysis, the p.d. across the electrolytic cell being 8.0 V. Calculate:
(*a*) the total quantity of electric charge that must be passed,
(*b*) the energy used.
(Relative atomic mass of copper = 63.6
Charge number of the copper ion = 2.)

9 The specific heat capacity of nitrogen at constant volume is 733 J kg^{-1} K^{-1}, and its relative molecular mass is 28.0. Calculate the average energy gained by a nitrogen molecule when the temperature of the gas is raised from 0 °C to 100 °C.

10 A battery of forty 2 V accumulators, each with internal resistance 0.1 Ω, is connected in series to be charged from a 100 V supply. If the charging current is 2 A, what resistance should be connected in the circuit?   (*Ox. Schol.*)

Other questions: Ch. 4: 1 Ch. 17: 3

B

11 An accumulator of e.m.f. 12 V and of internal resistance 1 Ω is to be charged from a 112 V d.c. supply.
(i) Draw a diagram of the circuit you would use.
(ii) Calculate the charging current had the accumulator been connected directly across the supply.
(iii) Calculate the value of the series resistance necessary to limit the charging current to 10 A.
(iv) With this resistance in the circuit, calculate the potential difference between the terminals of the accumulator.

State any assumptions you make in your calculations.   (*N*)

12 A battery of 24 accumulators in series is charged from 100 V d.c. mains. If the internal resistance of the *battery* is 2 Ω and the e.m.f. of each *cell* during charging is 2.5 V, what series resistance will be required to give a charging current of 10 A? What percentage of the energy taken from the mains will be converted into heat?   (*O & C*)

13 A box containing electrical equipment has two external terminals. One terminal is connected to one side of a 100 V electric supply. The other terminal is connected to the other side of the supply through a fixed resistance. An ammeter shows a constant current of 0.5 A through the box, and a voltmeter connected across the box terminals shows a p.d. of 75 V. The dissipation of power in the box is 25 W. Suggest what is in the box, and find the value of the external resistance.   (*W*)

14 Explain the terms *electromotive force* and

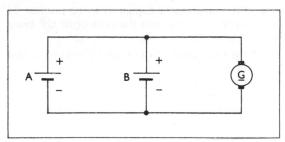

Fig. Q2

internal resistance of a cell. Describe how these quantities may be measured accurately.

Cell A of e.m.f. 1.6 V and internal resistance 0.1 Ω and cell B of e.m.f. 2.2 V and internal resistance 0.2 Ω are connected together as shown in the figure across a d.c. generator G of variable e.m.f. V and unknown polarity and whose internal resistance is 0.3 Ω. V is varied until the current through cell B is zero. At this value of V calculate:

(a) the current through cell A,

(b) the magnitude of V,

(c) the power generated,

(d) the energy dissipated per second in A and B, and

(e) the energy dissipated per second in the generator.

How may the differences between (d) + (e) and (c) be accounted for? (O & C)

Other questions: Ch. 1: 21 Ch. 2: 40 Ch. 4: 13
Ch. 9: 5 Ch. 10: 14(c) Ch. 16: 10

CHAPTER 4
Circuit Measurements

A

1 A certain thermocouple, which has a total resistance of 10 Ω, has one junction in melting ice and the other in steam. The e.m.f. between its ends, measured with a potentiometer, is 4.2 mV. What would be its reading when it is connected to a millivoltmeter which has a resistance of 50 Ω?

(O & C)

2 Explain how to use a potentiometer to measure the e.m.f. of a cell. What advantages does this method possess over a moving-coil voltmeter?

Two resistances, one of 800 Ω and the other of 1200 Ω, are joined in series and a current is passed through them from a cell of e.m.f. 2.00 V and negligible internal resistance. What value will be obtained for the potential difference across the 800 Ω coil (a) using a correctly calibrated voltmeter of resistance 1000 Ω, and (b) using a potentiometer?

(Cam. Forces)

3 Four resistances AB, BC, CD, DA are joined to form a Wheatstone bridge; a galvanometer is joined between B and D and a cell between A and C (with positive terminal joined to A). If AB = 4.0 Ω, BC = 5.0 Ω, CD = 6.0 Ω, and the bridge is balanced, what is the value of the resistance DA?

If this latter resistance is slightly increased, which way will the current flow through the galvanometer?

4 An electric current is passed through an ammeter and a standard coil of resistance 0.100 Ω, connected in series, and is adjusted intil the ammeter reads 5.00 A. The p.d. between the ends of the standard coil is balanced by the p.d. across 128.8 cm of a potentiometer wire. A standard Weston cell of e.m.f. 1.018 V is found to require 254.5 cm of the same potentiometer wire for a balance. What is the error in the ammeter reading? (O & C)

5 The resistance of the coil of a platinum resistance thermometer is 3.020 Ω at 0 °C, 4.172 Ω at 100 °C, and 13.20 Ω in a furnace. Deduce the temperature of the furnace as given by this thermometer.

(S)

Other question: Ch. 8: 2

B

6 Derive the relation between the resistances of the resistors forming a balanced Wheatstone bridge.

Four resistors AB, BC, CD and DA are joined with a cell and galvanometer to form a Wheatstone bridge. At temperature 0 °C each has resistance 10 Ω. AB and CD are made of iron wire while BC and DA are made of constantan wire. What change must be made in the resistance of BC in order to preserve the balance if the temperature of the whole rises to 100 °C?

(Temperature coefficient of resistance of iron
$= 5.0 \times 10^{-3}$ K^{-1};
temperature coefficient of resistance of constantan is negligible.)

(Cam. Forces)

7 Four 10 Ω coils AB, BC, CD, DA have a cell of e.m.f. 1 V and negligible internal resistance connected across AC.

(a) Calculate the change in the p.d. between B and D that occurs if an addition of 0.1 Ω is made to CD.

(b) The p.d. between B and D can be restored to the old value by connecting a resistor in parallel with one of AB, BC, or DA. Calculate the resistance of this resistor and state the position in which it must be connected. (Cam. Forces)

8 Explain how a potentiometer could be used to find the ratio of two resistances each known to be about 0.1 Ω. Draw a diagram of a suitable circuit and indicate on it appropriate values for the components.

The e.m.f. of a cell measured using a potentiometer is found to be 1.5 V. A voltmeter connected across the terminals of the cell reads 1.25 V. Explain the discrepancy and calculate the ratio of the resistance of the voltmeter to the internal resistance of the cell.

(O & C)

9 An accumulator of e.m.f. 2.0 V and negligible internal resistance is connected across a uniform wire of length 100 cm and resistance 5.0 Ω. The appropriate terminal of a cell of e.m.f. 1.5 V and internal resistance 0.90 Ω is connected to one end of the wire, and the other terminal of the cell is connected through a sensitive galvanometer to a slider on the wire. What length of the wire will be required to produce zero deflection of the galvanometer?

How will the balancing length change when a coil of resistance 1.0 Ω is placed (a) in series with the accumulator, (b) in parallel with the cell of e.m.f. 1.5 V?

(C)

10 Given a cell of e.m.f. 1.0 V as a reference standard, how would you measure accurately the e.m.f. of

(a) a thermocouple of e.m.f. about 5 mV; and
(b) a power pack of e.m.f. about 5 kV and internal resistance about 1 MΩ?

In each case give approximate values for the components in your circuit and estimate (with justification) the accuracy that you would expect to achieve.

(O & C)

11 A 2 V cell is connected in series with a resistance R and a uniform wire AB of length 100 cm and resistance 4 Ω. One junction of a thermocouple is connected to A, and the other through a galvanometer to a tapping key. No current flows in the galvanometer when the key makes contact with the mid-point of the wire. If the e.m.f. of the couple was 4 mV what was the value of R? If the same resistance R is now increased by 4 Ω, by how much would the balance point change?

(S)

12 Fig. Q3 shows part of a potentiometer circuit in which the wire AB is 1 m long and has a resistance of 2 Ω. The e.m.f. of the driver cell D is 2 V, and that of the standard cell S connected between C and B is 1.018 V. Calculate the values which the resistances R_1 and R_2 must have in order to produce a p.d. of 5 mV between the ends A and B of the potentiometer wire when no current is passing through the standard cell S.

A thermocouple of fine wire has a resistance of of 15 Ω and gives an e.m.f. of 40 μV for 1 K difference of temperature between the junctions.

(a) Find the length of the potentiometer wire which

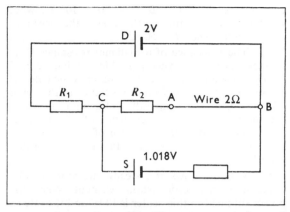

Fig. Q3

will balance the e.m.f. produced by the thermocouple when one junction is at 15 °C and the other at 100 °C.

(b) If a sensitive meter with a resistance of 25 Ω were used instead of the potentiometer, to measure the p.d. between the terminals of the thermocouple, what value would be obtained for the p.d. when the junctions are maintained at the same temperatures of 15 °C and 100 °C? On the basis of 40 μV K^{-1}, what temperature difference would be inferred from this value?

(O)

13 Describe and explain how to determine by experiment the relation between the e.m.f. E of a thermocouple and the temperature θ (in °C) of its hot junction, the cold junction being maintained at 0 °C.

The relation is sometimes given as (a) $E = \alpha\theta + \beta\theta^2$ and sometimes as (b) $E = k\theta^n$, where α, β, k, and n are constants. Indicate how you would plot your observations to test these equations and to determine values for the constants.

(L)

14 Write a short account of the variation with temperature of the electrical resistance of pure metals.

Give a brief description of a platinum resistance thermometer, and explain the principle of the electrical circuit in which it is normally used.

The resistance of the tungsten filament of an electric lamp was found to be 44.0 Ω at 0 °C and 65.7 Ω at 100 °C. When the lamp was running normally on 230 V mains it took 100 W. Calculate the temperature of the hot filament measured on the tungsten resistance scale.

(O & C)

15 If the resistance R_t of the element of a resistance thermometer at a temperature of t (in °C) on the ideal gas scale is given by $R_t = R_0(1 + At + Bt^2)$, where R_0 is the resistance at 0 °C and A and B are constants such that $A/B = -6.50 \times 10^3$ K, what will be the temperature on the scale of the resistance thermometer when $t = 50\,°C$? (*L*)

16 The resistance of a platinum resistance thermometer is 10.54 Ω when placed in melting ice, and 14.69 Ω when placed in boiling water. Calculate (*a*) the temperature coefficient of resistance of the metal of the coil, (*b*) the temperature of a liquid in which the thermometer has a resistance of 11.30 Ω, (*c*) the minimum detectable change of temperature if the least change of resistance that can be detected is 0.01 Ω. (*C*)

17 A liquid flows at a steady rate along a tube containing an electric heating element. When the power supplied to this heater is 10 W and the rate of flow of the liquid is 50 g min⁻¹, the outflow temperature of the liquid is 5 K higher than the inflow temperature. When the rate of flow is doubled, the power supplied to the heater must be increased to 19 W for the outflow temperature to be unchanged. If the inflow temperature is the same in each experiment, calculate a value for the specific heat capacity of the liquid. (*N*)

18 Electrical energy is supplied at a rate of 55 W to a heating coil immersed in a liquid, causing the liquid to boil. The vapour is condensed and collected at a steady rate of 8.52 g min⁻¹. When the power is changed to 34 W, the steady rate at which the liquid is collected becomes 4.97 g min⁻¹. Calculate the specific latent heat of vaporization of the liquid and the heat loss per minute. Discuss the advantages and disadvantages of this experimental method. (*N*)

19 A copper vessel of mass 0.1 kg contains 0.2 kg of water. Initially, the vessel and its contents are at the temperature of the air, which is 15 °C. It is then heated by an immersion heater which supplies 80 W, and after some time reaches a steady temperature of 95 °C.

(*a*) Find the initial rate of rise of temperature when the immersion heater is switched on.

(*b*) After the steady temperature has been reached, the heater is switched off. Find the initial rate of fall of temperature, and explain your reasoning.

(*c*) Assuming that, for temperatures up to 100 °C, the rate of loss of heat to the surroundings by all processes is proportional to the temperature excess over the surroundings, find the power which must be supplied just to maintain the temperature at 100 °C.

(*d*) Find the rate at which water will be boiled away if the power is increased to 160 W.

(Take the specific heat capacity of water to be 4.2×10^3 J kg⁻¹ K⁻¹, that of copper to be 400 J kg⁻¹ K⁻¹, and the specific latent heat of vaporization of water to be 2.3×10^6 J kg⁻¹.) (*O*)

Other questions: Ch. 2: 29, 32

C

20 One arm of a Wheatstone bridge network of equal resistances of 1000 Ω is replaced by a 1 V cell. The galvanometer resistance is 50 Ω and the driver cell is 2 V. The positive terminals of both cells are connected to the same point. If neither of the cells has appreciable internal resistance, calculate the current through the galvanometer and each of the cells. (*Ox. Schol.*)

21 P, Q, R and S are the resistances, taken in cyclic order, in the four arms of a Wheatstone bridge. P and Q are the ratio arms and each is approximately 1 Ω; S is a standard resistance, 1.001 50 Ω. To balance the bridge it is necessary to shunt Q with 2000 Ω. When R and S are interchanged and the shunt across Q removed, balance is restored when P is shunted with 1000 Ω. Obtain a value for the resistance of R. (*L*)

22 A Wheatstone bridge is made up from four resistances each of 100 Ω, a galvanometer also of resistance 100 Ω and a 2 V battery of negligible internal resistance. If one of the resistances is used as a thermometer and has a temperature coefficient of resistance of 4×10^{-3} K⁻¹, find the least change of temperature detectable if the galvanometer is sensitive to a current of 10^{-6} A. (*Cam. Schol.*)

CHAPTER 5
Magnetic Forces

Additional data required:
Charge of an electron $= -1.60 \times 10^{-19}$ C

A

1 A conductor 2 cm long, carrying a current of 8 A, lies at right-angles to a magnetic field in which the flux density is 1.0 T. Calculate the force exerted on the conductor and give a diagram to show its line of action and direction. (*Cam. Forces*)

2 One end of a simple rectangular wire-loop current balance is inserted into a solenoid. A force of 3.0×10^{-3} N is found to act on this end when a current of 2.0 A is flowing in it. If the length of conductor forming the end of the wire-loop is 0.10 m, what is the flux density in the solenoid?

3 A straight wire 1.0 m long carries a current of 100 A at right-angles to a uniform magnetic field of 1.0 T. Find the mechanical force on the wire and the power required to move it at 15 m s⁻¹ in a plane at right-angles to the field. (*S*)

4 A magnetized needle is allowed to oscillate in turn over a glass sheet and then over a sheet of copper. Describe and explain what is observed.

(*Cam. Oversea*)

5 What is the *electromagnetic moment* of a plane rectangular coil of 20 turns, measuring 20 cm by 10 cm, carrying a current of 0.5 A? What is the moment of the couple acting on this coil when it is placed with its plane parallel to a uniform magnetic field of flux density 4.0×10^{-2} T? What is the moment of the couple if the coil is turned through an angle of (a) 60°, (b) 90°, from its first position?

6 What is the force acting on an electron moving with a velocity of 2.0×10^7 m s^{-1} at right-angles to a uniform magnetic flux density of 1.5×10^{-2} T?

7 The current in a certain coil grows initially at a rate of 25 A s^{-1} when a steady p.d. of 5.0 V is applied to it. What is its self-inductance? If the mutual inductance between this coil and another coil nearby is 0.40 H, what is the e.m.f. induced in the latter coil?

8 What is the energy required to establish a current of 1.5 A in a coil of inductance 1.6 H?

9 The mutual inductance between two adjacent air-cored coils is 0.20 H. A current of 2.5 A in one coil is reduced to zero at a uniform rate in one millisecond. What is the e.m.f. induced in the second coil?

10 A current of 5.0 A flowing in a flat circular coil of 25 turns is found to produce a magnetic flux through the core of the coil of 3.0×10^{-5} Wb. Calculate the inductance of the coil.

Other questions: Ch. 6: 1 Ch. 9: 1, 4 Ch. 12: 1, 2 Ch. 13: 12

B

11 A closed tube of square cross-section measures 1 cm × 1 cm × 10 cm. It is filled with mercury and placed in a magnetic field of 1 T transverse to the axis of the tube. A current of 1000 A is passed between opposite sides of the tube in the direction at right-angles to the field. What is the pressure difference between the ends?

(Cam. Schol.)

12 A rectangular coil consisting of 100 turns, each 2.50 cm by 1.50 cm, is suspended so that it may turn about an axis through its centre parallel to one of its sides and perpendicular to a uniform magnetic field. When no current flows through the coil its plane is parallel to the field. When a current of 5.00 mA is passed through the coil it comes to rest with its plane making an angle of 30° with the direction of the field. If the torsion constant of the suspension is 1.44×10^{-4} N m rad^{-1}, calculate the flux density of the field.

(Cam. Oversea)

13 A small compass needle is freely pivoted so that it can swing in a horizontal plane. When a bar magnet is placed with its axis on the line passing through the centre of the compass perpendicular to the magnetic meridian, the needle is deflected through an angle of 35°. A second magnet is then placed with its axis on the same line and moved until the deflection is reduced to zero. If one of the magnets is now turned over end for end, what deflection will be produced?

14 An aeroplane with a wingspan of 25 m is flying horizontally at a speed of 900 km h^{-1}. Calculate the p.d. between the wingtips if the vertical component of the earth's field is 4.0×10^{-5} T. If a wire is connected between the wingtips, discuss whether a current will flow in it.

15 A fan having four blades each 20 cm long rotates at 3000 revolutions per minute about a horizontal axis in the magnetic meridian at a place where the horizontal component of the earth's flux density is 2.0×10^{-5} T. Calculate the induced e.m.f. (a) between the tip of a blade and the axle, (b) between the tips of diametrically opposite blades, (c) between the tips of two blades at right-angles to one another.

(Cam. Forces)

16 A horizontal metal disc of radius 10.0 cm is rotated about a central vertical axis at a region where the value of the earth's magnetic flux density is 5.3×10^{-5} T and the angle of dip 70°. A sensitive galvanometer of resistance 150 Ω is connected between the centre of the disc and a brush pressing on the rim. Assuming the resistance of the disc to be negligible, what will be the current through the galvanometer when the disc is rotated at 1500 revolutions per minute? If the system is frictionless, calculate the power required to maintain the motion.

(C)

17 A closed circular loop of wire of radius 5 cm is placed with its plane at right-angles to a uniform magnetic field in which the flux density is changing at the rate of 1.0×10^{-2} T s^{-1}. Calculate (a) the e.m.f. induced in the loop, (b) the energy dissipated in the loop in 10 s if its resistance is 2 Ω. (The self-inductance of the loop may be ignored.)

Explain why there is a tension in the loop while the magnetic field is changing. How does it vary with time? *(O & C)*

18 Calculate the r.m.s. value of the e.m.f. produced by a coil having 50 turns of wire each of area 30 cm^2, when the coil is rotated at a uniform speed of 2100 revolutions per minute about an axis perpendicular to a uniform magnetic field of flux density 0.80 T. *(Cam. Oversea)*

19 A 12 V lamp and a coil of negligible resistance but appreciable inductance are connected in series across (a) 12 V d.c. mains, (b) 12 V a.c. mains.

In case (*b*) the lamp is appreciably less bright than in case (*a*) and when a laminated soft iron core is pushed into the coil the lamp goes out in case (*b*) but is unaffected in case (*a*). Explain these observations. What would you expect to happen in case (*b*) if the iron core was absent and the frequency was gradually increased without altering the p.d.?

(*O & C*)

20 A coil of self-inductance 10 H in series with a wire of negligible resistance which fuses at a current of 10 A is connected to a 100 V d.c. mains supply. How long after making the connection will the wire fuse?

Describe the changes of current and p.d. before and after fusing.   (*Ox. Schol.*)

21 A coil of self-inductance 10 H is joined to a battery of e.m.f. 20 V, the total circuit resistance being 5 Ω. Calculate:

(*a*) The rate of growth of current at the instant at which the circuit is closed.

(*b*) The final value of the current.

(*c*) The mean e.m.f. induced if the circuit is broken in such a way that the current falls to zero in 0.1 s.

(*Cam. Forces*)

22 Describe the construction of a simple form of alternating current transformer.

If the secondary coil is on open circuit explain, without calculation, the effect on the current flowing in the primary of (*a*) a fall in the supply frequency, (*b*) a reduction in the number of primary turns.

Calculate the current which flows in a resistance of 3 Ω connected to a secondary coil of 60 turns if the primary has 1200 turns and is connected to a 240 V a.c. supply, assuming that all the magnetic flux in the primary passes through the secondary and that there are no other losses.   (*O & C*)

23 The primary and secondary coils of a transformer have negligible resistance and are wound so that all the magnetic flux due to one coil is linked by the turns of the other coil. The current through the primary coil is increased at a uniform rate from zero to 10 A in 2.5 s. With the secondary coil open-circuited, a steady potential difference of 0.5 V appears across the secondary, and a steady potential difference of 0.04 V across the primary. Find (i) the self-inductance of the primary coil, (ii) the ratio of the number of turns on the primary and secondary coils, (iii) the self-inductance of the secondary coil.

(*O & C*)

24 A current of maximum value 1 A, alternating at 50 Hz with pure sine-wave form, flows in the primary of a mutual inductance of 0.20 H. Find the maximum flux through the secondary of the mutual inductance, and the maximum e.m.f. induced in the secondary.   (*N*)

25 A flat circular coil of 2.0 cm diameter forms part of a circuit of 60 Ω resistance. There are 80 turns in the coil and when it is thrust suddenly into the space between the pole pieces of a large electromagnet so that its plane is perpendicular to the field, 123 μC of electric charge flows round the circuit. Calculate the field strength of the magnet.   (*L*)

26 A ballistic galvanometer gives a throw of 10 cm when a charge of 10^{-6} C is rapidly passed through it. It is used to determine the field between the poles of an electromagnet by connecting it in series with a search coil and resistance to make the total resistance of the galvanometer circuit 1000 Ω. The search coil has 10 turns of wire wound on a cylindrical former 1 cm in diameter. This is placed in the unknown field with its plane perpendicular to the field and, after the galvanometer has come to rest, the coil is quickly moved to a position of zero field. The throw of the galvanometer is 9 cm. What is the field in the gap of the electromagnet? (Galvanometer damping may be neglected.)   (*C*)

Other questions: Ch. 6: 7, 8, 9, 10, 11, 12, 15, 16, 17, 18 Ch. 8: 7 Ch. 10: 14(*a*) Ch. 12: 6, 13 Ch. 13: 13, 14 Ch. 17: 26

C

27 A coil whose self-inductance is 0.05 H and whose resistance is 100 Ω is connected to a battery. If a switch is thrown which disconnects the battery and replaces it by a short circuit, how long will it take for the current to fall to one-quarter of its initial value?   (*S*)

28 A superconducting solenoid (which has no resistance), immersed in liquid helium, is connected to a 2 V battery of negligible internal resistance. After 50 s the current carried is 5 A, and the field in the solenoid has reached a very high value. If the solenoid now 'goes normal' so that a large resistance appears and its stored energy is suddenly converted to heat, what volume of helium gas at s.t.p. will boil off?

(Specific latent heat of boiling of helium
$$= 2.2 \times 10^4 \text{ J kg}^{-1},$$
Relative molecular mass of helium = 4,
Molar volume of a gas at s.t.p.
$$= 2.24 \times 10^{-2} \text{ m}^3 \text{ mol}^{-1})$$
(*Cam. Schol.*)

Other questions: Ch. 6: 19, 20 Ch. 12: 31, 33 Ch. 15: 4

CHAPTER 6
Magnetic Fields

Additional data required:
Magnetic constant $\mu_0 = 4\pi \times 10^{-7}$ H m^{-1}
(permeability of a vacuum)
Acceleration due to gravity $g = 9.8$ m s^{-2}

A

1 A solenoid of length 30 cm is uniformly wound with 400 turns of wire. If a current of 1.5 A is flowing in it, what is the flux density of the magnetic field at its centre? If the solenoid is of square cross-section (2.0 cm by 2.0 cm), what is the total magnetic flux through its centre?

2 A long single-layer solenoid is wound with insulated wire of diameter 1.6 mm, the turns being laid down closely touching one another. What current is required to produce at its centre a field of flux density 5.5×10^{-3} T?

3 A current of 700 A for a certain piece of equipment is led into and out of it by means of two long flat parallel metal strips 0.50 m wide. If the separation of the strips is small compared with their width, what is the magnetic flux density (a) between the strips, (b) just above the top one?

4 What is the flux density at a distance of 0.10 m in air from a long straight conductor carrying a current of 6.5 A? Hence calculate the force per metre on a similar parallel conductor distant 0.10 m from the first and carrying a current of 3.0 A. Explain how the expression for the force between two such conductors is used to define the ampere. (C)

5 Calculate the force per unit length between two parallel long straight wires 2 cm apart in air, each carrying a current of 5 A. Show on a diagram the direction of the force acting on each wire. (O)

6 Find an approximate value for the inductance of a uniformly wound solenoid 50 cm long and of diameter 2 cm if there are 2000 turns and no magnetic material is present. (L)

Other questions: Ch. 5: 2, 7, 10

B

7 Draw a diagram to show the magnetic field due to a long solenoid carrying a current. Write down an equation for the flux density at its centre and at its ends, explaining your units. Describe an experiment you would perform to verify that the ratio of these flux densities is as predicted by your formulae.

A metal wire 10 m long lies east–west on a wooden table. What p.d. would have to be applied to the ends of the wire, and in what direction, in order to make the wire rise from the surface? Assume that the electrical connections to the wire cause no appreciable restraint.

(Density of the metal $= 1.0 \times 10^4$ kg m^{-3},
Resistivity of the metal $= 2.0 \times 10^{-8}$ Ω m,
Horizontal component of the earth's field
$= 1.8 \times 10^{-5}$ T.) (C)

8 A narrow solenoid 50 cm long, uniformly wound with 1000 turns, carries a current of 4.0 A. A square coil of side 2.0 cm having 40 turns is suspended inside the solenoid. Calculate the maximum couple exerted on the coil if the current passing through it is 0.25 A, and give a diagram showing the position of the coil when the couple has this value.

Discuss whether the couple on the coil would have the same mean value in the same position if the solenoid current were 4.0 A a.c. and the coil current 0.25 A a.c. of the same frequency. (*Cam. Forces*)

9 A flat circular coil consisting of 150 turns of mean diameter 50 cm is fixed with its plane vertical and parallel to the magnetic meridian and a current of 2 A is passed through it. A flat rectangular coil of 100 turns measuring 2 cm by 3 cm is situated at the centre of the larger coil with its longer sides vertical and its plane perpendicular to the meridian. Calculate the couple acting about a vertical axis on the small coil when a current of 4 mA is passed through it. (L)

10 A circular coil of radius 15 cm was mounted with its plane horizontal and another circular coil of 100 turns each of radius 2 cm was pivoted at its centre so that it could turn about a diameter of the large coil. The small coil carried a light counterbalanced pointer fixed at right-angles to its plane. The coils were connected in series and a current passed through them. A rider of mass 0.04 g was adjusted on the pointer to keep the plane of the small coil vertical. The following results were obtained for d, the distance from rider to pivot and N, the number of turns in the large coil.

d/cm	14.7	13.4	12.0	10.6	9.3	8.0
N	100	90	80	70	60	50

Represent these results graphically and use the graph to calculate the current in the coils and the strength of the earth's vertical component.

State clearly any formulae used in the calculation, and list the probable sources of error in this experiment. (S)

11 A coil of 50 turns of mean radius 5 cm is placed with its plane vertical in the earth's magnetic meridian and a short bar magnet is suspended at its centre so that it is free to rotate in a horizontal plane. Calculate the current required in the coil to cause a steady deflection of the magnet through 45° from the meridian. If the couple acting on the deflected magnet due to the current is 10^{-6} N m, calculate the electromagnetic moment of the magnet.

(Assume the horizontal component of the earth's magnetic field = 2.0×10^{-5} T.)

12 A small magnet, suspended with its axis horizontal so as to be able to rotate freely about a vertical axis, is situated at the centre of a long horizontal solenoid, the axis of which lies at right-angles to the magnetic meridian. If the solenoid has 2000 turns per metre, determine the value of the current passing through it which would cause the magnet to rotate through 50°.

(The horizontal component of the earth's magnetic field = 1.8×10^{-5} T.)   (N)

13 Two horizontal parallel straight conductors each 20 cm long are arranged in air one vertically above the other, and connected in series so that the current flowing through one is in the opposite direction to that flowing through the other. The lower conductor is fixed while the upper one is free to move vertically in guides under its own weight, the conductors always remaining parallel. If the upper conductor weighs 1.20 g, what is the approximate value of the current that will maintain the conductors at a distance of 0.75 cm apart? Neglect the effect of the earth's magnetic field.   (Cam. Oversea)

14 Two long straight wires lie parallel to each other a distance 5 cm apart. If one carries a current of 2 A and the other a current in the opposite direction of 3 A, find the force each wire exerts on the other, and the magnetic field at a point distant 5 cm from each.   (Ox. Schol.)

15 An air-cooled solenoid 100 cm long, diameter 2.00 cm, is uniformly wound with 1000 turns of wire in which a direct current of 2.00 A flows. Find the magnetic field strength at the middle of the solenoid. A short secondary coil of 200 turns is wound over the middle section of the solenoid. Find the e.m.f. induced in the secondary coil if the current in the solenoid is uniformly reduced to zero and then raised to its former value, but in the opposite direction, in 0.500 s.   (L)

16 A long solenoid consists of a single layer of turns of wire in the form of a helical spring. A small circular coil is placed coaxially inside the solenoid near its mid point, and connected to a centre-zero galvanometer. Explain carefully why the galvanometer deflects when the current in the solenoid is switched on.

If the galvanometer deflects through 10 divisions to the right when the current is switched on in this way, what would happen in each case if (a) the current were suddenly reversed, or (b) the circular coil were rapidly moved to one end of the solenoid, keeping it coaxial, or (c) the solenoid were suddenly compressed to two-thirds of its original length, the current remaining constant?   (O & C)

17 A small solenoid consisting of 100 turns of wire and having a cross-sectional area of 4 cm² is placed inside a larger solenoid which has 20 turns per cm and carries a current of 1 A. The small solenoid is connected to a ballistic galvanometer and the total resistance in the circuit is 500 Ω. What charge will pass through the galvanometer when the current in the larger solenoid is switched off?   (S)

18 A ballistic galvanometer of resistance 15.0 Ω and sensitivity 5.0 divisions per microcoulomb is connected in series with a resistance of 100 Ω and a secondary coil of 500 turns and of resistance 50 Ω. This coil is wound round the middle of a long solenoid of radius 3.0 cm having 10 turns per cm and carrying a current of 0.60 A. Assuming no damping, calculate the deflection produced in the galvanometer when the current in the solenoid is switched off.   (L)

Other questions: Ch. 5: 16, 23

C

19 A solenoid with its axis perpendicular to the magnetic meridian is 1 m long, has 1500 turns, and carries a current of 2 A. A copper disc 4 cm in diameter is rotated at a uniform speed of 300 revolutions per minute about a thin axle which lies along the axis of the solenoid, so that the disc is well inside the uniform solenoid field. Calculate the p.d. between the axle of the disc and its rim.

A lead with a rubbing contact with the axle of the disc is taken through a sensitive galvanometer to a contact A on a length of low resistance wire W connected in series with the solenoid. A second lead from a rubbing contact on the rim of the disc is taken to a contact B which slides along the wire W. The distance between A and B is adjusted until no current flows in the galvanometer. Calculate the resistance of the wire between A and B.   (O)

20 A 50 Hz alternating current of 1 A passes through a plane circular coil of radius 10 cm and 100 turns. Find the e.m.f. induced in a coplanar circular coil of 10 turns, each of radius 0.5 cm, with its centre at the centre of the larger coil.   (Ox. Schol.)

21 Explain the meaning of the terms *self-inductance* and *mutual inductance*. Calculate an approximate value for the self-inductance of a solenoid 40 cm long and 4 cm in diameter consisting of 100 turns. Is your value higher or lower than the true value? Give reasons.

A second, very short solenoid consisting of 200 turns 2 cm in diameter is mounted symmetrically inside and along the axis of the first. What is the mutual inductance of the two coils?

The outer solenoid is connected to an a.c. generator which maintains a r.m.s. potential difference of 1 V at a frequency of 10 kHz between its ends, the inner solenoid being left on open circuit. Calculate (a) the current in the outer solenoid assuming that its resistance is negligible, (b) the potential difference between the ends of the inner solenoid.

Describe and explain qualitatively the effect on the current in the outer coil if the inner coil is connected to a circuit and a current allowed to flow in it.
(O & C)

22 A coil of wire of 100 turns forms the perimeter of a rectangle with sides 10 cm by 200 cm. Estimate the magnetic flux density at the centre of the rectangle when a current of 1 A flows around the coil, explaining any approximations you make. A second small coil with five turns of area 1 cm² is placed at the centre of and coplanar with the rectangular coil. Calculate approximately (a) the mutual inductance between the two coils, (b) the r.m.s. value of the electromotive force induced in the small coil when an alternating current of r.m.s. value 1 A and frequency 50 kHz is passed through the rectangular coil.
(O & C)

CHAPTER 7
Magnetic Materials

Additional data required:
Magnetic constant $\mu_0 = 4\pi \times 10^{-7}$ H m^{-1}
(permeability of a vacuum)

A

1 A narrow solenoid 50 cm long having an iron core is wound with 1600 turns of wire. When a current of 2 A is passed the flux density in the iron is found to be 0.8 T. Calculate the relative permeability of the iron in these conditions.
(Cam. Forces)

2 An iron ring of mean diameter 30 cm and of cross-sectional area 4 cm² is to be magnetized by a coil uniformly wound round its circumference. If the current in the coil is to be 1.5 A, how many turns of wire are required to produce a total flux density of 3.2×10^{-4} Wb in the iron? Take the relative permeability of the iron to be 1000 under these conditions.

B

3 Compare and contrast the magnetic properties of ferromagnetic, paramagnetic and diamagnetic substances, and give a qualitative account of the processes which occur in each of these three classes of magnetic material when subjected to an external magnetic field.
(L)

4 A primary coil of 600 turns is wound uniformly on an iron ring of mean radius 8.0 cm and cross-sectional area 4 cm². A secondary coil of 5 turns is wound on the top of the primary, but insulated from it, and is connected with a ballistic galvanometer, the total resistance of the secondary circuit being 400 Ω. What quantity of electric charge will be discharged through the galvanometer when a current of 3 A is reversed in the primary coil, assuming that the relative permeability of the iron is 500?
(O & C)

5 An iron ring has a cross-section of 3 cm² and a mean diameter of 25 cm. An air gap of 0.4 mm has been made by a cut across the section of the ring. The ring is wound with a coil of 200 turns through which a current of 2.0 A passes. If the total magnetic flux is 2.1×10^{-4} Wb find the relative permeability of the iron.
(S)

Other questions: Ch. 6: 11 Ch. 12: 7

CHAPTER 8
Measuring Instruments

A

1 Two moving-coil galvanometers, P and Q, are alike in all respects except that P's coil has 10 turns of resistance 2 Ω and Q's coil has 100 turns of resistance 180 Ω. Compare (a) their current sensitivities, (b) their voltage sensitivities.
(O & C)

2 Explain, giving a circuit diagram, how you would use a potentiometer method to test the calibration of an ammeter of range up to 2 A.

When checked by this method, a certain moving-coil ammeter appeared to read approximately 2% low at all parts of the scale. What possible causes can you suggest for this?
(Cam. Forces)

3 A milliammeter with a full-scale deflection of 10 mA has a resistance of 7.3 Ω. What resistance would be necessary, and how would it be connected, so that the instrument could be used (a) as an ammeter up to 1.5 A, (b) as a voltmeter up to 150 V?
(O & C)

4 Give a labelled diagram which shows the construction of a moving-coil ammeter.

If such an instrument gives a full-scale deflection for a current of 15 mA and has a resistance of 5 Ω, how would you adapt it so that it could be used (i) as a voltmeter reading to 1.5 V, (ii) as an ammeter reading to 1.5 A? In case (i) above what p.d. would be indicated if the instrument were connected to the terminals of a cell of e.m.f. 1.40 V and internal resistance 5 Ω?   (W)

5 An ammeter of 0.02 Ω resistance has a linear scale and requires a current of 0.5 A to produce a full-scale deflection. How can it be adapted to operate as an ammeter reading to 10 A?

Show that the scale will still be linear.   (S)

6 The coil of a milliammeter has a resistance of 100 Ω and a potential of 100 mV between the terminals produces a full-scale deflection. Describe and explain quantitatively how the meter could be made to read (a) currents up to 1 A, (b) p.d.'s up to 10 V.
                    (O & C)

Other questions: Ch. 4: 4 Ch. 5: 5

B

7 A galvanometer with a linear scale has a coil consisting of 50 turns of wire wound on a former 2 cm square, mounted in a magnetic field of flux density 0.3 T. A current of 10 mA produces a deflection of 5°. Calculate the couple required to produce a twist of 1 radian in the suspension.   (O & C)

8 Two galvanometers, identical in all other respects, are fitted by the makers with coils in the one case of 50 turns and 5 Ω resistance, and in the other of 600 turns and 720 Ω resistance. Which will give the greater deflection when connected to a battery of 1.5 V e.m.f. and 80 Ω internal resistance?
                    (O & C)

9 Describe the structure and explain the action of a moving-coil galvanometer. Discuss the factors which contribute to high sensitivity in this type of galvanometer.

A galvanometer of resistance 95 Ω, shunted by a resistance of 5 Ω, is joined in series with a resistance of 2×10^4 Ω and an accumulator of e.m.f. 2.0 V. What is the sensitivity of the galvanometer if it shows a deflection of 50 divisions?   (L)

10 An accumulator of e.m.f. 2 V and negligible internal resistance is connected across two resistors AB and BC in series. The resistance of AB is 1000 Ω and that of BC is 1 Ω. A sensitive moving-coil galvanometer of resistance 200 Ω is connected across BC, and registers a deflection of 35 divisions. What is the current corresponding to a deflection of one division?   (O)

11 The galvanometer in the circuit shown (Fig. Q4) has a resistance of 20 Ω.

The terminals XY are connected to a thermopile generating an e.m.f. of 0.50 V with an internal resistance of 20 Ω.

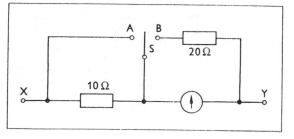

Fig. Q4

Calculate the current flowing in the galvanometer

(i) when the switch S is in position A,
(ii) when the switch S is in position B.

Calculate the potential difference across the terminals of the thermopile.

(iii) when the switch S is in the position A,
(iv) when the switch S is in the position B.

This circuit is sometimes labelled 'divide by two' when used in a multi-range meter. Explain the particular advantage of this form of shunt.   (C)

Other questions: Ch. 2: 26, 27, 28 Ch. 5: 12 Ch. 6: 8, 9, 10

C

12 An ammeter reads high by 5%. The resistance of its coil is 10 Ω. What resistances are required to make it read correctly and still present a resistance of 10 Ω?   (Ox. Schol.)

Other question: Ch. 13: 30

CHAPTER 9
Generators and Motors

A

1 A rectangular coil is pivoted about an axis perpendicular to a uniform magnetic field. Draw a graph showing how the induced e.m.f. in the coil varies with its angular position when the coil is rotated at a uniform speed. If the peak value of the induced e.m.f. is 200 V, calculate (a) its r.m.s. value, (b) its instantaneous value when the plane of the coil is at 60° to the direction of the magnetic field.
                    (C)

2 A resistance is sometimes connected so that it is in series with a d.c. motor when the motor is started, but completely cut out of the circuit when the motor has achieved its full running speed. Explain the purpose of this arrangement.   (Cam. Oversea)

3 Describe one simple type of d.c. motor and explain its action. Why is a series-wound d.c. motor specially suitable for electric traction?

(*Cam. Forces*)

4 State an equation giving the force F exerted on a wire of length l carrying a current I at right-angles to a magnetic field of flux density B. Give the units of all the quantities in your equation and draw a diagram showing the direction of the force in relation to field and current.

The armature of a motor contains a rectangular coil of area 20 cm² having 50 turns and carrying 2.5 A.

(*a*) Calculate the flux density of the field in which the coil is placed if the maximum torque on it is 0.30 N m, and draw a diagram showing the position of the coil when this torque is developed.

(*b*) Calculate the back e.m.f. in the coil in this position if it is rotating at 2400 revolutions per minute. (*Cam. Forces*)

Other question: Ch. 5: 2

B

5 What is meant by saying that the difference of potential between two points is 1 V?

The following are connected in series and joined to a direct-current supply:

(*a*) a 10 Ω resistor,

(*b*) an accumulator having an internal resistance 0.1 Ω, the positive terminal of the accumulator being connected to the positive of the supply,

(*c*) a series-wound d.c. electric motor, the total resistance of which is 2 Ω,

(*d*) an ammeter.

The ammeter reads 5 A, while the p.d. across the accumulator, measured with a high-resistance voltmeter, is 13.5 V, and that across the motor is 35 V.

Calculate the rate at which electrical energy is converted in (*a*), (*b*) and (*c*), and find the rate of production of heat in each case. (*Cam. Forces*)

6 Describe in outline the mode of working of (*a*) an a.c. generator, (*b*) a d.c. generator. How does the current taken by a shunt-wound motor automatically adjust itself to the external load?

A shunt-wound motor takes 15 A from 60 V mains. If the resistance of the armature is 0.2 Ω and that of the field coil is 30 Ω, calculate the armature current and back e.m.f. (*Cam. Schol.*)

7 A shunt-wound motor has field windings of resistance 50 Ω, and an armature of resistance 2 Ω. When connected across a 100 V supply, it takes a current of 10 A. Calculate (*a*) the back e.m.f.

EMP—N

generated in the armature, (*b*) the power available for mechanical use. (*O*)

8 What is meant by the *back e.m.f.* of a motor? What factors determine its magnitude?

A toy train has an electric motor with a permanent magnet. It runs steadily on level track at 0.50 m s⁻¹ while taking 2.0 A from a constant 12 V d.c. main supply. The resistance of its armature is 1.0 Ω. When the train runs into an obstruction and is stopped, the current rises. Explain why, and calculate the value of this larger current.

After being released, the train runs uphill at a steady speed of 0.10 m s⁻¹. What current does it now take? (*C*)

9 With the aid of a labelled diagram describe the essential features of a shunt-wound electric motor.

When a shunt-wound motor runs 'light' (i.e. off-load) at 12 revolutions per second on a 240 V d.c. supply, the armature current is 0.3 A. When driving a mechanism (i.e. on-load) the armature current increases to 16.0 A, the potential difference across the motor remaining the same. If the resistance of the armature is 3 Ω, calculate the number of revolutions per second that the motor makes on-load. (*A.E.B.*)

Other questions: Ch. 3: 14 Ch. 5: 16, 18 Ch. 12: 8, 24

C

10 Show that the mechanical power supplied by a series-wound electric motor to which a constant e.m.f. E is applied is a maximum when the motor is running at such a speed that the back e.m.f. is $\frac{1}{2}E$. (*N*)

11 A 250 V shunt-wound motor has an armature resistance of 0.2 Ω, and runs at 500 revolutions per minute on no-load, taking an armature current of 5 A. How is the energy dissipated? Calculate the speed and estimate the power output when running on-load and taking an armature current of 50 A. How will the energy losses differ from the case of no-load? (*L*)

12 A shunt-wound motor takes 22.5 A from the 100 V mains and runs at 1250 revolutions per minute. Calculate the back e.m.f. if the resistances of the armature and field circuits are 0.4 Ω and 40 Ω respectively. The field circuit resistance is then increased to 60 Ω. Find (*a*) the new back e.m.f., and (*b*) the new steady speed of the motor, assuming that the torque is the same as before and that the flux is proportional to the field current. (*C*)

Other question: Ch. 2: 40

CHAPTER 10
Capacitors

A

1 What charge is carried on the plates of a 100 μF capacitor charged to a potential difference of 200 V? If this capacitor is discharged in 4.0 s, what is the average discharge current?

2 A 30 μF capacitor is charged and then insulated. During the next minute the p.d. between the plates falls by 4.0 V. What is the average leakage current between the plates?

3 A steady charging current of 50 μA is supplied to the plates of a capacitor and causes the p.d. between them to rise from 0 to 5.0 V in 20 s. What is the capacitance?

4 An electroscope has a capacitance of 20 pF and its leaves diverge 25 divisions when charged to a potential of 600 V. If in this position the divergence of the leaves decreases at the rate of one division in a minute owing to imperfect insulation, find the leakage current and the resistance of the insulation. You may assume that the divergence of the leaves is proportional to their potential.   (N)

5 A capacitor of capacitance 100 μF is charged to a p.d. of 1000 V. It is discharged through a coil placed in a calorimeter of total mass 5 g and average specific heat capacity 2.1×10^3 J kg^{-1} K^{-1}. Calculate the rise of temperature in the calorimeter.
(Cam. Schol.)

6 A 20 μF capacitor is charged by joining to a 3000 V supply, and is then discharged. Calculate the mean power of the discharge, if the charge remaining after 10 μS is negligible.   (Cam. Forces)

7 If a steady potential difference of 250 V is maintained across the combination of a capacitor of capacitance 3 μF in series with one of 0.5 μF, calculate (a) the potential difference across each capacitor, (b) the charge on each capacitor, (c) the energy stored in each capacitor.   (A.E.B.)

8 Three capacitors of capacitances 2 μF, 3 μF and 6 μF respectively are joined (a) in parallel, (b) in series. What is the combined capacitance in each case?

9 How can three capacitors of capacitances 3 μF, 6 μF and 9 μF respectively be arranged to give a system of capacitance 11 μF?   (Cam. Schol.)

10 A 2 μF capacitor is required to work at 1000 V d.c. If the capacitor had to be replaced, but the only units available were of 2 μF with permissible d.c. rating of 400 V, how would you proceed?   (Cam. Forces)

11 Two capacitors of capacitances 0.5 μF and 0.3 μF are joined in series. What is their capacitance? What value of capacitance joined in parallel with this combination would give a capacitance of 0.5 μF? What will be the energy of the system when connected to a 240 V d.c. supply?   (W)

12 A capacitor is connected to a battery and to a galvanometer through a vibrating reed in such a way that the capacitor is fully charged by the battery and then fully discharged through the galvanometer, 20 times per second. If the e.m.f. of this battery is 120 V, and the galvanometer registers an apparently steady current of 0.6 mA, what is the capacitance of the capacitor?   (O & C)

13 A capacitance of 1.00×10^{-8} F is fixed between the terminals of an electrometer in order to perform the following experiment. An unknown capacitor is charged from a 25 V battery and then joined across the terminals of the electrometer; this process is repeated 10 times without discharging the electrometer in between. The electrometer reading is then found to be 0.28 V. Calculate the capacitance of the unknown capacitor.

Other questions: Ch. 2: 20 Ch. 11: 4, 5 Ch. 12: 3

B

14 Three identical components P, Q, R are joined to a battery as shown.

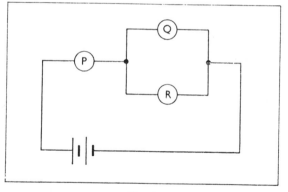

Fig. Q5

Answer the following, giving your reasons:

(a) If P, Q, and R are solenoids, how are the intensities of the magnetic fields inside them related to one another?

(b) If P, Q, and R are resistors, how are the rates at which heat is produced in them related to one another?

(c) If P, Q, and R are electrolytic cells, how are the masses of the same element deposited in them in the same time related to one another?

(d) If, P, Q, and R are capacitors, how are the potential differences across them related to one another?

Describe how you would check your statement by experiment in *one* of the cases.   (*Cam. Forces*)

15 Explain what is meant by *capacitance*, and define the *farad*. Fifty similar 1 μF capacitors are charged by joining them separately to a battery of e.m.f. 600 V. By a suitable mechanism they are then joined in series, with the positive plate of one to the negative of the next and so on. A 1 MΩ resistor is then joined across the ends of the series chain.

(a) Calculate the initial current through the resistor.

(b) Calculate the total energy dissipated in the resistor from the moment of connection until the current ceases.

(c) Is a capacitor in the middle of the chain discharged in this process? If so, what happens to the charges on its plates?   (*Cam. Forces*)

16 A 3.0 μF capacitor is charged by joining it to an 800 V battery through a non-inductive resistor. How much energy is taken from the battery and how much is stored in the capacitor? What happens to any difference between these amounts? The charged capacitor is joined to a tube containing gas at low pressure, which conducts only so long as the p.d. across it is 200 V or more. What quantity of electric charge will pass through the tube, and how much energy will be lost by the capacitor?   (*Cam. Forces*)

17 Define *capacitance* and deduce an expression for the energy stored in a charged capacitor.

A capacitor of capacitance 10.0 μF is momentarily connected across a 100 V d.c. supply. Calculate the charge on the capacitor and the energy stored. A second capacitor of capitance 5.0 μF is now switched in parallel with the first. Calculate the final potential difference across the capacitors and the energy now stored in the system. Comment on the change of energy.   (*Cam. Oversea*)

18 A capacitor of capacitance 4 μF, charged to a potential difference of 100 V, shares its charge with another capacitor of capacitance 2 μF charged to a potential difference of 400 V. Calculate

(a) the charge on each capacitor after sharing,
(b) the loss of energy as a result of the sharing.
      (*A.E.B.*)

19 The capacitance fixed between the terminals of an electrometer is 1000 pF. When a certain capacitor is charged by a small battery and then connected across the electrometer, the instrument gives a scale reading of 0.30. When the battery is joined directly across the electrometer the scale reading is 0·80. What is the capacitance of the capacitor? What fraction of the charge on the capacitor is delivered to the electrometer in this experiment?

20 Derive an expression for the energy of a capacitor of capacitance C when the potential difference between its terminals is V.

A multi-plate air capacitor of capacitance 3.0×10^{-10} F is connected in a circuit with a battery of e.m.f. 120 V, a sensitive galvanometer and vibrating switch, in such a way that the capacitor is fully charged by the battery then fully discharged through the galvanometer 100 times a second. Draw a diagram of the circuit and calculate (a) the value of the apparently steady current through the galvanometer, (b) the mean rate at which the battery supplies energy, (c) the mean rate at which energy is obtained from the capacitor.

Describe briefly how you would use an arrangement of this kind to measure the relative permittivity of oil.   (*N*)

21 A large capacitor was connected in series with a high-tension battery, a resistance of 1 MΩ, a microammeter and a plug key. Readings of the microammeter taken at the instant the key was put in, and at ten-second intervals afterwards, were as follows:

Time/s	0	10	20	30	40	50
Current/μA	100	61	37	22	13.5	8.2

Time/s	60	70	80	90	100	120
Current/μA	4.9	3.0	1.8	0.7	0.4	very small

Find the p.d. of the battery and, from a suitable graph, a value for the capacitance of the capacitor.   (*L*)

22 A capacitor of capacitance 1 μF is charged to 200 V. If its leakage resistance is 10^8 Ω, in what time will the p.d. fall to 100 V when the capacitor is isolated?   (*Cam. Schol.*)

23 The potential difference V across a capacitor, which had previously been charged, was measured every ten seconds; the results are tabulated below. At time $t = 15.0$ s, a resistor of 1.30×10^6 Ω was permanently connected across the capacitor.

t/s	0	10	20	30	40	50	60
V/V	10.0	10.0	8.87	6.98	5.49	4.32	3.40

(i) explain the inherent difficulty in measuring this potential difference V. What kind of instrument would you use?

(ii) Find the capacitance of the capacitor.   (*C*)

Other questions: Ch. 11: 15, 16, 17, 19 Ch. 12: 28(*b*)

C

24 What do you understand by the term *electrical energy*?

A capacitor of capacitance C is charged to potential difference V_0 in the following three ways:

(*a*) it is connected to a source of potential difference the value of which is slowly increased from zero to V_0;

(*b*) it is connected to a source of constant potential difference V_0 by means of a fixed resistance;

(*c*) it is connected directly to a source of constant potential difference V_0.

Discuss quantitatively what happens to the energy given out by the source in each case. The internal resistance of the source may be neglected.

(*Cam. Schol.*)

25 Obtain from first principles an expression for the electrical energy stored in a charged capacitor.

Show that, if the capacitor shares its charge with a previously uncharged capacitor of the same capacitance, the total electrical energy of the system is halved. How can this be accounted for?

After a 100 μF capacitor has been charged to 200 V, its plates are insulated and then joined by a 1 MΩ resistor of heat capacity 0.02 J K^{-1} which is at 15.0 °C. Assuming that no heat escapes, find (*a*) the rate of rise of temperature of the resistor 50 s after connection, (*b*) the temperature of the resistor 50 s after connection.

(*O*)

26 A device to develop short pulses across a 100 Ω resistor consists of a 0.01 μF capacitor which is connected alternately to a 100 V supply for 0.01 s and to the 100 Ω resistor for 0.01 s. The time during which the capacitor is connected neither to the supply nor to the load resistor is negligible.

What is the shape of the pulse across the load resistor? What is its peak amplitude? What is the average current through the load resistor?

(*Ox. Schol.*)

27 How would you measure three of the following:

(*a*) a resistance of 10^{11} Ω;

(*b*) a resistance of 10^{-3} Ω;

(*c*) the conductivity of 0.1 mol l^{-1} sulphuric acid solution;

(*d*) a capacitance of 100 μF;

(*e*) the capacitance of two small coins separated by a piece of mica 1 mm thick? (Relative permittivity = 6.)

State clearly the reason why you have selected the particular method you have used in each case.

(*Ox. Schol.*)

Other question: Ch. 11: 25

CHAPTER 11
Electric Fields

Additional data required:
Electric constant $\varepsilon_0 = 8.85 \times 10^{-12}$ F m^{-1} (permittivity of a vacuum)
Charge of an electron $e = -1.60 \times 10^{-19}$ C
Mass of an electron $m_e = 9.1 \times 10^{-31}$ kg
Acceleration due to gravity $g = 9.8$ m s^{-2}

A

1 Two parallel square metal plates (0.5 m by 0.5 m) are fixed 20 mm apart and are maintained at a potential difference of 1000 V. (*a*) What is the electric field strength in the gap? (*b*) What force would act on a small metallized pith ball in the gap carrying a charge of 2.0×10^{-10} C?

2 A parallel-plate capacitor is made up of eleven metal plates of area 5 cm by 3 cm, separated by sheets of mica of relative permittivity 6 and thickness 0.2 mm. Calculate its capacitance. (*O*)

3 Three parallel identical sheets of metal, each of area 0.15 m^2, are spaced apart by two large uniform sheets of insulator 2.0 mm thick and of relative permittivity 3.0. What is the capacitance of the capacitor formed (*a*) by the centre sheet as one electrode and the two outer sheets (joined together) as the other, (*b*) by taking the two outer sheets as the electrodes, the centre sheet being left insulated?

4 A metal sphere 4 m in diameter, is charged to a potential of 3 MV by a Van de Graaff machine. Calculate the heat generated when the sphere is earthed through a long resistance wire. (*O & C*)

5 A sphere of diameter 1 m is charged up to 200 V. How many times per second will it have to be discharged and recharged in order to give a mean discharge current of 1 μA? (*Cam. Schol.*)

6 Calculate the radius of a water drop which would just remain suspended in the earth's electric field of 300 V m^{-1} when charged with 1 electron.

(*S*)

7 A vacuum tube contains two plane parallel electrodes 7.5 mm apart. If a p.d. of 150 V is maintained between them, what is the electric field strength in the gap? What is the force acting on an electron in the gap?

An electron is emitted at negligible velocity from the negative electrode. How long does it take to cross the gap?

8 Give an account of a method by which the charge associated with an electron has been measured.

Calculate the potential difference necessary to be maintained between two horizontal conducting

plates, one 5.0 mm above the other, so that a small oil drop, of mass 1.31×10^{-14} kg with two electrons attached to it, remains in equilibrium between them. Which plate would be at the positive potential?
(L)

9 A charged oil drop of mass 10^{-14} kg is observed to remain stationary in the space between two horizontal charged plates, the electrical forces being sufficient to counteract its weight. Find the charge on the drop if this occurs when the p.d. between the plates is 3000 V and their distance apart 1 cm. *(Cam. Forces)*

10 A model Van de Graaff generator has a hollow sphere of radius 0.5 m mounted on a column of resistance 6×10^9 Ω. The belt conveys charge to the sphere at the rate of 10^{-6} A.

(a) What is the steady potential of the sphere?

(b) What is the surface density of charge on the sphere?

(c) What is the electric potential gradient close to its surface? *(O)*

11 Two small conducting spheres, each of mass 10 mg, are suspended from the same point by non-conducting strings of length 10 cm. They are given equal and similar charges, until the strings are equally inclined at 30° to the vertical. Calculate the charge on each sphere. *(N)*

12 If there is an electric field of strength 300 V m^{-1} at the surface of the earth, what is the charge per square kilometre on the earth's surface?
(O & C)

Other questions: Ch. 13: 2, 3, 4, 5 Ch. 17: 6

B

13 A 1 μF parallel-plate capacitor that can just withstand a p.d. of 6000 V uses a dielectric having relative permittivity 5, which breaks down if the electric field strength in it exceeds 3×10^7 V m^{-1}. Find (a) the thickness of dielectric required, (b) the effective area of each plate, (c) the energy stored per unit volume of dielectric. *(Cam. Forces)*

14 In the figure, C is a parallel-plate capacitor having two plates, each of area 10 cm^2, separated by a ceramic material of thickness 0.1 mm and relative permittivity 1200; R is a 10 MΩ resistance, and the battery has a constant e.m.f. of 60 V.

At a certain instant after the switch S has been closed, the microammeter registers 3.2 μA. Find the potential difference between the plates of C, and the charge on each plate, at this instant. *(O)*

15 It is found that a poor 50 μF capacitor (parallel plates) fails to retain its charge due to the dielectric layer having become imperfect and pos-

sessing finite resistance. During a test in which the capacitor was charged and then disconnected, the potential difference across its terminals was observed to fall to half its initial value in 10 minutes. If the relative permittivity of the layer is known to be 6, what is the resistivity of the layer? *(Cam. Schol.)*

16 A parallel-plate capacitor is filled with a 'leaky' dielectric of resistivity 1.50×10^{12} Ω m and relative permittivity 4.0. It is joined across a battery of e.m.f. 100 V and then isolated. What is the p.d. across the capacitor 60 s later?

17 A capacitor consisting of two parallel circular plates each of diameter 20 cm placed 1 mm apart in air is permanently connected to an electrostatic voltmeter of capacitance C. The system is charged to a potential difference of 80 V and is then isolated. When the plate separation is increased to 5 mm the voltmeter reading rises to 300 V. Explain the change in potential difference and calculate the value of C.
(O & C)

18 Explain what is meant by the terms *electric field strength* and *electric potential*.

Four infinite conducting plates A, B, C and D, of negligible thickness, are arranged parallel to one another so that the distance between adjacent plates is 2 cm. The outer plates A and D are earthed and the inner ones B and C are maintained at potentials of +20 V and +60 V respectively. Draw a graph showing how the potential between the plates varies as a function of position along a line perpendicular to the plates. What is the magnitude and direction of the electric field between each pair of adjacent plates?

By using Gauss's theorem, or otherwise, determine the charge density on plates B and C. If the inner plates were isolated and then connected together, what would be their potential? *(O & C)*

19 Describe the experiments you would perform in order to determine how the capacitance of a parallel-plate capacitor varies with the area of overlap of the plates, and the separation of the plates.

A parallel-plate air capacitor has plates each of

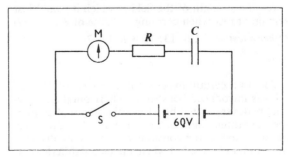

Fig. Q6

area 1.0×10^{-2} m² separated by 2×10^{-3} m. If the p.d. between the plates is 200 V, calculate (a) the charge, (b) the stored energy, and (c) the force between the capacitor plates.   (O)

20 In Millikan's oil drop experiment the drops are observed to move (under the action of constant forces due to gravity and any uniform electric field that may be present) with constant velocity. Explain why their velocity, rather than their acceleration, is constant.

In this experiment an oil drop was prevented from falling or rising by applying a potential difference of 5750 V between parallel horizontal plates 1.5 cm apart. Assuming that the drops each carried one electron charge, what was the radius of the drop?
(Density of oil = 0.92×10^3 kg m⁻³.)   (O & C)

21 Without giving the full theory or full experimental details, explain the general principle of Millikan's determination of the value of the electronic charge e. Why is this determination important?

In an oil drop experiment, two drops A and B both remain stationary at the same time in the electric field. When the field is switched off, the times taken to fall the same distance are 50 s for A and 38.2 s for B. Which drop has the greater charge? If the larger charge is $12e$, what is the smaller one?
                                                          (O)

22 Explain the physical principles involved in Millikan's method for determining the value of the electronic charge e. How did he establish that this is indeed the fundamental 'atom' of electric charge?

A uniform electric field of strength 5×10^5 V m⁻¹ acts vertically downwards. A small oil drop carrying a charge $8e$ and situated in the field falls with a uniform speed 2×10^{-4} m s⁻¹; when the charge on the drop is increased to $11e$, it moves upwards in the field with a uniform speed of 1×10^{-4} m s⁻¹. Find

(a) the mass of the drop;

(b) the charge it carries when it just remains at rest in the field of 5×10^5 V m⁻¹;

(c) the strength of the field in which it would just remain at rest when carrying a charge of $8e$.   (O)

Other questions: Ch. 13: 15 Ch. 17: 7

C

23 In a certain experiment a very small drop of oil was introduced between two horizontal and parallel plates separated by a vertical distance of 20 mm, the apparatus being open to the air. The drop was observed to fall at a constant velocity over a distance of 15 mm in 15.1 s, but when a potential difference of 7500 V was applied across the plates, it was found to

rise, covering the same distance in times of 65.4 s, 65.4 s, 52.2 s, 52.2 s in four successive measurements. Given that under the conditions of the experiment the viscosity of air was 1.830×10^{-5} N s m⁻², and the density of oil 0.923×10^3 kg m⁻³, deduce all you can from the observations.

(A sphere of radius r moving with steady velocity v through a fluid of viscosity η experiences a retarding force $F = 6\pi\eta rv$.)

                                                   (*Cam. Schol.*)

24 A parallel-plate air capacitor has two square plates, each of area 1×10^{-3} m² and separated by a distance of 4 mm. The lower plate is earthed, and the upper plate is raised to a potential of 4 kV and then insulated. Calculate (i) the surface density of charge on each plate, (ii) the electric field strength between the plates, and (iii) the energy stored in the capacitor.

Without touching either plate of the capacitor, an insulated square sheet of aluminium 2 mm thick and of area 1×10^{-3} m² is now inserted between the capacitor plates so that it is parallel to them and does not project beyond the region between them. Calculate the new value of the energy stored in the system. Explain your calculations and account for the energy change.

The insulated aluminium sheet is now withdrawn from between the plates of the capacitor, and the lower earthed plate is moved sideways in its own plane until the area of overlap is reduced to 5×10^{-4} m². Find the new value of the electric field strength between the capacitor plates, and explain your calculation.

(Edge effects may be neglected throughout.)   (O)

25 Define *capacitance*. Suggest suitable dimensions for a parallel-plate air capacitor, of capacitance 10^{-10} F, for use in the laboratory, and describe how you would measure its capacitance experimentally.

The area of either plate of a parallel-plate air capacitor is 0.1 m², and their separation is 2×10^{-3} m. The plates are permanently connected across a 120 V battery. Find (a) the charge on each plate; (b) the energy stored in the capacitor; (c) the change in the energy stored in the capacitor when the separation of the plates is altered to 5×10^{-3} m; (d) the corresponding change in the total energy of the system formed by the capacitor and the battery; and hence (e) the mechanical work done in increasing the separation of the plates from 2×10^{-3} m to 5×10^{-3} m.
                                                          (O)

26 A parallel-plate air capacitor is made from two discs, each of radius 8 cm, placed 1 cm apart. The upper plate is earthed and hung from a balance. The lower plate is initially at earth potential and is then raised to 1.5×10^4 V. What is the change in the apparent weight of the upper plate? Ignore edge effects.   (*Cam. Schol.*)

27 A 1 μF capacitor consisting of two parallel plates separated by 0.1 mm is charged to a potential of 100 V. What is the force between the plates?
(*Cam. Schol.*)

28 Derive an expression for the capacitance per unit length between two coaxial cylindrical shells of radii r_1 and r_2 separated by a dielectric of relative permittivity ε_r.

A cable consists of a wire of radius 1 mm embedded coaxially in a dielectric cylinder of $\varepsilon_r = 2.25$ sheathed in metal braid. If the capacitance of 1 m of the cable is 6×10^{-11} F, what is the radius of the sheath?

This dielectric has the property that any part of it which is subjected to an electric field greater than 10^7 V m^{-1} is permanently damaged and becomes a conductor. What is the maximum p.d. which can be applied between the wire and the braid without damaging the dielectric? Discuss whether the whole of the dielectric is damaged when a p.d. slightly greater than this maximum is applied.   (*O & C*)

CHAPTER 12
Alternating Currents

Additional data required:
Velocity of light $c = 3.00 \times 10^8$ m s^{-1}

A

1 An a.c. supply, which may be assumed to be of a sinusoidal type, is applied to (*a*) a resistance, (*b*) a capacitor, (*c*) a large self-inductance of negligible resistance. Draw graphs to illustrate the variation of the applied potential difference and the resulting current with time in each case.

What is the effect on the current in each case of (i) increasing the frequency of the supply, (ii) increasing the resistance, the capacitance, and the inductance respectively. Give reasons for your answers.   (*W*)

2 An iron-cored inductance of negligible resistance is connected across the 230 V, 50 Hz mains supply. An alternating current ammeter in series reads 0.5 A. Find the value of the inductance.

How do you account for the fact that the power dissipation in the inductance is zero?   (*O*)

3 An 8 μF capacitor is placed across the 200 V r.m.s. a.c. 50 Hz electric supply. Calculate the r.m.s. current which flows in the circuit. What is the peak value of the p.d. across the capacitor?
(*Cam. Schol.*)

4 A tuning coil of inductance 180 μH is to be connected in series with a capacitor to form a cir-

cuit which resonates at 720 kHz. What value of capacitance is required?   (*A.E.B.*)

5 Derive an expression for the resonant frequency of a series tuned circuit.

A 150 μH inductor is tuned by a parallel variable capacitor having maximum and minimum capacitances 500 pF and 20 pF, respectively. If stray capacitances amount to 40 pF, calculate the maximum and minimum frequencies to which the circuit can be tuned.   (*Cam. Forces*)

Other questions: Ch. 2: 14, 15, 16, 23, 24 Ch. 9: 1 Ch. 13: 6, 7, 9

B

6 Describe fully how you would calibrate a moving-iron ammeter to read r.m.s. values of sinusoidal currents alternating at 50 Hz.

Such an instrument is used to measure the current flowing in an air-cored solenoid connected to a 50 Hz alternating p.d. State, with reasons, what changes in current you would expect to observe if (*a*) a laminated soft iron cylinder, (*b*) a copper cylinder, were inserted in the solenoid.

Why should the ammeter not be placed near the solenoid when measuring the current?   (*O & C*)

7 The loss due to hysteresis in the iron core of a transformer is 20 W under normal load conditions, the frequency of the a.c. supply being 50 Hz. What change, if any, do you expect in this loss (*a*) if the current taken from the secondary winding is doubled by reducing the resistance joined to the secondary, (*b*) if the transformer is used with a supply of frequency 100 Hz, and the primary p.d. is doubled? Neglect any effect due to resistance of the windings.
(*Cam. Forces*)

8 An alternating e.m.f. is represented by the equation $E = E_0 \sin \omega t$. Explain the physical significance of E_0 and ω. Express the variations in E in terms of the periodic time and of the frequency.

An osglim lamp lights when the p.d. applied to it exceeds 170 V and goes out when the p.d. drops to 140 V. If such a lamp is connected to a 240 V a.c. supply, for what fraction of the time is the lamp lit?
(*W*)

9 The e.m.f. of an a.c. generator is 100 V r.m.s. Its internal impedance is 50 Ω and is independent of frequency. It is connected in turn to (*a*) a pure inductance, (*b*) a pure capacitance, and (*c*) a resistance of 50 Ω in series with a pure capacitance. Calculate the r.m.s. values of the current flowing and the p.d. across the generator in each case when the frequency f is very low, i.e. $f \ll R/L$ and $1/RC$ and also when it is very high, i.e. $f \gg R/L$ and $1/RC$.
(*O & C*)

10 Explain what is meant by the *power factor* of a circuit. Why is it desirable that the power factor of an installation should not differ much from unity?

A lamp taking 0.40 A at 200 V is supplied from 250 V 50 Hz a.c. mains using a series capacitor. Calculate (*a*) the power factor of the circuit, (*b*) the capacitance of the capacitor, (*c*) the peak p.d. across the capacitor.

In actual practice the power factor is found to be a little higher than the calculated value. Suggest an explanation. (*Cam. Forces*)

11 A coil of inductance 0.50 H and resistance 220 Ω is joined to an alternating supply of frequency 50 Hz. What is its power factor?

12 A non-inductive circuit of resistance 5.0 Ω carries sinusoidal alternating current of r.m.s. value 4.0 A and frequency 50 Hz.

(i) Explain the significance of *r.m.s.* in this statement.

(ii) What is the peak value of the current?

(iii) What is the instantaneous current flowing 6×10^{-4} s after it changes direction?

(iv) How much heat is developed in the circuit in 1 minute?

(v) How much heat would have been developed in 1 minute if part of the circuit had been made into a coil of inductance 0.010 H, and the same potential difference applied? (*L*)

13 A transformer takes a current of 5.0 A from 210 V mains when the secondary supplies 18.0 A through a resistive load, the p.d. across the load being 50 V. In these circumstances the power factor of the primary circuit is 0.99. The resistance of the primary winding is 2.0 Ω, and that of the secondary winding 0.10 Ω. Calculate (*a*) the efficiency of the transformer at this load, (*b*) the power loss in the iron core. (*Cam. Forces*)

14 If the impedance of a moving-coil loudspeaker is 15 Ω and it is to be matched to an output power valve for which the optimum load is 2500 Ω, calculate the turns ratio of the coupling transformer required. (*A.E.B.*)

15 A coil with an inductance of 10 mH and resistance 10 Ω is connected across an a.c. supply of 10 V r.m.s. at a frequency of 200 Hz. Calculate the current in the circuit. (*A.E.B.*)

16 A capacitor of capacitance 2 μF in series with a resistance of 1000 Ω is connected across a 50 Hz alternating supply of e.m.f. 240 V r.m.s. Calculate (*a*) the current through the circuit, (*b*) the potential difference across the capacitor. (*A.E.B.*)

17 The instantaneous e.m.f. *E* of a source is represented by $E = 300 \sin 100 \, \pi t$ volts, where *t* represents time in seconds. If this e.m.f. is applied to a choke of resistance 10 Ω and self-inductance 100 mH, what is (*a*) the maximum value of the current, (*b*) the r.m.s. value of the current through the choke? (*A.E.B.*)

18 A non-inductive resistor of resistance 20 Ω and a coil are connected in series across a source of potential difference of r.m.s. value 8 V alternating at 50 Hz. The r.m.s. potential differences across the resistor and coil are 4 V and 6 V respectively. Find the current in the circuit and, by calculation or by drawing a phasor diagram on graph paper, determine the resistance and inductance of the coil. (*N*)

19 A coil carries a steady current of 50 mA when the steady p.d. across its ends is 2.0 V. When the steady p.d. is replaced by a p.d. alternating at 50 Hz and of magnitude 4.1 V r.m.s., the current flowing through the coil is 100 mA r.m.s.

Calculate (*a*) the resistance of the coil, (*b*) the self-inductance of the coil, and (*c*) the phase difference between the alternating current and the alternating p.d. (*N*)

20 A coil is joined in series with a non-inductive resistor of resistance 800 Ω across a potential difference of 100 V alternating at 50 Hz. The potential difference across the ends of the coil is found to be 45 V and across the resistor 80 V. Find the inductance and resistance of the coil by drawing a phasor diagram on graph paper. (*N*)

21 A 10 Ω resistor and a coil having resistance and self-inductance are joined in series with a 50 Hz a.c. supply, the r.m.s. p.d. between the terminals of which is 91 V. The r.m.s. p.d.'s across the resistor and coil are found to be 70 V and 35 V respectively. Calculate values for the following quantities associated with the coil; its impedance, its resistance, its reactance, its self-inductance, its power factor and the power dissipated in it.

Explain the action of a choking coil. (*L*)

22 Explain the significance of the symbols in the equation for an alternating current $I = I_0 \sin 2\pi ft$. Draw sketch graphs of *I* against *t* and I^2 against *t* for one complete cycle. From your graphs or otherwise derive a relation between the r.m.s. current I_{rms} and I_0.

A resistor of resistance 1000 Ω and a capacitor of capacitance 2 μF are connected in series across the 200 V 50 Hz supply. Calculate the impedance of this combination and find the potential difference across (*a*) the resistor, (*b*) the capacitor, (*c*) the supply, at the instant when the current has its maximum value. (*O & C*)

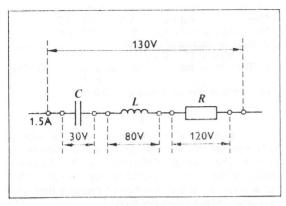

Fig. Q7

23 A capacitor C, an inductance L, and a resistance R are connected in series across a 50 Hz alternating supply, and the current flowing through the circuit is recorded as 1.5 A while the p.d. across the whole circuit is 130 V. What do these readings represent?

A voltmeter connected in turn across the terminals of C, L and R separately registers 30 V, 80 V, and 120 V respectively (Fig. Q7). Explain, with the help of a diagram, the phase relationship between these p.d.'s, and show how they can be reconciled with the reading of 130 V across the whole circuit.

Calculate the capacitance of C and the inductance of L. *(O)*

24 Explain how a sinusoidal alternating p.d. may be generated by a rotating coil, and obtain an expression for the p.d. developed at time t.

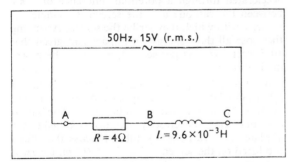

Fig. Q8

For the circuit of Fig. Q8, carrying alternating current of frequency 50 Hz, calculate, correct to two significant figures,

(a) the reactance of the inductor L,
(b) the impedance of the circuit between A and C,
(c) the r.m.s. value of the circulating current,
(d) the r.m.s. potential difference between A and B and between B and C,

(e) the power dissipated in the circuit,
(f) the phase difference between the current and supply p.d. *(O)*

25 A circuit contains a coil of inductance L and resistance R in series with a capacitor of capacitance C. Deduce from first principles an expression for the magnitude Z of the impedance of this circuit in terms of L, R, C and the frequency f of the a.c. supply.

If, in this circuit, L is 0.5 mH, R is 10 Ω and C is 1.0 μF, find the magnitude of the impedance if the frequency of the supply is 5000 Hz.

Find also the resonant frequency of a series circuit with these component values. *(A.E.B.)*

26 A capacitor of capacitance 300 pF and an inductive coil of self-inductance 150 μH and resistance 6 Ω are available. The power factor of the capacitor may be assumed to be zero. Calculate

(a) the impedance of these two components when connected in series across an a.c. supply of frequency 2 MHz;
(b) the resonant frequency of these two components in a series tuned circuit;
(c) the value of the additional capacitance it would be necessary to connect in series with these two components to double the resonant frequency.
(Assume $\pi^2 = 10$.)

27 A variable tuning capacitor has a maximum capacitance of 500 pF and a minimum capacitance one-tenth of this value. If it is used in series with a fixed inductor coil to form a tuned circuit, what must be the inductance of this coil if the circuit is to be resonant, at the maximum setting of the capacitor, to a radio-frequency signal of wavelength 600 m? Calculate also the resonant wavelength for this circuit at the minimum setting of the capacitor.

If the inductor coil has a resistance of 5 Ω, what will be the magnification of p.d. provided by the circuit when tuned to a frequency of 1 MHz? *(A.E.B.)*

28 Calculate the following:

(a) the resonant frequency of a series circuit consisting of an inductance of 2 mH in series with a capacitance of 1000 pF;
(b) the time constant of a capacitor of 0.15 μF across which there is a resistor of 250 kΩ;
(c) the Q-factor at a frequency of 100 kHz of a coil which has an inductance of 100 μH and a resistance of 3 Ω;
(d) the impedance at 50 Hz of a capacitance of 2 μF in series with a resistance of 1 kΩ. *(A.E.B.)*

29 A coil of self-inductance 200 μH and resistance 5 Ω is connected in series with a capacitor of capacitance 300 pF. Calculate

(a) the resonant frequency of this circuit,

(b) the impedance of this circuit at a frequency half the resonant frequency,

(c) the amplification of p.d. provided by this circuit at resonance.   (A.E.B.)

Other questions: Ch. 5: 18, 19, 22, 23, 24 Ch. 6: 8

C

30 A d.c. test applied between two terminals on a sealed box shows an open circuit. When an alternating p.d. is applied a current flows which is proportional to the p.d. and has a value of 0.024 A for an applied p.d. of 12 V at 50 Hz, and 0.028 A at 12 V and 100 Hz.

Describe an arrangement of components inside the box which would be consistent with these readings, and give their values.

Describe and explain one further experiment which could be used to test your answer without opening the box.   (O & C)

31 The primary of an a.c. transformer has a large self-inductance and a low resistance; the mutual inductance between primary and secondary is 0.3 H. If a 50 Hz current of 2 A r.m.s. flows in the primary, what will be the r.m.s. p.d. developed in the secondary on open circuit? How will the phase of this p.d. be related to the phase of the p.d. across the primary?   (O & C)

32 Write an account of tuned circuits, including the series acceptor circuit and the parallel rejector circuit.

Explain briefly the importance of tuned circuits in radio reception.

Calculate the dynamic resistance at resonance of a rejector circuit consisting of a capacitor of capacitance 300 pF and negligible power factor in parallel with a coil of inductance 150 μH and resistance 5 Ω.

Evaluate approximately the resonant frequency of this circuit.   (A.E.B.)

33 A current of 2 A passing through a circuit of self-inductance 200 mH and negligible resistance is suddenly interrupted by opening a switch. Sparking is liable to occur if the p.d. across the switch exceeds 1000 V. Show how a capacitor may be used to prevent sparking and calculate the capacitance that will suffice. Justify carefully the method of calculation you adopt.   (Cam. Schol.)

Other questions: Ch. 6: 21, 22

CHAPTER 13
Electrons and Ions

Additional data required:
Planck constant $h = 6.63 \times 10^{-34}$ J s

Velocity of light $c = 3.00 \times 10^8$ m s^{-1}
Charge of an electron $= -1.60 \times 10^{-19}$ C
Mass of an electron $m_e = 9.1 \times 10^{-31}$ kg
$\therefore$ Specific charge of an electron
$\quad = -1.76 \times 10^{11}$ C kg^{-1}
Unified atomic mass constant $m_u = 1.66 \times 10^{-27}$ kg
$\qquad\qquad\qquad\qquad\qquad = 1$ u
Acceleration due to gravity $g = 9.8$ m s^{-2}

A

1 A milliammeter connected in series with a hydrogen discharge tube indicates a current of 10^{-3} A. The number of electrons passing the cross-section of the tube at a particular point is 4.0×10^{15} per second. Find the number of protons that pass that same cross-section per second.   (L)

2 A potential difference of 15 V is maintained between two electrodes in a vacuum tube. An electron is emitted with negligible velocity from the negative electrode; calculate its velocity when it reaches the positive electrode.

3 Describe a method by which the charge per unit mass of an electron has been determined.

An electron is accelerated from rest through a potential difference of 200 V. Find the velocity that it acquires.   (L)

4 Describe and give the theory of an experiment to determine the value of the specific charge of an electron.

If this value is actually -1.76×10^{11} C kg^{-1}, calculate the velocity of an electron of energy 10^4 eV moving in a region of zero potential.   (L)

5 Electrons in a certain cathode-ray tube are accelerated through a potential difference of 2 kV between the cathode and the screen. Calculate the velocity with which they strike the screen. Assuming they lose all their energy on impact and given that 10^{12} electrons pass per second, calculate the power dissipation.   (O & C)

6 A small step-down mains transformer and a 6 V accumulator are joined in turn to a 6 V bulb and are found to light it with equal brilliance. The bulb is disconnected and each source in turn is joined to the Y-plates of a cathode-ray tube. Discuss the effects produced on the screen.   (Cam. Oversea)

7 Describe the principles of a cathode-ray oscilloscope.

When a p.d. of 100 V d.c. is applied to the Y-plates of a C.R.O. the spot is deflected 1 cm. A sinusoidal alternating p.d. is then applied to the Y-plates and a suitable time base to the X-plates. The wave thus presented is 5 cm in height from trough to crest. What is the r.m.s. value of the current taken when this same p.d. is applied to a resistance of 1000 Ω?   (S)

8 Describe a cathode-ray tube and explain the principles of its operation. How would you use it to display the current vs. p.d. characteristic of a semiconductor diode?   (*O & C*)

9 A p.d. alternating at 50 Hz is connected across the Y-plates of a cathode-ray oscilloscope. Sketch and explain the forms of the traces on the oscilloscope screen when a linear time base of frequency (i) 10 Hz, (ii) 100 Hz is connected across the X-plates. What would be the effect of disconnecting the time base and connecting the X- and Y-plates in parallel?   (*N*)

10 What is the energy of 1 quantum of sodium D light (wavelength 5.89×10^{-7} m)?

A 200 W sodium vapour street light has an efficiency of 30% (i.e. 30% of the supplied energy is emitted as the D light). How many quanta of light does it emit per second?

11 The ionization potential of hydrogen is 13.6 V.

(*a*) Calculate the energy and speed of the slowest electron that can ionize a hydrogen atom when it collides with it.

(*b*) Calculate the longest wavelength of electromagnetic radiation that could produce ionization in hydrogen.

(*c*) The lowest two excited states of a hydrogen atom are 10.2 eV and 12.0 eV above the ground state. Calculate *three* wavelengths of radiation that could be produced by transitions between these states and the ground state.

12 The mass of the atom of the neon isotope ^{20}Ne is 3.2×10^{-26} kg, and the charge on a singly ionized atom is 1.6×10^{-19} C. What is the flux density of a magnetic field in which singly charged ions of ^{20}Ne travelling at 10^6 m s^{-1} describe an arc of radius of curvature 0.36 m?

Find also the radii of curvature of the arcs described in this field by (*a*) singly charged ^{22}Ne ions and (*b*) doubly charged ^{20}Ne ions when the velocity is 10^6 m s^{-1} in each case.   (*O*)

Other questions: Ch. 3: 2, 3 Ch. 5: 6 Ch. 11: 7
Ch. 16: 3, 4

B

13 A beam of electrons accelerated by a potential difference of 1000 V in a cathode-ray tube passes between a pair of electrostatic deflecting plates which produce a field of 5×10^4 V m^{-1} perpendicular to the initial direction of the beam. What magnetic field, superimposed on the electric field and acting over the same region, will allow the beam to pass undeviated? How should the magnetic field be

orientated? Describe how this principle has been used to measure the velocity of a beam of electrons.   (*O & C*)

14 In one type of cathode-ray tube, electrons are accelerated in a narrow beam through a potential difference of 3000 V. Calculate the velocity with which the electrons are then moving.

If the electrons then pass into a uniform magnetic field orientated at right-angles to the beam, calculate the flux density B needed to deflect the beam so that the electrons begin to move along a circular arc of radius 10 cm, given that the force acting on a charge e moving with velocity v normally to the field is *Bev*.   (*O & C*)

15 An electron is travelling horizontally after having been accelerated from rest by a p.d. of 100 V. How far would it fall under gravity in travelling 100 m in a vacuum?   (*C*)

16 At a certain setting of the controls, the Y-deflection sensitivity of a cathode-ray tube is 5 mm V^{-1}, the period of the time base is 1 ms, and the trace is a horizontal line along 00 when no p.d. is applied to the Y-terminals. A 10 Ω resistor, part of a circuit which contains a d.c. generator and an a.c. generator in series, is connected across the Y-plates as in Fig. Q9(i), and the resulting trace is as in Fig. Q9(ii), which shows also the 1 cm squares of the scale on the screen.

(*a*) What is the frequency of the alternating p.d.?

(*b*) What is the magnitude of the d.c. potential difference across the resistor?

(*c*) What is the power converted in heating the resistor?   (*O*)

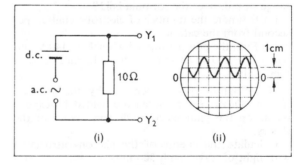

Fig. Q9

17 What is meant by the statement that the sensitivity of a certain cathode-ray tube for vertical deflection is 20 V cm^{-1}?

With the help of a diagram explain the factors in the design of the tube which govern this sensitivity.

How would you use a cathode-ray oscilloscope to do the following experiments?

(i) Observe the waveform of the p.d. of the a.c. supply mains (approximately 240 V r.m.s. at 50 Hz).

(ii) Check the calibration of an audio oscillator of variable frequency at a scale reading of about 1000 Hz, the frequency of the supply mains being assumed to be exactly 50 Hz.   (*O & C*)

18 The current I through a given diode for various values of the p.d. V between anode and cathode is given in the following table

V/V	0	50	100	150	200	250	300
I/mA	0	8	30	55	70	76	78

Sketch a characteristic curve for this tube and explain its general shape.

When a diode with this characteristic is connected in series with an anode resistor to a 100 V d.c. supply, a current of 8 mA is observed to flow. To what value must the supply p.d. be raised in order to increase the current to 30 mA?   (*O & C*)

19 For a thermionic diode with its cathode maintained at a fixed temperature the following table gives the values for the current I through the diode at various values of the positive anode potential V with respect to the cathode:

$I/\mu A$	120	320	600	910	1300
V/V	20	40	60	80	100

$I/\mu A$	1500	1600	1600	1600	1600
V/V	120	140	160	180	200

(*a*) Plot a graph of I against V; explain its form in general terms.

(*b*) Over part of the range covered by these readings, the relation between I and V is of the form $I = 1.3\ V^n$. What is the value of n, and for what range of p.d. does the formula hold?

(*c*) Estimate the number of electrons emitted per second from the cathode.

(*d*) Find the power supplied when $V = 140$ V and when $V = 200$ V; account for the difference between these values.   (*O*)

20 What do you understand by the *quantum theory*? Describe the evidence provided by experiments on the photoelectric effect in favour of the theory.

Calculate the energy of the photons associated with light of wavelength 300 nm.   (*C*)

21 Discuss the facts of the photoelectric effect.

The maximum wavelength that can cause emission of electrons from a metal plate is 5×10^{-7} m. Find the maximum velocity with which electrons are emitted if light of wavelength 4×10^{-7} m is incident on its surface.   (*S*)

22 If a sodium surface in vacuum is illuminated with a monochromatic beam of ultra-violet light with a wavelength of 2×10^{-7} m, what is the maximum velocity of emission of the electrons if the work function of the sodium is 2.46 V?   (*A.E.B.*)

23 Describe two types of photocell and indicate their application with reference to their characteristics.

The maximum wavelength that can cause electron emission from a metal is 5.2×10^{-7} m. Calculate the work function of the metal and the maximum speed of electrons emitted by light of wavelength 4×10^{-7} m.   (*A.E.B.*)

24 Explain Einstein's application of the quantum theory of radiation to the understanding of photoelectric emission.

Light of wavelength 4.3×10^{-7} m is incident on (*a*) a nickel surface of work function 5 V, (*b*) a potassium surface of work function 2.3 V. Determine by calculation if electrons will be emitted and, if so, the maximum velocity of the emitted electrons in each case.   (*A.E.B.*)

25 How would you attempt to measure the maximum velocity with which photoelectrons leave a metal surface irradiated by monochromatic light?

Explain how this velocity depends on the wavelength of the light, the intensity of the light and the nature of the metal.

When the incident light is monochromatic and of wavelength 600 nm, the kinetic energy of the fastest electrons is 2.56×10^{-20} J. For light of wavelength 400 nm, it is 1.92×10^{-19} J. Use this information to estimate the value of the Planck constant h and the work function of the metal.

26 When ultra-violet light of wavelength 1.7×10^{-7} m falls in a vacuum on the clean surface of a well-insulated sheet of a certain metal the metal attains a potential of $+6.6$ V relative to its surroundings. For wavelengths of 3.0×10^{-7} m and 4.5×10^{-7} m the corresponding potentials are $+2.4$ V and $+1.0$ V respectively. What theory can be advanced to explain these data?

Explain what is meant by *threshold wavelength* and *work function* and use the above data to determine numerical values for these quantities and for the Planck constant.

Show how the theory put forward explains (i) the effect of increasing the intensity of the light falling on the metal surface, (ii) the effect of illuminating the surface simultaneously with light of all three wavelengths, the surface being initially at earth potential in each case.   (*O & C*)

27 A heated filament provides the electrons in a high-vacuum cathode-ray tube. A potential difference of 2000 V is maintained between the anode and cathode. The stream of electrons, after falling through this potential difference, passes for 2 cm between two parallel metallic plates, separated by a distance of 1.2 cm and maintained at a potential difference of 120 V. The path of the electrons on entry into the space between the two deflecting plates is parallel with the plates. Find the change in direction of the electron stream in its passage between the plates. Edge effects are to be neglected. (W)

28 A parallel beam of electrons, accelerated through a potential difference of 1000 V, is injected horizontally half-way between two horizontal conducting plates 4 cm apart charged to a difference of potential of 100 V. Find the horizontal distance from the point of injection to the point of impact of the electrons on one of the plates.

Describe in general terms the effect on the beam of a horizontal magnetic field at right-angles to the beam.

Find the value of such a field if the path of the beam remains horizontal. (Ox. Schol.)

29 Explain what is meant by the statement that neon has isotopes of mass number 20, 21, and 22.

Outline how a beam of positive neon ions may be produced.

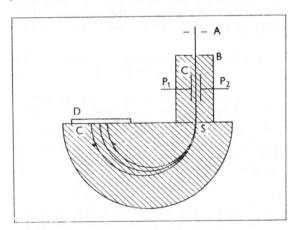

Fig. Q10

In the apparatus shown in the sketch, a beam of singly-charged $^{20}_{10}Ne$, $^{21}_{10}Ne$ and $^{22}_{10}Ne$ ions passes through the slits A and B and enters the evacuated chamber C. A uniform magnetic field is applied throughout C. The beam traverses the electrostatic deflection plates P_1 and P_2, between which there is an electric field of strength 1.50×10^5 V m^{-1}.

Show, by means of a sketch, the polarity of the plates P_1 and P_2 and the direction in which the magnetic field must be applied for the ions to follow the path shown.

If the magnetic field is of flux density 0.60 T find the velocity of the ions which emerge from S.

The ions of one isotope form a trace on the photographic plate D at a distance of 0.190 m from the slit S. Find the mass number of the isotope. (C)

Other question: Ch. 17: 26

C

30 Discuss the constructional details and the principle of operation of a thermionic diode. Measurements of V_a, the p.d. across the diode, and I_a, the anode current, gave the following results:

V_a/V	0	10	20	30	40	50
I_a/mA	0	2	6	10	17	25

V_a/V	60	70	80	90	100
I_a/mA	35	47	63	80	100

Give a qualitative explanation of the general shape of the curve and calculate approximately, by means of a graphical method, the current indicated by a moving-coil meter of negligible resistance connected in series with the diode and an alternating p.d. of 50 V r.m.s. at 50 Hz. (O & C)

31 Show that a charged particle moving with constant speed at right-angles to a uniform magnetic field in a vacuum travels in a circular path and find an expression for the time taken to complete one revolution.

What is the path of a charged particle in a uniform magnetic field if its initial velocity is not at right-angles to the field?

An electron of speed 2×10^7 m s^{-1} is emitted from a point on the axis of a very long solenoid wound with 10 turns per cm and carrying a current of 2.5 A. If the initial velocity makes an angle α with the solenoid axis, find at what distance from its starting point the particle next crosses the axis, and show that this distance is almost independent of α if α is small. (O & C)

32 Make a clear labelled diagram of a cathode-ray tube using electrostatic deflection and explain the function and mode of action of each part. The deflector plates in a given tube are 5 cm long and 0.5 cm apart. Calculate the sensitivity of the instrument (in V cm^{-1}) if the potential of the gun is -2000 V and the distance between the centre of the deflector plates and the screen is 25 cm. What factors do you consider will affect this sensitivity when the potential difference between the deflector plates oscillates at high frequency?

(Ox. Schol.)

CHAPTER 15
Radio

Additional data required:
Velocity of electromagnetic waves in air
= 3.00×10^8 m s^{-1}

B

1 Draw the diagram and explain the action of a diode detector circuit suitable for the demodulation of a 400 kHz signal carrying speech modulation.

Calculate a suitable value for the circuit time constant.

Why is automatic gain control desirable in a communications receiver? Explain briefly how a.g.c. is achieved.   (*Cam. Forces*)

2 Describe how you would demonstrate an electromagnetic stationary wave, assuming you have a generator of 3 cm waves. The generator should not be described but you should mention the principle of the detector.

Assuming that electromagnetic waves are efficiently launched from an aerial when a stationary wave is set up within it, estimate the shortest possible length of aerial to radiate electromagnetic waves:

(i) of 3 cm wavelength, in air;
(ii) of frequency 50 Hz, in air;
(iii) of frequency 2×10^4 Hz, from an aerial where the effective wave velocity is 2×10^8 m s^{-1}.
                                                     (*C*)

3 Two loudspeakers face each other at a separation of about 100 m and are connected to the same oscillator, which gives a signal of frequency 110 Hz. Describe and explain the variation of sound intensity along the line joining the speakers. A man walks along the line with a uniform speed of 2.0 m s^{-1}. What does he hear?

(Speed of sound = 330 m s^{-1}.)   (*C*)

Other questions: Ch. 12: 4, 5, 14, 26, 27, 28, 29 Ch. 16: 6

C

4 A moving-coil loudspeaker has a cylindrical coil of diameter 2.0 cm with 40 turns, mounted in a radial magnetic field of flux density 1.2 T. If the effective mass of the moving system is 4.0 g, what is the amplitude of movement when a sinusoidal current of 0.10 A r.m.s. at 1000 Hz flows through the coil? Neglect elastic forces due to the suspension.
                                                     (*C*)

5 Explain what is meant by *interference of electromagnetic waves*. Why is it necessary that two trains of *light* waves must come from the same source if interference is to be observed? Discuss whether this is also necessary in the case of radio waves.

Radio waves pass from a transmitter to a receiver by two paths which differ in length, the length of the shorter path being 120 km. When the frequency used is changed from 2.5 to 2.4 MHz the received signal passes through 8 minima of intensity, being at maximum intensity at the initial and final frequencies. Calculate the length of the longer path.
                                                     (*Cam. Forces*)

6 An observer at the equator wishes to study the 5 m radio waves emitted by a star overhead. He sets up two identical aerials 1 km apart on an east–west line and arranges for the signals they pick up from the star to be combined in a single tuned receiver. As the earth rotates the combined signal is found to oscillate in strength. Explain this effect and calculate the period of oscillation.   (*Cam. Schol.*)

7 A radio receiver, situated at the top of a cliff of height 50 m overlooking the sea, is observing electromagnetic radiation of wavelength 10 cm coming from a transmitter carried by a balloon. Interference occurs between the direct beam and that reflected in the surface of the sea (taken to be plane). The balloon is released at a point 10 km from the foot of the cliff and rises at a constant rate of 2 m s^{-1}.

(*a*) What time elapses between successive minima in the signal at the receiver when the balloon starts to rise?

(*b*) How would you use this effect to estimate the height of the transmitter at any time?   (*Ox. Schol.*)

8 Two vertical radio aerials are spaced one-quarter of a wavelength apart and are fed with signals from the same transmitter but with a phase difference of 90°. How does the intensity of the transmitted radiation vary in a horizontal plane? What would be the effect of increasing the number of aerials, maintaining a constant distance and phase difference between one and the next?   (*Ox. Schol.*)

9 Two vertical line sources of radiation, ABC and DEF, are placed with their mid-points B and E in the same horizontal plane and 0.5 m apart. Each source radiates uniformly in this plane. The intensity of the radiation received along a line GH, parallel to BE, at the same horizontal level and distant 5 m from BE, is measured and is found to be zero at O where the perpendicular bisector of BE meets GH, with a series of maxima and sharp minima on either side of O. The first minima on either side of O are 30 cm from O.

What deductions can be made concerning the radiators ABC and DEF?

When ABC is turned through 90°, to lie along BE, the intensity at points along GH shows negligible variation, but when DEF is rotated also in the same way the maxima and minima re-appear. What further information does this give?   (O & C)

Other question: Ch. 12: 32

CHAPTER 16
X-rays

Additional data required:
Planck constant $h = 6.63 \times 10^{-34}$ J s
Velocity of light $c = 3.00 \times 10^8$ m s^{-1}
Charge of an electron $= -1.60 \times 10^{-19}$ C
Mass of an electron $m_e = 9.1 \times 10^{-31}$ kg
Faraday constant $F = 9.65 \times 10^4$ C mol^{-1}
Specific heat capacity of water
  $= 4.2 \times 10^3$ J kg^{-1} K^{-1}

A

1 In what circumstances are X-rays produced? State *briefly* what you know about the nature and properties of these rays.

Draw a labelled diagram of a modern form of X-ray tube. Why is tungsten commonly used as the target material in such a tube? How would you expect the intensity and penetrating power of the X-rays from a tube to be altered by an increase in (*a*) the filament heating current, (*b*) the high-tension p.d. across the tube?   (*Cam. Oversea*)

2 Draw a labelled diagram of an X-ray tube. Explain its mode of operation, and indicate the electrical arrangements that are needed in order to work it from the normal mains supply.

If the potential difference applied across the tube is 5000 V and the current through it is 2 mA, calculate the number of electrons striking the target per second, and the speed at which they strike it.   (*O*)

3 Describe, with a labelled diagram, an X-ray tube of the high vacuum (Coolidge) typé which employs a thermionic filament.

An X-ray tube is operated with a potential difference of 10 kV between the target (anti-cathode) and the filament. What will be the minimum wavelength of the X-rays generated by this tube?   (*A.E.B.*)

4 Calculate the minimum wavelength of the X-rays produced at the anode of a diode electron tube which is operated at a potential difference of 100 V with respect to the cathode.   (*A.E.B.*)

5 The spacing between the layers of atoms parallel to the cleavage face of calcite is 3.03×10^{-10} m.

A monochromatic beam of X-rays is found to be strongly reflected from a crystal of calcite when the glancing angle it makes with the cleavage face is 14° 42′; strong reflection does not occur for any smaller angle than this. Calculate the wavelength of the X-rays. At what other values of the glancing angle might strong reflection occur?

Other questions: Ch. 13: 5, 10, 11

B

6 State briefly the methods that are available for the production and detection of electromagnetic waves of the following wavelengths:

(i) 10^{-10} m,   (ii) 5×10^{-7} m,   (iii) 1 cm.
          (*Cam. Forces*)

7 In a commercial X-ray tube, a current of 15 mA of electrons strikes a target which is at a potential of 100 kV positive with respect to the filament. What flow of water is required to keep the target cool assuming that all the heat is conducted away by the water, and that its rise in temperature should not be greater than 30 K? (You may assume that the entire kinetic energy of the electrons is converted into heat in the target.)   (*O & C*)

8 Draw a labelled diagram of a high-vacuum X-ray tube. Sketch curves showing the manner in which the intensity of the X-rays varies with the wavelength for a given target with the target potential as the parameter.

What is meant by the characteristic X-ray for an element? Distinguish between those factors which give rise to the continuous X-ray spectrum and the characteristic line spectrum for a target metal like molybdenum.   (*A.E.B.*)

9 Give an account of the production and chief properties of X-rays. How can the wavelength of X-rays be determined?

The wavelength of the K_β line in the characteristic X-ray spectrum of molybdenum is 6.3×10^{-11} m. What is the glancing angle in the first order with a crystal whose grating spacing is 2.8×10^{-10} m?

Explain why the continuous spectrum of the radiation from an X-ray tube has a short-wavelength limit, the value of which depends on the potential difference across the tube. What is the value of this limit for a p.d. of 40 kV?   (*O*)

10 Describe how X-rays are produced and give an account of their significance in modern physics.

The diffraction of X-rays by a crystal of rock-salt, NaCl, shows that in the crystal lattice the sodium and chlorine ions occupy alternate corners of cubes of side 2.8×10^{-10} m. Given that the density of rock-salt is 2.16×10^3 kg m^{-3}, calculate the charge of an electron.

(Relative molecular mass of NaCl = 58.5.)

(*Cam. Schol.*)

Other questions: Ch. 1: 19, 21 Ch. 13: 20 Ch. 15: 2

CHAPTER 17
Nuclear Physics

Additional data required:
Velocity of light $c = 3.00 \times 10^8$ m s^{-1}
Charge of an electron $= -1.60 \times 10^{-19}$ C
Avogadro constant $L = 6.02 \times 10^{23}$ mol^{-1}
Electric constant $\varepsilon_0 = 8.85 \times 10^{-12}$ F m^{-1}
   (permittivity of a vacuum)
Unified atomic mass constant $m_u = 1.66 \times 10^{-27}$ kg
                                    $= 1$ u

Assume:
   1 year $= 3.15 \times 10^7$ s

A

1 Describe and explain simple experiments which demonstrate the distinctive properties of the radiations emitted by radioactive substances.

Write brief accounts of TWO *industrial* (*not medical*) uses to which radioactive substances may be put.

(*N*)

2 '$^{24}_{11}$Na is a *radioactive isotope* of sodium, which has a *half-life* of 15 hours and disintegrates with the emission of *β-particles* and *γ-rays*. It emits *β*-particles that have energies of 4.2 *MeV*.'

Explain the meanings of the five terms that are italicized in the statement above.

(*L*)

3 (*a*) What is the number of electrons in 1.0 kg of hydrogen.

(*b*) Show that most other light elements contain about half as many electrons per kg as hydrogen.

(*c*) How many electrons are there in 1.0 kg of uranium-238?

(Atomic number of uranium = 92.)

4 Write brief notes on *three* of the following: (*a*) *β*-particles, (*b*) neutrons, (*c*) protons, (*d*) gamma radiation.

Discuss the reaction represented by

$$^{14}_{7}N(\alpha, p)^{17}_{8}O$$

explaining the meaning of the subscript and superscript numbers.

(*L*)

5 In an experiment with a high-energy beam, hydrogen atoms each of mass 1.67×10^{-27} kg strike a target with a velocity of 2×10^7 m s^{-1}. If 10^{15} atoms arrive each second, and the target is a lump of brass of 0.50 kg thermally insulated, find how long it will take for the temperature of the brass to rise by 100 K.

(Specific heat capacity of brass
$$= 0.38 \times 10^3 \text{ J kg}^{-1} \text{ K}^{-1}.)$$

(*O & C*)

6 Compare the velocities attained by a proton and an α-particle each of which has been accelerated from rest through the same potential difference. (*L*)

Other questions: Ch. 1: 20 Ch. 3: 3, 4, 5

B

7 The *atomic number* and the *mass number* of aluminium are 13 and 27 respectively. What is the significance of these statements in relation to the structure of the aluminium atom?

A solid aluminium sphere has a radius of 0.10 m. Find (*a*) the total number of electrons in the sphere, (*b*) the fraction of these which are removed when the sphere is raised to a positive potential of 100 V.

(Density of aluminium $= 2.7 \times 10^3$ kg m^{-3}.)

(*N*)

8 List the chief properties of α-radiation, β-radiation and γ-radiation.

Describe a type of Geiger-Müller tube that can be used to detect all three types of radiation. If you were supplied with a radioactive preparation that was emitting all three types of radiation, describe and explain how you would use the tube to confirm that each of them was present.

A radon ($^{222}_{86}$Rn) nucleus, of mass 3.6×10^{-25} kg, decays by the emission of an α-particle of mass 6.7×10^{-27} kg and energy 8.8×10^{-13} J.

(*a*) Write down the values of the mass number A and the atomic number Z for the resulting nucleus.

(*b*) Calculate the momentum of the emitted α-particle.

(*c*) Find the velocity of recoil of the resulting nucleus. (*O*)

9 An α-particle is emitted with a kinetic energy of 2.87 MeV from a stationary $^{209}_{83}$Bi nucleus. Find the approximate velocity with which the residual thallium nucleus recoils. (*C*)

10 What conclusions as to the properties of α-particles can be drawn from an examination of their cloud-chamber tracks in air and in helium? Explain how these conclusions are reached.

How would you expect the range in a given gas of α-particles from a given source to depend on the pressure?

A certain isotope has a half-life of 8800 years, emitting α-particles of energy 5 MeV, with a range of about 30 mm in air at atmospheric pressure.

(*a*) Explain the term *half-life*.

(*b*) Calculate the velocity with which the particles are emitted.

(*c*) Assuming an average value for the ionization potential of the gases in the air to be 15 V, estimate the average number of ion pairs produced per millimetre of an α-particle's path.

(Take the mass of an α-particle to be 6.7×10^{-27} kg.) (*O*)

11 Describe how the nature of α-particles has been established experimentally.

The half-life of the nuclide polonium-210 is about 140 days. During this period the average number of α-emissions per day from a mass of polonium initially equal to 1 μg is about 1.2×10^{13}. Assuming that one emission takes place per atom and that the approximate density of polonium is 10^4 kg m^{-3}, estimate the number of atoms in 1 m³ of polonium. (*N*)

12 Iodine-131 has a half-life of 8 days. A source containing this isotope has an initial activity of 2.0 Ci.

(*a*) What is the activity of the source after 24 days?

(*b*) What time elapses before the activity of the source falls to 1.0 μCi?

13 The half-lives of two radioactive isotopes X and Y are 25 minutes and 40 minutes respectively. Initially a sample of the isotope X has the same activity as a sample of the isotope Y. Compare the number of radioactive atoms present in the two samples.

Compare the number of radioactive atoms present in the two samples, and also the activities of the two samples, at a time 2 hours later. (*O*)

14 Radon is a monatomic gas of mass number 222 and with a radioactive constant equal to 2.1×10^{-6} s^{-1}. Calculate the number of α-particles emitted per second by 1 g of radon at s.t.p. when free from disintegration products. (*S*)

15 Calculate the mass of caesium-137 that has an activity of 5.0 μCi.

(Half-life of caesium-137 = 30 years, 1 Ci is a disintegration rate of 3.7×10^{10} s^{-1}.)

16 What is meant by the terms (*a*) *atomic number*, and (*b*) *relative atomic mass*? What are *isotopes*?

How would you attempt to discover whether or not a given sample of a material is radioactive?

What changes in atomic number and relative atomic mass occur when a nucleus emits (*a*) an alpha particle, (*b*) a negative beta particle, and (*c*) a positive beta particle?

A certain sample of pure lead is known to consist initially of only the isotope of mass number 210.

After some years it is found to contain another isotope of mass number 206. Give possible explanations of this phenomenon. How might it be possible to discover which of your explanations is correct?

(Lead has atomic number 82; neighbouring elements and their atomic numbers are: gold 79, mercury 80, thallium 81, bismuth 83, polonium 84 and astatine 85.) (*O & C*)

17 Technetium-99 (atomic number 43, chemical symbol Tc) decays by emission of a negative β-particle into ruthenium (chemical symbol Ru). Write down the nuclear equation representing this process.

A sample containing 0.100 μg of technetium is found to emit β-particles at a rate of 135 s^{-1}. Calculate the half-life of technetium.

18 A compartment on a Geiger-Müller tube is filled with a solution containing 1.00 g of carbon extracted from one of the Dead Sea scrolls. The count rate recorded is 1000 per hour. When a similar solution containing 1.00 g of carbon extracted from a living plant is used instead, the count rate is 1200 per hour. Without any solution in the compartment the background count rate is found to be 300 per hour. Estimate the age of the scroll, if the half-life of carbon-14 is 5600 years.

19 Give an account of the types of radioactive emissions found in nature and explain how you would distinguish between them.

A piece of timber has been recovered from an archaeological excavation and it is required to find its approximate age by measuring the radioactivity of the carbon-14 contained therein. For this purpose it may be assumed that the proportion of carbon-14 in the natural carbon of living wood is everywhere and at all times the same and that it begins to decay at death. If the number of disintegrations observed from 5 g of carbon prepared from the specimen is 21 per minute, how old is the specimen?

(The ratio of carbon-14 to natural carbon in living wood is 1.25×10^{-12} and the half-life of carbon-14 may be taken to be 5600 years. The mass number of natural carbon is 12.) (*N*)

20 Uranium-234 is formed as one of the decay products of uranium-238. The half-lives of the two isotopes are 2.5×10^5 years and 4.5×10^9 years, respectively. What proportion (by mass) of a sample of natural uranium would you expect to find in the form of uranium-234?

21 State the nature of *alpha*, *beta* and *gamma* radiations.

Suggest a method by which you could ascertain whether a beam of particles from a radioactive source consists of α-particles alone, β-particles alone, or a mixture of α-particles and β-particles. State what apparatus you would require.

The radioactive isotope of polonium, ^{218}Po, decays with a half-life of 3 minutes. It emits α-particles and produces a radioactive isotope of lead, ^{214}Pb, which emits β-particles and has a half-life of 27 minutes. On the same axes draw sketch graphs to show how, starting with pure ^{218}Po, (a) the α-particle emission-rate, and (b) the amount of ^{214}Pb present, would vary with time. (O & C)

22 The ratio of the mass of lead-206 to the mass of uranium-238 in a certain rock is measured to be 0.45. Assuming that the rock originally contained no lead-206, estimate its age.
(Half-life of uranium = 4.5×10^9 years.)

23 One of the decay products of thorium-232 is the monatomic radioactive gas radon-220. Calculate the mass of radon-220 in radioactive equilibrium with 25 g of thorium-232. What volume would this quantity of radon occupy at s.t.p.?
(Half-life of radon-220 = 54 s,
Half-life of thorium-232 = 1.4×10^{10} years,
Molar volume of a gas at s.t.p.
$$= 2.24 \times 10^{-2} \text{ m}^3 \text{ mol}^{-1}.)$$

24 Discuss the assumption on which the law of radioactive decay is based.
What is meant by the *half-life* of a radioactive substance?
A small volume of a solution which contained a radioactive isotope of sodium had an activity of 1.2×10^4 disintegrations per minute when it was injected into the bloodstream of a patient. After 30 hours the activity of 1.0 cm^3 of the blood was found to be 0.50 disintegrations per minute. If the half-life of the sodium isotope is taken as 15 hours, estimate the volume of blood in the patient. (N)

25 When iron is irradiated with *neutrons* an *isotope* of iron is formed. This isotope is *radioactive* with a *half-life* of 45 days. Give the meanings of the terms printed in italics.
A steel piston ring of mass 16 g was irradiated with neutrons until its activity due to the formation of this isotope was 10 μCi. Ten days after the irradiation the ring was installed in an engine and after 80 days' continuous use the crankcase oil was found to have a total activity of 1.85×10^3 disintegrations per second. Determine the average mass of iron worn off the ring per day assuming that all the metal removed from the ring accumulated in the oil and that 1 Ci is equivalent to 3.7×10^{10} disintegrations per second. (N)

26 A charged particle moves in a circle under the influence of a magnetic field of flux density 0.6 T. Show that the frequency of revolution is independent of the velocity of the particle, and find the value of this frequency for a proton.

Describe briefly any way in which this constant frequency property has found practical application.
(Specific charge of the proton =
$$9.57 \times 10^7 \text{ C kg}^{-1}.) \quad (Ox. Schol.)$$

27 (a) What quantity of energy (in J) would have a mass of 1.00 kg?
(b) Express this quantity in eV.
(c) At what speed is the mass of a particle twice its rest mass?
(d) What is the p.d. required to accelerate (i) an electron, (ii) an α-particle, to this speed?
(e) What is the smallest quantity of energy (in eV) that a γ-ray must have in order to give rise to an electron–positron pair?
(f) What is the wavelength of such a γ-ray?
(g) If a γ-ray of half this wavelength produces an electron–positron pair and its energy is equally shared between the particles, what is the energy (in eV) and speed of the particles?
(Mass of an electron $m_e = 9.1 \times 10^{-31}$ kg,
Mass of an α-particle = 6.64×10^{-27} kg,
Planck constant $h = 6.63 \times 10^{-34}$ J s.)

28 Phosphorus-32 decays by β-emission into sulphur-32, the energy liberated in each disintegration as kinetic energy of the particles being 1.7 MeV. The relative atomic mass of sulphur-32 is 31.972 1. Calculate the relative atomic mass of phosphorus-32.

C

29 Describe an experiment which suggests that atoms have nuclei. Explain how the results of the experiment indicate the existence of nuclei in atoms.
α-particles with a speed 3.0×10^6 m s^{-1} are fired at a block of gold. Assuming that the gold atoms are rigidly fixed in the block, calculate the closest distance of approach between an α-particle and a gold nucleus.
What law are you assuming still holds at these small distances? What path are you assuming for the incident α-particle?
(Atomic number of gold = 79.) (C)

30 In an experiment to investigate how the radiation from a radioactive source diminished with distance the following data were obtained:

x/cm	3	4	5	6	7	8
count rate/s^{-1}	43.3	30.3	22.6	16.9	13.4	11.2

x was the separation between the window of the Geiger tube and the outside of the perspex disc in which the source was encapsulated. The background count was measured to be 150 counts in 5 minutes.
State fully the law which you would expect the above data to confirm if the radiation consisted of gamma rays only. Explain possible ways of treating

the data, each of which should lead to a linear graph. Use ONE of these ways and the above data to plot a graph, and comment briefly on the result.

Give detailed discussion of the possible reasons why, in such an experiment, the experimentally observed intensity of radiation might diminish with distance at a greater or lesser rate than is theoretically expected. (*N*)

31 How many α-particles are emitted altogether by an atom of uranium-238 as it decays progressively into lead-206?

A quantity of uranium ore contains 1.00 kg of uranium, which can be assumed to be in equilibrium with its decay products. Estimate the mass of helium generated per year in the ore.

(Half-life of uranium-238 = 4.5×10^9 years.)

Answers

Answers are given to the number of significant figures justified by the accuracy of the data. Where an answer to only 2 significant figures is appropriate a more exact answer is sometimes added alongside in brackets to help a student check his arithmetic. In some cases the final digit is written as a suffix to help indicate accuracy. Thus '2.7$_4$ V' is intended to show that the answer has an uncertainty of perhaps 4 or 5 in the second place of decimals. An answer given as '2.74 V' would mean that the uncertainty is not more than 1 in that place.

Chapter 1   Atoms and Electric Currents (Questions p. 355)

1 5 A

4 225 C; 1.4×10^{21}

7 1.8×10^4 J

10 1 : 16

13 $6.2_5 \times 10^{-4}$ m s^{-1}

16 3.5×10^5 J

19 2(.5) $\times 10^{-10}$ m

21 $1.99_2 \times 10^{-26}$ kg; 5.03×10^{25} kg^{-1}; 8.7×10^{-30} m^3

2 3 V

5 500 A; (a) 20 V, (b) 60 kW

8 200 MW; 1.08×10^{12} J

11 2.0×10^{28} m^{-3}

14 85 cm h^{-1}

17 7.0 V

20 2.9 : 1 (by mass)

3 15 W

6 240 J; (a) 720 J, (b) 240 J

9 75 C; 1.25 A

12 7.8×10^7

15 7×10^8 V; 10^4

18 7 min; 38 min

22 5.6×10^{-13} kg m s^{-1}

Chapter 2   Resistance (Questions p. 356)

1 25 Ω

4 0.25 mA; 1.25 W

6 (i) 3 V, 0.8 Ω, 1.25 A; (ii) 1.5 V, 0.2 Ω, 0.83 A

8 6

10 (a) 8.3 Ω, (b) 11.7 Ω, (c) 12.5 Ω, (d) 1.7 Ω

12 12.0 m

16 (a) 1%, (b) 1000 V; 1.41×10^5 V

18 500 W, 1000 W and 2000 W

20 2.0×10^{15} Ω m; 5.0×10^{-16} S m^{-1} (or Ω^{-1} m^{-1})

22 1.52×10^{-4} K^{-1}

25 1.5 V; 0.5 Ω

28 31 V; 10 V

31 2.6×10^{-2} mm

34 6.6×10^{-3} m s^{-1}

37 2 : 3

40 6.4 A; 2.0 A, charging

2 0.5 Ω

5 (i) 960 Ω, (ii) 484 Ω

9 0.2 A; 1.4 V

14 2.83 A

17 0.61 Ω

19 6.3 m

23 2.2 A (2.24)

26 17 Ω; 1.4 V (1.365)

29 (a) 6.7 V, (b) 8.0 V, (c) 0

32 5.0 mV; 5.2 mA; 383 Ω

35 2.4×10^3 W m^{-2}

38 14 A

41 0.39 m s^{-1}

3 4 V; 1.6 W

7 (a) 5 Ω, (b) 1.2 Ω

11 7.5 V, 1.0 Ω; 7.35 V

15 707 V

21 80 Ω; 1.18 cm

24 11.3 A; 314 rad s^{-1}; 4 kW

27 80 V; 64.5 V

30 27.79 Ω

33 $3.6_3 \times 10^{-4}$ m s^{-1}

36 18.7$_5$ kW

39 1.24; 0.48

Chapter 3   Electrolysis and Thermoelectricity (Questions p. 359)

1 3.2×10^{-19} C; 2.5×10^{19}; 0.4 A

3 0.024 kg; 3.6×10^{24}

5 $4.1_5 \times 10^{-15}$ mol; 9.9×10^{-14} kg

7 18 min

10 6 Ω

12 2 Ω; 40%

14 (a) 6 A, (b) 4.0 V, (c) 24 W, (d) 3.6 J s^{-1}; 0, (e) 10.8 J s^{-1}

2 8.9×10^5 C kg^{-1}

4 3.0×10^{-5} mol; 1.8×10^{19}

6 5.6 A

8 3.03×10^6 C; 2.43×10^7 J

9 3.41×10^{-21} J

11 (ii) 100 A, (iii) 9 Ω, (iv) 22 V

13 25 V battery + 100 Ω resistance; 50 Ω

Chapter 4   Circuit Measurements (Questions p. 360)

1 3.5 mV

4 0.15 A low

2 (a) 0.540 V, (b) 0.800 V

5 883 °C

3 4.8 Ω; from B to D

6 increase to 22.5 Ω

7 2.49 mV; 1000 Ω in parallel with AB   **8** 5

9 75 cm; (a) 90 cm, (b) 39(.5) cm   **11** 996 Ω; 0.2 cm

12 392.8 Ω, 405.2 Ω; (a) 68.0 cm, (b) 3.125 mV, 53.1 K   **14** 2235 °C

15 50.4 °C   **16** (a) $3.9_4 \times 10^{-3}$ K^{-1}, (b) 18.3 °C, (c) 0.2_4 K

17 2.2×10^3 J kg^{-1} K^{-1} (2.16)   **18** 3.5×10^5 J kg^{-1}

19 (a) 0.091 K s^{-1}, (b) 0.091 K s^{-1}, (c) 85 W, (d) 3.3×10^{-5} kg s^{-1} (3.26)

20 0; 1 mA; 2 mA   **21** 1.000 75 Ω **22** 0.05 K

Chapter 5   Magnetic Forces (Questions p. 362)

1 0.16 N   **2** 1.5×10^{-2} T   **3** 1.0×10^2 N; 1.5 kW

5 0.2 A m^2; 8.0×10^{-3} N m; 4.0×10^{-3} N m; 0   **6** 4.8×10^{-14} N

7 0.20 H; 10 V   **8** 1.8 J   **9** 500 V

10 150 μH   **11** 1.0×10^5 N m^{-2} **12** 0.402 T

13 54(.5)°   **14** 0.25 V   **15** (a) 0.12_6 mV, (b) zero, (c) zero

16 0.26_1 μA; $1.0_2 \times 10^{-11}$ W   **17** (a) 79 μV, (b) 3.1×10^{-9} J   **18** 18.7 V

20 1.0 s   **21** (a) 2 A s^{-1}, (b) 4 A, (c) 400 V   **22** 4 A

23 (i) 10 mH, (ii) 12.5, (iii) 1.6 H (1.56)   **24** 0.2 Wb; 63 V

25 0.29_4 T   **26** 1.15 T   **27** 0.7 m s

28 6.25×10^{-2} m^3

Chapter 6   Magnetic Fields (Questions p. 365)

1 2.5×10^{-3} T; 1.0×10^{-6} Wb   **2** 7.0 A   **3** 1.76×10^{-3} T; 0

4 1.3×10^{-5} T; 3.9×10^{-5} N   **5** 2.5×10^{-4} N m^{-1} **6** 3(.2) mH

7 1.1 kV   **8** 4.0×10^{-5} N m   **9** 1.8×10^{-7} N m

10 1.00_3 A; 3.7×10^{-5} T   **11** 32 mA; 7.1×10^{-2} A m^2 **12** 8.3 mA

13 47 A   **14** 2.4×10^{-5} N m^{-1} (repulsion); 1.06×10^{-5} T

15 2.51×10^{-3} T; 6.3×10^{-4} V   **16** (a) 20 div L, (b) 5 div L, (c) 5 div R

17 0.20 μC   **18** 32(.3) div

19 24 μV (23.7); 1.2×10^{-5} Ω (1.18)   **20** 0.15_5 mV

21 40 μH ($4\pi^2$); higher; 20 μH ($2\pi^2$); 0.40 A; 0.50 V

22 8.0×10^{-4} T; (a) 0.4 μH, (b) 0.13 V

Chapter 7   Magnetic Materials (Questions p. 367)

1 100   **2** 400   **4** 22(.5) μC

5 2500 (2470)

Chapter 8   Measuring Instruments (Questions p. 367)

1 1 : 10; 9 : 1   **3** (a) 0.049 Ω in parallel, (b) 1.5×10^4 Ω in series

4 (i) 95 Ω in series, (ii) $5.0_5 \times 10^{-2}$ Ω in parallel; 1.33 V

5 $1.0_5 \times 10^{-3}$ Ω in parallel   **6** (a) 0.100 Ω in parallel, (b) 9900 Ω in series

7 6.9×10^{-7} N m rad^{-1} **8** the 2nd, (40 : 51)

9 10 div μA^{-1} **10** 0.28_4 μA

11 (i) 12.5 mA, (ii) 6.2_5 mA, (iii) 0.25 V, (iv) 0.25 V

12 either 200 Ω in parallel with meter, and 0.48 Ω in series with both or 0.5 Ω in series with meter, and 210 Ω in parallel with both

Chapter 9   Generators and Motors (Questions p. 368)

1 (a) 141 V, (b) 100 V   **4** (a) 1.2 T, (b) 30 V

5 total energy conversion: (a) 250 W, (b) 67.5 W, (c) 175 W

heat production: (a) 250 W, (b) 2.5 W, (c) 50 W

6 13 A; 57(.4) V   **7** 84 V; 670 W (672)

8 12 A; 10 A   **9** 9.6_4 s^{-1}

11 5 W in windings, 1245 W by friction; 480 rev min^{-1} (482); 10.8 kW; 500 W in windings, 1200 W by friction

12 92 V; (a) 88 V, (b) 1800 rev min^{-1} (1794)

Chapter 10   Capacitors (Questions p. 370)

1 ± 20 mC; 5.0 mA   **2** 2.0 μA   **3** 200 μF
4 8×10^{-12} A; 7.5×10^{13} Ω **5** 4.8 K   **6** 9×10^6 W
7 (a) 36 V, 214 V, (b) $1.0_7 \times 10^{-4}$ C, (c) $1.9_2 \times 10^{-3}$ J, $11(.5) \times 10^{-3}$ J
8 (a) 11 μF, (b) 1 μF   **9** 3 μF and 6 μF in series, with 9 μF in parallel with both
10 3 groups in parallel, each of 3 in series   **11** 0.19 μF (0.1875); 0.31 μF; $1.4_4 \times 10^{-2}$ J
12 1.06 mm apart; 0.25 μF   **13** 11.2 pF
14 The effects in P are twice those in Q or R, except in (b) where it is 4 times
15 (a) 3×10^{-2} A, (b) 9 J, (c) yes   **16** 1.92 J; 0.96 J; 1.8×10^{-3} C; 0.90 J
17 1.00×10^{-3} C; 5.00×10^{-2} J; 66.7 V; 3.33×10^{-2} J
18 (a) 8×10^{-4} C and 4×10^{-4} C, (b) 6×10^{-2} J   **19** 600 pF; 0.62_5
20 (a) 3.6 μA, (b) 4.3×10^{-4} W, (c) 2.2×10^{-4} W   **21** 100 V; 20 μF
22 69 s   **23** 32.1 μF   **25** (a) 0.74 K s^{-1}, (b) 78 °C
26 100 V; 50 μA

Chapter 11   Electric Fields (Questions p. 372)

1 (a) 5.0×10^4 V m^{-1}, (b) 1.0×10^{-5} N   **2** 4×10^{-9} F (3.98)
3 (a) 4.0×10^{-9} F, (b) 1.0×10^{-9} F   **4** 1000 J
5 90 s^{-1} **6** $1.0_5 \times 10^{-7}$ m
7 2.0×10^4 V m^{-1}; 3.2×10^{-15} N; 2.1×10^{-9} s   **8** 2.01×10^3 V; top
9 3.3×10^{-19} C   **10** (a) 6 kV, (b) 1.06×10^{-7} C m^{-2}, (c) 12 kV m^{-1}
11 8.0×10^{-9} C (7.94)   **12** 2.65×10^{-3} C km^{-2}
13 (a) 0.2 mm, (b) 4.5 m^2, (c) 2.0×10^4 J m^{-3} **14** 28 V; 3.0 μC (2.97)
15 1.6×10^{13} Ω m   **16** 32.3 V   **17** 25(.2) pF
18 -1.0×10^3 V m^{-1}; -2.0×10^3 V m^{-1}; $+3.0 \times 10^3$ V m^{-1}; $-8.8_5 \times 10^{-9}$ C m^{-2}; 4.4×10^{-8} C m^{-2};
     40 V   **19** 8.85×10^{-9} C; 8.85×10^{-7} J; 4.4×10^{-4} N
20 1.2×10^{-6} m (1.18)   **21** B; 8e
22 (a) 8×10^{-13} kg, (b) 10e, (c) $6.2_5 \times 10^5$ V m^{-1}
23 The charge changes half-way through from 21 electrons to 22
24 $\pm 8.85 \times 10^{-6}$ C m^{-2}; 1.0×10^6 V m^{-1}; 1.77×10^{-5} J; 8.85×10^{-6} J; 2×10^{-6} V m^{-1}
25 (a) $\pm 5.3 \times 10^{-8}$ C, (b) 3.2×10^{-6} J, (c) -1.9×10^{-6} J, (d) $+1.9 \times 10^{-6}$ J, (e) 1.9×10^{-6} J
26 0.20 N   **27** 50 N   **28** 8 mm; 10(.4) kV

Chapter 12   Alternating Currents (Questions p. 375)

2 1.4_6 H   **3** 0.50_3 A; 283 V   **4** 270 pF
5 1.68 MHz; 0.559 MHz   **7** (a) no change, (b) doubled   **8** 0.70
9 low f : (a) 2 A, 0; (b) 0, 100 V; (c) 0, 100 V;
     high f : (a) 0, 100 V; (b) 2 A, 0; (c) 1 A, 50 V
10 (a) 0.80, (b) 8.5 μF, (c) 212 V   **11** 0.81
12 (ii) 5.7 A, (iii) 1.06 A, (iv) 4.8×10^3 J, (v) 3.4×10^3 J
13 (a) 86(.5)%, (b) 58 W   **14** 13 : 1   **15** 0.62_3 A
16 0.128 A; 203 V   **17** (a) 9.1_0 A, (b) 6.4_3 A   **18** 0.2 A; 7.5 Ω; 9.25 mH
19 40 Ω; 29 mH; 12.7$^{\bullet}$ **20** 1.4_0 H; 98(.5) Ω
21 5.0 Ω; 2.2 Ω; 5.0 Ω (4.98); 1.6×10^{-2} H; 0.44; 108 W
22 1880 Ω; (a) 150 V, (b) 0, (c) 150 V   **23** 1.6×10^2 μF; 0.17 H
24 (a) 3.0 Ω, (b) 5.0 Ω, (c) 3.0 A, (d) 12 V; 9.0 V, (e) 36 W, (f) 37° (tan^{-1} 0.75)
25 19 Ω; 7.1 kHz   **26** (a) $1.6_3 \times 10^3$ Ω, (b) 0.74_5 MHz, (c) 100 pF
27 203 μH; 190 m; 255   **28** (a) 1.13×10^5 Hz, (b) $3.7_5 \times 10^{-2}$ s, (c) 21, (d) 1880 Ω
29 (a) 0.65 MHz, (b) 1220 Ω, (c) 245   **30** 10.7 μF in series with 400 Ω
31 190 V (188.5); in phase   **32** 1.0×10^5 Ω; 0.75 MHz   **33** 0.8 μF across switch

Chapter 13   Electrons and Ions (Questions p. 378)

1 $2.2_5 \times 10^{15}$ s^{-1} **2** 2.3×10^6 m s^{-1} **3** 8.4×10^6 m s^{-1}
4 5.93×10^7 m s^{-1} **5** $2.6_5 \times 10^7$ m s^{-1}; 3.2×10^{-4} W

7 0.17_7 A   **10** 3.37×10^{-19} J; 1.8×10^{20} s^{-1}
11 (a) 2.18×10^{-18} J; 2.19×10^6 m s^{-1}, (b) 9.14×10^{-8} m, (c) $1.03_5 \times 10^{-7}$ m; 1.22×10^{-7} m;
   6.9×10^{-7} m   **12** 0.56 T; (a) 0.39_6 m, (b) 0.18 m
13 2.66×10^{-3} T   **14** 3.45×10^7 m s^{-1}; 1.96×10^{-3} T
15 1.4×10^{-9} m   **16** (a) 3 kHz, (b) 2 V, (c) 0.6 W   **18** 288 V
19 (b) 1.5; 0 to 100 V, (c) 1.0×10^{16} s^{-1}, (d) 0.224 W; 0.320 W   **20** 6.6×10^{-19} J
21 4.68×10^5 m s^{-1} **22** 1.15×10^6 m s^{-1} **23** 2.4 V; 5.0×10^5 m s^{-1}
24 (a) no, (b) 4.6×10^5 m s^{-1} **25** $6.6_6 \times 10^{-34}$ J s; 3.07×10^{-19} J (or 1.92 eV)
26 7.0×10^{-7} m; 1.8 V; 6.7×10^{-34} J s   **27** $2.9°$
28 17.9 cm; 1.33×10^{-4} T   **29** 2.5×10^5 m s^{-1}; 22 **30** $24(.5)$ mA
31 0.23 m   **32** 16 V cm^{-1}

Chapter 15 Radio (Questions p. 382)

2 (i) 7.5 mm, (ii) 1500 km, (iii) 2.5 km   **3** 4 maxima every 3 s
4 2.7×10^{-6} m   **5** 144 km
6 69 s   **7** 5 s

8 intensity $\propto \cos^2 \left[\dfrac{\pi}{4}(1 - \cos\theta) \right]$ at an angle θ with the plane of the aerials; the maximum would be narrower

9 wavelength $= 3.0$ cm; sources in antiphase; equal amplitude; radiation is plane polarized

Chapter 16 X-rays (Questions p. 383)

2 $1.2_5 \times 10^{16}$ s^{-1}; 4.2×10^7 m s^{-1} **3** 1.24×10^{-10} m
4 1.24×10^{-8} m   **5** $1.53_7 \times 10^{-10}$ m; $30°\ 30'$; $49°\ 36'$
7 0.71 kg min^{-1} **9** $6.45°\ (6°\ 27')$; 3.1×10^{-11} m
10 1.6×10^{-19} C (1.56)

Chapter 17 Nuclear Physics (Questions p. 384)

3 6.0×10^{26}; 2.3×10^{26} **5** 57 s   **6** $1.41:1$
7 (a) 3.3×10^{27}, (b) 2.1×10^{-18} **8** (a) 218; 84, (b) 1.09×10^{-19} N s, (c) 3.1×10^5 m s^{-1}
9 2.4×10^5 m s^{-1} **10** (b) 1.55×10^7 m s^{-1}, (c) 1.1×10^4 mm^{-1}
11 3.4×10^{28} m^{-3} (3.36) **12** (a) 0.25 Ci, (b) 170 d (167)
13 $0.625:1$; $0.179:1$; $0.287:1$ **14** 5.7×10^{15} s^{-1} **15** 5.7×10^{-11} kg
17 9.9×10^4 y   **18** 2030 y old   **19** 8.9×10^3 y
20 5.5×10^{-5} (5.47) **22** $2.7_2 \times 10^9$ y   **23** 2.9×10^{-18} kg; 2.9×10^{-19} m^3
24 6×10^{-3} m^3 **25** 4.0 mg d^{-1} **26** 9.15×10^6 s^{-1}
27 (a) 9.0×10^{16} J, (b) $5.6_2 \times 10^{35}$ eV, (c) 2.60×10^8 m s^{-1}, (d) (i) $5.1_2 \times 10^5$ V, (ii) 1.87×10^9 V,
   (e) 1.02×10^6 eV, (f) 1.21×10^{-12} m, (g) $5.1_2 \times 10^5$ eV; 2.60×10^8 m s^{-1}

28 $31.973\ 9$ **29** 1.2×10^{-12} m   **31** 8; $2.0_7 \times 10^{-11}$ kg y^{-1}

Appendix

Hints on using a slide-rule

It is best not to buy too elaborate an instrument. The essential scales are the A, B, C and D scales and the reciprocal scale down the middle of the slide. Sine and tan scales are useful if they are on the *front* of the instrument. It is also useful to have the special scale for sines and tangents of small angles; and cube and log scales are also of some value. But log–log scales find little application in school work. Too elaborate an instrument is both confusing and unnecessarily expensive.

The slide-rule is designed for the rapid calculation of expressions involving multiplication and division. Most calculations may be done with either the A and B pair of scales or the C and D pair. However, the divisions in the latter pair are more widely spaced, so that greater accuracy can be obtained with these. Also the reciprocal scale in the middle of the slide is designed to be used in conjunction with the C and D scales, and enables many short-cuts in calculations to be made. The A and B scales should therefore ordinarily be used only in calculations involving squares and square-roots. The accuracy obtainable with a good 10″ slide-rule is about 0.2%.

The basic expression that may be worked out with a single movement of the slide is one of the form

$$\frac{a \times c}{b}$$

Other simple expressions should be brought to this form by mentally adding 'ones' to the top or bottom lines. For example, think of

$$a \times c \quad \text{as} \quad \frac{a \times c}{1}$$

and

$$\frac{a}{b} \quad \text{as} \quad \frac{a \times 1}{b}$$

It is then not necessary to memorize separate rules for multiplication and division. The procedure is best described by considering examples:

(1) $$\frac{1.92 \times 5.45}{2.71}$$

Set the cursor to 1.92 on the D scale.
Move the slide to bring 2.71 on the C scale under the cursor.
Move the cursor to 5.45 on the C scale.
Read off the answer (3.86) on the D scale under the cursor.

(2) $$6.3 \times 1.4$$

Set the cursor to 6.3 on the D scale.
Move the slide to bring 1 on the C scale under the cursor.
Move the cursor to 1.4 on the C scale.
Read off the answer (8.82) on the D scale under the cursor.

(3) $$\frac{9.5}{7.6}$$

Set the cursor to 9.5 on the D scale.
Move the slide to bring 7.6 on the C scale under the cursor.
Move the cursor to 1 on the C scale.
Read off the answer (1.25) on the D scale under the cursor.

The significant figures of a calculation are not affected by the position of the decimal point in the numbers involved. The slide-rule only deals with the *significant figures* in a calculation. For instance, the point on the scale marked '2.5' stands equally for 0.25, 2500 2.5×10^{13}, etc. The position of the decimal point in the answer must be decided by a separate approximate calculation, which can usually be done in a moment in the head.

It follows also that the '10' can be used as a '1' whenever it is convenient to do so. Thus in the last example above, if the numerator were 6.5 instead of 9.5, the '1' on the C scale would come off the end of the D scale. The answer must then be read below the '10' on the C scale instead. Consider likewise the example:

(4) $$4.8 \times 6.3$$

First set the cursor to 4.8 on D.
If now we move the slide to bring 1 on C under the cursor, 6.3 on C is well beyond the end of the D scale. Therefore the 10 on C must be brought under the cursor instead. The answer (30.2) is then found below 6.3 as usual.

To some extent calculations can be saved from 'going off the end of the scale' by choosing carefully the order in which the figures are taken. But sometimes it is unavoidable. The procedure is then to interchange the positions of the '1' and '10' of the C scale, and carry on as before. The next example shows the technique.

(5) $$\frac{425 \times 134}{6.75}$$

Set the cursor to 4.25 on D.
Bring 6.75 on C under the cursor. Now 1.34 on the C scale is beyond the left-hand end of the D scale. We therefore interchange the positions of the '10' and '1' as follows. Set cursor to 10 on C. Bring 1 on C under the cursor. Then read off the answer (8440) on D under 1.34 on C. (An approximate calculation shows that the result is somewhere near 8000.)

The basic slide-rule technique is readily extended to handle more elaborate calculations. Thus expressions of the form

$$\frac{a \times c \times e}{b \times d}$$

may be worked out with *two* movements of the slide (provided we can avoid 'going off the end of the scale'); and

$$\frac{a \times c \times e \times g}{b \times d \times f}$$

requires only *three* movements of the slide (with the same proviso). Note that there must always be *one more* number on the top line than on the bottom. As before, 'ones' (or 'tens') are mentally inserted in the expression to make this so. The procedure is shown in the next example.

(6)
$$\frac{8\pi}{2.65 \times 3.86 \times 4.23}$$

Set cursor to 8 on D. Bring 2.65 on C under cursor.
Set cursor to π on C. Bring 3.86 on C under cursor.
Set cursor to 1 on C. Bring 4.23 on C under cursor.
Set cursor to 1 on C. Read off the answer (0.581) on D.

The technique may be analysed as follows:

1. Take the numbers from the top line and bottom line alternately.
2. The D scale is used only for the *first number* and the *answer*. All other numbers are found on the C scale.
3. Move the *cursor* for *top line* figures, and the *slide* for *bottom line* figures. That is, we start the calculation by moving the cursor to a top line number on the D scale, and then move the slide and cursor alternately.

The use of the reciprocal scale

Apart from the obvious use of such a scale in giving reciprocals, its main purpose is to enable us to cut down the number of movements of the slide in many calculations. In this way accuracy is increased, since each movement of the slide must introduce a small error, and these errors accumulate in a long calculation. The ideal expression for slide-rule work is one in which there is one more number on the top line than on the bottom. When this is not so, we must usually insert extra 'ones' as needed; but an alternative technique is to *transfer* a number from one side of the fraction bar to the other by using the reciprocal scale. Thus any number on the top line can be treated as its reciprocal on the bottom line—or vice versa. For instance in the calculation of example 6 given above two extra 'ones' had to be inserted in the top line. Instead we could transfer the 4.23 to the top line and think of the expression as

$$\frac{8\pi \times \dfrac{1}{4.23}}{2.65 \times 3.86}$$

Set cursor to 8 on D. Bring 2.65 on C under cursor.
Set cursor to π on C. Bring 3.86 on C under cursor.
Set cursor to 4.23 on the reciprocal scale.
Read off the answer (0.581) on D.

The method is thus quicker and saves a movement of the slide.

By this means almost any expression involving three numbers can be worked out with only one movement of the slide. Here is another example:

(7)
$$7.86 \times 14.7 \times 273$$

This is re-arranged mentally as

$$\frac{7.86 \times 273}{1/14.7}$$

Set cursor to 7.86 on D.
Bring 14.7 on reciprocal scale under cursor.
Set cursor to 273 on C.
Read off the answer (3.15×10^4) on D.

The reciprocal scale can also be used to avoid the annoyance of 'going off the end of the scale'. It will be noticed that there is no possibility of this happening with a simple division of one number by another. Any denominator (on the C scale) can always be brought opposite any numerator (on the D scale). Thus in a calculation like that in example 4 above, all danger of this particular nuisance obtruding itself may be avoided by treating the expression as

$$\frac{4.8}{1/6.3}$$

Set cursor to 4.8 on D. Bring 6.3 on reciprocal scale under cursor. Then read off the answer (30.2) under the 'one' of the C scale.

Calculations involving squares and square-roots should be planned likewise to involve the minimum movement of the slide. Here are two examples:

(8)
$$8\sqrt{2}$$

Set cursor to 2 on the A scale ($\sqrt{2}$ is then under the cursor on the D scale). Bring 8 on the reciprocal scale under cursor. Read off answer (11.31) on D under the 'one' of the C scale.

(9) What is the area of a circle of diameter 5.83 cm?

$$\text{The area} = \frac{\pi(5.83)^2}{4}$$

Set cursor to π on the A scale. Bring 4 on the B scale under the cursor. Set cursor to 5.83 on C, and read off the answer (26.7 cm²) on A.

(Many slide-rules have special marks on the cursor to enable a few commonly recurring types of calculation like this one to be done by a single setting of the cursor. It is worth consulting the maker's instruction booklet to find out what is possible in this way.)

As the student gains familiarity with the slide-rule many other short-cuts will be discovered. We conclude with an example to illustrate another variation of technique.

(10)
$$\frac{1}{4\pi}$$

Set cursor to 4 on D. Bring π on reciprocal scale under cursor. The number under the 'one' of the C scale is now 4π. The reciprocal of this is then to be found on the C scale above the 'ten' of the D scale (0.0796).

Although accuracies of 0.2% can be obtained if a slide-rule is used with great care, it is by no means always necessary to do this. If the figures in the calculation involve errors of 1% or more, it is a foolish waste of time to struggle to adjust the cursor or slide to a hair's breadth. A fairly slapdash slide-rule technique is then quite adequate.

Mathematical short-cuts

When the angle θ is given in radians,

1
$$\sin \theta = \theta - \frac{\theta^3}{6} + \ldots$$

2
$$\tan \theta = \theta + \frac{\theta^3}{3} + \ldots$$

For sufficiently small angles both these expansions reduce to

3
$$\tan \theta \approx \sin \theta \approx \theta \text{ (in radians)}$$

The error in ignoring the other terms of the expansions is less than 1% for angles up to about $10°$.

4
$$1 - \cos \theta = \frac{\theta^2}{2} - \frac{\theta^4}{24} + \ldots$$

5
$$e^x = 1 + x + \frac{x^2}{2} + \ldots$$

6
$$\log_e (1 + x) = x - \frac{x^2}{2} + \frac{x^3}{3} - \ldots$$

The last two expansions are of value only when x is small. However, it is never necessary to resort to tables of $\log_e y$ or e^x, if a set of common log tables or a slide-rule with a log scale is available. Thus

7
$$\log_e y = \log_e 10 . \log_{10} y = 2.303 . \log_{10} y$$

The multiplier 2.303 is worth memorizing.

Also, if $\quad y = e^x$
then $\quad x = \log_e y = 2.303 \log_{10} y$

8
$$\therefore \ y = e^x = \text{antilog} \left(\frac{x}{2.303} \right)$$

The *binomial expansion* is of great value in many calculations in physics.

9
$$(1 + x)^n = 1 + nx + \frac{n(n-1)}{2} x^2 + \ldots$$

When x is small, the terms in x^2 and higher powers of x may be ignored. Here are some examples showing how this may be applied:

(i)
$$(1 + x)^2 = 1 + 2x + \ldots$$
e.g. $(1.0003)^2 = 1.0006$

(ii)
$$\sqrt{(1 + x)} = (1 + x)^{\frac{1}{2}} = 1 + \tfrac{1}{2}x + \ldots$$
e.g. $\sqrt{0.996} = (1 - 0.004)^{\frac{1}{2}} = 0.998$

(iii)
$$\frac{1}{1 + x} = (1 + x)^{-1} = 1 - x + \ldots$$
e.g. $\dfrac{1}{1.0003} = 0.9997$

VALUES OF PHYSICAL CONSTANTS

Quantity	Symbol	Value*	
speed of electromagnetic waves (in a vacuum)	c	$2.997\ 925 \times 10^8$ m s^{-1}	$\approx 3.00 \times 10^8$ m s^{-1}
magnetic constant permeability of a vacuum	μ_0	$4\pi \times 10^{-7}$ H m^{-1}	$\approx 1.256 \times 10^{-6}$ H m^{-1}
electric constant permittivity of a vacuum	$\varepsilon_0 \left(= \dfrac{1}{\mu_0 c^2} \right)$	$8.854\ 19 \times 10^{-12}$ F m^{-1}	$\approx 1/(36\pi \times 10^9)$ F m^{-1}
charge of a proton or electron	e	$\pm\,1.602\ 19 \times 10^{-19}$ C	
unified atomic mass constant	m_u	$1.660\ 53 \times 10^{-27}$ kg	$= 1$ u
rest mass:			
of proton	m_p	$1.672\ 61 \times 10^{-27}$ kg	$= 1.007\ 276\ 6$ u
of neutron	m_n	$1.674\ 92 \times 10^{-27}$ kg	$= 1.008\ 665\ 2$ u
of electron	m_e	$9.109\ 6 \times 10^{-31}$ kg	$= 5.485\ 9 \times 10^{-4}$ u
specific charge:			
of proton	e/m_p	$+9.579\ 0 \times 10^7$ C kg^{-1}	
of electron	e/m_e	$-1.758\ 80 \times 10^{11}$ C kg^{-1}	
electromagnetic moment:			
of proton	μ_p	$1.410\ 62 \times 10^{-26}$ A m^2	
of electron	μ_e	$9.284\ 9 \times 10^{-24}$ A m^2	
Planck constant	h	$6.626\ 2 \times 10^{-34}$ J s	
Faraday constant	F	$9.648\ 7 \times 10^4$ C mol^{-1}	
Avogadro constant	L	$6.022\ 2 \times 10^{23}$ mol^{-1}	
molar gas constant	R	8.314 J K^{-1} mol^{-1}	
molar volume at s.t.p.	V_m	$2.241\ 4 \times 10^{-2}$ m^3 mol^{-1}	

* based on the values given by Taylor, Parker and Langenberg, *Reviews of Modern Physics*, **41**, 375, 1969.

Index